AF615221

HANDBOOK OF MICROWAVE AND OPTICAL COMPONENTS

VOLUME 4

HANDBOOK OF MICROWAVE
AND OPTICAL COMPONENTS

Editor-in-Chief

KAI CHANG
Department of Electrical Engineering
Texas A&M University
College Station, Texas 77843

HANDBOOK OF MICROWAVE AND OPTICAL COMPONENTS

Volume 4

Fiber and Electro-Optical Components

Edited by

Kai Chang
Department of Electrical Engineering
Texas A&M University
College Station, Texas

A Wiley-Interscience Publication

JOHN WILEY & SONS, INC.

New York · Chichester · Brisbane · Toronto · Singapore

Library of Congress Cataloging in Publication Data:

Fiber and electro-optical components./edited by Kai Chang.
p. cm.—(Handbook of microwave and optical components; v. 4)

"A Wiley-Interscience publication."
Includes bibliographical references.
ISBN 0-471-61365-7
1. Electrooptical devices. 2. Fiber optics. I. Chang, Kai, 1948– . II. Series.

TA1750.F53 1991
621.381'33 s—dc20 90-12108
[621.381'33] CIP

Printed in the United States of America

10 9 8 7 6 5 4 3 2 1

HANDBOOK OF MICROWAVE AND OPTICAL COMPONENTS

VOLUME 1 MICROWAVE PASSIVE AND ANTENNA COMPONENTS

1 Transmission Lines
2 Transmission-Line Discontinuities
3 Filters, Hybrids and Couplers, Power Combiners, and Matching Networks
4 Cavities and Resonators
5 Ferrite Control Components
6 Microwave Surface Acoustic Wave Devices
7 Quasi-Optical Techniques
8 Components for Surveillance and Electronic Warfare Receivers
9 Microwave Measurements
10 Antennas I: Fundamentals and Numerical Methods
11 Antennas II: Reflector, Lens, Horn, and Other Microwave Antennas of Conventional Configuration
12 Antennas III: Array, Millimeter Wave, and Integrated Antennas
13 Antennas IV: Microstrip Antennas

VOLUME 2 MICROWAVE SOLID-STATE COMPONENTS

1 Molecular Beam Epitaxy and Its Application to Microwave and Optical Devices
2 Mixers and Detectors
3 Multipliers and Parametric Devices
4 Semiconductor Control Devices: PIN Diodes
5 Semiconductor Control Devices: Phase Shifters and Switches
6 Transferred Electron Devices
7 IMPATT and Related Transit-Time Diodes
8 Microwave Silicon Bipolar Transistors and Monolithic Integrated Circuits
9 High-Electron-Mobility Transistors: Principles and Applications
10 FETs: Power Applications
11 FETs: Low-Noise Applications

VOLUME 3 OPTICAL COMPONENTS

1 Optical Wave Propagation
2 Infrared Techniques
3 Optical Lenses
4 Optical Resonators
5 Spatial Filters and Fourier Optics
6 Semiconductor Lasers
7 Solid-State Lasers
8 Liquid Lasers
9 Gas Lasers

VOLUME 4 FIBER AND ELECTRO-OPTICAL COMPONENTS

1 Optical Fiber Transmission Technology
2 Optical Channel Waveguides and Waveguide Couplers
3 Planar Optical Waveguides and Waveguide Lenses
4 Optical Modulation: Electro-Optical Devices
5 Optical Modulation: Acousto-Optical Devices
6 Optical Modulation: Magneto-Optical Devices
7 Optical Detectors
8 Liquid Crystals: Materials, Devices, and Applications

CONTENTS

PREFACE

In the past two decades, we have witnessed a rapid development in high-frequency spectra above the microwave frequency. Components and subsystems have been built in microwave, millimeter-wave, infrared, and optical spectra for applications in communications, radar, remote sensing, remote control, sensors, navigation, surveillance, electronic warfare, radiometers, medicine, plasma research, imaging, computers, signal processing, industry heating, fabrication and processing, astronomy, and other fields. Extensive use of these spectra creates an urgent need for a practical handbook to assist practicing professionals in their daily work.

This handbook is intended to serve as a compendium of principles and design data for practicing microwave and optical engineers. Although it is expected to be most useful to engineers actively engaged in designing microwave and optical systems, it should also be of considerable value to engineers in other disciplines who have a desire to understand the capabilities and limitations of microwave and optical systems. Many handbooks for low-frequency electronics are available; a good handbook covering microwave and optical components is nonexistent. This handbook represents the most comprehensive treatment of both microwave and optical engineering that has appeared in book form to date.

To achieve these goals, this handbook covers almost all important components in microwave, millimeter-wave, submillimeter-wave, infrared, and optical frequency spectra. Theoretical discussions and mathematical formulations are given only where essential. Whenever possible, design results are presented in graphic and tabular form; references are given for further study. The book provides, in practical fashion, a wealth of essential principles, methods, design information, and references to help solve problems in high-frequency spectra.

The handbook was organized into two major parts with four volumes:

Part I. Microwave Components

Vol. 1 Microwave Passive and Antenna Components

Vol. 2 Microwave Solid-State Components

Part II. Optical Components

Vol. 3 Optical Components

Vol. 4 Fiber and Electro-Optical Components

Each chapter is written as a self-contained unit. It has its own table of contents and list of references. Some overlap is inevitable among chapters but has been kept to a minimum. It is hoped that this comprehensive handbook will offer the type of detailed information necessary for use in today's complex and rapidly changing high-frequency engineering.

The authors who have contributed chapters to this handbook have done an excellent job of condensing mountains of material into readable accounts of their respective areas.

The emphasis throughout has been to provide an overview and practical information of each subject.

I would like to express my special appreciation to my friend, Mr. Algie Lance, who was writing his chapter while terminally ill. He submitted his manuscript on time and passed away just after the completion of his chapter. I would also like to thank all members of the editorial board for their advice and suggestions, Dr. Felix Schwering for organizing the antenna chapters, and Mr. George Telecki, our Wiley editor, for his constant encouragement. I wish especially to thank my wife, Suh-jan, for her assistance in typing and managing this project.

KAI CHANG

College Station, Texas

CONTRIBUTORS

Alan E. Craig†
U.S. Naval Research Laboratory
Washington, DC

Uzi Efron
Research Laboratories
Hughes Aircraft Company
3011 Malibu Canyon Road
Malibu, CA 90265

Talal K. Findakly
Research Division
Hoechst Celanese Corporation
86 Morris Avenue
Summit, NJ 07901

H. D. Law
PCO, Inc.
20200 Sunburst Street
Chatsworth, CA 91311

Chinlon Lin
Bell Communications Research (Bellcore)
331 Newman Springs Road
Red Bank, NJ 07701

S. Thaniyavarn
Boeing High Technology Center
P.O. Box 24969
Seattle, WA 98124-6269

Chen S. Tsai
Department of Electrical Engineering and Institute for Surface and Interface Science
University of California
Irvine, CA 92717

Shi-Kay Yao
Optech
4071 Towhee Drive
Calabassas, CA 91302

P. K. L. Yu
Department of Electrical and Computer Engineering
University of California, San Diego
La Jolla, CA 92093-0407

† Present address: Air Force Office of Scientific Research, Bolling Air Force Base, Washington, DC 20332-6448.

1

OPTICAL FIBER TRANSMISSION TECHNOLOGY

Chinlon Lin

Bell Communications Research (Bellcore)
Red Bank, New Jersey

1.1 INTRODUCTION

The idea of using low-loss optical glass fiber waveguides for long-distance optical transmission was first proposed by C. K. Kao in 1966. The realization of low-loss glass optical fibers was first achieved by Corning in 1970; in the very same year, semiconductor diode lasers operating continuously at room temperature were obtained by Bell Labs. The combination of a low-loss compact optical transmission medium and a miniature, directly current-modulated diode laser as the optical source for an optical signal transmitter paved the way to a revolution in communications technology. For the past 20 years or so, there were dramatic progresses made in the field of optical fiber transmission technology in laboratories around the world. Furthermore, the technology has moved from laboratory experiments to actual system applications in the outside world. Optical fiber transmission systems are now widely used in carrying real-life communications traffic throughout the optical-fiber-based long-haul digital telecommunications network.

Clearly optical fiber transmission technology has emerged as the telecommunication system technology of the 1980s and will remain the key communications technology for a long time to come. The key advantages of the optical fiber transmission technology are now well known:

1. small size and light weight;
2. low transmission loss;
3. very high bandwidth possible;
4. immunity to electromagnetic interference;
5. optical cable flexibility and ruggedness;
6. optical fiber's electrical isolation;
7. signal security within the fiber waveguide;
8. potentially low cost per bandwidth–distance product;
9. optoelectronic devices capable of very high speed operation;
10. technology base shared with other emerging optoelectronic information systems such as laser printers, optical disks, and optical fiber sensors.

Because of these unique advantages, optical fiber transmission systems are widely used not only in the long-haul telephone transmission network but also in specialized data transmission applications within a ship, an aircraft, an industrial plant, an electric power company's monitoring network, along the railroads, and so on. In the near future, the use of fibers for intrabuilding- or intercampus-type local area data communications as well as for broadband video services will certainly see a significant growth. An all-optical-fiber interconnected Broadband Integrated Services Digital Network (BISDN) providing two-way transmission of telephone, data, and high-quality audio and video signals will likely be implemented, eventually making the information age a reality.

This chapter presents a basic review of the essential aspects of optical fiber transmission technology, including optical fibers, optoelectronic devices, and optical components used in optical fiber communication systems. Basic and essential practical information is presented and summarized without the theoretical derivations; relevant emerging technology trends are also briefly discussed. For detailed in-depth theoretical study as well as comprehensive discussions, readers should consult existing books [1–4] and advanced literature [5].

1.2 OPTICAL FIBERS FOR TRANSMISSION

Before the first proposal on using optical glass fiber waveguides for long-distance telecommunications, optical fibers have high losses and have been used for short-distance light guiding. The key advance thus was the realization of very low transmission loss in glass fiber waveguides, which enabled the use of optical fibers for long-distance communication applications. Both the optical waveguiding and the material transmittance (loss due to attenuation, scattering, etc.) of a glass optical fiber are therefore equally important. The next few sections discuss the basic waveguiding properties including the transmission loss and dispersion characteristics.

MULTIMODE STEP INDEX FIBER

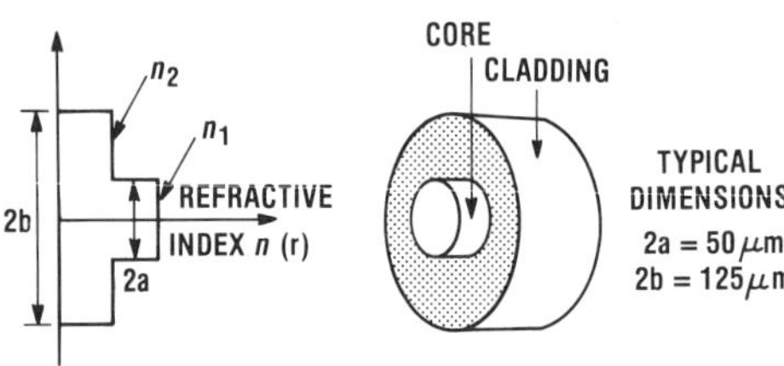

SINGLE-MODE STEP INDEX FIBER

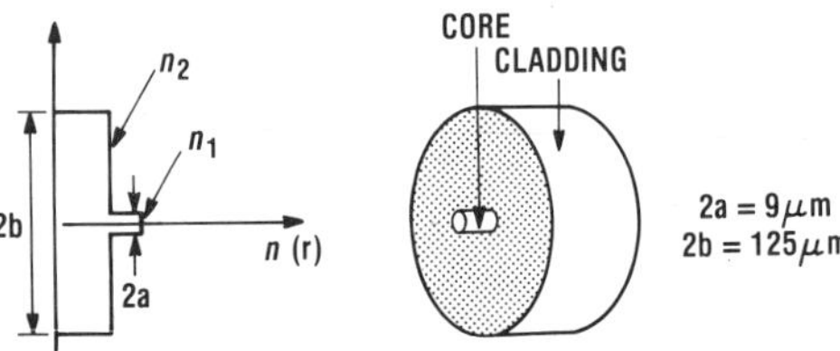

Figure 1.1 Refractive index profiles of the core and the cladding of a multimode step index fiber and a single-mode step index fiber. The core diameter is $2a$ and the cladding diameter is $2b$.

1.2.1 Optical Fiber Waveguides

The basic structure of an optical fiber waveguide is a cylindrical glass fiber with two cylindrical layers: the interior region, the core, of refractive index n_1, is surrounded by an outer region, the cladding, of refractive index n_2, where $n_2 < n_1$. Typically $n_1 - n_2 \ll 1$; the index difference is usually achieved by a slightly different doping concentration in the host material (e.g., SiO_2 host with small concentration of GeO_2 as dopant). Figure 1.1 shows two such typical step index fiber structures and the corresponding refractive index variations (profiles) along the radial direction. For a typical single-mode fiber the core diameter is about 9 μm, the cladding diameter is 125 μm, and the index difference, $\Delta n = n_1 - n_2$, is about 0.0035. For multimode fibers, the core diameter could be 50, 65, 85 μm, or larger depending on the fiber design. Figure 1.1 shows a multimode fiber with 50 μm core diameter and 125 μm cladding diameter, with the index difference being $\Delta n = 0.016$, which is higher than that of the single-mode fiber case.

The rigorous theoretical approaches [2,3,6] involve solving Maxwell's equations for the electromagnetic wave propagation in these cylindrically symmetric dielectric waveguide structures. Simplifications in the analysis can be made by noting that the index differences are usually very small ($\Delta n \ll 1$). Nevertheless, an electromagnetic wave propagation analysis is still rather complicated. A physical intuitive understanding of most of the basic propagation properties in optical fiber waveguides could be obtained with a simple ray picture based on geometrical optics of total internal reflection between two dielectric media of different refractive indices. Figure 1.2 illustrates such a case. From Fig. 1.2*a*, Snell's law for refraction leads to

$$n_1 \sin \theta_1 = n_2 \sin \theta_2 \tag{1.1}$$

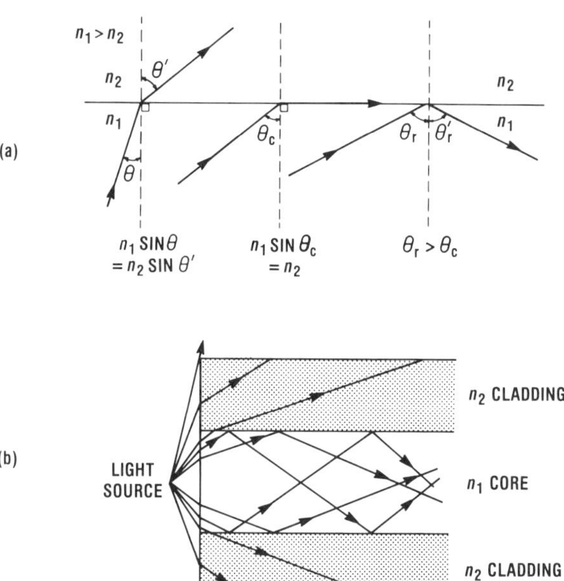

Figure 1.2 (*a*) Refraction of light at the interface between media of refractive index n_1 and n_2, for $n_1 > n_2$. Total internal reflection occurs when incident angle θ is $\geqslant \theta_c$, the critical angle. (*b*) A simple geometrical optics picture of guiding in an optical fiber by total internal reflections at the core–cladding boundaries.

The expression of the critical angle θ_c for total internal reflection (the angle of incidence above which the light ray is totally internally reflected rather than refracted into the second medium) is

$$n_1 \sin\theta_c = n_2 \sin(\tfrac{1}{2}\pi) = n_2 \qquad (n_1 > n_2)$$
$$\theta_c = \sin^{-1}\frac{n_2}{n_1} \tag{1.2}$$

Figure 1.2*b* shows that for incident light rays with different incident angles at the core–cladding boundary, some of them, with an incident angle less than the critical angle for total internal reflection, will be reflected back and thus guided in the core, while some of them will be refracted into the cladding and not be guided in the core. Usually there is a light-absorbing jacket region outside the cladding, so the light scattered out of the cladding will be absorbed. This simple ray picture of total internal reflection can thus explain the basic light-guiding principle.

However, to go one step further to explain the existence of fiber waveguide "modes" based on simple ray optics is not simple. A better description is obtained directly from electromagnetic wave propagation analysis. Such analysis results [2,3,6] show that light rays in the optical fiber waveguide can only propagate with a discrete set (and not a continuum) of incident angles, corresponding to the discrete set of waveguide modes, each mode having a specific spatial intensity distribution in the transverse plane [2,3,6]. The number of waveguide modes an optical fiber can support depends on the fiber's waveguide parameters such as core diameter, core–cladding refractive index difference, and the light signal wavelength. Optical fibers that can support many waveguide modes are called *multimode fibers*; optical fibers that are designed to support only one mode (the fundamental waveguide mode) are called *single-mode fibers*.

Before we go to the next section, let us first define a few useful parameters. From the optical fiber waveguide dimension a (core diameter), core index n_1, and cladding index n_2, several commonly used parameters in describing optical fiber waveguides can be defined:

1. The *relative* core–cladding index difference Δ is given by

$$\Delta = \frac{n_1 - n_2}{n_1} = \frac{\Delta n}{n_1} \tag{1.3}$$

2. The numerical aperture NA, a measure of the light-capturing capability of the optical waveguide, is given by

$$\mathrm{NA} = n_1 \sin(\tfrac{1}{2}\pi - \theta_c) = n_1 \cos\theta_c = (n_1^2 - n_2^2)^{1/2} \simeq n_1(2\Delta)^{1/2} \tag{1.4}$$

3. The normalized frequency V (also called the V-number) is given by

$$V = \frac{2\pi a}{\lambda}(n_1^2 - n_2^2)^{1/2} \simeq \frac{2\pi a}{\lambda} n_1(2\Delta)^{1/2} \tag{1.5}$$

where λ is the wavelength of the light. These parameters, defined based on the fiber waveguide structural parameters, are very useful in various discussions on fiber waveguide propagation properties.

1.2.2 Multimode and Single-Mode Optical Fibers

A typical multimode fiber has a large core diameter (50–200 μm) and a large index difference between the core and the cladding ($\Delta n = n_1 - n_2 = 0.01$–$0.03$). In this case there is a large number of discrete waveguide modes that can propagate in the fiber.

It can be shown [2,3,6] that for a large-core, multimode step index fiber with a large number of modes M, M can be approximately expressed by

$$M \cong \frac{V^2}{2} = \frac{1}{2}\left(\frac{2\pi a}{\lambda}\right)^2 (n_1^2 - n_2^2) = 2\pi \frac{A_c}{\lambda^2} (\text{NA})^2 \tag{1.6}$$

where $A_c = \pi a^2$ is the core area. Thus, in multimode step index fibers, the large index differences and large core diameters lead to a large NA and a large V-number. Therefore, the number of modes that can propagate in the fiber is large. For example,

$$n_1 = 1.46 \qquad n_2 = 1.45 \qquad 2a = 50\ \mu\text{m} \qquad \text{NA} = 0.17$$

$$V = 20 \quad \text{at} \quad \lambda = 1.27\ \mu\text{m} \qquad M \simeq \frac{V^2}{2} = 200$$

The main difference in the waveguide characteristics between the single-mode and the multimode optical fiber waveguides is that the former can only support one waveguide mode while the latter can support propagation of many waveguide modes. This is due to the fact that multimode fibers have larger core diameters and larger refractive index differences between the core and the cladding, while single-mode fibers have much smaller core diameters and index differences that only one waveguide mode can propagate. This in turn leads to a very significant difference in transmission bandwidth and dispersion properties.

It can be shown [2,3,6] that in a *step index fiber* the incident light will propagate in only one waveguide mode (the fundamental mode) when the core diameter and the refractive index difference between the core and the cladding are small enough such that the normalized frequency V (the V-number) is

$$V = \frac{2\pi a}{\lambda} (n_1^2 - n_2^2)^{1/2} < 2.405 \tag{1.7}$$

This small V-number ($V < 2.405$) is to be compared with a typical V-number of 20 or more for a large-core, large-index-difference multimode fiber.

The single-mode condition described in the preceding can also be expressed in terms of the wavelength of the propagating light, λ, with respect to the fiber's *cutoff wavelength* λ_c,

$$\lambda_c = \frac{2\pi a(n_1^2 - n_2^2)^{1/2}}{2.405} = \frac{V}{V_c} \lambda \tag{1.8}$$

In this case of $\lambda > \lambda_c$, only one mode can be guided in the fiber waveguide (actually two if counting the two possible polarizations). The optical fiber is thus a single-mode fiber waveguide *for light with optical wavelength longer than the cutoff wavelength of the single-mode fiber waveguide.* For example, if a single-mode fiber has a core diameter $2a = 10\ \mu$m and index difference $n_1 - n_2 = 0.003$, then the condition $V < 2.405$ leads to

TABLE 1.1 Examples of SM and MM Fiber Parameters

	SM Fiber	MM Fiber
$2a$	10 μm	50 μm
n	0.003	0.01 (NA = 0.17)
V-number at 1300 nm	<2.4	20
Number of modes	1	200

a cutoff wavelength

$$\lambda_c = \frac{\pi(10)(2 \times 1.46 \times 0.003)^{1/2}}{2.405} = 1.22\ \mu\text{m}$$

With this optical fiber waveguide, an InGaAsP laser or light-emitting diode (LED) at 1.3 μm will propagate in the fiber in the fundamental mode only because $\lambda > \lambda_c$, so the fiber is a single-mode fiber at 1.3 μm wavelength.

It is important to note that the same 10-μm-core-diameter fiber in the preceding example will support higher-order modes (thus becoming a multimode fiber that supports a few higher-order modes) at the He–Ne laser wavelength of 0.63 μm or the GaAlAs laser wavelength of 0.8 μm, because in either case the corresponding V-number exceeds 2.405. In terms of the fiber's single-mode cutoff wavelength, both optical sources have wavelengths *shorter than* the fiber's cutoff wavelength ($\lambda < \lambda_c$) and thus will propagate in the fiber as a mixture of higher-order modes. For an optical fiber to be single mode at these shorter wavelengths, a smaller core diameter (5–6.5 μm) is needed to satisfy the same condition of $V < 2.405$. The definition of the single-mode fiber is thus closely tied to the cutoff wavelength and is not independent of the optical source wavelength.

It should be pointed out that in actual single-mode fibers, the practical cutoff wavelength has to be experimentally measured in the fiber cable, because the actual cutoff wavelength depends on the cabling, the fiber/cable length, and the bend radius and could be shorter than the theoretical cutoff wavelength calculated based on (1.7) or (1.8), which is true only for step index single-mode fibers. Table 1.1 illustrates the fiber core diameters and index differences for a typical single-mode (SS) fiber (uncabled) with its cutoff wavelength near 1220 nm, designed for 1300-nm transmission systems. For comparison, parameters for a multimode (MM) fiber are also shown.

1.2.3 Graded Refractive Index Profiles and Modal Dispersions

The refractive index difference between the core and the cladding region is responsible, as we already discussed, for the total internal reflection and hence waveguiding in the fiber core region. The index step shown in Fig. 1.1 is only one of the possible index profiles for the core and the cladding regions; in reality, a truly step index profile with sharp boundaries is not always obtained.

There are a variety of refractive index profiles possible for both multimode and single-mode optical fibers; the refractive index profiles together with the doping material properties determine the majority of the optical fiber waveguiding characteristics. Different refractive index profiles have been proposed and realized to achieve various special optical waveguide characteristics. The most well known example is the nearly parabolic graded index profile in multimode fibers for minimizing the intermodal delay distortions (modal dispersion). The other important example is the triangular refractive index profile in single-mode fibers for dispersion shifting (shifting the minimum chromatic dispersion wavelength to the lowest loss region of 1550 nm), which will be discussed in a later section.

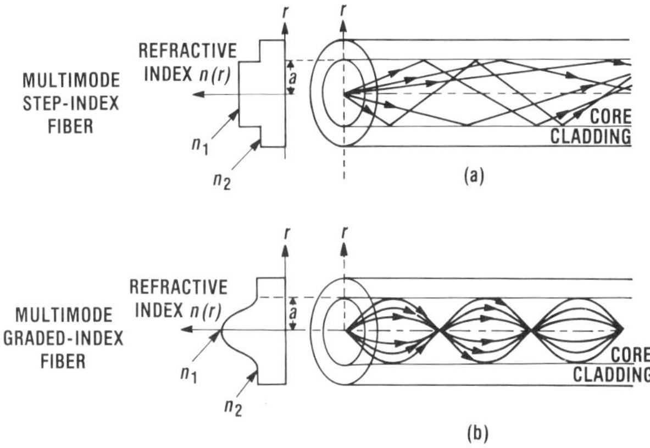

Figure 1.3 The refractive index profile and ray path in a multimode step index fiber and those in a multimode graded index fiber. Note the multipath delay difference in the step index fiber is large (large modal dispersion), while that in an ideal graded index fiber is minimized.

Figure 1.3 shows the graded index multimode fiber profile as compared with the step index multimode fiber profile. In a typical step index multimode fiber, the different rays with different angle of incidence will have different travel path lengths in the optical fiber waveguide; this will result in a time-of-arrival difference, also called modal delay spread between the different modes, which causes intermodal delay distortions (modal dispersion). The maximum modal delay spread $\Delta T_{\max}$ is that between the ray traveling along the fiber axis (the fastest ray) and the ray traveling at the maximum angle bouncing off the core–cladding interface (the slowest ray), as shown in Fig. 1.4 [1–4]:

$$\begin{aligned}\Delta T_{\max} &= t_{2,\max} - t_1 = \left.\frac{n_c L}{c\cos\theta}\right|_{\max} - \frac{n_c L}{c} \\ &= \frac{n_c L}{c}\left(\frac{n_c - n}{n}\right) = \frac{L}{c}\frac{\Delta n\, n_c}{n} = \frac{L}{2cn}(\mathrm{NA})^2 \end{aligned} \tag{1.9}$$

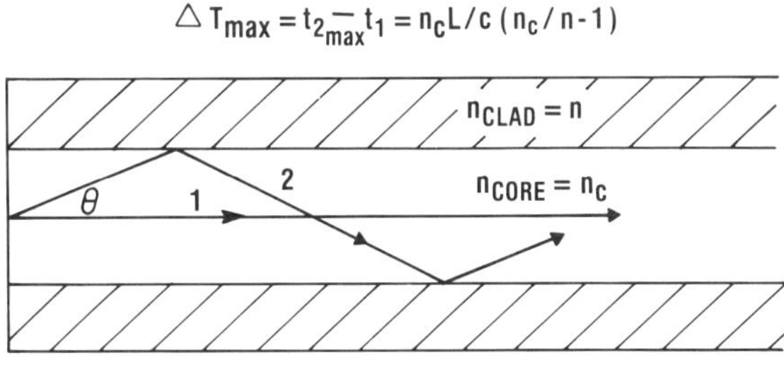

Figure 1.4 The optical path difference between the fastest ray (on axis) and the slowest guided ray leading to modal delay spread in the step index multimode fiber.

Note here that $n_c = n_1$ (core) and $n = n_2$ (cladding) and c is the speed of light. The maximum delay spread due to modal dispersion is therefore proportional to the index difference Δn and to the square of the NA. Thus, in general, the larger the index difference (or equivalently the larger the numerical aperture), the larger the maximum modal delay spread. For example, if $n_1 - n_2 = 0.01$, NA = 0.17, the maximum modal delay spread amounts to 50 ns/km. This means for 10-Mb/s optical signals (minimum pulse spacing 100 ns) the delay spread in a 2-km length of this fiber would be 100 ns, causing the intersymbol interference of the adjacent signal pulses. Such a large modal delay spread, due to modal dispersion of the fiber, would limit the use of this fiber to <10 Mb/s and 2-km long-distance applications. A higher bit rate or longer distance transmission would be seriously degraded in such a multimode fiber.

To reduce the modal delay spread and improve the performance, instead of using a step-index-profile multimode fiber, graded-index-profile multimode fibers can be fabricated to reduce the modal delay spread (modal dispersion), as shown in Fig. 1.3. With the graded index profile shown, the outer rays, traveling farther from the core center axis and longer overall path, see a lower refractive index (to speed up the propagation time), and the rays closer to the core axis, traveling a shorter distance, see a larger refractive index and travel at a slower speed. In this way the graded index profile is designed to equalize the travel times of all the different rays (modes) so that the overall delay spread is minimized.

The graded index profile in a multimode graded index fiber is often expressed by

$$n(r) = \begin{cases} n_1\left[1 - 2\Delta\left(\dfrac{r}{a}\right)^{\alpha}\right]^{1/2}, & r < a \quad \text{(core)} \\ n_1(1 - 2\Delta)^{1/2} = n_2, & r \geqslant a \quad \text{(cladding)} \end{cases} \tag{1.10}$$

where α is the graded index profile parameter; $\alpha = 2$ corresponds to a parabolic profile; $\alpha = 1$ corresponds to a triangular profile, while the step index profile is the special case of $\alpha = \infty$. Extensive theoretical studies [2,3,6] have shown that when α is very close to 2, the fiber has the optimum refractive index profile for minimizing the modal delay spread and thus achieving the smallest modal dispersion multimode fiber propagation. This ideal profile is called a nearly parabolic profile.

Without getting into detailed discussion, the following important facts concerning multimode graded index fibers and modal dispersion characteristics can be summarized [1–4,6]:

1. Actual graded index profile multimode fibers have their profiles made by small step index increments attempting to approximate the ideal near-parabolic index profile; the resultant bandwidth is usually much less than that expected of the ideal profile. Consequently, while ideal profile theoretically can yield 100-GHz-km bandwidths (bandwidth–distance product), the best experimental results are in the 4–6-GHz-km range, while the majority of typical commercial multimode graded index fibers have much smaller bandwidths, in the 400-MHz-km to 1.5-GHz-km range, depending on the NA, the dopants, and the exact index profile.

2. The refractive index profile in a graded index multimode fiber is usually designed to approach the optimum profile *at a given operating wavelength.* Due to "profile dispersion" (wavelength dependence of the index difference), a graded index multimode fiber optimized to have a high bandwidth at one wavelength (e.g., 850 nm) may have a low bandwidth at a different wavelength (e.g., 1300 nm), because the optimum α is wavelength dependent, that is, $\alpha = \alpha(\lambda)$. Figure 1.5 shows examples of such "bandwidth spectra" of

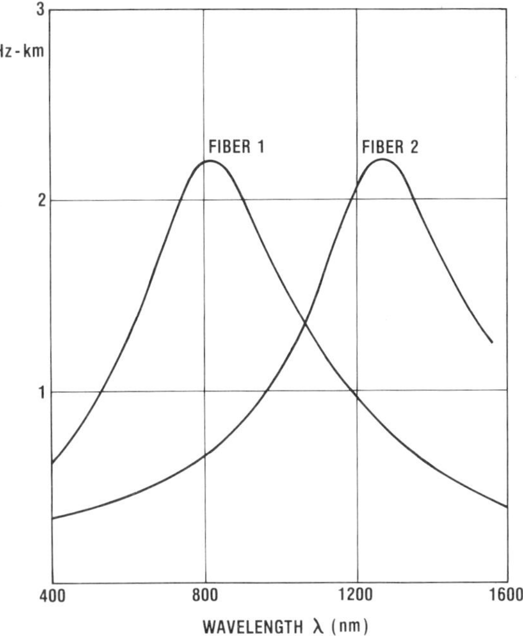

Figure 1.5 Bandwidth spectra of two graded index multimode fibers, fiber 1 was optimized for 800 nm and fiber 2 for 1300 nm.

two graded index multimode fibers [7]. This spectral dependence of the bandwidth, which is sharply peaked near the optimized wavelength only, makes multimode fibers unattractive for optical fiber systems requiring upgrading by WDM (wavelength division multiplexing).

3. The phenomenon of mode mixing in a multimode fiber leads to a situation where the bandwidth of a concatenated multimode fiber is a nonlinear function of fiber link length, causing some complication in the design of long-link-length multimode fiber transmission systems.

In spite of these serious bandwidth limitations, several early long-haul multimode optical fiber transmission systems were implemented (usually at a lower bit rate such as 45–90 Mb/s) in the 1970s. However, after the 1980s, single-mode fiber transmission systems have been exclusively used for long-distance and high-bit-rate optical communication links because multimode fiber-based systems not only have the above-mentioned bandwidth limitations but also are found to have the "modal noise" problem [2,3] caused by time-varying, fluctuating intensity patterns due to the coherent interference of the fiber modes associated with external disturbances and/or displaced fiber joints. The presence of modal noise is a serious transmission limitation for long-distance transmission systems using multimode fibers; the use of low-coherence optical sources such as LEDs or wide-spectral-width multi-longitudinal-mode laser diodes could reduce the modal noise, but at the expense of increased chromatic dispersion penalty.

Today multimode fibers are principally being used in short-distance (< 1 km typically) transmission, although there were longer-distance multimode fiber telecommunication systems installed before 1980. With a cladding diameter of 125 μm, multimode fibers typically have a core diameter of 50, 62.5, or 85 μm, with an NA of 0.2, 0.275, and 0.26, respectively; the choice depends on specific application requirements. In most cases, especially for short-distance, local area network (LAN) communication between computers, the primary concern is usually the power budget, coupling and branching losses, and not the bandwidth or dispersion limitations. For example, with commercial large-NA graded index multimode fibers, the modal dispersion limitation is not serious for a few 100-Mb/s, $\leqslant$1-km transmission applications, as the fiber bandwidth is typically in the 200–600-MHz-km range. The high LED light-coupling efficiency, the low fiber jointing (connectors and splices) losses, and low branching losses (through fiber couplers, splitters, etc.) of large-core, large-NA graded index multimode fibers are unique advantages that make it possible to use low-cost optical sources and components for large volumes of LAN applications.

1.2.4 Doping and Optical Transmission Loss Spectra

The refractive index profile, whether a step index or a graded index difference between the core and the cladding region, is achieved in practice by doping the host material (in our case silica, SiO_2) with a dopant that either raises or reduces the effective refractive index of the doped region. For example, GeO_2 doping will raise the refractive index above that of the pure silica, while B_2O_3 doping reduces the index. Depending on the dopant, the doped region (core or cladding) will have a loss characteristic modified from that of the pure silica. Figure 1.6 illustrates the effect of various dopants on the fiber transmission loss; note some dopants cause the infrared (IR) absorption edge to rise at a shorter wavelength. Similarly, water, not an intentional dopant but practically unavoidable, contributes to the OH^- loss peaks near 1.4, 1.25, and 0.95 μm [1–4]. Likewise, other impurities in the fiber preform fabrication process may cause localized absorption peaks.

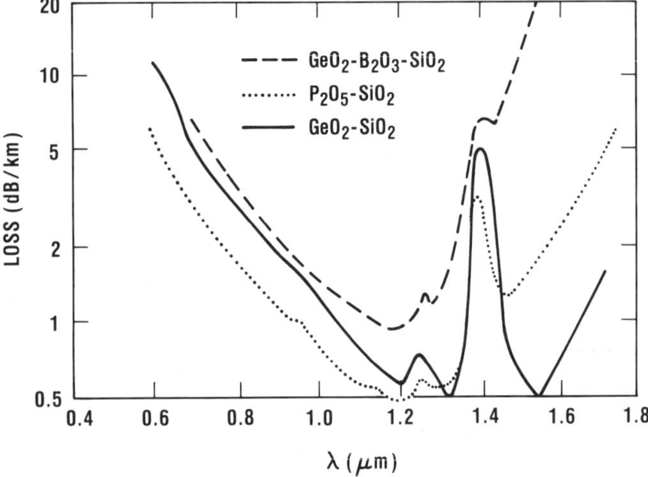

Figure 1.6 Effect of various dopants on the silica glass fiber loss spectra.

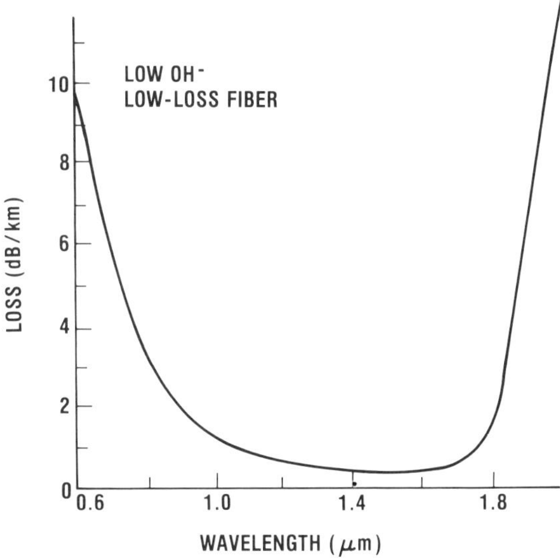

Figure 1.7 Loss spectrum of a very low OH content silica glass fiber, showing a single low-loss transmission window.

Various fiber preform fabrication and fiber drawing processes have been developed for making low-loss glass fibers with the desired refractive index profiles, geometrical parameters, and mechanical properties but with least impurities or OH absorption. For a discussion on the various fiber fabrication processes such as modified chemical vapor deposition (MCVD), outside vapor phase oxidation (OVPO), vapor axial deposition (VAD), and plasma-activated chemical vapor deposition (PCVD), see Refs. 3 and 4.

With the advances in the past 18 years, fiber fabrication and manufacturing technology have now reached a mature stage; commercially available optical silica glass fibers now have in general very low losses. Figure 1.7 shows the loss spectra of a low-OH, low-loss silica glass optical fiber. Typical low-loss fibers can easily have a loss of about 2.4 dB/km at 0.8 μm, 0.4 dB/km at 1.3 μm, and 0.25 dB/km near 1.56 μm. Since fiber loss, or attenuation coefficient α_l in decibels per kilometer, is defined by

$$P(z) = P_0 10^{-\alpha_l z/10} \qquad P_0 = P(z=0)$$
$$\alpha_l = \frac{1}{L}\left[10\log_{10}\frac{P_0}{P(L)}\right] \tag{1.11}$$

where L is the fiber length. To appreciate the high transparency of optical glass fibers, note that a loss of 0.25 dB/km means 94.4% of optical transmission after 1 km of glass fiber, an extremely low loss indeed.

While the rising edge in the IR region beyond about 1.7 μm is due to the IR absorption of the host and the dopant glass, the loss curve in the visible–near IR region actually is very close to the intrinsic loss limit due to the Rayleigh scattering associated with microscopic compositional and density fluctuations of the glass [2–6]. Figure 1.8 illustrates the contributions of various loss mechanisms including ultraviolet (UV) and IR absorption (intrinsic, not due to impurity), waveguide imperfection, and Rayleigh scattering [3,4] and the experimental loss spectra that include the contributions from impurity

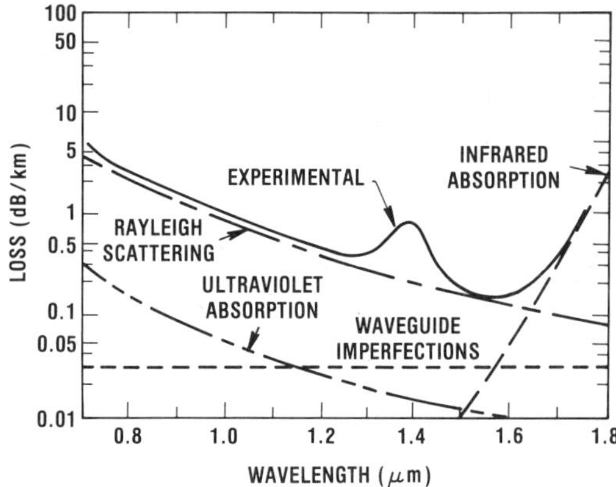

Figure 1.8 Figure showing the relative contributions of the various attenuation mechanisms (shown in dashed and dotted lines) to the loss spectra (solid curve) of a low-loss silica glass fiber.

and OH absorptions. Because Rayleigh scattering is inversely proportional to λ^4, it is smaller in the IR region. The loss minimum is thus somewhere near the 1.6-μm region bounded by the decreasing Rayleigh scattering loss curve and the rising IR absorption edge. A minimum loss of 0.16 dB/km near 1.58 μm is possible.

In typical low-loss silica glass optical fibers, the transmission window spans from 0.6 to 1.7 μm, with a water peak around 1.4 μm. Although with a low-water-peak fiber the *entire transmission window is useful* for telecommunication applications, due to the availability of laser diode and LED sources, most of the present telecommunication optical fiber systems operate in the transmission loss windows near 0.8, 1.3, and 1.55 μm. In addition to telecommunication applications, optical fibers are also useful for a variety of very short distance transmission applications such as light delivery and data transmission in an automobile and an airplane, high-power IR and UV laser power delivery, medical applications in a hospital, or robotic material processing applications in a factory.

In a discussion of fiber loss, it should also be mentioned that in addition to low-loss silica-glass-based fibers, there are medium-loss (e.g., 20–40 dB/km) multicomponent glass fibers and higher-loss plastic-based optical fiber waveguides [2,3], usually for large-core multimode-fiber short-distance transmission applications. In fact, some plastic-based fibers have acceptably low (a few 100 dB/km) loss windows in the visible, allowing the use of inexpensive visible LEDs for some of the very short distance (e.g., automobile) applications previously mentioned.

1.2.5 Single-Mode Fiber Dispersion and Dispersion Control

In single-mode optical fibers, there is no group delay spread due to the different modes (modal dispersion) as only one mode is supported. The dispersion limitation in single-mode fibers is due to the glass material's wavelength dispersion (material dispersion) and is associated with the finite (nonzero) spectral width of the optical signals. This is often also referred to as chromatic dispersion, except in fiber waveguides, for which chromatic dispersion includes, in addition to material dispersion, the (usually small) waveguide dispersion due to the wavelength dependence of the propagation constant of the single mode. Actually chromatic dispersion exists in multimode fibers too, but it is usually small compared with modal dispersion, except when very wide spectral width LED sources are used in the fiber transmission. In single-mode fibers the chromatic dispersion becomes the dominant bandwidth limitation; the study, the measurement, and the improvement of chromatic dispersion characteristics have been important aspects of high-bandwidth optical fiber technology research and development [2–8].

Material dispersion in glass materials is due to the variation of the material's refractive index with wavelength. As a result, for a light signal with a finite spectral width $\Delta\lambda_s$ different spectral components of the signal travel at different speeds in the glass, causing a group delay difference due to chromatic dispersion (such as in a glass prism). The group velocity is determined by the group index n_g, that is, $v = c/n_g$, where the group index is related to the refractive index n by

$$n_g = n - \lambda \frac{dn}{d\lambda} \tag{1.12}$$

The most commonly used material dispersion parameter M is defined as

$$M(\lambda) = \frac{d\tau}{d\lambda} = \frac{-1}{c}\frac{dn_g}{d\lambda} = \frac{\lambda}{c}\frac{d^2n}{d\lambda^2} \tag{1.13}$$

For silica glass, material dispersion curves cross zero in the wavelength region near 1300 nm, the exact zero crossing depending on the dopants and the doping concentration [9]. This is the region of zero or minimum material dispersion.

After an optical signal pulse has propagated in a single-mode fiber, there is a group delay spread $\Delta\tau$ due to the different travel speeds of the various spectral components in the optical signal source, where $\Delta\tau$ is given by

$$\Delta\tau = M(\lambda)\,\Delta\lambda_s\,L \tag{1.14}$$

where M is a more general chromatic dispersion parameter (usually in ps/nm-km), including both the material dispersion and the waveguide dispersion, discussed later; $\Delta\lambda_s$ is the rms spectral width of the optical signal (in nanometers), and L is the single-mode fiber length (in kilometers). Therefore, it is desirable to have a small-spectral-width optical source such as a narrow-spectral-width laser diode operating near the minimum chromatic dispersion wavelength region (M small) to have the minimum dispersion limitation and higher bandwidth transmission. For example, with a conventional single-mode fiber designed for 1300-nm operation, the chromatic dispersion is typically less than 2 ps/nm-km in the spectral range of 1280–1330 nm. On the other hand, the dispersion is as large as 110 ps/nm-km near 800 nm and 20 ps/nm-km near 1550 nm, limiting the transmission bandwidth distance to a value much less than that achievable near the minimum dispersion region.

Since silica glass fibers have lowest transmission loss in the 1550-nm spectral region rather than in the 1300-nm region, it is desirable to have a fiber with minimum dispersion also in the 1550-nm region. This requires the ideas and techniques for dispersion shifting in single-mode fiber design by controlling the waveguide dispersion. In optical fibers, the wavelength dependence of the propagation constant within an individual mode gives rise to a small spectral-dependent delay difference; this is called waveguide dispersion. Since a single-mode glass fiber is a glass fiber waveguide, the material dispersion effect and the waveguide dispersion effect both need to be considered. The sum is called total chromatic dispersion, or simply, the chromatic dispersion of the single-mode fiber, with the same parameter symbol $M(\lambda)$ used to designate the chromatic dispersion. Waveguide dispersion is usually small compared with material dispersion except in the region of zero material dispersion. In the region near zero material dispersion, the wavelength dependence is such that waveguide dispersion can be used to modify the resultant total dispersion characteristics, such as shifting the zero crossing point to a new wavelength.

The dispersion-shifted single-mode fiber was first realized by

1. using GeO_2 doping to increase the material dispersion minimum wavelength and
2. tailoring the waveguide dispersion in a step index single-mode fiber.

The latter was achieved by operating the single-mode fiber at a small V-number, achieved by designing a smaller core diameter with a larger index difference for the single-mode fiber. The combination of a small V-number and a larger index difference led to the desired waveguide dispersion characteristics for dispersion shifting [10]. In this first experimental demonstration, the dispersion minimum was successfully shifted to the desired 1550-nm region. However, the increased germanium doping caused a higher Rayleigh scattering and hence a higher transmission loss.

Subsequent research and development efforts based on the same idea of waveguide dispersion tailoring have come up with the triangular-shaped index profile single-mode fibers [11–13] that achieved dispersion shifting to the 1550-nm region with a smaller doping requirement (thus keeping the loss lower than if the step index profile is used). Presently commercially available dispersion-shifted single-mode fibers are based on the triangular index profile design and its variations.

To distinguish between the conventional single-mode fibers (minimum dispersion near 1300 nm) and the dispersion-shifted single-mode fibers (minimum dispersion near 1550 nm), we shall use C-SMF to designate the former and DS-SMF for the latter in the rest of this chapter.

It should be pointed out that whether it is C-SMF or DS-SMF, operating at the "zero"-dispersion wavelength does not mean there is no dispersion effect. Theoretical analyses [14–16] have shown that at the minimum dispersion wavelength, only the first-order dispersion vanishes, but the higher-order dispersions may still cause pulse distortions that may lead to intersymbol interference. This represents the ultimate dispersion limitations of single-mode fibers (if there is no polarization dispersion [16]). For the majority of present applications, however, this is of little concern, because these limitations are significant only at 10–100 Gb/s bit rates and over substantial transmission distances [16].

Figure 1.9 shows the chromatic dispersion characteristics of the two types of single-mode fibers: C-SMF and DS-SMF. Note that in addition to the different zero crossing wavelengths, the slopes of the chromatic dispersion curves near the zero crossing point [$S(\lambda_0)$] are also different; while the dispersion slopes for both types of fibers are less than 0.1 ps/km-nm^2, the DS-SMF has a more gradual slope, which is advantageous for multiwavelength WDM applications. Near the minimum dispersion wavelength region the single-mode fiber bandwidth-distance product is $\geqslant$1000 GHz-km [16]. This is a practically unlimited bandwidth, considering that the wide low loss window allows multiwavelength WDM operation. This is in sharp contrast to the $<$10 GHz-km bandwidth of multimode fibers.

In terms of practical applications, it should be mentioned that the C-SMF fibers have been widely used with 1300-nm laser diode optical transmitters in the existing digital

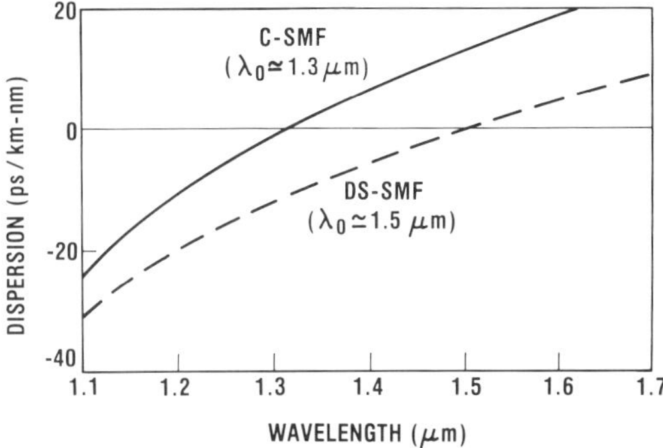

Figure 1.9 Chromatic dispersion as a function of wavelength for the C-SMF (conventional single-mode fiber, $\lambda_0 = 1.3\ \mu m$) and the DS-SMF (dispersion-shifted single-mode fiber, $\lambda_0 = 1.5\ \mu m$).

telephone trunk and interoffice transmission network, typically carrying 140 Mb/s, 565 Mb/s, and 1.12 Gb/s traffic over 10–40-km distances. In contrast, DS-SMF fibers, which have become commercially available only in the past few years and typically require 1550-nm laser sources, have only begun to be used in a few long-distance transmission systems. In a later section the different system implementation strategies, the device and fiber requirements, and potential applications will be discussed.

In addition to the conventional step index single-mode fiber and the triangular profile dispersion-shifted single-mode fiber, there are also single-mode fibers with various different refractive index profiles in both the core and the cladding regions that help achieve certain desired propagation characteristics. There are a large number of single-mode fiber designs using various index profiles and dopants. A few examples are shown in Fig. 1.10. These include W-profile fibers and doubly clad (DC) and quadruply clad (QC) fibers designed to tailor the waveguide dispersion characteristics such that the total chromatic dispersion either has two zero crossings (two dispersion minima) or has low dispersions (e.g., less than ± 2 ps/nm-km) over a broad spectral range (e.g., from 1200 to 1600 nm). These are called *broadband low dispersion* or *dispersion-flattened* single-mode fibers [17]. They are still in the research and exploratory development stage; if these broadband

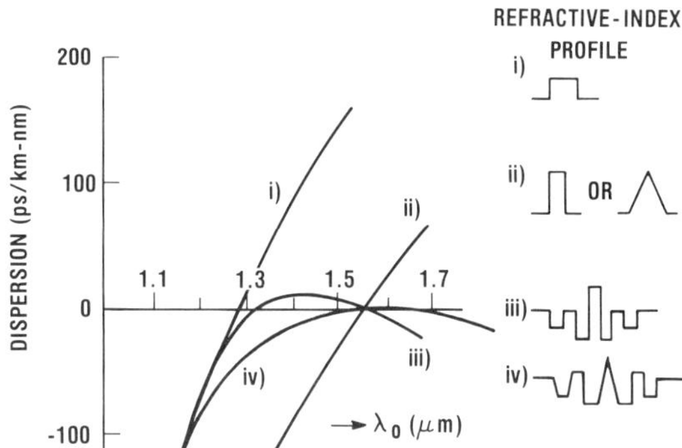

Figure 1.10 Dispersion versus wavelength of four types of single-mode fibers with different refractive index profiles: (i) conventional step index; (ii) dispersion-shifted, step index (small core, high index step) or triangular profile; (iii, iv) dispersion-flattened or broadband low-dispersion fibers, with multiple-cladding layers.

low-dispersion fibers can be made inexpensively with low loss and low dispersion over the 1200–1600-nm spectral range, they would be ideal for broadband WDM applications. Economics and future system needs will determine the practical applications of such advanced single-mode fibers.

1.2.6 Spot Size, Mode Field Diameter, and Gaussian Beam in Single-Mode Fibers

One important parameter of single-mode optical fiber waveguides is the mode field diameter, also known as spot size, of the optical field propagating in the single-mode optical fiber waveguide as the fundamental mode (also called LP_{01} mode, see Refs. 2–6). It has been shown [18] that the optical field's intensity distribution in the single-mode optical fiber waveguide can be closely approximated, in many cases except for very small V-numbers or unusual index profiles, by a Gaussian intensity distribution:

$$P(r) = P(0)e^{-2(r/\omega_0)^2} \tag{1.15}$$

where the intensity of light is at its maximum at the core center ($r = 0$) and drops to 13.5% of its maximum (i.e., the $1/e^2$ point) at $r = \omega_0$. Here, ω_0 is the mode field radius and $2\omega_0$ is the mode field diameter, or the spot size of the Gaussian intensity distribution. In general, the mode field diameter is not the same as the single-mode fiber core diameter $2a$. Typically for step index single-mode fibers the mode field diameter $2\omega_0$ is between 1 and 1.5 times the fiber core diameter $2a$ [18], depending on the ratio of operating (source) wavelength λ to the cutoff wavelength λ_c.

It is important to note that the optical field in a single-mode fiber waveguide extends well into the cladding region and is not confined to the core region. This of course points out the limitation of the simple ray picture of waveguiding by total internal reflection at the core–cladding boundary as described in Section 1.2.2. Maxwell equations for the electromagnetic waves propagating in a cylindrically symmetrical single-mode optical fiber waveguide have to be solved to obtain the mode field distributions in the core and the cladding regions [18]. Such solutions show that the optical energy of the propagating mode in the single-mode optical fiber that is contained in the core region is only about

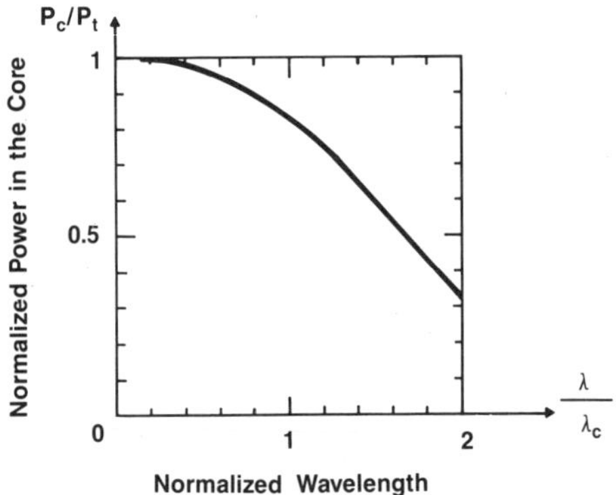

Figure 1.11 The fundamental mode (LP_{01} mode) power contained in the core, P_c, normalized to the total guided power, P_t, as a function of normalized wavelength λ/λ_c (λ_c is the cutoff wavelength of the single-mode fiber): $\lambda/\lambda_c = 1$ corresponds to $V = 2.4$; $\lambda/\lambda_c < 1$ corresponds to $V > 2.4$, when the higher-order mode can be guided.

80% when the V-number [from Eq. (1.7) the V-number is a normalized wavelength] is at 2.405, that is, right at the single-mode fiber cutoff condition ($\lambda = \lambda_c$). Figure 1.11 shows the optical power contained in the core of a single-mode optical fiber for different λ/λ_c or V-numbers. The power in the core decreases with a decreasing V-number or an increasing ratio of λ/λ_c. When a high percentage of optical energy is propagating in the core region of the waveguide, it is a stronger guiding case. On the other hand, if a high percentage of the energy of the propagating optical wave extends well into the cladding, it is a weaker guiding case. This happens when the optical source wavelength is at such a long wavelength that the corresponding V-number is too small. For example, when $V = 1.2$ from Eq. (1.7) [corresponding to $\lambda = 2\lambda_c$ from Eq. (1.8)], it can be shown that the mode field diameter $2\omega_0$ is about 3 times the fiber core diameter $2a$ [18], which means there is weaker guiding with a substantial amount of energy ($>60\%$) in the cladding. In such a case the fiber is more sensitive to bending- and microbending-induced losses [18]. Typically, it is desirable to operate with a V-number in the range of 1.8–2.4 for a given single-mode fiber. Equivalently, this corresponds to having the optical source wavelength between λ_c and $1.33\lambda_c$ (cutoff wavelength) to reduce sensitivity to bending and microbending while maintaining a reasonable mode field diameter or spot for ease of fiber coupling and jointing.

1.2.7 Optical Fiber Jointing: Splices and Connectors

In optical fiber fabrication, glass preforms are first made by proper deposition of core and cladding materials (host and dopants), with the appropriate refractive index profiles and core–cladding diameter ratios. The preforms are then heated and drawn into thin optical fibers and coated with appropriate coatings in a tall drawing tower. Typical practical fiber length is in the range of 2–10 km, although much longer fibers can be drawn from large preforms. In practical applications, it is often necessary to have the ability to splice and connect the fibers so that different fiber lengths can be jointed together permanently or disconnected and reconnected as needed.

There are several types of splices and splicing techniques [2–5] for permanent or semipermanent fiber jointing as needed in initial system installation or subsequent fiber repair. There are fusion splices and mechanically bonded splices [5]. Fusion splicing machines with fully automated active optical alignment and splicing and with TV camera monitoring of the splicing are available commercially. Fusion splicing techniques use arc fusion or flame fusion of the glass materials to achieve low-loss and high-strength permanent joints. Mechanical splices are also nearly permanent joints except they use bonding materials such as epoxy and index gels and precision mechanical alignment fixtures such as silicon V-grooves for alignment. Both single-fiber and multifiber (such as in a fiber ribbon or array) fusion splicing and mechanical splices are available. Note good cleaved fiber ends are needed for low-loss splicing.

Connectors are intended for connecting and disconnecting the fiber links at the connecting points for system installation, rerouting, and service and maintenance purposes, which may or may not be frequent, although repeated connections are expected. There are many types of optical connectors, but basically they are either fiber butt-joint connectors or lensed expanded-beam connectors [3,4]. In butt-joint connectors, the fibers are aligned with each other and the fiber ends are butt jointed; the ends may or may not be in physical contact. Biconic connectors and ferrule connectors (FCs) belong to this category and require precision mechanical alignment of the butted fibers. In lensed expanded-beam connectors, collimating lenses are attached to the fiber ends, and the connectors are actually a high-precision imaging system plus stable mechanical fixtures for the optics. The optical fiber alignment tolerances are increased by the magnification of the optical system. Due to the nature of the collimated beam, lensed expanded-beam connectors have several unique features and are useful for device coupling applications such as in-line filters and isolators. However, these connectors are more complicated in optical design.

While insertion loss may be the most important parameter for optical jointing such as splices and connectors, optical reflection at the joint interfaces is becoming more and more of great concern. It is now recognized that even a low level of reflection can induce many undesirable effects in a high-speed optical fiber transmission systems (to be discussed in a later section); for future coherent transmission systems the reflection effect is even more severe. To avoid the reflection-induced noises and optical fiber transmission system penalties, low-loss fusion splices, angled mechanical splices, and low-reflection connectors should be used whenever possible. For optical connectors, several commercially available single-mode optical fiber connectors are either polished to have a convex fiber end surface to guarantee physical contact (such as in FC–PC connectors, where PC stands for physical contact) or polished to have an angle (5°–12°) so that reflection is out of the fiber core direction. In lensed expanded-beam connectors, good AR (antireflection) coatings on the lenses are needed to achieve low reflection. These special connectors have low reflections, with a return loss of 30 dB (less than 0.1% reflection) or higher (a higher return loss corresponding to a lower reflection level). This level of return loss (30–40 dB) is found to be desirable for advanced optical fiber transmission systems.

With good fusion splices between identical single-mode fibers (fibers from the same manufacturer and with the same glass composition and fiber parameters), the splice loss is typically less than 0.1 dB. Good mechanical splices have losses in the range of 0.1–0.2 dB. For single-mode fibers of different design or manufacture, the key parameter is

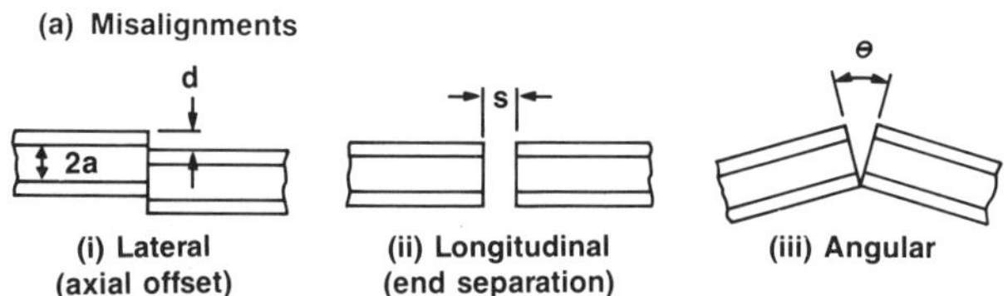

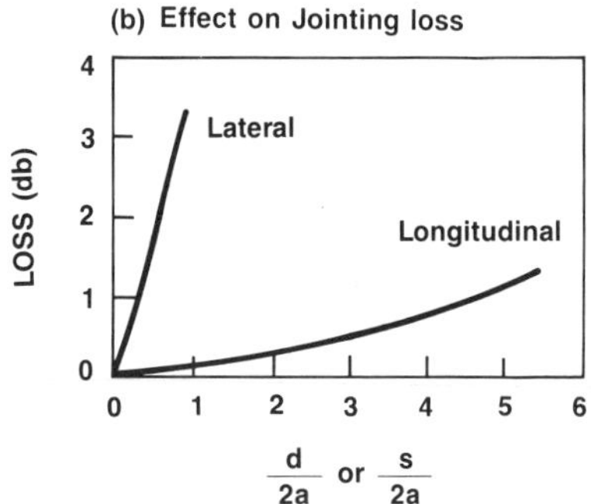

Figure 1.12 (*a*) Mechanical misalignments at a fiber-to-fiber joint (connector or splice) due to lateral, longitudinal, or axial offset. (*b*) Curves showing the sensitive effect of axial offset or end separation on the fiber joint loss.

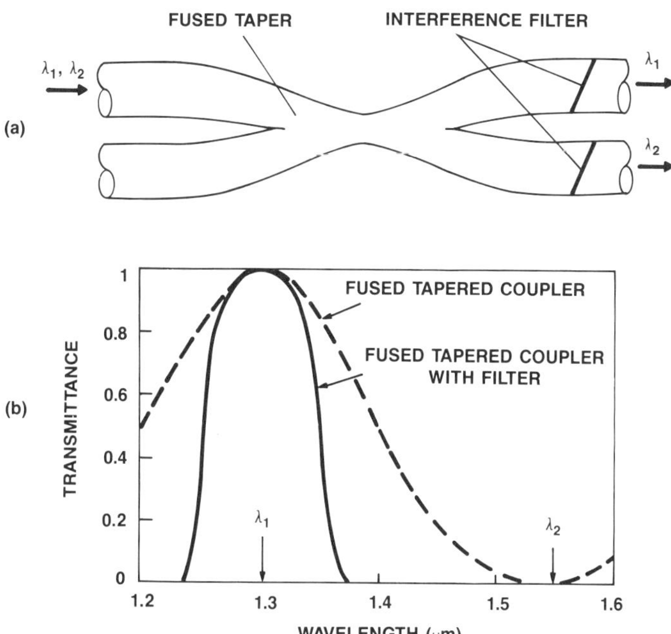

Figure 1.13 Example of a simple WDM device based on fused biconical tapered single-mode fiber coupler plus bandpass interference filters for multiplexing and demultiplexing of 1.3 and 1.55-μm spectral region laser sources onto the same single-mode fiber.

the mode field diameter, not the core diameter. Additional splice loss will be incurred if the mode field diameters are different; for example, if one fiber has a mode field diameter of 10 μm while the other has a mode field diameter of 8.7 μm, the additional loss due to mode field mismatch is 0.1 dB. Therefore, it is important to have mode field diameter match for very low loss jointing.

With splices, "splice organizers" are typically used, in which the extra lengths of fibers are coiled and "organized" and properly protected. It is important to use a large enough coil radius (or diameter) for conventional single-mode fibers (with cutoff wavelength and mode field diameter designed primarily for operation at 1300 nm wavelength) to minimize possible bending losses in the 1550-nm-wavelength region if upgrading to include 1550-nm operation is contemplated.

In general, in addition to mode field diameter mismatch between fibers, other common causes of fiber joint losses are due to mechanical misalignments: fiber angular tilt angle θ, axial or lateral offset d, and longitudinal end separation s [3,4,18]. Figure 1.12 illustrates the different types of fiber end misalignments and effect of lateral and longitudinal misalignments on the fiber jointing loss. For single-mode fibers, the jointing loss is not a very sensitive function of end separation s but increases quadratically with the angular tilt angle θ and lateral offset d [18]. In practice, the state of the art in fiber jointing technology today is such that single-mode fiber fusion splice loss of $\leqslant 0.1$ dB, mechanical splice loss of 0.1–0.2 dB, and single-mode fiber connector loss of 0.1–0.3 dB can all be readily obtained with properly prepared splices and properly designed and assembled connectors.

1.2.8 Other Passive Optical Components

In addition to optical fiber splicers and connectors, there are other passive optical components, such as optical attenuators, optical fiber couplers for power mixing and splitting, optical filtering components for spectral separation and combination, and optical fiber taps for power distribution or line monitoring. These can be used in a typical optical fiber transmission system to enhance the system flexibility. For example, optical attenuators are often used to reduce the received optical power to within the dynamic range of the receiver (to avoid saturation and nonlinear distortion), because the optical transmitter– receiver pair could have been designed for a fiber link with a loss higher than the specific one in use. Optical fiber couplers can also be used as optical power splitters (e.g., 50–50% coupler or 10–90% power splitting for distribution bus). Also, as will be discussed in Section 1.3.5, WDM components can be used to multiplex two or more optical source wavelengths for transmission over a single fiber to achieve higher total transmission capacity over an existing fiber. Figure 1.13 illustrates a WDM component for multiplexing 1.3- and 1.55-μm laser diodes; the component is based on a simple fused biconical tapered fiber coupler (two fibers fused together sideways along the core region). The optical interference filters are added to improve the isolation between the two optical channels and therefore reduce the crosstalk in a WDM transmission system.

For all passive optical components used in optical fiber transmission systems, in addition to optical characteristics such as insertion loss, spectral dependence, reflection, and durability, there are other concerns, such as long-term stability in uncontrolled environments, cost of technology and installation, and ease of maintenance in mass system deployments; this is especially true for future subscriber loop applications.

1.3 OPTICAL FIBER TRANSMISSION TECHNOLOGY

1.3.1 Importance of Optical Transmission Technology in Telecommunications

Transmission systems based on optical fiber communication technology have revolutionalized the field of communications where there is a change toward a totally digital network for a variety of new functions and features based on the concept of BISDN. The BISDN promises to provide services including high-speed data, low-speed telemetry, high-quality audio, high-definition TV distribution, and two-way video, in addition to the POTS (plain old telephone service) we have today on paired copper wires.

In the following sections we will discuss the basics of optical transmission technology including the choices of fibers, passive optical components, and active devices as well as some of the essential features and limitations of these transmission technologies.

1.3.2 Optical Fiber Transmission Systems

Figure 1.14 shows the schematic block diagram of an optical fiber transmission system. There are four important basic building blocks in a typical optical fiber transmission system:

1. transmission medium—the optical fiber/cable;
2. active optical device modules—the laser or LED optical transmitters and photodiode-based optical receivers;

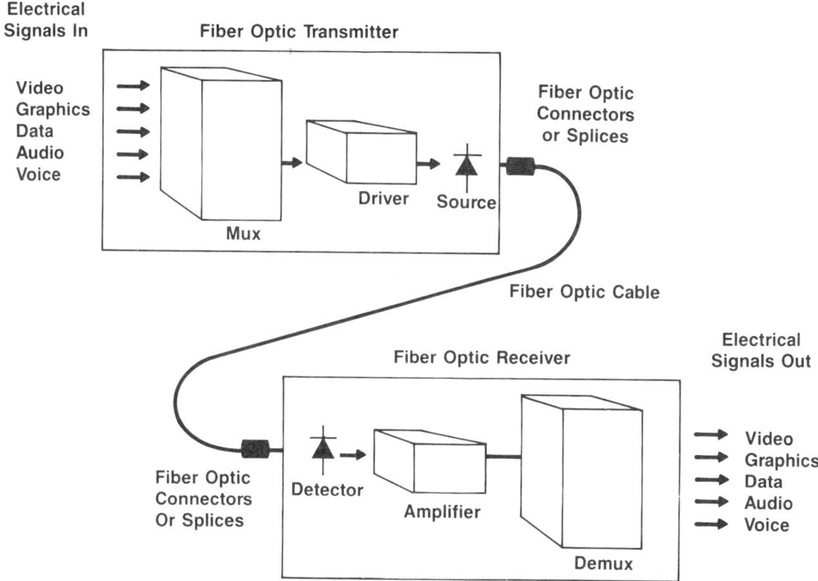

Figure 1.14 Schematic diagram of an optical fiber transmission system showing the fiber-optical transmitters, receivers, and transmission fiber with fiber jointing components.

3. passive optical components—connectors, splices, couplers, attenuators, WDM components, and so on; and
4. electronics—electronic multiplexers/demultiplexers, supervisory and maintenance circuits, and so on.

The electronic part of the system is not unique to fiber transmission technology. In contrast, the optoelectronic parts of the system are unique to optical fiber transmission technology. Since we have already discussed fiber characteristics and some of the passive components in the previous sections, the following sections will discuss active devices and the fiber/device choices for different system applications. Note that in an optical fiber transmission system, in general, the characteristics of the active optoelectronic devices and the transmission fibers together determine the most essential features of the particular system. However, the system flexibility is made possible only with the use of various passive interconnection and branching components such as splices, connectors, optical attenuators, fiber couplers, and WDM components.

1.3.3 Multimode Optical Fiber Transmission Systems

Most of the earlier optical fiber transmission systems installed were based on multimode optical fiber technology using, first, short-wavelength (0.8-μm) AlGaAs semiconductor laser diodes and then, long-wavelength (1.3-μm) InGaAsP semiconductor laser diodes [19]. However, compared with single-mode optical fiber systems, multimode optical fiber systems have two distinct disadvantages, as already mentioned in Section 1.2.3:

1. limited bandwidth because modal dispersion is large and
2. modal noise, which causes significant system penalty.

Modal noise, due to time-varying modal-distribution-related interference effects, can cause serious system penalty. It is difficult to predict because it depends on the location and characteristics of the interconnections in the multimode fiber link and optical source coherence, requiring extra care in system design, deployment, and maintenance.

Mainly due to these factors, multimode optical fiber systems are typically limited to shorter-distance, lower-bit-rate optical transmission applications. In addition to more complex profile fabrication control and system design, multimode fibers have higher losses than single-mode fibers because of the increased Rayleigh scattering due to higher dopant concentration in the core. Thus, for long-haul transmissions, multimode fibers have distinct disadvantages both in loss and in bandwidth, although the coupling, splicing, and connector losses are lower. Early fiber transmission system installations (pre-1980) used multimode fibers because of the availability of multimode fiber transmission technology. Shortly after 1980, though, there was a clear trend toward using single-mode fiber systems in all long-haul applications as the technology moved toward higher bit rates and longer repeater spacing systems [20].

Nevertheless, there is a strong interest in multimode fiber-based, low-bit-rate, short-distance intrabuilding optical systems because modal dispersion and modal noise do not pose significant problems in these applications, in which the coupling and branching efficiency of large-core multimode fibers are desired. Together with inexpensive LEDs or low-cost lasers, multimode fiber-based systems are finding significant demands for high-speed data transmission in LAN applications.

1.3.4 Single-Mode Fiber Transmission Technology

Single-mode fiber transmission technology is clearly the choice for high-bit-rate, long-distance communication systems. Since low-loss optical fiber transmission windows are in the 1.3-μm region and the broad low-loss window is in the 1.5–1.6-μm spectral range, these two wavelength regions are used for long-distance single-mode fiber transmission. The minimum loss achievable is about 0.37 dB/km near 1.3 μm and 0.16 dB/km near 1.58 μm for high-quality silica glass single-mode optical fibers.

Assuming, for example, a loss budget of 30 dB (considering optical transmitter power, receiver sensitivity, and system margin) for the fiber link alone, the allowable transmission distance is about 80 km near 1.3 μm and 160 km near 1.56 μm. This illustrates the clear advantage of using the lowest-loss transmission window with 1.56-μm-wavelength laser sources in single-mode silica glass optical fibers.

Loss-limited versus Dispersion-limited Transmission. As can be seen from Figure 1.10, the chromatic dispersion of a regular, conventional single-mode fiber (C-SMF) in the lowest-loss 1.56-μm region is typically 17 ps/nm-km, or more than 10 times that in the 1.3-μm low-dispersion region. The intersymbol interference due to the fiber's chromatic dispersion can limit the high-bit-rate optical signal transmission to a distance much shorter than the loss-limited distance. Since the single-mode fiber's zero-dispersion wavelength is typically around 1.3 μm, optical sources operating in the 1.3-μm loss window region experience much smaller pulse broadening or other dispersion-induced penalty as compared with the 1.56-μm region operation. As a consequence, the dispersion-limited distance in 1.56-μm spectral region transmission is much shorter than that for the 1.3-μm spectral region.

To take advantage of the lowest fiber loss near 1.56 μm without introducing a large dispersion penalty, one can either reduce chromatic dispersion $M(\lambda)$ or reduce the optical

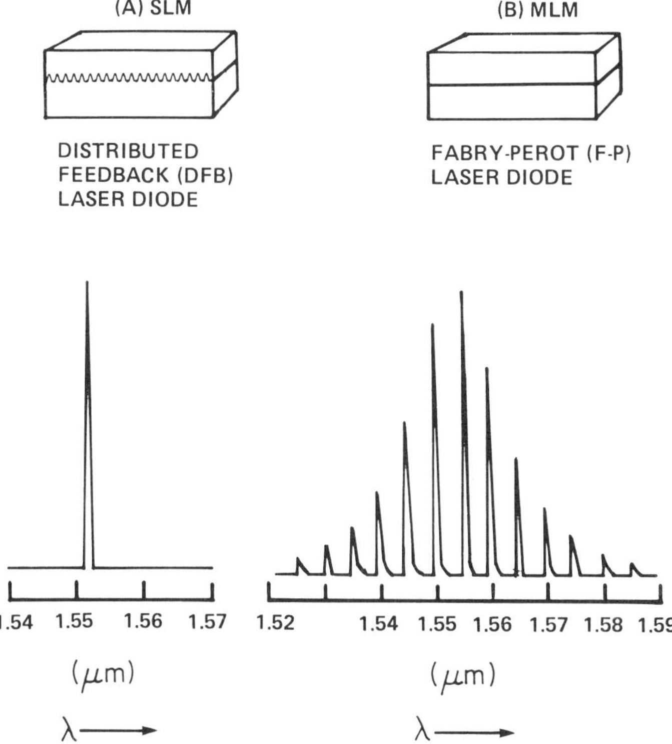

Figure 1.15 Spectra of MLM Fabry–Perot (FP) laser diode as compared with the SLM DFB structure laser diode (long-distance transmission).

spectral width $\Delta\lambda_s$ in Eq. (1.14) by

1. using a dispersion-shifted single-mode fiber (DS-SMF) with its zero-dispersion wavelength shifted to the 1.56-μm spectral region or
2. using a narrow-spectral-width (small $\Delta\lambda_s$) single-longitudinal-mode (SLM) distributed feedback (DFB) laser diode [21] as the optical source.

Either of these two approaches will reduce the chromatic dispersion-related transmission penalties. Figure 1.15 illustrates the spectral characteristics of a SLM DFB semiconductor laser diode as compared with the regular multi-longitudinal-mode (MLM) Fabry–Perot (FP) cavity laser diodes. The broad spectral width (large $\Delta\lambda_s$) of the MLM laser diodes contributes to a large dispersion penalty due to pulse broadening in the transmission fiber, as described by Eq. (1.14) and due to the MLM laser's mode partition noise (MPN), to be discussed later.

Optical Source–Fiber Combinations for Single-Mode Fiber Systems. Table 1.2 shows some examples of possible source–fiber combinations using different sources (MLM FP laser diodes vs. SLM DFB laser diodes, 1.3 μm versus 1.56 μm wavelength) and fibers (regular 1.3-μm zero-dispersion fiber, C-SMF, versus 1.56-μm dispersion-shifted fiber, DS-SMF) for long-distance single-mode optical fiber transmission systems.

TABLE 1.2 Examples of Laser Source–Fiber Combinations for Long-Haul System Applications

Source Characteristics		Typical Fiber Characteristics		
Laser Type	Wavelength λ_s (nm)	Fiber Type	Zero Dispersion	Loss at λ_s (dB/km)
1. MLM F-P LD	1300	C-SMF	1310-nm region	0.40
2. SLM DFB LD	1300	C-SMF	1310-nm region	0.40
3. SLM DFB-LD	1560	C-SMF	1310-nm region	0.21
4. MLM F-P LD	1560	C-SMF	1310-nm region	0.21
5. MLM F-P LD	1560	DS-SMF	1550-nm region	0.24
6. SLM DFB-LD	1560	DS-SMF	1550-nm region	0.24

Note: (i) Combinations 1 and 5 are multi-longitudinal-mode (MLM) lasers operating at the fibers' zero-dispersion wavelength; 2 and 6 are single-longitudinal-mode (SLM) lasers operating at the fibers' zero-dispersion wavelength; combinations 3 and 4 are SLM and MLM lasers, respectively, operating at a high-dispersion region.
(ii) Most of the existing single-mode fiber systems use combinations 1 and 2, at various bit rates, and may use 3 for upgrading these existing systems by WDM. Combinations 4 and 5 are used in much longer repeater spacing or repeaterless island-hopping applications, with 4 limited to low bit rate (e.g., less than 100 Mb/s). Combination 6 is being considered for the next generation of undersea single-mode fiber transmission systems.

The majority of today's single-mode fiber transmission systems use C-SMF fiber (zero-dispersion, 1.31 μm) and FP-type MLM laser diodes operating near 1.3 μm. However, 1.3 μm SLM DFB lasers are beginning to be used in gigabit-per-second systems. These DFB laser-based single-mode fiber systems represent the first use of SLM (sometimes called single-frequency) laser diodes in commercial optical fiber communication applications. There are also WDM single-mode fiber systems using both the 1.3-μm FP MLM laser diodes and 1.56-μm SLM DFB laser diodes in conventional single-mode fibers (C-SMF) with the zero-dispersion wavelength at 1.3 μm.

The main use of 1.56-μm single-mode fiber systems with MLM laser diode sources are typically for long-repeater-spacing but lower-bit-rate (less than gigabit-per-second) applications. The long-repeater spacing is of particular interest for repeaterless, island-hopping, undersea optical fiber systems [22]. For the next generation of undersea optical fiber systems, 1.56-μm single-mode fiber transmission systems based on DS-SMF or SLM DFB lasers are being developed as such systems promise greatly reduced number of undersea repeaters required [22].

Obviously if one designs a system using *both* a 1.56-μm SLM DFB laser diode source *and* DS-SMF fibers with the minimum dispersion wavelength near 1.56 μm, the loss and dispersion limitations would be simultaneously minimized. The transmission distance/capacity will thus be maximized if we assume that the same optical source power and the receiver sensitivity are available at 1.56 μm as at 1.3 μm. Such DFB laser/DS-SMF systems operating at 1.56 μm are just beginning to be considered for some special transmission applications [23].

Recently, terrestrial single-mode optical fiber transmission systems that operate above 1 Gb/s bit rate have been developed and have become available for commercial applications. Most of these commercial gigabit-per-second, single-mode transmission systems use SLM DFB lasers operating at 1300 nm, although 1300-nm MLM LD (laser diode) is also used in some cases. The use of MLM LD instead of SLM DFB LD for gigabit-per-second system transmission would require a very tight selection of multimode laser spectral width and center wavelength to match very closely the fiber's dispersion minimum, thus minimizing dispersion broadening and MPN. This could mean a less flexible system design or less margin for device degradation.

Beyond the now commercially available 1.12- and 1.7-Gb/s optical single-mode fiber transmission systems, recent advances of very high speed semiconductor laser diodes with a small-signal 3 dB bandwidth up to the 10–15-GHz range [24,25] has made possible transmission experiments at high gigabit-per-second rates. Laboratory research demonstration of high-speed modulation and transmission experiments at 2.4, 8, 10, and 16 Gb/s [26–28] have been reported; very wide bandwidth (> 20-GHz) *pin* photodetectors have also been developed [29]. Therefore, it seems that in the 1990s, multi-gigabit-per-second, high-speed optical transmission technology may become available for a variety of applications, including long-haul telecommunications and short-distance ultrahigh-speed data transmission between supercomputers.

1.3.5 Wavelength Division Multiplexing in Single-Mode Optical Fiber Systems

While the development of high-capacity single-mode optical fiber transmission systems continues to move in the direction of higher-speed, high-gigabit-per-second bit rates, to upgrade an existing transmission link capacity, one can either go to a higher transmission bit rate or use additional optical channels sharing the same optical fiber for increased bandwidth transmission. The optical WDM approach [30] uses several simultaneous optical channels transmitting through the same optical fiber. With WDM schemes, for an established route where fiber cable installation has been completed long ago, one can upgrade the system capacity by adding more optical channels without installing new fiber cable. It appears that upgrading by going to higher speeds or higher bit rates (i.e., time division multiplexing) is likely to be considered first, because it is a more straightforward approach if the higher-speed system is already available. However, when the economic considerations are such that adding another optical channel at the same bit rate is better cost justified, or if the higher-bit-rate system equipments are simply not yet available (e.g., limited by the high-speed electronics required), then WDM may be deployed. This will depend on the relative maturity and cost of WDM technology versus the high-speed optoelectronics technology.

Figures 1.16*a* and 1.16*b* show two typical WDM system configurations. Figure 1.16*a* shows a typical application, that is, two-channel transmission in one direction, in a single fiber. Figure 1.16*b* shows the case for bidirectional transmission in the same fiber with lasers at different wavelengths. Although only two optical channels (laser wavelengths) are shown in the figures, the number of channels can be as high as 10–20 depending on the channel spacing, the laser spectral properties, and the WDM device characteristics. High-density WDM will require SLM DFB laser diodes.

The technology of WDM devices has been the subject of research and development for years [30,31]. Several multimode fiber transmission systems have used WDM technology to increase the total capacity, usually at low transmission bit rates. Single-mode WDM optical fiber transmission systems operating at two to four wavelengths in the 1.3and 1.55-μm region have also been developed in the laboratory [31] for multichannel transmission at higher bit rates. There was no significant use of WDM technology until recently when multiplexing 1.3 and 1.55 μm in an existing 1.3-μm single-mode transmission link was deployed for upgrading.

Wavelength division multiplexing can be accomplished by using various dispersive optical components such as dielectric interference filters, fused biconical tapered fiber couplers with wavelength-dependent transmission (Fig. 1.13), gratings, and micro-optical components such as graded refractive index (GRIN) rod lenses with wavelength-dependent coatings [30,31]. Important practical requirements of WDM components include

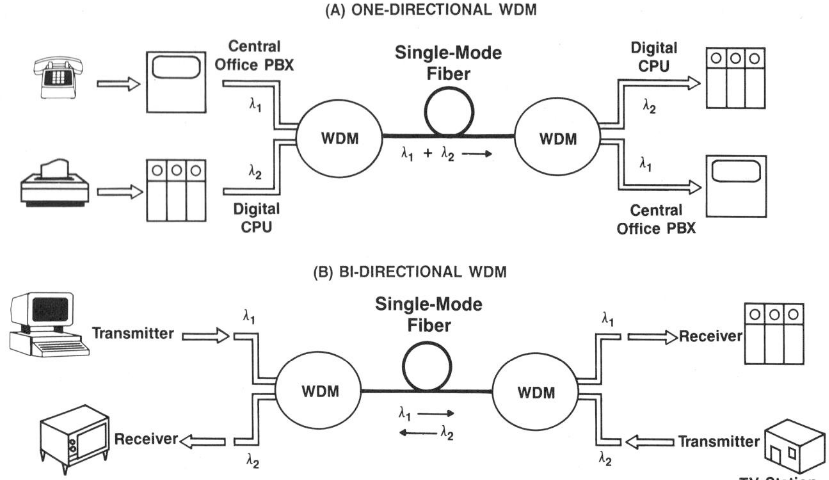

Figure 1.16 The WDM for increasing the transmission capacity by using more than one optical wavelength channel: (*a*) one-directional WDM; (*b*) bidirectional WDM. Although only 2 optical channels are shown, in principle, 10–50 optical channels per fiber are possible.

low loss, low crosstalk between optical channels (different wavelength optical sources), low reflection, high stability under operating environmental conditions, small size, and low cost. Specific applications will determine the WDM technology of choice. For example, multiplexing 2 channels such as 1.3 and 1.56 μm could be accomplished readily with simple fused, biconical, tapered fiber couplers and optical dielectric interference filters. To multiplex 10–20 channels within either the 1.3- or the 1.56-μm spectral region, on the other hand, may require the use of a more bulky grating multiplexer/demultiplexer with GRIN lenses or equivalent collimating optics for input/output coupling of multiwavelength channels unless a star coupler is used to mix the optical signal for distribution and tunable narrow-band optical filters are used before the optical detector in the receiver.

It is important to note that WDM technology offers not only the overall bandwidth/bit rate upgrading for long-haul optical fiber transmission systems but also the needed system/network architectual flexibility for future subscriber loop broadband distribution applications. A WDM system, with 4–16 optical channels, for example, allows different broadband services (digital or analog, high-bit-rate or medium-bandwidth, NTSC TV or high-definition TV) to be provided over different optical channels over the same fiber to the customer, depending on the service needs. The high-density multichannel WDM technology requires closely spaced optical channels (2–10 nm); this in turn requires the use of multiwavelength optical sources (such as DFB lasers at different wavelengths) and optical channel selection filters (such as fixed-wavelength optical filters, tunable optical etalons, or FP interferometers). There are currently significant research and development efforts in these areas, which will lead to the realization of multichannel high-density WDM technologies and their applications in future optical systems and networks.

1.4. TRANSMISSION LIMITATIONS

As discussed in the preceding, the most obvious transmission limitations are due to fiber transmission loss and fiber dispersion. However, in addition to transmission loss and dispersion of the fibers, in a single-mode fiber transmission system there are other im-

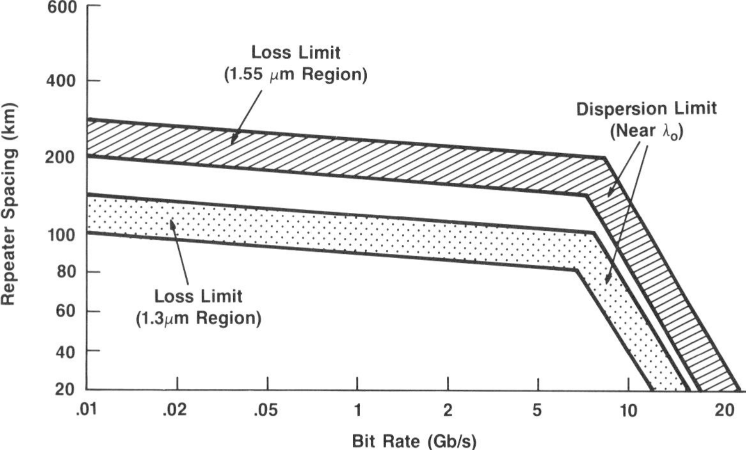

Figure 1.17 Repeater spacing versus bit rate in single-mode optical fiber transmission, with boundaries (limits) due to fiber loss and dispersion near λ_0.

portant system degradation factors associated with the interaction of the optoelectronic devices and components with the transmission media that can cause system impairments. In this section we discuss these limitations and a few of the important factors—*mode partition noise* in MLM lasers, *chirping* in SLM lasers, *reflection-induced noise*, and so on—and the corresponding system penalties.

Figure 1.17 shows the general features of single-mode fiber transmission distance versus bit rate bounded by these limiting factors. The individual factors are discussed in the following sections.

1.4.1 Transmission Loss and Dispersion

Loss-limited system operation means that the power budgeting consideration limits the transmission distance. For example, if

$$P \quad \text{(transmitter output power)} = P_t \quad \text{(dBm)}$$

$$P \quad \text{(receiver sensitivity)} = P_r \quad \text{(dBm)}$$

and if there is no dispersion-related penalty or limit, then the maximum transmission distance is

$$L_m = \frac{(P_t - P_r)\ \text{(in dB)}}{\text{average fiber link loss (in dB/km)}} \tag{1.16}$$

The fiber link loss includes fiber loss, splice loss, and the loss due to connectors, WDM components, fiber couplers, and any other interconnecting components. For example, assuming we have a single-mode fiber system operating at 1300 nm with $P_t = 3$ dBm, $P_r = -34$ dBm, and an average fiber link loss of 0.41 dB/km (C-SMF at 1300 nm), the maximum loss-limited transmission fiber length is $L_m = 37/0.41$ km $= 90$ km. Here, the system margin is not considered.

Dispersion-limited transmission occurs when the power budget allows for transmission over a length L_m but the achievable transmission distance is shorter than the L_m from loss calculation alone because the dispersion-induced signal pulse broadening $\Delta\tau$ is causing significant intersymbol interference between transmitted neighboring pulses. This is usually the case with higher-bit-rate systems with a large overall dispersion. Significant intersymbol interference happens when the pulse broadening $\Delta\tau$ is larger than half of the bit period T:

$$\Delta\tau \geq \tfrac{1}{2}T \tag{1.17}$$

where the pulse broadening $\Delta\tau$ is given by Eq. (1.14).

For example, suppose the loss-limited transmission distance is 90 km, as in the preceding example. Now suppose the particular system is to operate at 2 Gb/s (500 ps pulse period) using an MLM laser diode with an rms spectral width of 4 nm at a wavelength of 1280 nm (assuming the laser wavelength does not match the fiber's minimum dispersion wavelength exactly). Since the fiber dispersion M is about 2 ps/nm-km, the dispersion limits the transmission distance to

$$L \cong \frac{500 \text{ ps}/2}{2 \text{ ps/nm} \times 4 \text{ nm}} \text{ km} = \frac{250}{8} \text{ km} = 31 \text{ km}$$

which of course is almost one-third of the 90-km length from the loss consideration alone. As a rule of thumb, loss-limited transmission is usually obtained at relatively low bit rates (<150 Mb/s) where the dispersion effects are negligible.

1.4.2 MLM Laser Mode Partition Noise

In FP cavity MLM laser diodes, there is a certain degree of power distribution fluctuation among the various longitudinal modes. From theoretical analysis, in MLM lasers the statistical nature of the buildup of the various longitudinal modes leads to a fluctuating power distribution among the modes from one optical pulse to the next in a modulating pulse train. This pulse-to-pulse variation of the optical power distribution means more power is in longitudinal mode A for one pulse while for the next pulse there is more power in longitudinal mode B. When coupled with the chromatic dispersion of a long single-mode fiber, the pulse-to-pulse fluctuating optical power distribution will cause pulse-to-pulse signal waveform fluctuations because of the wavelength-dependent delays for different longitudinal modes. This in turn leads to a bit error rate "floor" [32] that is independent of optical power received. This is called mode partition noise, or MPN [32,33].

If there were no mode partitioning, that is, if the longitudinal-mode power distribution profile were constant from pulse to pulse, then the system penalty would be due to the pulse broadening alone, the broadening being associated with the fixed (not pulse-to-pulse varying) spectral width and the fiber dispersion, as was calculated in the example in the last section. Serious MPN can dominate over the chromatic dispersion broadening in limiting the system transmission [32,33]. Indeed, this has been the case in several system experiments using MLM laser diodes, where a floor in the bit error rate versus received optical power curve is observed. Figure 1.18 shows a schematic, qualitative diagram of such a bit error rate versus received optical power curve. The pulse broadening due to dispersion causes a dispersion penalty in terms of the power increase needed to get the same bit error rate, while the bit error rate floor is a floor in the bit error rate due to the MPN and is independent of the received optical power.

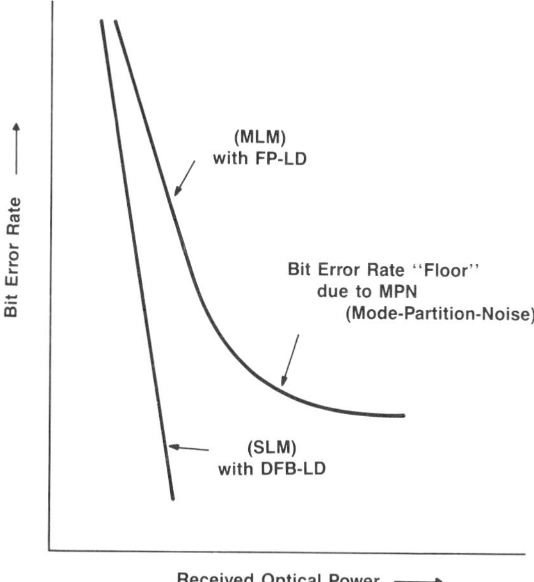

Figure 1.18 Qualitative diagram showing the effect of the MPN of a MLM laser diode in a single-mode optical fiber transmission system. The bit error rate floor, which cannot be reduced by simply increasing optical signal power, limits the transmission bit error rate. No such floor due to MPN is observed in a good SLM DFB laser diode.

Because of the MPN limitations, for gigabit-per-second systems, MLM laser diodes should probably only be used near the minimum-dispersion region of the single-mode fiber, except for short-distance transmission. The SLM laser diodes are undoubtedly the primary optical sources of choice for gigabit-per-second long-haul single-mode fiber systems.

1.4.3 Chirping in SLM Lasers and System Implications

It has been recognized very early that the broadening of the individual longitudinal modes of a laser diode under fast modulation is due to the transient phenomenon of chirping [34,35]. Chirping refers to a time-dependent frequency (or wavelength) shift during the transient turn on (and off) in a laser diode under fast excitation. The wavelength shift is caused by the time-varying refractive index excursion associated with the carrier density change, which is a natural consequence of nonconstant excitation, such as in short-pulse pumping or high-speed digital modulation. Chirping results in a dynamic line width broadening of the individual longitudinal modes, which is broadened from the continuous-wave (cw) line width of less than 0.001 nm to about 0.2 nm, depending on the biasing and modulation conditions. Since in direct-detection optical transmission systems the laser diode transmitters are directly current modulated, chirping and the resultant line broadening are commonly observed in both MLM and SLM laser diodes. However, in MLM laser diodes, the overall spectral envelope broadening is more important than the individual longitudinal mode broadening, so the latter effect (due to chirping) is usually neglected.

In good SLM lasers, since there is only one dominant longitudinal mode, chirp broadening becomes the most important factor other than the side-mode suppression

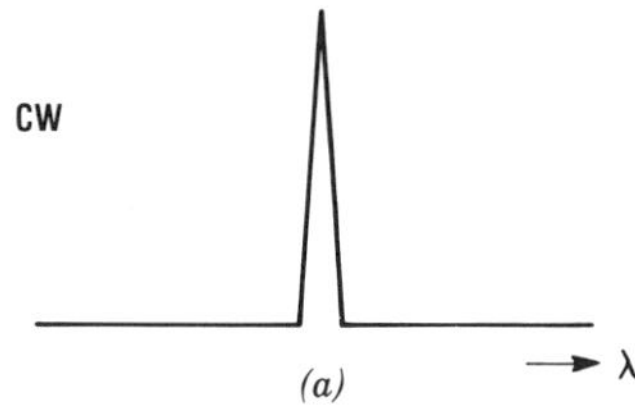

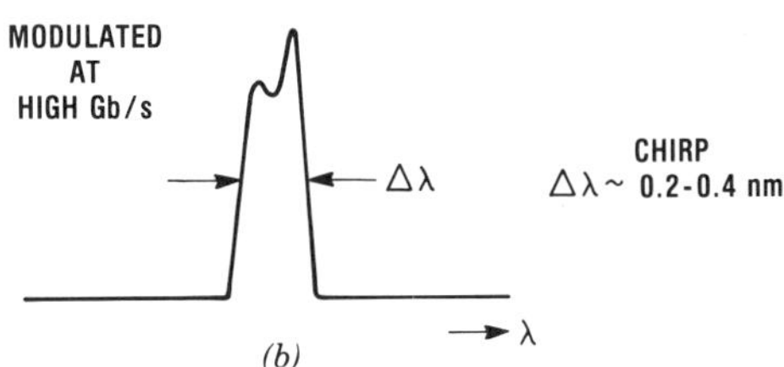

Figure 1.19 Typical SLM DFB laser diode spectra (*a*) in cw operation and (*b*) under high-speed Gb/s modulation showing the dynamic line broadening due to chirping when modulated.

ratio. Figure 1.19 shows an example of chirp-broadened DFB laser longitudinal mode under high-speed modulation; for comparison, the cw (no modulation) spectra is also shown. This chirped, dynamically broadened line width could impose a significant transmission limitation in multi-gigabit-per-second single-mode fiber transmission systems using SLM DFB laser diodes operating at a dispersive wavelength. If there were no side modes, the limitation is only due to the chirped width of the only longitudinal mode. For transmission near the zero-dispersion wavelength, there is negligible dispersion penalty. However, if the dispersion is, say, 18 ps/nm-km (e.g., with a 1560-nm DFB laser in a C-SMF), then 0.2 nm of chirped laser line width gives rise to 3.6 ps/km of pulse broadening, which for 100 km of low-loss C-SMF transmission amounts to a total pulse broadening of about 360 ps, limiting the transmission bit rate to about 2 Gb/s. So, indeed, this is a significant limitation. Chirp limitation has been observed in several gigabit-per-second system transmission experiments where a high dc bias was sometimes needed to minimize the chirp-related penalty.

It is possible to reduce the chirp and the chirp-induced penalty in a dispersive transmission by using, for example, an external optical modulator to modulate the laser rather than using dc modulation. Perhaps the simplest solution is to operate the laser wavelength near the zero-dispersion region so that the chirped line width contributes very little to the pulse broadening. However, in a multiwavelength WDM system, there will be some SLM lasers operating in the dispersive regions of the spectrum, and it would be difficult to reduce the chirp penalty for all the laser wavelengths. A broadband low-dispersion single-mode fiber is useful in such applications.

In the meantime, chirping will continue to be an important limitation in SLM-DFB-LD-based gigabit-per-second C-SMF single-mode fiber transmission systems. For this reason, adjusting the laser transmitter bias and modulation conditions to optimize the overall system performance (which could correspond to reduced chirp, increased side-mode suppression, high speed, low pulse patterning, but low extinction ratio) is usually required as part of the system design and testing task.

1.4.4 Reflection-induced Noise and Penalty

Depending on the phase and magnitude of optical reflection back into a laser cavity, the laser output characteristics can be disturbed and modified. Reflection-induced noise in semiconductor laser diodes has received much attention, but by nature reflection effects on the laser oscillation behavior are complicated and difficult to quantify.

In direct-detection optical fiber transmission systems, reflection from, for example, the laser transmitter fiber pigtail end and the near end connectors has been observed to introduce noises and system penalties [36]. One can measure directly the induced system penalty for a given reflection feedback noise. Such measurement is most important in single-mode fiber systems using SLM laser diodes. The reflection effect on MLM laser diode characteristics may be just as significant, but the system consequence is less significant in MLM-LD-based systems unless the reflection introduces unpredictable mode jumps.

The experimental results of a 1-Gb/s NRZ (nonreturn to zero) transmission experiment with a 1550-nm DFB laser diode in a conventional single-mode fiber (C-SMF) with zero dispersion at 1310 nm showed that the measured power penalty increased with the increase in the reflection level [36]. Under the particular experimental conditions, a reflection level of −19 dB was found to cause a 1 dB penalty; a reflection level of −10 dB caused more than a 4-dB penalty. When the optical feedback was large, an increase in noise and jitter and degradation of response speed were observed in the received signal. The effect of the reflection was also dependent on the modulation condition; a larger modulation signal improved the tolerance to reflections.

In general, a reflection level of −30 dB (0.1% reflection) is probably satisfactory for most applications; in contrast, a reflection level of −12 dB (8% reflection typical of an air gap in noncontacting connectors or other components) or higher will probably always cause a significant system penalty. Figure 1.20 illustrates qualitatively the effect of different levels of reflections on the system bit error rate.

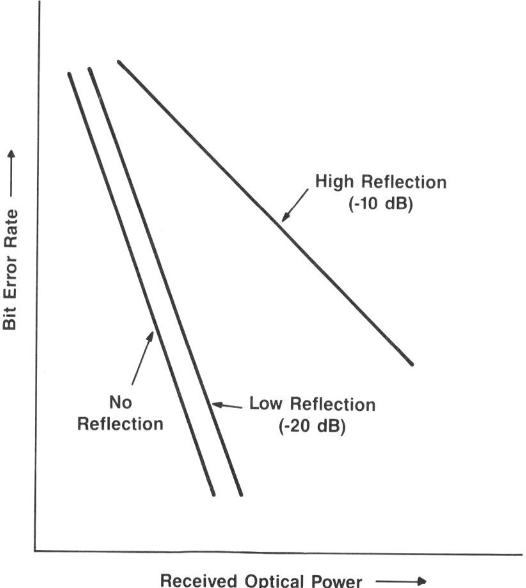

Figure 1.20 Bit error rate versus received optical power diagram showing qualitatively the effect of reflection level. High-level reflection induces power penalty.

To make sure there is little reflection-induced noise, some recent gigabit-per-second systems use an optical isolator (based on Faraday rotation of the optical polarization in a magnetic field) within the SLM DFB laser package and use low-reflection physical contact connectors to minimize and prevent reflection penalties. Clearly as the high-speed systems get into high-gigabit-per-second ranges with the increased use of SLM DFB laser diodes, the issues of reflection tolerance in SLM lasers, optical isolator isolation requirement, low-reflection connectors, splices and other branching components, and the effect of multiple reflections will need to be addressed in system engineering.

1.5 FUTURE DIRECTIONS IN OPTICAL TRANSMISSION TECHNOLOGY

Since single-mode fibers have practically unlimited bandwidth [16] and very low losses, optical transmission system technology is really not limited by this excellent transmission medium. Rather, the single-mode fiber technology is relatively mature and waiting to be fully utilized. In contrast, the optoelectronics technology and the electronic integrated circuit (IC) technology, impressive as they are today, still cannot fully utilize the remarkable transmission medium—single-mode fiber—to its full potential. Future directions in optical transmission technology are discussed briefly.

1.5.1 Very High Speed and High-Density WDM Systems

The rapid progress in both the high-speed optoelectronic and IC technology have been very impressive. The speed limits of the laser diodes and photodetectors seem to be beyond at least 20 Gb/s; the current limitations seem to be mainly in the electronic IC technology. The VHSIC (very high speed IC) technology based on the III–V compound semiconductors and novel structures is under intense research and development and seems to hold great promise for the future. The wide low-loss window of single-mode fibers promises tens or hundreds of simultaneous optical channels, each capable of operating at 10–20 Gb/s bit rates, transmitting through the same fiber by large-channel-number, high-density WDM [37], constituting an ultrahigh-transmission capacity system.

1.5.2 Optoelectronic Integrated Circuit Technology

By combining various optical devices and electronic circuits on a single chip to perform complex functions on a monolithic chip or a hybrid chip, optoelectronic IC (OEIC) technology [38] promises to be the equivalent of microelectronic technology in the optoelectronic area in terms of potential applications and impacts on the optoelectronic information transmission and information storage/processing/handling applications. Because of the great importance of OEIC technology, there are significant research and development efforts in various research laboratories worldwide; it is almost certain that in time OEIC technology will be at the center stage of the optoelectronic technology.

1.5.3 Coherent Optical Transmission Technology

Based on coherent optical modulation and demodulation (optical heterodyne/homodyne) techniques, one can significantly increase the optical receiver sensitivity over the present direct-detection optical transmission technology as well as achieve very high density multichannel (50–1000 or more channels in principle) optical frequency division multiplexing (OFDM) transmission [39]. The practical deployment of such coherent optical transmission technology depends on future development of stabilized tunable narrow-line-width single-frequency lasers for optical sources and local oscillators, polarization control techniques, coherent receiver designs, and so on. Coherent optical fiber transmission systems will most likely be used first in long-haul island-hopping applications; for multichannel OFDM applications the successful deployment may have to depend on the development of OEIC coherent transmitters and receivers.

1.6 FUTURE APPLICATIONS: FIBER TO THE HOME AND INTELLIGENT BUILDINGS

So far we have described the important essential aspects of the single-mode fiber transmission technology for actual system applications. In the past 10 years or so, the emphasis was mainly on deploying the single-mode optical fiber transmission systems for the digital long-haul transmission network and on multimode optical fiber transmission technology for business or campus environment local area computer data network.

In the next decade and beyond, the new wave of optical fiber transmission applications will be in the subscriber loop part of the network. This includes both the "loop feeder" (typically 10 km or shorter) and the "distribution" portions of the subscriber network. The distribution portion of the subscriber loop is typically between 1 and 3 km long, with the majority of distribution distance being less than 2 km. Thus, the fiber loss and dispersion considerations take on a different significance; in fact, other than the consideration that single-mode fibers should be used for once-for-all installation and nearly unlimited future upgrading, the fiber loss and dispersion are not as important as the network architectural considerations, coupling and distribution losses, maintenance and service concerns, and so on.

There are a number of studies and field trials being conducted worldwide [40,41] on fiber-to-the-home and fiber-to-the-curb systems and networks to determine the service, architecture, economics, and maintenance/installation issues as well as evolution strategies for deploying the optical fiber transmission technology in subscriber loops to homes and intelligent buildings.

The narrow-band and broadband information could be transmitted to the subscriber's home using various possible architectures and strategies. Figure 1.21 [40] shows a simplified schematic of a single-mode fiber-based transmission approach in which video

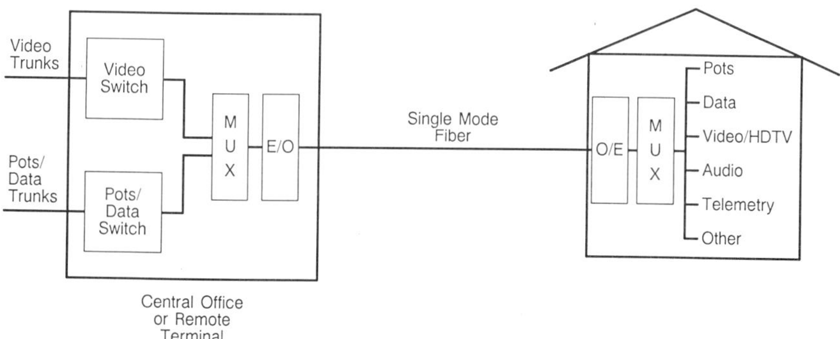

Figure 1.21 Single-mode optical fiber transmission to the subscriber's home providing the narrow-band and broadband information services.

information (e.g., 600 Mb/s), POTS, and data at a lower bit rate are electronically multiplexed for downstream transmission over a single fiber to the home. Certainly many other versions are possible; for example, two fibers per subscriber instead of one fiber per subscriber, WDM instead of electronic multiplexing, or broadcast video in addition to switched video distribution.

A number of technological, architectural, economic, and legal issues need to be addressed and resolved before single-mode fiber transmission to the subscriber's home can be widely deployed. Nevertheless, with progress in these fiber-to-the-home field trials and future progress in associated optoelectronic and electronic technologies, a broadband integrated services digital network (BISDN) can be envisioned (noting that analog video distribution can be mixed with digital transmission). Such a BISDN, providing a variety of services including POTS, telemetry and remote metering and energy control, high-speed data, high-quality audio and TV distribution, a video library and video-on-demand including HDTV technology, and so on, will probably be realized eventually. This B-ISDN-based broadband optical network will then be at the center stage of the new information society, which may have significant societal implications. All this is made possible by the dramatic advances in the past 15 years of optoelectronics and optical fiber transmission technology.

REFERENCES

1. S. D. Personick, *Fiber Optics: Technology and Applications*, Plenum, New York, 1985.
2. J. Gowar, *Optical Communication Systems*, Prentice-Hall, Englewood Cliffs, NJ, 1984.
3. G. Keiser, *Optical Fiber Communications*, McGraw-Hill, New York, 1983.
4. J. Senior, *Optical Fiber Communications: Principles and Practice*, Prentice-Hall, Englewood Cliffs, NJ, 1985.
5. See, e.g., papers published in *IEEE/OSA Journal of Lightwave Technology*, and *Technical Digests of Optical Fiber Communications (OFC) Conferences.*
6. M. J. Adams, *An Introduction to Optical Waveguides*, Wiley, New York, 1981.
7. L. G. Cohen, P. K. Kaiser, and Chinlon Lin, *Proc. IEEE*, Vol. 68, p. 1203, 1980.
8. See, e.g., *Technical Digest, OFC'85*, pp. 92–96, San Diego, CA, 1985.
9. J. W. Fleming, *Electron. Lett.*, Vol. 14, p. 326, 1978.
10. L. G. Cohen, Chinlon Lin, and W. G. French, *Electron. Lett.*, Vol. 15, p. 334, 1979.
11. M. A. Saifi et al., *Opt. Lett.*, Vol. 7, p. 43, 1982.
12. B. J. Ainslie et al., *Tech. Digest, 9th ECOC* (European Conference on Optical Communications), Geneva, p. 53, 1983.
13. C. M. Lemrow and V. A. Bhagavatula, *Laser Focus*, p. 82, March 1985.
14. D. N. Payne and W. A. Gambling, *Electron. Lett.*, Vol. 11, p. 176, 1975; F. P. Kapron, *Electron. Lett.*, Vol. 13, p. 96, 1977.
15. K. Jurgensen, *Appl. Opt.*, Vol. 18, p. 1259, 1979.
16. D. Marcuse and Chinlon Lin, *IEEE J. Quantum Electron.*, Vol. QE-17, p. 869, 1981.
17. L. G. Cohen, W. L. Mammel, and S. J. Jang, *Electron. Lett.*, Vol. 18, p. 1023, 1982.
18. L. B. Jeunhomme, *Single-Mode Fiber Optics: Principles and Applications*, Marcel Dekker, New York, 1983.
19. T. Li, *IEEE J. Select. Areas Commun.*, Vol. SAC-1, p. 356, 1983.
20. D. C. Gloge and K. Ogawa, *Tech. Digest, OFC'85*, p. 84, 1985.
21. K. Kobayashi and I. Mito, *IEEE J. Lightwave Tech.*, Vol. LT-3, p. 1202, 1985.
22. R. E. Wagner, *IEEE J. Lightwave Tech.*, Vol. LT-2, p. 1007, 1984.
23. L. C. Blank, L. Bickers, and S. D. Walker, *IEEE J. Lightwave Tech.*, Vol. LT-3, p. 1017, 1985.

24. C. B. Su et al., *Appl. Phys. Lett.*, Vol. 46, p. 344, 1985.
25. J. E. Bowers et al., *Appl. Phys. Lett.*, Vol. 47, p. 78, 1985.
26. A. H. Gnauck et al., *Tech. Digest, OFC'85*, Paper PD-2, 1985.
27. A. H. Gnauck et al., *Tech. Digest, OFC'86*, Paper PDP-9, 1986.
28. Chinlon Lin and J. E. Bowers, *Electron. Lett.*, Vol. 21, p. 906, 1985; A. Gnauck and J. E. Bowers, *Electron. Lett.*, Vol. 23, p. 801, 1987.
29. S. Y. Wang and D. M. Bloom, *Electron. Lett.*, Vol. 19, 1983; J. E. Bowers and C. A. Burrus, *IEEE J. Lightwave Tech.*, Vol. LT-5, p. 1339, 1987.
30. G. Winzer, *IEEE J. Lightwave Tech.*, Vol. LT-2, p. 369, 1984.
31. H. Ishio, J. Minowa, and K. Nosu, *IEEE J. Lightwave Tech.*, Vol. LT-2, p. 448, 1984.
32. Y. Okano, K. Nakagawa, and T. Ito, *IEEE Trans. Commun.*, Vol. COM-28, p. 238, 1980.
33. K. Ogawa, *IEEE J. Quantum Electron.*, Vol. QE-18, p. 849, 1982.
34. Chinlon Lin, T. P. Lee, and C. A. Burrus, *Appl. Phys. Lett.*, Vol. 42, p. 141, 1983.
35. S. Yamamoto et al., *IEEE J. Lightwave Tech.*, Vol. LT-5, p. 1518, 1987.
36. M. Shikada et al., *Tech. Digest, OFC'87*, Paper TuB4, p. 46, 1987.
37. S. Sasaki, H. Nakano, and M. Maeda, *Proc. ECOC'86*, 1986.
38. K. Nosu, H. Toba, and K. Iwashita, *IEEE J. Lightwave Tech.*, Vol. LT-5, p. 1301, 1987.
39. S. R. Forrest, *IEEE J. Lightwave Tech.*, Vol. LT-3, p. 1248, 1985.
40. P. Kaiser, *Tech. Digest, ECOC'85*, Vol. II, p. 125, 1985.
41. B. Catania, *IEEE J. Lightwave Tech.*, Vol. LT-4, p. 699, 1986.

2

OPTICAL CHANNEL WAVEGUIDES AND WAVEGUIDE COUPLERS

Talal K. Findakly

Research Division
Hoechst Celanese Corporation
Summit, New Jersey

This chapter deals with the basic structures of integrated optical circuitry, namely, optical channel waveguides and waveguide couplers. The treatment is intended to provide aid and understanding of these components to the designer of such circuits. The properties, behavior, and varieties of planar and channel waveguide structures are presented first. This is followed by a treatment of directional couplers from the standpoint of design and operation for practical applications.

2.1 OPTICAL PLANAR WAVEGUIDES

Integrated optical waveguides are simply structures that confine and guide optical waves due to an induced refractive index increase in the guiding region with respect to the surrounding regions. Such waveguides are typically formed at or near the surface of the substrate material by a variety of fabrication techniques. Channel waveguides confine the light in three dimensions, two transverse and one longitudinal, in contrast with the more general form of planar waveguides in which the light is confined in two directions, one transverse and one longitudinal. Since channel waveguides are derived from planar waveguides by providing the extra transverse direction confinement, it is instructive to look first at planar waveguides in order to better understand channel waveguides.

A planar waveguide is typically a thin, flat layer whose refractive index is higher than the two regions that come into immediate contact with it. These regions typically comprise the substrate material and a cover layer, which is often air, but could be any layer of lower refractive index. A typical illustration of such a structure is shown in Fig. 2.1. The index profile of planar guiding layers can be either uniform or graded, depending on the method used in forming such layers.

For the designer and maker of integrated optical waveguides, certain information is needed in order to utilize these waveguides for the application intended. Usually, this information includes the number of modes desired for a particular state of polarization, the amount of index change required for a given layer thickness and vice versa, and the

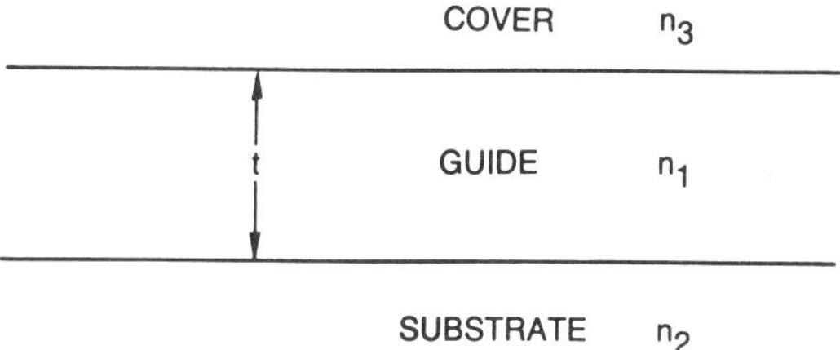

Figure 2.1 Planar optical waveguide.

effect of the cover layer on such parameters. This information is accurately obtained from solutions to the well-known Maxwell equation leading to a general dispersion equation.

The derivation of the characteristic equation for uniform planar waveguides is briefly given next for TE modes. For a waveguide structure comprising a film of thickness t and refractive index n_1, a substrate of index n_2, and a cover region of index n_3, with $n_1 > n_2, n_3$, as shown in Fig. 2.1, the electric field components are oscillatory in the film and decaying outside of it, leading to solutions of the form

$$E_y = \begin{cases} Ae^{\gamma_3 x} & x \leqslant 0 \quad (2.1) \\ Be^{ik_x x} + Ce^{-ik_x x} & 0 \leqslant x \leqslant t \quad (2.2) \\ De^{-\gamma_2(x-t)} & x \geqslant t \quad (2.3) \end{cases}$$

where $A, B, C, D, \gamma_2, \gamma_3$, and k_x are constants to be determined. The boundary conditions on the electric field component E_y require continuity at $x = 0$ and $x = t$, yielding

$$A = B + C \tag{2.4}$$

$$D = Be^{ik_x t} + Ce^{-ik_x t} \tag{2.5}$$

Furthermore, the magnetic field component H_z obtained from the wave equation

$$\frac{\partial}{\partial x} E_y = i\omega\mu H_z \tag{2.6}$$

must also be continuous at $x = 0$ and $x = t$, leading to

$$\gamma_2 A = ik_x(B - C) \tag{2.7}$$

$$-\gamma_3 D = ik_x(Be^{ik_x t} - Ce^{-ik_x t}) \tag{2.8}$$

Equations (2.4), (2.5), (2.7), and (2.8) can be used to solve for the characteristic equation by setting the determinate to zero:

$$\begin{vmatrix} -\gamma_2 & ik_x & -ik_x & 0 \\ -1 & 1 & 1 & 0 \\ 0 & ik_x e^{ik_x t} & ikxe^{-ik_x t} & \gamma_3 \\ 0 & e^{ik_x t} & e^{-ik_x t} & -1 \end{vmatrix} = 0 \tag{2.9}$$

yielding the following characteristic equation:

$$k_x t = \tan^{-1}\frac{\gamma_2}{k_x} + \tan^{-1}\frac{\gamma_3}{k_x} + m\pi \tag{2.10}$$

The eigenvalue constants k_x, γ_1, and γ_2 are obtained from the wave equation and are related to the propagation wave vector β as follows:

$$k_x = \sqrt{k_0^2 n_1^2 - \beta^2} \tag{2.11}$$

$$\gamma_2 = \sqrt{\beta^2 - k_0^2 n_2^2} \tag{2.12}$$

$$\gamma_3 = \sqrt{\beta^2 - k_0^2 n_3^2} \tag{2.13}$$

where $k_0 = 2\pi/\lambda$ and $\beta = (2\pi/\lambda)n_m$, with n_m being the refractive index of the guided mode. In order to simplify the presentation of the dispersion properties of these waveguides, it is convenient to use two normalized parameters that relate to the physical parameters of the structure, namely, a normalized thickness and a normalized propagation wave vector or refractive index. Therefore, we define the normalized thickness as

$$V = k_0 t\sqrt{n_1^2 - n_2^2} \tag{2.14}$$

and the normalized index of the guided mode as

$$b = \frac{n_m^2 - n_2^2}{n_1^2 - n_2^2} \tag{2.15}$$

where k_0 is the wave number in free space, t is the guiding layer thickness, n_1 is the guiding layer index, and n_2 is the substrate index. Since these structures might be asymmetric, that is, the indices of the substrate and cover are different, an asymmetry factor is introduced for the TE and TM modes as

$$a_{\mathrm{TE}} = \frac{n_2^2 - n_3^2}{n_1^2 - n_2^2} \tag{2.16}$$

$$a_{\mathrm{TM}} = \frac{(n_1^4/n_3^4)(n_2^2 - n_3^2)}{n_1^2 - n_2^2} \tag{2.17}$$

The eigenconstants k_x, γ_2, and γ_3 can now be redefined in terms of V, b, and a as follows:

$$k_x = \frac{V}{t}\sqrt{1 - b} \tag{2.18}$$

$$\gamma_2 = \frac{V}{t}\sqrt{b} \tag{2.19}$$

$$\gamma_3 = \frac{V}{t}\sqrt{b + a} \tag{2.20}$$

Following these normalized definitions, the dispersion equation (2.10) becomes

$$V\sqrt{1 - b} = \tan^{-1}\sqrt{\frac{b}{1 - b}} + \tan^{-1}\sqrt{\frac{b + a}{1 - b}} + m\pi \tag{2.21}$$

From this equation, the cutoff frequency of the mth mode as well as the number of modes can be determined as follows:

$$V_m = V_0 + m\pi \tag{2.22}$$

where V_0 is the cutoff condition for the fundamental mode given by

$$V_0 = \tan^{-1}\sqrt{a} \tag{2.23}$$

For symmetric waveguides, $a = 0$, and therefore $V_0 = 0$. The solution to Eq. (2.21) is depicted graphically in Fig. 2.2a. From this figure, the desired point of operation (number of modes allowed) as well as the degree of optical confinement (especially in the case of single-mode operation) are chosen first. Knowledge of the V-number allows for the choice of either the thickness of the layer or the index difference ($\Delta n = n_1 - n_2$), from which the second parameter is easily calculable. The b value also determines the effective index

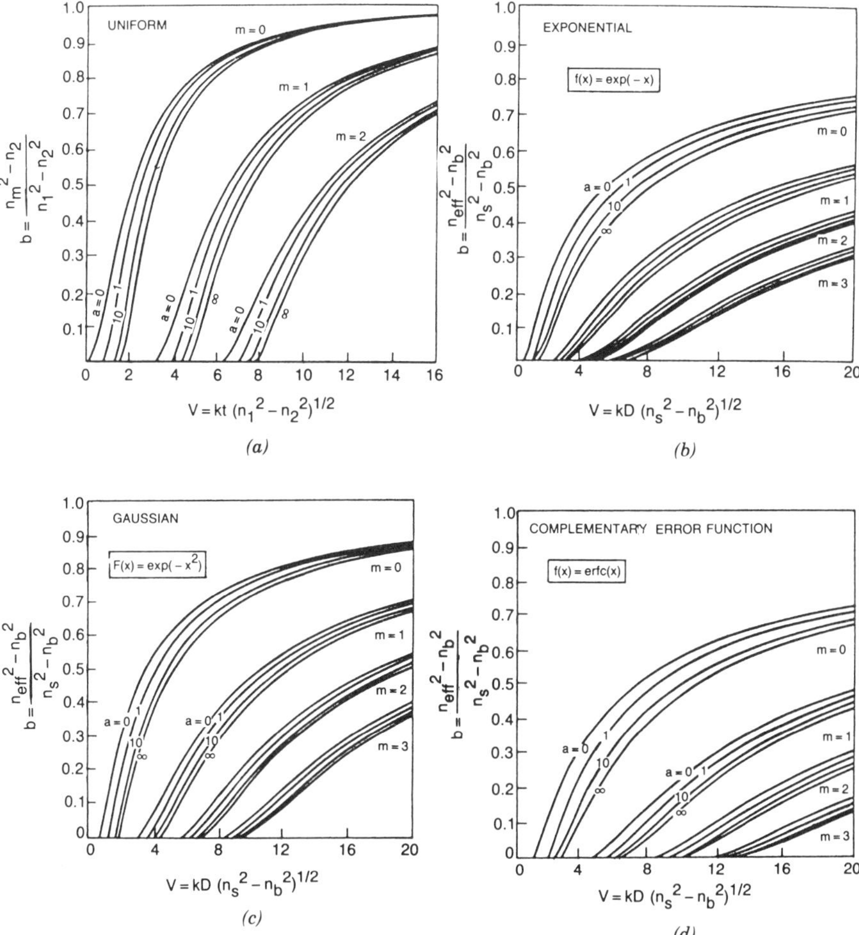

Figure 2.2 Normalized dispersion curves of planar waveguides for (a) uniform, (b) exponential, (c) Gaussian, and (d) complementary error function index profiles.

of the guided modes represented for small index difference approximately by

$$n_m = n_2 + b(n_1 - n_2) \tag{2.24}$$

Knowledge of the V and b parameters also allows for the calculation of the eigennumbers (k_x, γ_2, and γ_3) inside and outside the waveguide as given by Eqs. (2.12)–(2.20).

For guiding layers with nonuniform refractive index distribution, the dispersion relation as well as the field distributions can be obtained from solution to the wave equation with the substitution of the particular index distribution $n(x)$:

$$\frac{d^2E_y}{dx^2} = [\beta^2 - n^2(x)\, k_0^2]E_y \tag{2.25}$$

for TE modes and

$$\frac{d^2H_y}{dx^2} = [\beta^2 - n^2(x)\, k_0^2]H_y \tag{2.26}$$

for TM modes.

Such solutions are usually difficult to obtain analytically for complicated index distributions but can be obtained by numerical methods. In practice, fabrication methods employing diffusion techniques lead to a slowly varying index distribution. The relationship between impurity concentration and refractive index distribution justifies the use of Gaussian and error function distributions for diffused waveguides as predicted from diffusion theory. This has been found to be close to actual distributions obtained in the fabrication of $Ti:LiNbO_3$ and glass waveguides, for example. Accordingly, it is useful to generate modal dispersion curves for such distributions. Figures 2.2*b*, *c*, and *d* correspond to normalized dispersion curves obtained numerically for graded index waveguides with exponential, Gaussian, and complementary error function distributions, respectively, represented by the following index distributions:

Exponential: $$n(x) = n_s + \Delta n \exp\left(-\frac{x}{D}\right) \tag{2.27a}$$

Gaussian: $$n(x) = n_s + \Delta n \exp\left[-\left(\frac{x}{D}\right)^2\right] \tag{2.27b}$$

Error function: $$n(x) = n_s + \Delta n \operatorname{erfc} \frac{x}{D} \tag{2.27c}$$

It can be noticed from Figs. 2.2*b*, *c*, and *d* that the asymmetry in the index profile (maximum at the superstrate interface and decaying to a steady state within the substrate) leads to a cutoff condition for the fundamental mode even when the symmetry parameter $a = 0$, in contrast with the uniform index distribution (Fig. 2.2*a*) where the fundamental mode has no cutoff at $a = 0$.

2.2 OPTICAL CHANNEL WAVEGUIDES

Two-dimensional optical confinement can be achieved in a variety of ways. In all cases, the refractive index within the channel waveguide must be larger than that in all the regions around its perimeter. The five basic structures of integrated optical channel waveguides are shown in Fig. 2.3. The strip-loaded waveguide shown in Fig. 2.3*a* consists of

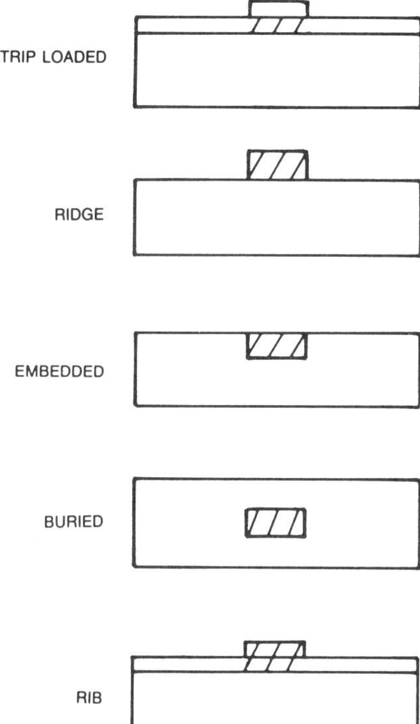

Figure 2.3 General configurations of integrated optical channel waveguides.

a planar film deposited on a substrate of lower index. The channel confinement is provided by depositing a narrow superstrate strip film whose index is higher than air but lower than that of the film. Due to this, the region in the film below the superstrate strip has a higher effective index than the side regions covered with air, and therefore light is confined underneath the strip. The ridge waveguide shown in Fig. 2.3*b* is simply a narrow film deposited on a substrate of lower refractive index, with air covering the top. Embedded channel waveguides are formed by diffusing impurities into a substrate such that the index in the diffused region is higher than the substrate, thus forming a channel guide bound by the substrate on three sides and by air on the fourth, as shown in Fig. 2.3*c*. Buried waveguides are formed when the channel area of higher index is driven into the substrate and is therefore surrounded symmetrically by regions of the same refractive index, as shown in Fig. 2.3*d*. Finally, a rib waveguide may be formed by depositing a planar film layer of higher index than the substrate and then removing part of the film on both sides of a narrow channel, as shown in Fig. 2.3*e*, thus forming a waveguide underneath the rib area. Variations on these structures may be generated, for example, by using a cover layer instead of air.

The formation of such waveguide circuitry requires the use of photolithographic techniques for the definition of the channel. Fabrication techniques include deposition (thermal evaporation, electron beam evaporation, sputtering, spin coating, CVD deposition, etc.), epitaxial growth (sputtering, melting, liquid phase epitaxy, vapor phase epitaxy, metal organic chemical vapor deposition, molecular beam epitaxy, etc.), and modification (out-diffusion, in-diffusion, ion exchange, proton exchange, ion implantation, etc.).

The exact analysis of channel waveguides is more difficult than that of planar waveguides. This is primarily due to the increase in complexity in the boundary conditions where field matching around the waveguide perimeter has to be satisfied. Rigorous and approximation methods have been employed to determine the dispersion properties of rectangular channel waveguides. The more rigorous analysis employed circular harmonic field expansions [1], wave vector optimization for field match [2], and variational methods [3], all of which require computerized numerical computation since closed-form solutions are not easily obtained. The most accurate results have been reported using circular harmonic computer analysis [1]. This method is based on the expansion of the electromagnetic fields in terms of a series of circular harmonics in the form of Bessel functions multiplied by trigonometric functions where the electric and magnetic fields are matched at the boundaries, yielding a set of equations solvable by a computer. Approximation methods have also been introduced to reduce the computational complexities involved in the rigorous solutions. An approximate analytical method for well-guided structures where most of the energy is confined in the channel waveguide assumes oscillatory field solutions inside the channel and decaying fields outside of it both in the horizontal and vertical regions surrounding the channel with the decaying constants being independent of each other [4,5,6]. Accordingly, two characteristic equations are derived for each axis and are linked to each other by their relation to the propagation constant in the channel waveguide. The accuracy of these solutions is very good for waveguides well above cutoff. A second approximate but simple method known as the effective index method utilizes the division of the channel waveguide structure into three regions looking like a slab and sandwiched between two side regions. An effective index for each region is found using the simple planar waveguide analysis. From that, the step index planar waveguide treatment is applied to determine the wave vector of the new slab waveguide [6–8]. This method also yields accurate results for waveguides well above cutoff. The accuracy improves with larger width-to-depth aspect ratio. The simplicity and accuracy of this method makes it attractive for a quick prediction of the approximate dispersion properties of a variety of channel waveguide structures. Due to the usefulness of this method, an example is cited to illustrate the approach. Consider the ridge waveguide shown in Fig. 2.4, where the channel waveguide is composed of a raised ridge of width W, thickness t, and uniform index n_1 on a substrate of index n_2. The thickness of the film on both sides of the channel is t_2, and the structure is bound on top by a region of index n_3. The first step is to determine an effective index for the three regions a, b, and c corresponding to the channel and its sideway surroundings, respectively, as though they were all planar regions. With the help of Fig. 2.2a, the V-numbers for the respective regions are first found:

$$V_a = k_0 t_i \sqrt{n_1^2 - n_2^2} \tag{2.28a}$$

$$V_b = V_c = k_0 t_2 \sqrt{n_2^2 - n_2^2} \tag{2.28b}$$

To find the effective indices in each region, the b-numbers are found from Fig. 2.2a for the appropriate asymmetry factor a, as defined in Eqs. (2.16) and (2.17). The corresponding effective indices are found from knowledge of the b-numbers, which for small $n_1 - n_2$ are approximately given by

$$N_a^2 = n_2^2 + b_a(n_1^2 - n_2^2) \tag{2.29}$$

$$N_{b,c}^2 = n_2^2 + b_{b,c}(n_1^2 - n_2^2) \tag{2.30}$$

The same exercise is applied to the new slab waveguide. The V-number of the channel waveguide is now given by

$$V_g = k_0 W \sqrt{N_a^2 - N_b^2} = k_0 W \sqrt{(n_1^2 - n_2^2)(b_a - b_b)} \tag{2.31}$$

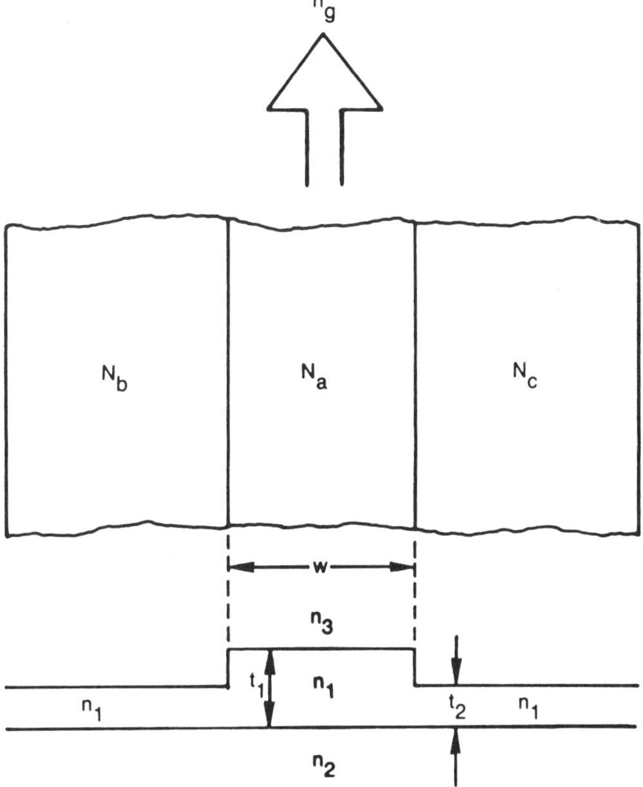

Figure 2.4 Equivalent index representation of channel waveguides.

Due to the symmetry of this structure, $a = 0$, and hence the b-number of the waveguide (b_g) is found from the $a = 0$ curve, yielding the effective index of the channel waveguide:

$$n_g = \sqrt{n_2^2 + b_g(b_a - b_b)(n_1^2 - n_2^2)} \tag{2.32}$$

For a small index difference between the guide and the substrate (Δn), Eq. (2.23) is approximated to

$$n_g \cong n_2[1 + b_g(b_a - b_b)\,\Delta n] \tag{2.33}$$

While exact solutions to the dispersion relations of channel waveguides involve complicated boundary conditions that require lengthy numerical computations, it is possible to obtain approximate solutions in approximated closed forms by simplification of the boundary conditions. By assuming the waveguide to be of a rectangular geometry, as shown in Fig. 2.5, oscillatory field solutions are assumed in the waveguide region of index n_1 in both the x and y directions, and decaying solutions are assumed in the four regions abounding the waveguide on its four sides denoted by indices n_2, n_3, n_4, and n_5. For well-guided waveguides, most of the power is confined to region 1, while a small portion propagates in regions 2–5, and even smaller power resides in the excluded areas bound by the corner intersections, thus yielding small error in boundary condition

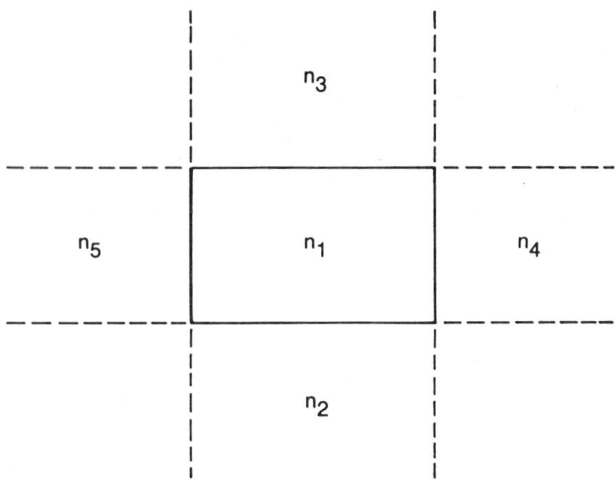

Figure 2.5 Cross section and boundaries of channel waveguides.

matching. By separation of variables, two characteristic equations are obtained in the x and y directions in much the same way as obtained in the slab waveguides as described earlier, yielding, for TE-like modes, denoted as E^{y}_{MN},two equations similar to Eq. (2.10):

$$k_x W = \tan^{-1}\frac{\gamma_4}{k_x} + \tan^{-1}\frac{\gamma_5}{k_x} + M\pi \tag{2.34}$$

$$k_x t = \tan^{-1}\frac{\gamma_2}{k_x} + \tan^{-1}\frac{\gamma_3}{k_x} + N\pi \tag{2.35}$$

with

$$\gamma_{4,5} = \left[\left(\frac{V_{xi}}{W}\right)^2 - k_x^2\right]^{1/2} \tag{2.36}$$

$$\gamma_{2,3} = \left[\left(\frac{V_{yi}}{t}\right)^2 - k_y^2\right]^{1/2} \tag{2.37}$$

$$V_x = \frac{2\pi}{\lambda} W\sqrt{n_1^2 - n_i^2} \tag{2.38}$$

$$V_{yi} = \frac{2\pi}{\lambda} t\sqrt{n_1^2 - n_i^2} \tag{2.39}$$

and

$$k_z^2 = k_1^2 - k_x^2 - k_y^2 \tag{2.40}$$

For well-guided modes, k_x and k_y can be approximated by

$$k_x \cong \frac{m\pi}{W}\left(1 + \frac{1}{V_{x4}} + \frac{1}{V_{x5}}\right)^{-1} \tag{2.41}$$

$$k_y \cong \frac{n\pi}{t}\left[1 + \left(\frac{n_2}{n_1}\right)^2 \frac{1}{V_{yz}} + \left(\frac{n_3}{n_1}\right)^2 \frac{1}{V_{y3}}\right]^{-1} \tag{2.42}$$

From the preceding, an approximate closed-form solution of the propagation number or the effective index of the guided mode is obtained as follows:

$$n_{\text{eff}} \cong \left\{ n_1^2 - \frac{M\lambda}{2W}\left(1 + \frac{1}{V_{x4}} + \frac{1}{V_{x5}}\right)^{-2} - \frac{N\lambda}{2W}\left[1 + \frac{n_2}{n_1}\frac{1}{V_{y2}} + \left(\frac{n_3}{n_1}\right)^2 \frac{1}{V_{y3}}\right]^{-2} \right\}^{1/2}$$

A more exact solution is obtained by solving Eqs. (2.34) and (2.35) without using the approximate values for k_x and k_y given in Eqs. (2.41) and (2.42). Such solutions yield accurate results except near cutoff conditions, which are not of great interest in most cases.

Normalized dispersion curves for the various forms of channel waveguides shown in Fig. 2.3 are presented in Figs. 2.6–2.11. These charts were obtained by numerical solutions based on the various methods described earlier. Figure 2.6 shows the normalized waveguide index of the lowest-order mode for a strip-loaded waveguide (Fig. 2.3*a*) for a special case where $W = 2t$ as obtained by the effective index method. Figure 2.7 shows the normalized waveguide index of the ridge channel waveguide (Fig. 2.3*b*) for which $W = 2t$ and $W = 4t$ as obtained also by the effective index method. Figures 2.8 and 2.9 show the normalized waveguide index for embedded channel waveguides (Fig. 2.3*c*) for which $W = t, 2t, 4t$ as obtained by solutions to the transcendental Eqs. (2.34) and (2.35). Figure 2.10 shows the normalized waveguide index for buried channel waveguides (Fig. 2.3*d*)

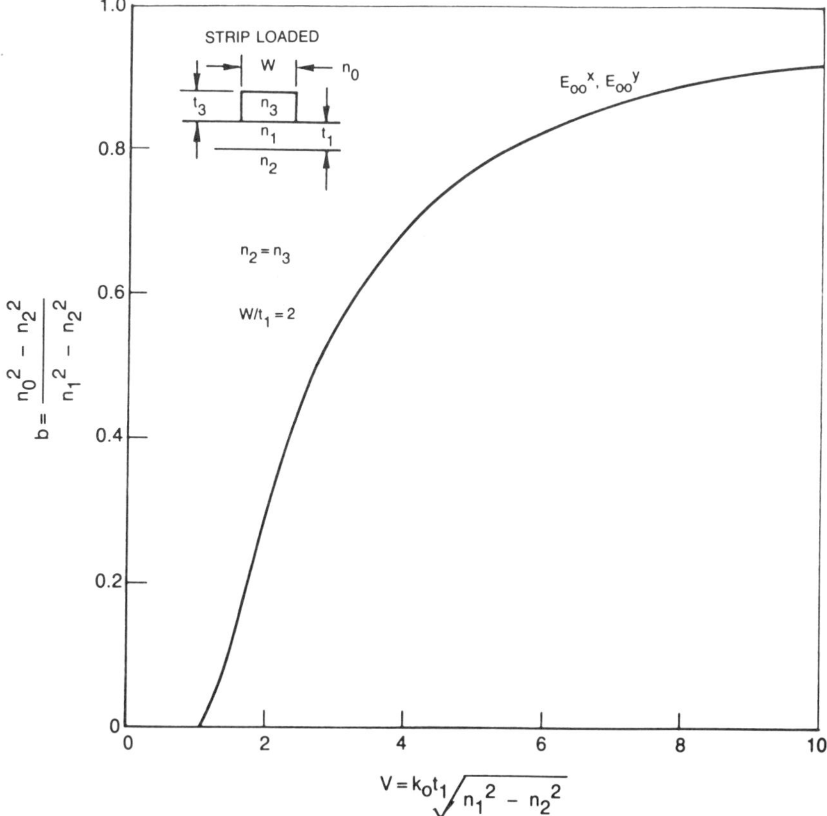

Figure 2.6 Normalized dispersion curves of strip-loaded channel waveguides. (From Ungar [2], © 1977, by permission of the Oxford University Press.)

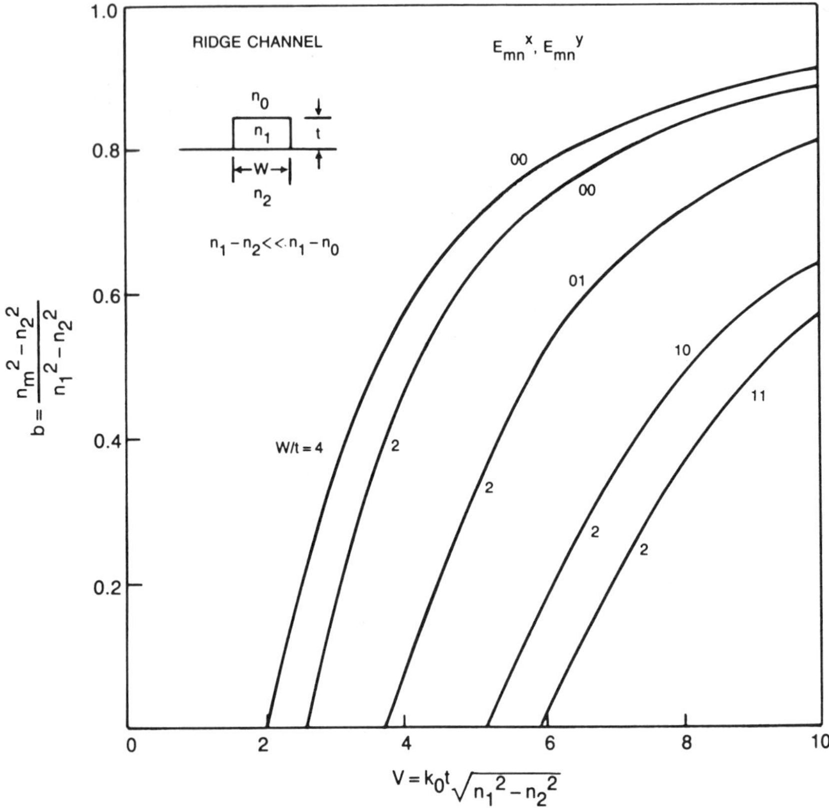

Figure 2.7 Normalized dispersion curves of ridge channel waveguides. (From Ungar [2], © 1977, by permission of the Oxford University Press.)

for which $W = 2t$ as obtained from exact solutions by the circular harmonic field expansion method. Figure 2.11 shows the normalized waveguide index of rib channel waveguides (Fig. 2.3*e*) for which the rib thickness is 70% of the film thickness and $W = 4t$ as obtained by the effective index method.

2.3 OPTICAL DIRECTIONAL COUPLERS

2.3.1 General Solution

Optical directional couplers consist of a set of closely spaced optical waveguides whose fields interact with each other by proximity coupling. Coupling is cumulative over a distance (known as the interaction length) along the propagation direction where the waveguides are in sufficient proximity to cause coupling. The coupling strength depends on parameters that can be categorized under three general conditions, namely, synchronism or wave number phase matching, optical confinement, and proximity. Synchronism of the wave vectors of the propagating optical waves in the coupled waveguide set is very important. The coupling strength is highest when the wave vectors of all the interacting waveguides are the same. Conversely, the coupling could be substantially reduced if sufficient imbalance between the wave vectors exists, regardless of proximity and optical confinement. Optical confinement is also important as it determines the extent of field

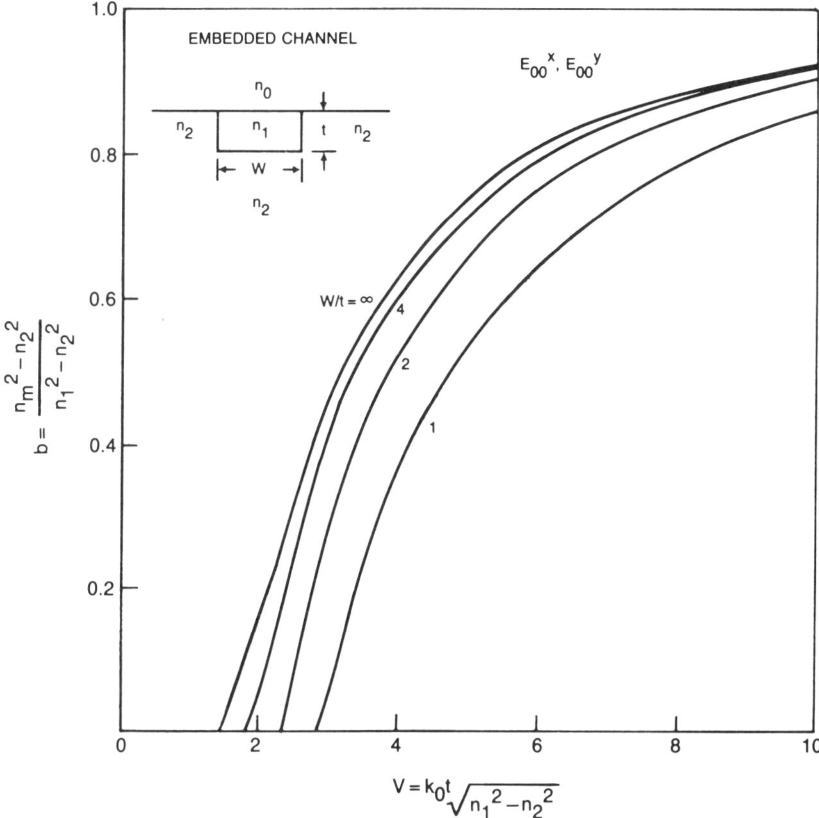

Figure 2.8 Normalized dispersion curves of embedded channel waveguides. (From Ungar [2], © 1977, by permission of the Oxford University Press.)

overlap between the interacting waveguides. The optical confinement at a particular wavelength is controlled by the waveguide size (width and height) and index difference with its surrounding (Δn), which is the make-up of the numerical aperture of the waveguide. Smaller optical confinement (small Δn and/or size) implies that a larger portion of the optical fields reside outside the waveguide, decaying in the transverse plane, yielding greater overlap between the fields of the adjacent waveguides, which increases the coupling strength. Stronger optical confinement (large Δn and/or size) yields shorter tails of the decaying fields outside the waveguide and therefore lower field overlap and coupling strength. Finally, the coupling strength is critically dependent on the spacing between the waveguide set. Smaller spacing yields higher coupling strength, and vice versa. In order to translate the preceding discussion into meaningful design parameters, a generalized formulation of the coupling problem is introduced next and then is narrowed down to practical cases of interest.

Consider a set of N coupled waveguides, as shown in Fig. 2.1. The optical fields in the various waveguides are represented by the following equations:

$$\frac{d}{dz}\begin{bmatrix} A_1(z) \\ A_2(z) \\ \vdots \\ A_n(z) \end{bmatrix} = -i \begin{bmatrix} \beta_1 & K_{12} & \cdots & K_{1n} \\ K_{21} & \beta_2 & & \\ \vdots & K_{ij} & & \vdots \\ K_{n1} & \cdots & \cdots & \beta_n \end{bmatrix} \begin{bmatrix} A_1(z) \\ A_2(z) \\ \vdots \\ A_n \end{bmatrix} \tag{2.43}$$

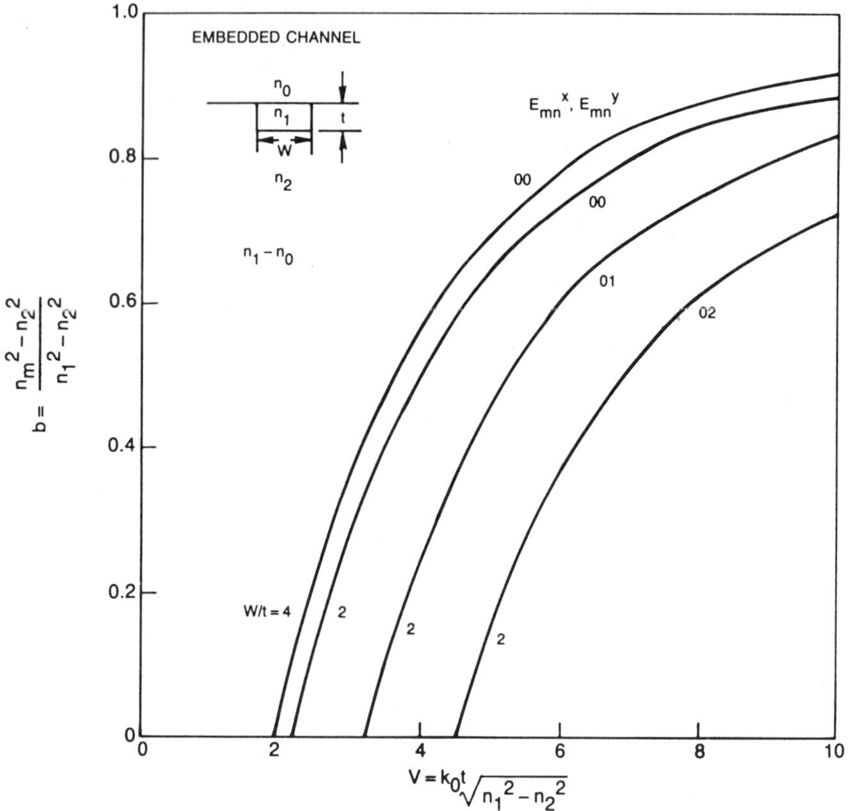

Figure 2.9 Normalized dispersion curves of embedded channel waveguides. (From Ungar [2], © 1977, by permission of the Oxford University Press.)

where

$$A_i(z) = a_i(z)e^{-i\beta \cdot z} \tag{2.44}$$

β_i is the wave vector of the ith waveguide, and K_{ij} is the coupling coefficient between the ith and jth waveguides. In most cases of practical interest, the waveguides are made identical, thus having the same β's, so that the coupling strength is maximized. Furthermore, the interaction between the nonadjacent waveguides is much smaller than the interaction between the adjacent waveguides due to the substantial increase in spacing. Therefore, it can be assumed that

$$K_{ij} \cong 0 \quad \text{for} \quad |i - j| > 1 \tag{2.45}$$

Finally, and assuming lossless conditions, the conservation of power, in normalized form, is defined as

$$\frac{d}{dz} \sum |Ai(z)|^2 = 1.0 \tag{2.46}$$

requiring

$$K_{ij} = K_{ji}^* = K \tag{2.47}$$

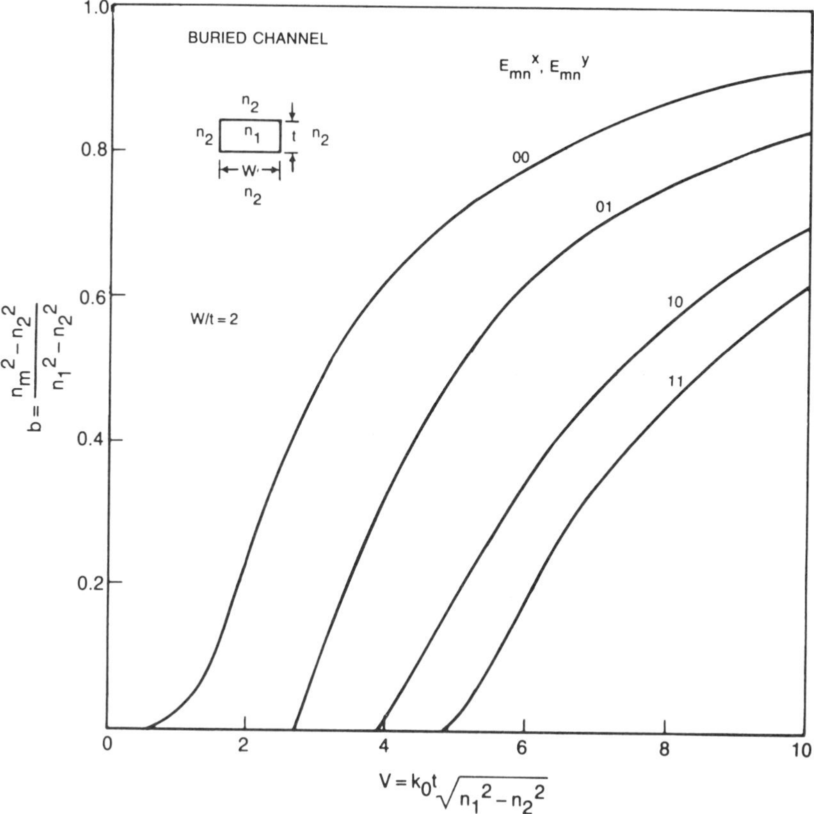

Figure 2.10 Normalized dispersion curves of buried channel waveguides. (From Goel [1]. Reprinted with permission. © 1969 AT&T.)

under these conditions, and for an equal and uniformly spaced set of waveguides, Eq. (2.1) can be rewritten as

$$\frac{d}{dz}\begin{bmatrix} A_1(z) \\ A_2(z) \\ \vdots \\ A_n(z) \end{bmatrix} = -i\begin{bmatrix} \beta & k & 0 & 0 \\ k & \beta & k & \vdots \\ 0 & k & \vdots & \vdots \\ \ddots & & & \beta \end{bmatrix}\begin{bmatrix} A_1(z) \\ A_2(z) \\ \vdots \\ A_n(z) \end{bmatrix} \tag{2.48}$$

Equation (2.48) can now be solved under the desired initial conditions to determine the power flow in the coupled waveguide set.

2.3.2 Waveguide Pair Directional Coupler

The two-channel directional coupler is an important element in integrated optical circuitry and has been extensively used to build passive couplers, 2 × 2 switches, and intensity modulators. For a two-waveguide system in which the waveguides are uniformly spaced

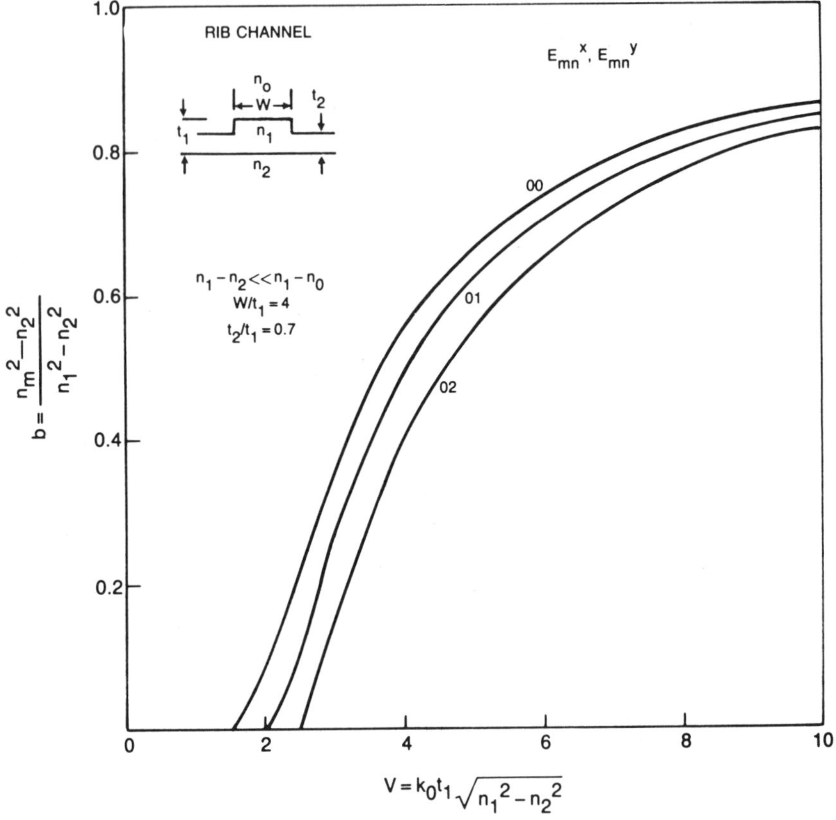

Figure 2.11 Normalized dispersion curves of rib channel waveguides. (From Ungar [2], © 1977, by permission of the Oxford University Press.)

throughout the interaction region, as shown in Fig. 2.12, Eq. (2.48), subject to conservation of power, yields the following solution:

$$\begin{bmatrix} a_1(z) \\ a_2(z) \end{bmatrix} = \begin{bmatrix} A & -iB \\ -iB^* & A^* \end{bmatrix} \begin{bmatrix} a_1(0) \\ a_2(0) \end{bmatrix} \tag{2.49}$$

where

$$A = \cos\theta + i\frac{\Delta\beta L}{\pi}\frac{\pi}{2}\frac{\sin\theta}{\theta} \tag{2.50}$$

$$B = \frac{\pi}{2}\frac{L}{l_c}\frac{\sin\theta}{\theta} \tag{2.51}$$

$$\theta = \frac{\pi}{2}\sqrt{\left(\frac{L}{l_c}\right)^2 + \left(\frac{\Delta\beta L}{\pi}\right)^2} \tag{2.52}$$

$$l_c = \frac{\pi}{2K} \tag{2.53}$$

$$\Delta\beta = \beta_1 - \beta_2 \tag{2.54}$$

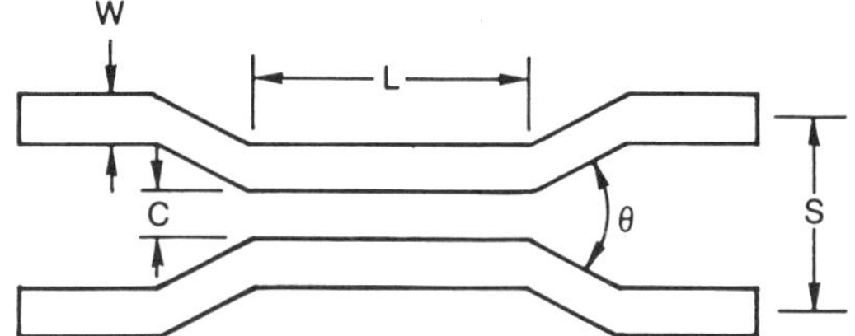

Figure 2.12 Integrated optical 2 × 2 parallel directional coupler.

where L is the interaction length and l_c is the coupling length. For identical waveguides, $\Delta\beta = 0$, and therefore,

$$A = \cos(Kz) \tag{2.55}$$

$$B = \sin(Kz) \tag{2.56}$$

The amplitudes in the two waveguides in Eq. (2.49) become

$$a_1(z) = a_1(0)\cos(Kz) - ia_2(0)\sin(Kz) \tag{2.57}$$

$$a_2(z) = a_1(0)\sin(Kz) - ia_2(0)\cos(Kz) \tag{2.58}$$

The power flow in the coupled waveguides is therefore

$$|a_1(z)|^2 = |a_1(0)|^2\cos^2 Kz + |a_2(0)|^2\sin^2 Kz - \mathrm{Im}[a_1(0)_{a_2}^*(0)]\sin^2 Kz \tag{2.59}$$

$$|a_2(z)|^2 = |a_1(0)|^2\sin^2 Kz + |a_2(0)|^2\cos^2 Kz - \mathrm{Im}[a_1(0)_{a_2}^*(0)]\sin^2 Kz \tag{2.60}$$

Depending on the initial conditions $[a_1(0)$ and $a_2(0)]$, the power in the two waveguides oscillates between maximum and minimum at intervals equal to the coupling length l_c, except when $a_1(0) = a_2(0)$, where no coupling or power transfer occurs. Under the condition when all the power is initially excited into one guide, say, for example, $|a_1(0)|^2 = 1.0$ and $|a_2(0)|^2 = 0.0$, the power flow becomes

$$|a_1(z)|^2 = \cos^2 Kz \tag{2.61}$$

$$|a_2(z)|^2 = \sin^2 Kz \tag{2.62}$$

The preceding treatment applies to parallel waveguide directional couplers in which the coupling is uniform throughout. When the coupling within the interaction region is not uniform, such as in nonparallel waveguides, the coupling coefficient is no longer constant with z. In this case, under lossless conditions Eq. (2.48) yields the following modified solutions for a two-identical-waveguide coupler:

$$a_1(z) = a_1(0)\cos\int K(z)\,dz - ia_2(0)\sin\int K(z)\,dz \tag{2.63}$$

$$a_2(z) = a_1(0)\sin\int K(z)\,dz - ia_2(0)\cos\int K(z)\,dz \tag{2.64}$$

The oscillatory behavior of the power exchange is now modified depending on the $K(z)$. In certain cases, the oscillatory behavior may have variable periodicity; in others it may not have any oscillatory behavior. The treatment of such cases is presented later in this section for special cases of practical interest.

The coupling coefficient K is a measure of the field overlap between the coupled waveguides. In the case of two adjacent identical waveguides, the coupling coefficient is defined as

$$K = -\frac{iw}{4} \varepsilon_0 \iint_{\infty}^{\infty} a_1(x_1 y) a_2^*(x, y) \, \Delta\varepsilon_y(x, y) \, dx \, dy \tag{2.65}$$

where the relative dielectric constant difference is $\Delta\varepsilon_y = n_g^2 - n_s^2$ within the waveguide, and $\Delta\varepsilon_y = 0$ outside the waveguide, with n_g and n_s the refractive indices of the guide and its surroundings, respectively. The integration of Eq. (2.65) over the entire cross section of the two-waveguide coupler yields the following expression for the coupling coefficient:

$$K = \frac{2k_x^2 \gamma \exp(-\gamma x)}{\beta(W + 2/\gamma)(k_x^2 + \gamma^2)} \tag{2.66}$$

where β is the wave vector, k_x and γ are the eigenconstants inside and outside the waveguides, respectively, W is the guide width, and c is the inner-edge channel separation. By using the normalized slab waveguide parameter notation, the coupling length can be represented in a simple form by utilizing the notation used in Eqs. (2.18) and (2.19):

$$l_c = \frac{\pi^2 n W^2 (V\sqrt{b} + 2)}{2\lambda V^2 b(1 - b)} \exp\left(\frac{V}{W}\sqrt{bc}\right) \tag{2.67}$$

where

$$V = k_0 W \sqrt{2n \, \Delta n} \tag{2.68}$$

In most cases of practical interest, a well-confined single-mode optical waveguide is desired. Good single-mode optical confinement is usually achieved at $V \simeq 3$, such that the guide is well above cutoff yet adequately below the cutoff of the next higher-order mode so that any electro-optically induced increase in the index does not push the guide into a two-mode regime. An approximate design chart for Ti:$LiNbO_3$ parallel directional couplers is presented in Fig. 2.13 at various wavelengths for guide widths and V-numbers of practical interest. The effect of optical confinement is evident by the longer coupling lengths for higher V-numbers and by the longer wavelengths. The spacing is also critical due to the exponential dependence of the coupling factor on spacing. It should be noted that as the spacing is decreased to zero, the formulation used in deriving the coupling efficiency by the perturbation method (which is based on weak coupling) loses accuracy as the situation transforms from weak to strong coupling.

When the waveguide pair are not parallel, the coupling coefficient is no longer constant with distance, and the solution to the coupled-mode equations yield the solutions given in Eqs. (2.63) and (2.64). The argument representing the coupling coefficient–distance product is now an integral of the coupling coefficient over distance. For a linearly changing waveguide spacing under the initial condition where one input guide is excited, the optical powers in the two guides may be obtained by integration over a distance where the spacing is linearly changing with distance to obtain

$$P_1(\infty) = \cos^2 \frac{K_0}{\gamma \tan \alpha} \tag{2.69}$$

$$P_2(\infty) = \sin^2 \frac{K_0}{\gamma \tan \alpha} \tag{2.70}$$

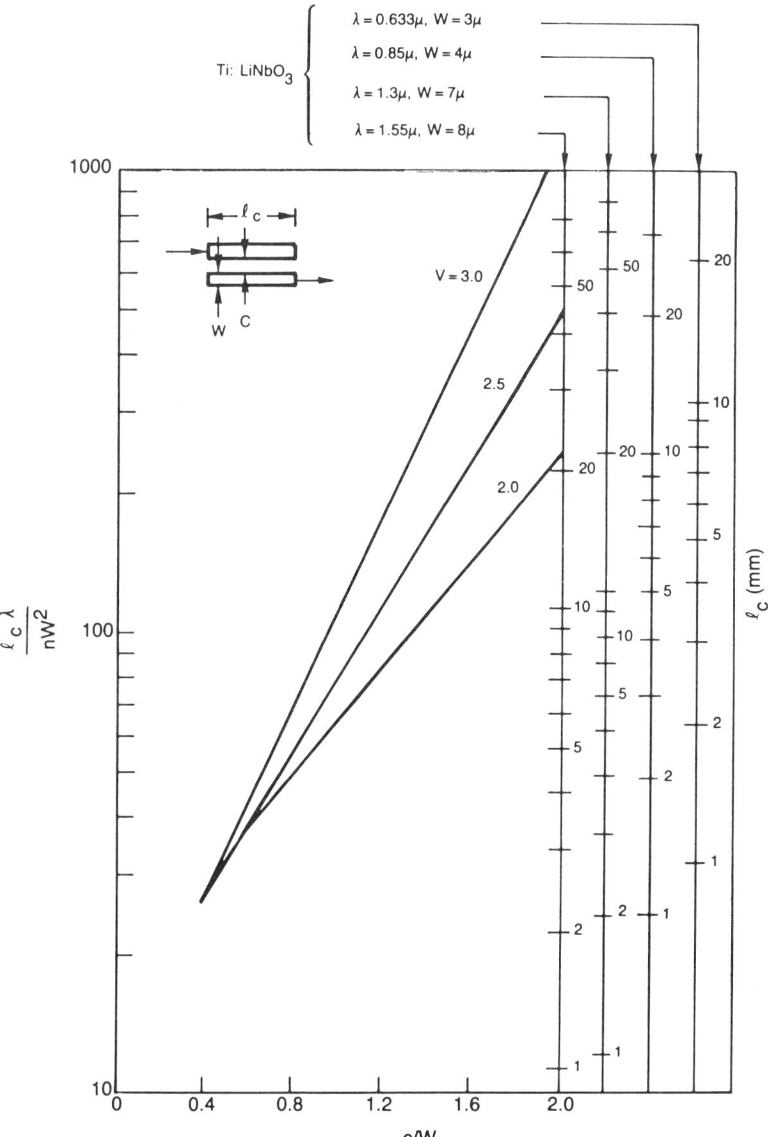

Figure 2.13 Calculated parameters for $Ti{:}LiNbO_3$ 2×2 directional couplers.

where α is the angle between the two guides, K_0 is the coupling coefficient at the smallest separation between the two guides as defined in Eq. (2.66), and γ is the decaying eigennumber outside the guides.

The behavior of such directional coupler configuration is shown in Fig. 2.14 for a single-mode $Ti{:}LiNbO_3$ coupler at various wavelengths of practical interest. Here, again, the waveguide was assumed to have a single-mode optical confinement at $V = 3$. In this case, the coupling efficiency represents the effect of coupling over infinite distance. Therefore, for very small angles, the power coupling is very sensitive to minor changes in angle

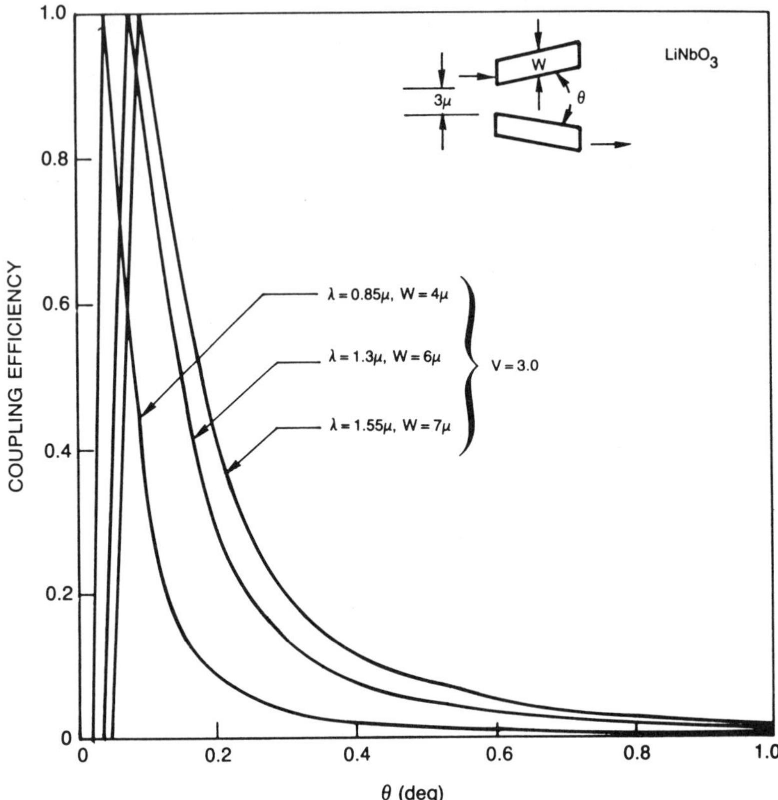

Figure 2.14 Coupling characteristics of nonparallel directional couplers.

as the power is exchanged many times between the two guides in much the same way as in parallel waveguides. As the angle increases, the region where a strong interaction exists becomes shorter and the number of times the power is exchanged decreases, up to a point where coupling is weak and only a small fraction of the power is allowed to couple to the second waveguide. The behavior is especially important in the design of parallel couplers where the input and output guides have to be separated away from proximity to a wider spacing to allow for fiber coupling, for example. In this case, small angles may cause significant coupling prior to the principal interaction region, thus modifying the assumed initial excitation conditions.

REFERENCES

1. J. E. Goel, "A Circular-Harmonic Computer Analysis of Rectangular Dielectric Waveguides," *Bell Syst. Tech. J.*, Vol. 48, p. 2071, 1969.
2. H. G. Unger, *Planar Optical Waveguides and Fibers*, Chapter 3, Oxford University Press, 1977.
3. C. B. Shaw, B. T. French, and C. Wagner, "Further Research on Optical Transmission Lines," Science Report No. 2, contract AF449(638)-1504 AD 625 501, Autonetics Report No. C7-929/501, pp. 13–44, 1976.

4. E. A. J. Marcatili, "Dielectric Rectangular Waveguide and Directional Coupler for Integrated Optics," *Bell Syst. Tech. J.*, Vol. 48, p. 2071, 1969.
5. E. A. J. Marcatili, *Bell Syst. Tech. J.*, Vol. 53, p. 645, 1974.
6. R. M. Knox and P. P. Toulios, in J. Fox, Ed., *Proceedings of MRI Symposium of Submillimeter Waves*, Polytechnic, Brooklyn, NY, 1970.
7. V. Rameswamy, *Bell Syst. Tech. J.*, Vol. 53, p. 697, 1974.
8. G. B. Hocker and W. K. Burns, "Mode Dispersion in Diffused Channel Waveguides by the Effective Index Method," *Appl. Opt.*, Vol. 113, 1977.

3

PLANAR OPTICAL WAVEGUIDES AND WAVEGUIDE LENSES

Shi-Kay Yao

Optech
Calabassas, California

Since the invention of lasers, immense progresses have been made in optoelectronics. Optical modulation, beam scanning, frequency mixing, and parametric oscillations have achieved tremendous advances and are employed in various systems applications such as optical sensors, communications, data storage, signal processing, and many other instruments. With the exception of fiber-optics, almost all the applications are implemented in bulk device forms and with a light beam of nearly Gaussian intensity distribution. However, since the introduction of the concept of integrated optics in the late sixties [1–11], the electro-optical community has been in vigorous pursuit of more efficient electro-optical modulators, switches, scanners, spatial light modulators, and nonlinear optical devices in a planar thin-film form. Furthermore, an important question has been raised on the possibility of constructing an optical system or subsystem on the surface of a single substrate using photolithography and planar fabrication processes similar to the integrated electronics circuits. Very compact, efficient, and environmentally stable optical devices could be produced economically using the integrated optical techniques.

Contrary to the one-dimensional nature of electronic systems, classical optical systems tend to be multidimensional. The advances of electro-optics and integrated optics follow both of these two directions. In the areas of optical communications, and optical sensors, the approaches have been analogous to electronic circuits with channel optical waveguides serving as wires. On the other hand, in the area of optical signal processing, multidimensional systems taking advantage of the parallel processing property of Fourier optics are employed most of the time. When reduced to the integrated optics form, a planar optical waveguide is used as a two-dimensional medium in which optical wavefronts are manipulated. Just like in classical optics, the most important components in the two-dimensional integrated optics are the planar optical waveguides (the optical medium), the waveguide optical lenses (the optical lens), and the spatial light modulators (the optical transparency).

In this chapter, a number of popular planar optical waveguide systems and waveguide optical lenses will be described. General theory, preparation techniques, physical properties, and measurement techniques will be discussed.

3.1 PLANAR OPTICAL WAVEGUIDES

3.1.1 Introduction

In the late 1960s researchers turned their attention to the intense laser light that can be trapped in an optical thin film and the optical prism film couplers as well as the optical grating film couplers that can bring the laser light into and out of such a thin film. In the meantime many dielectrical optical thin-film systems have been discovered to exhibit good optical waveguiding properties. These include films prepared on glass substrates, semiconductor wafers, and electro-optical crystal surfaces. Optical waveguide effective refractive indices of these waveguiding films are perturbed by additional processes such as thin-film overcoats and thickness modifications over selected areas.

In general, planar optical waveguiding films can be classified as either step index optical waveguides or graded index optical waveguides. Films deposited by the majority of conventional thin-film coating processes such as physical vapor deposition, chemical vapor deposition, vacuum sputtering, and epitaxial crystal growth processes tend to belong to the step index waveguide category. The optical waveguiding layer has a substantially uniform optical index of refraction that is higher than the optical index of refraction of the substrate material. Optical waveguiding is facilitated by the total internal reflection at the thin-film interfaces where a step in the optical index of refraction exists. On the other hand, graded index optical waveguides are made by diffusion type of processes such as in-diffusion, out-diffusion, or ion exchange processes. Again, the waveguiding layer must have a higher index of refraction than its surrounding layers if low waveguide losses are desired. At the graded index interface, the Poynting vector of the optical field is gradually turned back toward the direction where the optical index of refraction is higher. Graded index waveguides are extremely important because they can be fabricated onto dielectric crystal surfaces with a single-crystal waveguide body and because they are relatively tolerant to microscopic local defects.

3.1.2 General Theory of Dielectric Planar Waveguides

Ray Optics of Planar Waveguide. The simplest dielectric waveguide is the planar guide shown in Fig. 3.1, where a film of refractive index n_2 is sandwiched between a substrate and a cover material with lower refractive indices n_3 and n_1, respectively. Often the cover

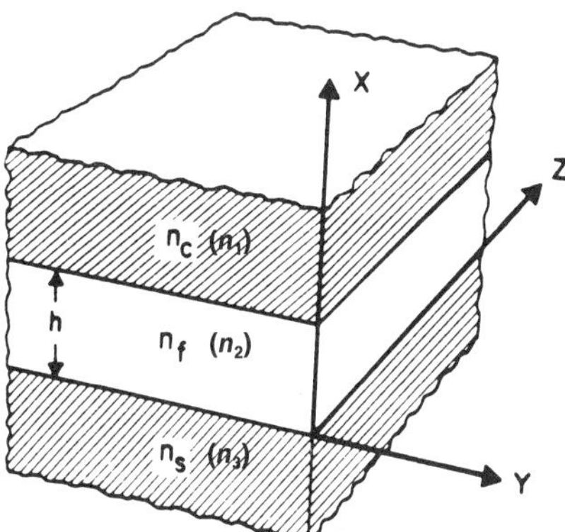

Figure 3.1 Cross section of a planar waveguide.

material is air, $n_1 = 1$. Due to total internal reflection at the film–substrate and film–cover interfaces, light can be confined in the film layer as guided optical waves. The conditions for waveguiding can be illustrated by considering the optical waves as rays propagating inside a thick slab of glass [9–11], as shown in Fig. 3.2. For example, at the air–film interface, the rays can be turned back completely when the incidence angle is greater than the so-called critical angle θ_c, which is given by

$$\sin \theta_c = \frac{n_1}{n_2} \tag{3.1}$$

In Eq. (3.1), n_2 is the refractive index of the medium in the input ray side of the interface. Note that the critical angle exists only when $n_2 > n_1$. Figure 3.2*b* illustrates such a condition in which radiation is allowed in the substrate side only. Further increases in the ray incidence angle beyond the critical angle at the film–substrate interface can cause the total internal reflection to occur at both the interfaces, as shown in Fig. 3.2*c*. Such rays can only propagate in the zig-zag manner within the film region as guided light.

In Fig. 3.2, the rays are shown to actually penetrate slightly into the lower index side of the total internal reflection interfaces while turning back toward the higher index side of the interface. This is due to the presence of evanescent waves on the lower index side of a total internal reflection interface for satisfying the boundary conditions. With the evanescent waves, the energy flow of light actually passes the interface during the total internal reflection and causes a lateral shift in the reflected ray relative to the incident ray. This is the Goos–Hanchen shift, which turned out to be an important element in the understanding of the flow of energy in dielectric waveguides in terms of the ray optics.

Refering to Fig. 3.2*c*, the zig-zag rays may be considered as two superimposed plane-wave components with wave normals that follow the zig-zag directions and are totally reflected at the film boundaries. These waves are coherent and monochromatic with wavelength λ. For a guided mode of the planar waveguide, the zig-zag model predicts

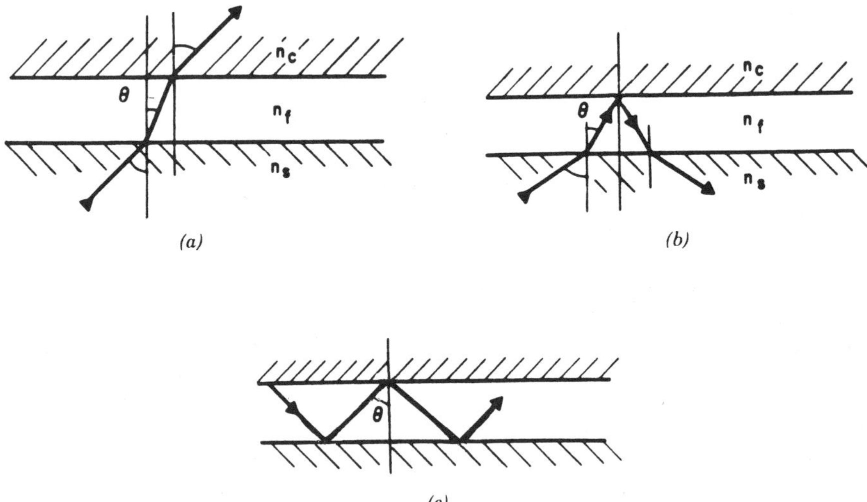

Figure 3.2 Zig-zag ray pictures for light waves in a planar waveguiding structure where $n_2 > n_3 > n_1$: (*a*) radiation mode; (*b*) substrate radiation mode; (*c*) guided modes.

a propagation constant

$$\beta = kn_2 \sin \theta \tag{3.2}$$

which is the projection of the wave vector of the plane waves, n_2k, in the direction of the waveguiding film. However, only a discrete set of angles that allow the reflected plane waves to interfere constructively will lead to acceptable guided modes. For a film of thickness h, the optical phase shift for each transverse passage through the film is simply $n_2k \cos \theta$. Letting $2\phi_1$ and $2\phi_2$ denote the phase shift on total internal reflections from the air–film interface and the film–substrate interface, respectively, the discrete angles for guided modes is given by the "transverse resonance condition" when the round-trip phase shift of the zig-zag rays equals multiples of 2π:

$$2n_2kh \cos \theta - 2\phi_1 - 2\phi_2 = 2m\pi \tag{3.3}$$

where m is an integer that identifies the mode number. Equations (3.2) and (3.3) are essentially the dispersion equation for the planar waveguide when the Goos–Hanchen shift angles as a function of the incidence angle θ are known.

From the preceding equations it is observed that

$$n_3k < \beta < n_2k \tag{3.4}$$

or in terms of an *effective waveguide index*, $N = n_2 \sin \theta$,

$$n_3 < N < n_2 \tag{3.5}$$

The dispersion diagram of an asymmetric planar waveguide is illustrated by Fig. 3.3. At the cutoff frequency, the effective waveguide indices assume the value of the lower bound, n_3, and as the waveguide thickness h increases (or as the wavelength decreases), the effective waveguide index approaches the upper bound of n_2 and more and more waveguide modes are allowed.

To obtain a more precise dispersion diagram for asymmetric planar waveguides in general, the waveguide parameters are rearranged to a number of normalized parameters [12]. The first is a normalized thin-film thickness parameter V such that the effect of

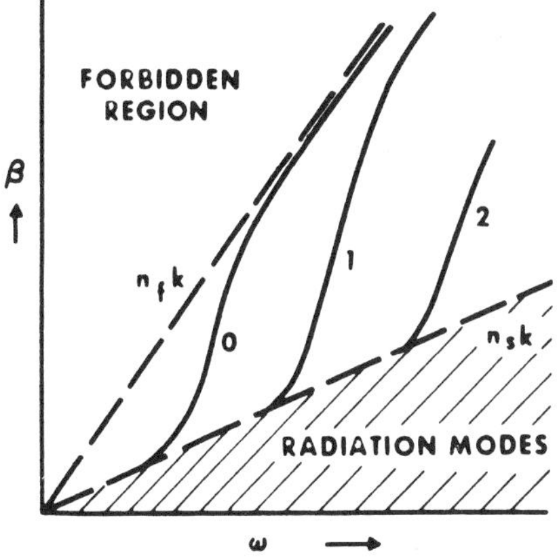

Figure 3.3 Typical ω–β diagram of a dielectric waveguide.

optical wavelength is included:

$$V = kh\sqrt{n_2^2 - n_3^2} \tag{3.6}$$

and then a normalized guide mode index b is related to the effective mode index N by

$$b = \frac{N^2 - n_3^2}{n_2^2 - n_3^2} \tag{3.7}$$

The index b is zero at cutoff and approaches unity far away from it. Typically, the index difference $(n_2 - n_3)$ is small. Thus,

$$N \approx n_3 + b(n_2 - n_3) \tag{3.8}$$

Finally, we introduce the asymmetry parameter for the waveguide structure as

$$a = \frac{n_3^2 - n_1^2}{n_2^2 - n_3^2} \tag{3.9}$$

for the TE modes, which ranges in value from zero for perfect symmetry $(n_3 = n_1)$ to infinity for strong asymmetry $(n_3 \neq n_1$ and $n_3 \approx n_2)$. Typical values of the asymmetry parameter range from 4 for spattered glass waveguide film over glass substrate to 881 for out-diffused waveguide film on lithium niobate crystal substrate. The asymmetry parameter for the popular titanium diffused waveguide on lithium niobate crystal substrate is 44. For TM modes the asymmetry parameter has larger values of from 27 for the glass waveguide to 21,206 for the out-diffused lithium niobate waveguide. In the form of normalized parameters, the generalized dispersion relation becomes, following Kogelnik and Ramaswamy,

$$V\sqrt{1-b} = m\pi + \tan^{-1}\sqrt{\frac{b}{1-b}} + \tan^{-1}\sqrt{\frac{b+a}{1-b}} \tag{3.10}$$

Figure 3.4 is the normalized dispersion diagram where the normalized effective mode index b is plotted as a function of the normalized frequency V for four values of asymmetry parameters a and for the first three modes, $m = 0, 1, 2$. In Eq. (3.10), the cutoff frequency for the mth mode, V_m, can be obtained by letting $b = 0$:

$$V_m = \tan^{-1}\sqrt{a} + m\pi \tag{3.11}$$

For a symmetric waveguide, $a = 0$, the fundamental mode cutoff frequency $V_0 = 0$. There is always a fundamental mode even for thickness approaching zero. On the other hand, for asymmetric waveguides, a minimum film thickness is needed in order to support any guided mode. From Eq. (3.11), the number of guided modes allowed in a waveguide is

$$m = \frac{2h}{\lambda}\sqrt{n_2^2 - n_3^2} \tag{3.12}$$

The cutoff conditions and dispersion diagrams for the TM mode are very similar to the TE modes, particularly when $n_2 - n_3$ is small. The main difference is in the definition

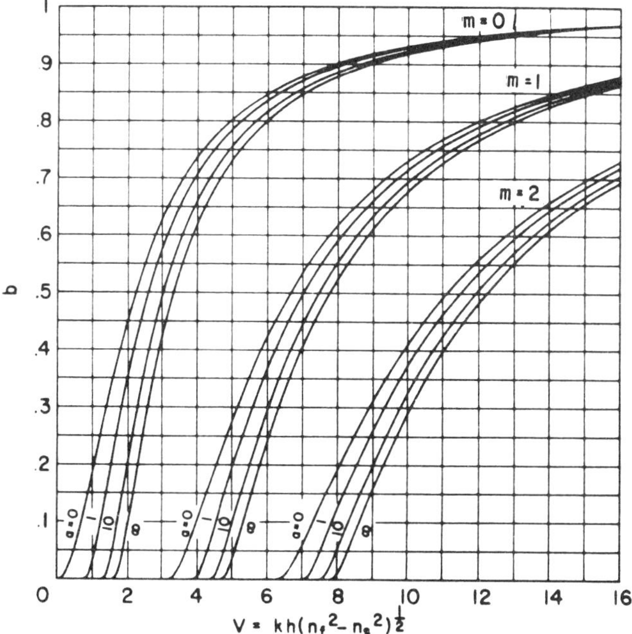

Figure 3.4 Normalized ω–β diagram of a planar slab waveguide showing the guide index b as a function of the normalized thickness V for various degrees of asymmetry. (After Ref. 12.)

of the asymmetry parameter a, which for the TM mode is

$$a_{\mathrm{TM}} = \frac{n_2^4}{n_1^4}\left(\frac{n_3^2 - n_1^2}{n_2^2 - n_3^2}\right) \tag{3.13}$$

Waveguide Modes for Step Index Guides. The waveguide modes are solutions of Maxwell's wave equation [9,13–15],

$$\nabla^2 E(r) + k^2 n^2(r) E(r) = 0 \tag{3.14}$$

subject to the continuity of the tangential components of E and H at the dielectric interfaces. Consider the step index planar waveguide illustrated in Fig. 3.1. The field components E_y of a TE mode propagating along the y direction must obey the equation

$$\partial^2 E_y / \partial x^2 = (\beta^2 - n^2 k^2) E_y \tag{3.15}$$

where the E_y for guided modes are, with $p^2 = \beta^2 - n_1^2 k^2$ and $q^2 = \beta^2 - n_3^2$,

$$E_y = \begin{cases} E_1 \exp[-p(x - h)] \quad \text{for } h < x & (3.16) \\ E_2 \cos(k_2 x - \phi_2) \quad \text{for } 0 < x < h & (3.17) \\ E_3 \exp(qx) \quad \text{for } x < 0 & (3.18) \end{cases}$$

From the boundary conditions, the phase shifts and dispersion relation V can be obtained:

$$\phi = \tan^{-1} \frac{p}{k_2} \tag{3.19}$$

$$= \tan^{-1} \frac{q}{k_2} \tag{3.20}$$

$$K_2 h - \tan^{-1} \frac{p}{k_2} - \tan \frac{q}{k_2} = m\pi \tag{3.21}$$

in agreement with the results from the zig-zag model. The relation between the field magnitude is

$$E_2^2(n_2^2 - N^2) = E_3^2(n_2^2 - n_3^2) = E_1^2(n_2^2 - n_1^2) \tag{3.22}$$

where $N = \beta/K$ is the effective mode index. Finally, the power carried by a mode per unit guide width is

$$P = -2 \int_{-\infty}^{\infty} E_y H_x \, dx = N \sqrt{\frac{\varepsilon_0}{m_0}} E_2^2 h_{\text{eff}} \tag{3.23}$$

where

$$h_{\text{eff}} = h + \frac{1}{q} + \frac{1}{p} \tag{3.24}$$

is the effective thickness of the waveguide.

For the case of TM modes, the boundary conditions require the continuity of H_y and E_z at the interfaces. Assuming the field solutions for the guided modes as

$$H_y = \begin{cases} H_1 \exp[-p(x-h)] & \text{for } h < x \quad (3.25) \\ H_2 \cos(k_2 x - \phi_2) & \text{for } 0 < x < h \quad (3.26) \\ H_3 \exp(qx) & \text{for } x < 0 \quad (3.27) \end{cases}$$

the phase shifts and the dispersion relation are

$$\phi_1 = \tan^{-1} \left(\frac{n_2}{n_1}\right)^2 \frac{p}{k_2} \tag{3.28}$$

$$\phi_2 = \tan^{-1} \left(\frac{n_2}{n_3}\right)^2 \frac{q}{k_2} \tag{3.29}$$

and

$$k_2 h - \tan^{-1} \left(\frac{n_2}{n_3}\right)^2 \frac{q}{k_2} - \tan^{-1} \left(\frac{n_2}{n_1}\right)^2 \frac{p}{k} = m\pi \tag{3.30}$$

The magnitude of these field components are related by

$$\frac{H_2^2(n_2^2 - N^2)}{n_2^2} = H_3^2(n_2^2 - n_3^2)\frac{(N/n_2)^2 + (N/n_3)^2 - 1}{n_3^2}$$
$$= H_1^2(n_2^2 - n_1^2)\frac{(N/n_2)^2 + (N/n_1)^2 - 1}{n_1^2} \tag{3.31}$$

The power per unit guide width carried by the TM mode is

$$P = 2\int_{-\infty}^{\infty} E_x H_y \, dx = \frac{N\sqrt{\frac{\mu_0}{\varepsilon_0}}H_2^2 h_{\text{eff}}}{n_2^2} \tag{3.32}$$

where the effective guide thickness for the TM mode is

$$h_{\text{eff}} = h + \frac{1}{q\left[\left(\frac{N}{n_2}\right)^2 + \left(\frac{N}{n_3}\right)^2 + 1\right]} + \frac{1}{p\left[\left(\frac{N}{n_2}\right)^2 + \left(\frac{N}{n_1}\right)^2 + 1\right]} \tag{3.33}$$

The dispersion diagrams are given in Fig. 3.4.

Waveguide Modes for Graded Index Guides. Many highly useful optical waveguides are made by diffusion processes that lead to waveguides with refractive indices varying gradually over the cross section. Solution of such graded index waveguides have been done by solving the wave equations for exact solutions for certain index profiles. However, it can also be solved either by the WKB approximation method or by taking the odd symmetric solutions of previously solved Schrödinger equations with symmetric potential profiles. For arbitrary profiles, a numerical method has been developed to obtain the waveguide mode properties.

For TE modes, the wave equations for the E_y component is

$$\frac{d^2E_y}{dx^2} = (\beta^2 - n^2k^2)E_y \tag{3.34}$$

which has the same form as the Schrödinger equation of quantum mechanics. The waveguide mode index $N^2 = \beta^2/k^2$ is equivalent to the energy level for each solution to the potential energy well of $n^2(x)$. However, most practical planar waveguides are highly asymmetric with a small gradient into the substrate and a large index step at the air–film interface. The small gradient means close approximation between TE and TM modes. The large step index at the air–film interface allows the assumption that the field in the air region ($x < 0$) is negligibly small. This in turn allows the use of known solutions for Schrödinger equations with odd symmetry that have zero field amplitude at $x = 0$ corresponding to the near-zero amplitude of the guided mode at the air–film interface. A number of examples are given in Ref. 15 and are listed here in Table 3.1, which lists the index formulas, approximate wave solutions, and effective mode indices [16–18]. The potential well profiles employed in the Schrödinger equation for these three index profiles are illustrated in Fig. 3.5. Of greater interest is the exponential profile [16,17] of Table

TABLE 3.1

	Parabolic Profile	sech^2 Profile	Exponential Profile
Index formula	$n(x) \approx n_2(1 - \frac{1}{2}x^2/x_0^2)$	$n(x) \approx n_3 + \Delta n\, \text{sech}^2(2x/h)$	$n(x) \approx n_3 + \Delta n \exp(-2\lvert x\rvert/h)$
Solution			
For $x \geqslant 0$	$E_y \simeq H_{2m+1}(\sqrt{2}x/w) \exp(-x^2/w^2)$	$E_y = U_{2m+1}(2x/h)\, \text{sech}^2(2x/h)$	$E_y = J_p(V \exp[-x/h])$
For $x < 0$	$E_y = 0$	$E_y = 0$	$E_y = 0$
Polynomial	Hermite polynomials: $H_1 = 2x$ $H_3 = 8x^3 - 12x$	Hypergeometric functions: $u_1 = \sinh(2x/h)$ $u_3 = \sinh(2x/h)[1 - \frac{2}{3}(s-2)\sinh^2(2x/h)]$	Bessel function of the first kind with noninteger order p
Normalized thickness	—	$V = kh\sqrt{2n_3\,\Delta n}$	$V = kh\sqrt{2n_3\,\Delta n}$
Effective index	$N_m^2 = n_2^2 - (2m + \frac{3}{2})(n_2\lambda/\pi x_0)$	$N_m^2 = n_3^2 + (s - 2m - 1)^2(\lambda/\pi h)^2$	$N_m^2 = n_3^2 + (P_m^2/4)(\lambda/\pi h)^2$
Comment	$w = \sqrt{\lambda x_0/\pi n_2}$ is the beam radius	$S = \frac{1}{2}(\sqrt{1 + V^2} - 1)$ specifies the maximum number of modes $m \leqslant (s-1)/2$	The order p is solved by finding the Bessel functions such that $J_p(V) = 0$

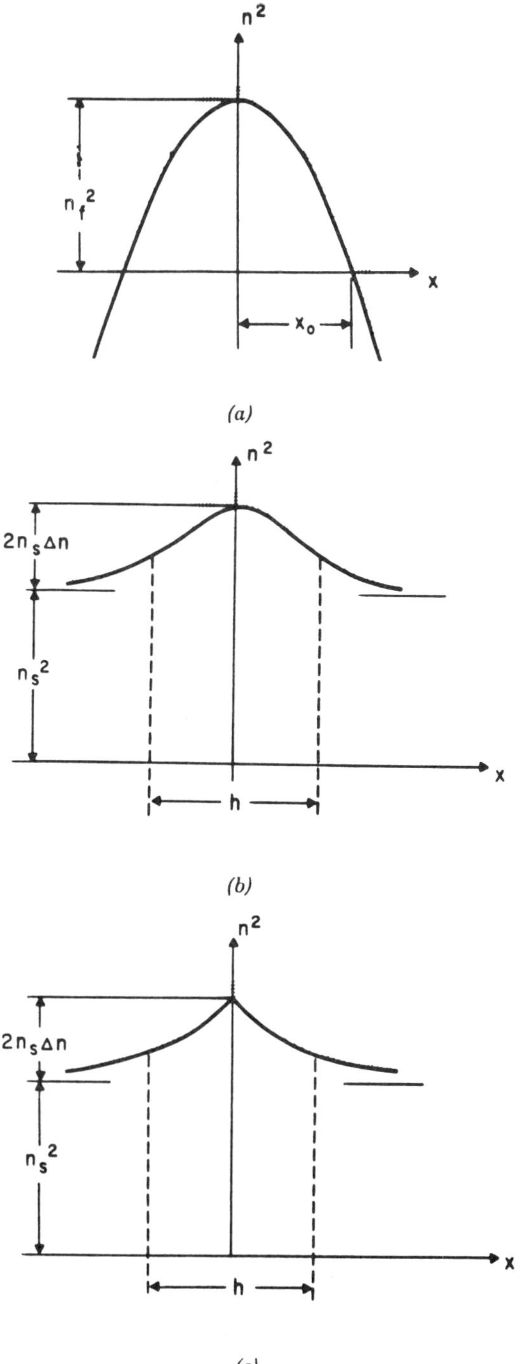

Figure 3.5 The potential well profile used to solve the wave equation for (*a*) parabolic index profile, (*b*) sech^2 index profile, and (*c*) exponential index profile.

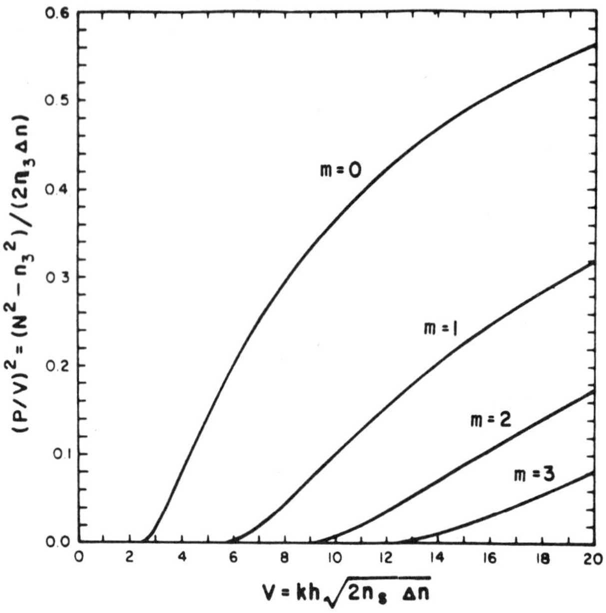

Figure 3.6 Dispersion relation for the first four modes of an exponential profile waveguide.

3.1, which is similar to many practical waveguide systems. The mode index is related to the parameter P by $(P_m/V)^2 = (N_m^2 - n_3^2)/(2n_3\,\Delta n) \approx (N_m - n_3)/\Delta n$. The dispersion relation for the first four modes of such a waveguide is given in Fig. 3.6.

Although these approximate solutions work very well with most asymmetric graded index waveguides, it is sometimes of interest to know the field values at the air–film interface. One then needs to match the boundary conditions by assuming an evanescent field in the air region [15] with the exponential decay constant given by, as usual,

$$l^2 = \beta^2 - n_1^2 k^2 \tag{3.35}$$

The y component of this field is related to the approximate mode solutions by

$$E_y(0) = \frac{1}{l}\left(\frac{dE_y}{dx}\right)x = 0 \tag{3.36}$$

Another example is for the silver ion exchange waveguide on glass that has a waveguide index given in Fig. 3.7 and is

$$n^2 = \begin{cases} n_3^2 - 2n_3\,\Delta n\left[\dfrac{x}{d} + b\left(\dfrac{x}{d}\right)^2\right] & \text{for } x > 0 \\ n_1^2 & \text{for } x < 0 \end{cases} \tag{3.37}$$

A rigorous solution can be obtained by means of a change of variable. Equations (3.34) and (3.37) can be rewritten in the form [19]

$$\frac{d^2E_y}{d\xi^2} + \left(v + \frac{1}{2} - \frac{\xi}{4}\right)E_y = 0 \tag{3.38}$$

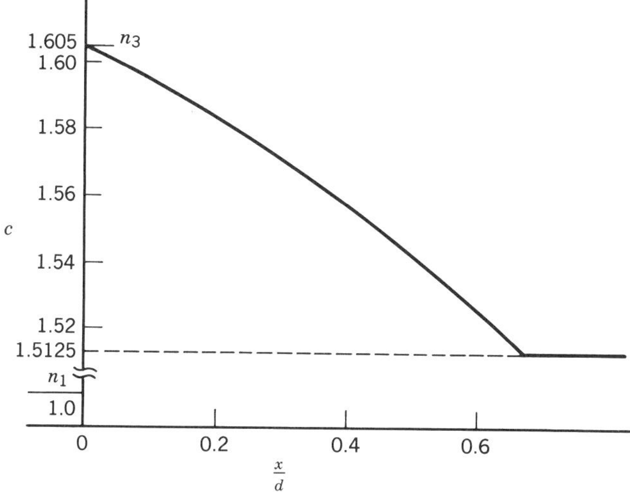

Figure 3.7 Refractive index profile of a planar waveguide formed by silver ion exchange in glass.

where

$$\xi = \left(\frac{8k^2 n_3 \Delta n\, b}{d^2}\right)^{1/4}(x + a) \tag{3.39a}$$

$$v = \left(\frac{8k^2 n_3 \Delta n\, b}{d^2}\right)^{-1/2}\left(k^2 n_3^2 + \frac{k^2 n_3 \Delta n}{b} - \beta^2\right)^{-1/2} \tag{3.39b}$$

and

$$a = \frac{d}{2b} \tag{3.39c}$$

The solutions of Eq. (3.38) are the *parabolic cylinder functions* $D_v(\xi)$ and $D_v(-\xi)$. For the field to vanish as ξ becomes large, only the $D_v(\xi)$'s are retained. Thus

$$E_y(x) = AD_v(\xi) \quad \text{for } x > 0 \tag{3.40a}$$

$$= B\exp(\sqrt{\beta^2 - n_1^2 k^2}\, x) \quad \text{for } x < 0 \tag{3.40b}$$

where A and B are arbitrary constants. If the discontinuity in the index at $x = 0$ is large (i.e., highly asymmetric), then the boundary conditions yield

$$D_v(\xi_0) = 0 \tag{3.41}$$

where

$$\xi_0 = \xi(x = 0) = \left(kd\sqrt{\frac{n_3 \Delta n}{2b^3}}\right)^{1/2} \tag{3.42}$$

The ξ_0 value determines the number of modes allowed in this waveguide. The discrete number of v values satisfy Eq. (3.41) leads to the discrete number of solutions. For

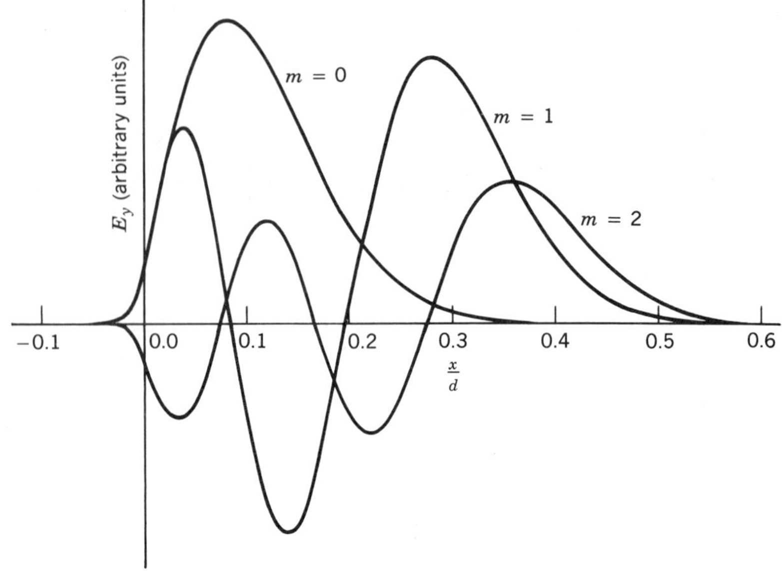

Figure 3.8 Modal patterns for the three lowest-order TE modes for a typical value of $k_0 d = 100$. (After Ref. 19.)

$n_3 = 1.605$, $n_1 = 1.0$, $\Delta n = 0.0925$, and for a typical value of $kd = 100$, the three lowest-order modes are given in Fig. 3.8.

The WKB Method. The WKB method, which has been thoroughly treated in quantum mechanics, can be applied to obtain approximate solutions of the wave equation (3.34) with slowly varying index profile $n(x)$ [20,21]. A trial value for the propagation constant β (or the mode index N) of the waveguide is selected. This leads to the determination of "turning points" x_{t1} and x_{t2}, as indicated in Fig. 3.9 and mathematically by

$$n(x_{t1}) = N = \beta/k \tag{3.43}$$

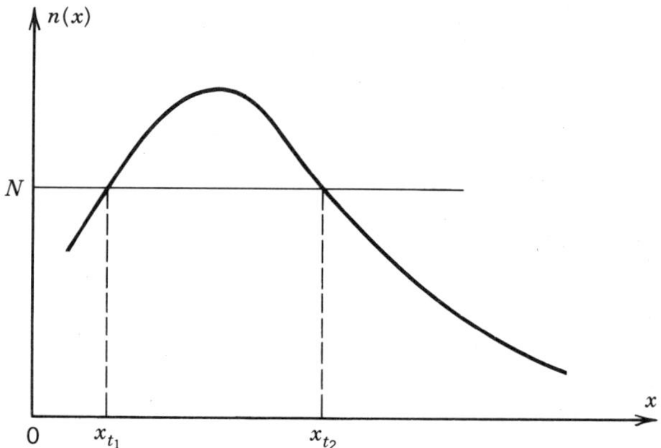

Figure 3.9 Turning points are where $n(x) = N$. The field is trapped between x_{t1} and x_{t2}.

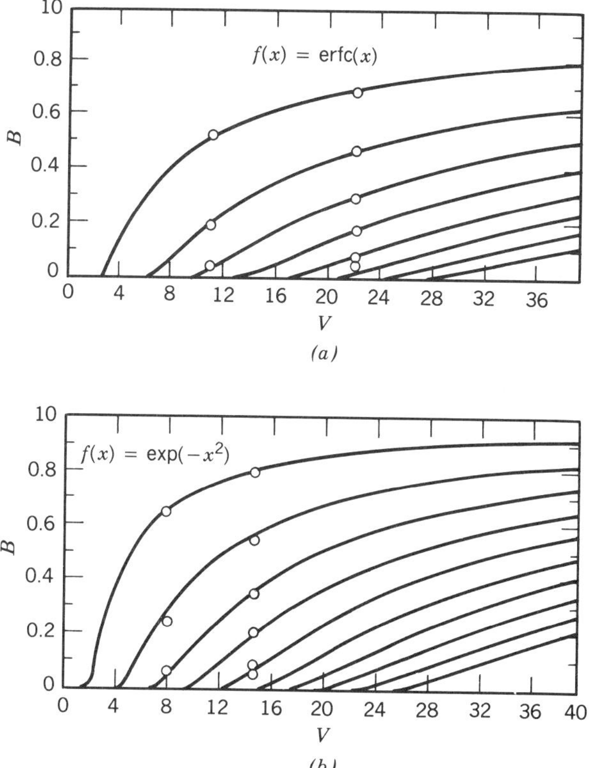

Figure 3.10 Normalized dispersion relation for waveguides with (*a*) a complimentary error function index profile and (*b*) a Gaussian index profile; $B \approx (N_m^2 - n_3^2)/2n_3\,\Delta n$ and $V \approx kd\sqrt{2n_3\,\Delta n}$.

The phase shift at the turning point is $-\pi/2$ each. Thus, if this value β satisfies the phase condition

$$\int_{x_{t1}}^{x_{t2}} \sqrt{n^2k^2 - \beta^2}\,dx = (m + \tfrac{1}{2})\pi \tag{3.44}$$

where m is the mode number, it is the propagation constant for the mth waveguide mode. For the mode, the WKB method predicts oscillatory field distribution between x_{t1} and x_{t2} and an exponentially decaying field outside this range. For the cases of highly asymmetric graded index waveguides, x_{t1} is the total internal reflection at the air–film interface. The phase shift at $x_{t1} = 0$ is given by Eq. (3.19) and is very close to $-\pi$. Thus

$$2\int_{0}^{x_{t2}} \sqrt{n^2k^2 - \beta^2}\,dx = (2m + \tfrac{3}{2})\pi \tag{3.45}$$

Equation (3.45) can be used to solve for the dispersion relation for each given index profile numerically. Figure 3.10 gives the dispersion curves for two highly useful index profiles [22]. The complimentary error function profile generally describes the diffusion process with infinite surface source. When surface source is depleted with a substantial "drive-in" diffusion process, the index profile will approach a Gaussian function. Figure

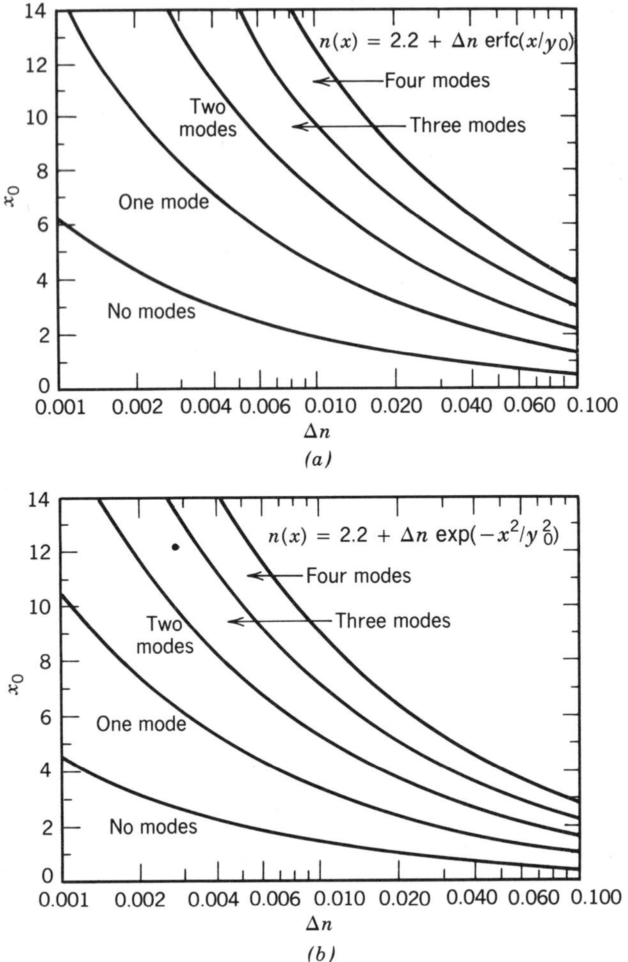

Figure 3.11 Allowed waveguide modes on $n_3 = 2.2$ substrate as function of surface index change Δn and v normalized diffusion depth d/λ for (*a*) an erf *c* index profile and (*b*) a Gaussian index profile.

3.11 gives the allowed mode number as a function of surface index change Δn and diffusion depth d for the two cases [22]. In these figures $B = (N_m^2 - n_3^2)/[\Delta n(2n_3 + \Delta n)] \approx (N_m^2 - n_3^2)/2n_3 \Delta n$, $V \cong kd\sqrt{2n_3 \Delta n}$, and $x_0 = d/\lambda$.

3.1.3 Waveguide Materials and Fabrication Techniques

As described in the previous sections, a planar optical waveguide is a dielectric thin film with its optical index of refraction higher than its substrate as well as cover regions. The index of refraction of the thin-film optical waveguide can be either homogeneous or graded with certain profiles. The index-of-refraction profile is primarily determined by the waveguide fabrication process. Since the name *integrated optics* was coined in 1968, many waveguide fabrication techniques have been proposed and used to form various optical waveguides on a variety of substrates. For instance, organic and inorganic films have been deposited either by physical deposition processes (thermal evaporation, electron beam evaporation, RF sputtering, spin or dip coating) or by chemical deposition processes. Single-crystal films have been grown on single-crystal substrates by various epitaxial processes. Surface layers with modified optical properties have been made on crystalline and noncrystalline substrates by means of ion exchange, diffusion, or ion implantation processes. The deposition process makes waveguides with physical properties unrelated to the substrate material. Such waveguide materials tend to be amorphous and are useful for passive device applications. On the other hand, the epitaxial processes and the surface modification processes tend to extend certain physical properties from the substrate into the waveguiding layer, making it useful for active device applications. To date, low-loss optical waveguides with propagation loss less than 1 dB/cm have been demonstrated with most processes on selected substrates.

In the meantime, the great variety of potential applications for integrated optical devices have spurred intensive investigation of a large number of waveguide as well as substrate materials over the past two decades. For instance, semiconductive substrates were studied in the hope of facilitating easy interface with microelectronic or optoelectronic components. Glass substrates were investigated with intended applications as advanced passive fiber-optical components. Electro-optical crystal substrates were explored for the development of optical switches, modulators, and optical signal-processing devices. Nonlinear optical materials were studied in the attempts to make high-efficiency optical harmonic generators. With feasibilities demonstrated, the practicality of each of these potential applications depends on the availability of matured and substrate-compatible process technologies that will provide the required physical properties to the waveguide structure.

In this section, waveguides reported to date are summarized in terms of fabrication technologies with emphasis on the ones that yield low-loss waveguides.

Deposited Thin-Film Waveguides. This section deals with physically or chemically deposited optical waveguides that exhibit either an amorphous structure or a polycrystal structure. The amorphous optical waveguides are passive devices that merely present themselves as light-guiding media. In general, such waveguides can provide very low optical propagation losses when properly prepared. They can serve the purposes of optical interconnecting, branching, combining, wavelength filtering, spatial filtering, and interfacing with optical sources or detectors. On the other hand, polycrystal optical waveguides tend to show considerably higher optical propagation losses. Much of the loss is due to the rough surface texture as the result of anisotropic growth characteristics of each crystallite, and such loss can be improved by surface polishing. Under certain deposition conditions, the crystallites of a polycrystal film can be made to orient all in the same direction. Such oriented film will show some electro-optical or piezoelectric properties and can be utilized to make active devices on a great variety of substrates. However, much work is needed to further reduce the waveguide loss in such oriented polycrystal films in order to fully realize its potential. As a result, to date, the majority of the demonstrated deposited thin-film optical waveguides are passive amorphous waveguides.

A rather straightforward means of waveguide fabrication is by simply depositing a layer of dielectric thin film of high index of refraction over an optically smooth and transparent substrate. This deposition can be done by spin or dip coating, evaporation, RF sputtering, and chemical vapor deposition. The spin or dip coating techniques are popular for the preparation of the delicate organic films while the evaporation, chemical vapor deposition, and the sputtering methods are commonly used for the preparation of inorganic films.

Organic Thin-Film Waveguides. Thin-film optical waveguides have been deposited on glass substrates from liquid solutions. After the solvents are evaporated, a solid film is left on the substrate surface. The liquid film can be coated by spinning the substrates that have been covered by liquid or by dipping into a solution bath. Film thickness is determined by the solid content, viscosity, and spinning or dip withdrawal speed. Various materials were used, including polyurethane, polystyrene, epoxy, photoresist, and organometallic solutions. Usually, coating is done at room temperature followed by curing at elevated temperature. Often the baking is done in several steps with gradual increase in temperature in order to obtain a smooth film surface. For the polymer films VTMS and HMDS, the coated monomer films have to be polymerized by exposure to an RF plasma environment. Table 3.2 lists the waveguide parameters and fabrication procedures of several organic waveguides reported in the literature. Waveguide losses less than 1 dB/cm at the He–Ne laser wavelength of 0.6328 μm can be easily obtained with all the materials except photoresists containing photosensitizer dyes. An interesting feature of these films is the possibility of adjusting the film index as well as film thickness by means of mixing or even subsequent heat treatment or UV exposure.

Most organic films are soft and susceptible to chemical attack. The coating technique also lacks the control in film thickness and uniformity needed for waveguide optical devices. However, due to the ease in fabrication, these optical waveguides may be useful for the fabrication of passive components in optical circuits such as wafer scale optical interconnections. In addition, the relatively low film index also make these films useful as overlay coating to reduce the waveguide surface scattering.

Also listed in Table 3.2 are two electro-optically active liquid waveguides, namely the MBBA nematic liquid crystal and nitrobenzene. Unfortunately, the usefulness of these liquid waveguides are hampered by the rather high propagation losses.

Inorganic Thin-Film Waveguides. Like solution-coated films, practically all the deposited inorganic films exhibit a step index profile. These waveguides tend to have higher loss than the organic waveguides but are much stronger and durable. The deposition process can be controlled to provide tight thickness as well as uniformity tolerances. Although most deposited films are passive, some are electro-optical. Others have demonstrated frequency doubling and even laser gain.

The deposition of dielectric thin films by evaporation have been known in the optics industry for a long time. However, most evaporated films are not of sufficient quality for optical waveguide applications due to surface roughness scattering and absorption. In Table 3.3, the ZnS film and the Planer CAS 10 are prepared by electron beam evaporation. The ZnS film exhibits high scattering loss unless it is hand polished.

The RF sputtering of dielectric films has been a popular technique for the deposition of low-loss waveguiding films. Note that most of the reported low-loss deposited waveguide films were prepared by the RF sputtering technique. The sputtering process is well understood, and precise control of process parameters is obtainable with commercial sputtering systems. By altering the ground condition, the sputtering can be performed in several modes including RF sputtering, RF sputter-etch, and biased RF sputtering. By admitting reactive gases such as oxygen or nitrogen to the sputtering chamber, reactive sputtering can modify the chemical composition of the deposited films.

Corning 7059 glass (a pyrex glass with composition of SiO_2 50.2%, BaO 25.1%, B_2O_3 13%, Al_2O_3 10.7%, and As_2O_3 0.4%) was the first reported low-loss optical waveguide

TABLE 3.2 Organic Thin-Film Waveguides

Material	Index	Loss (dB/cm)	Fabrication Method	Comments	References
Polyurethane					
9653-1	1.555	0.8	Solvent dry and bake	Solvent is toluene/MEK	23
LX500	1.573	4	Solvent dry and bake	Solvent is toluene/MEK	23
Epoxy (Araldite)	1.581	0.3	Solvent and bake	Ethanol	23
Photoresist	1.615	7	Solvent and bake, UV	Thinner	23
KPR	—	2.2	Solvent and bake, UV	Thinner, dye removed	23
AZ 1350		1.6	Solvent and bake	Dye removed	24
HMDS, $(CH_3)_3SiOSi(CH_3)_3$	1.488	Low loss	RF plasma, polymerize	n and thickness, reduce by heat	25–27
VTMS, $CH_2{=}CHSi(CH_3)_3$	1.532–1.4797	0.04	RF plasma, polymerize	n and thickness, reduce by heat, mix HMDS and VTMS to adjust n	25–27
PMMA or Glycidyle MA	1.515	0.1–0.2	—	UV increase n 1% and thickness	28–30
PMMA	1.49–1.56	Very low	Dip and bake	—	
Polystyrene	1.586 TE, 1.589 TM	Low loss	Dip and dry	Stress birefingence	28
Polyimide	1.7	Very low	Spin and bake	150°C, then 325–400°C	31
Nitrocellulose	—	Low	Langmür coating	Isoamyl acetate solution, spread on H_2O surface	32
MBBA	$1.52n_0$, $1.72n_e$	500	Sandwich	Liquid crystal, loss = 30–40 db/cm with field on	33, 34
Nitrobenzene	1.55	—	Sandwich	Electro-optical liquid	35–37

Note: Data is for 0.6328 μm optical wavelength.

TABLE 3.3 Inorganic Thin-Film Waveguides

Material	Index	Loss (dB/cm)	Fabrication Method	Comments	References
ZnO/glass	1.973	60, 20	Sputtering	Oriented crystal, after polishing	38
ZnO	—	7	Sputtering	Oriented	39
ZnO	1.98	0.01	Sputtering	On oxidized Si, laser annealed	40
ZnS	2.342	5	E-Beam	—	38
Planar CAS 10	1.469	1.2	E-Beam	Glass film	41
7059 glass	1.53–1.585	<1	Sputtering in oxygen	*n* depends on deposit rate	42–45
Ba–Si glass	1.48–1.62	<1	Sputter from silica and $BaCO_3$		46
Nd–Glass	—	0.5	Sputter from AO1838 target, anneal 500°C	Gain is 1/cm	47, 48
Aluminum oxide	1.66	40	Sputtering	—	49
Ta_2O_5	2.08–2.214	0.9	Sputter Ta in O_2, then heat 500°C	Amorphous	50, 51
Ta_2O_5	1.85–2.13	Low	Sputter in O_2–N_2 mix		52
Ta_2O_5/SiO_2	1.46–2.08	0.8	Ta from 0–100% adjusted *n*	Anneal, 450°C for 12 h.	51

Nb_2O_5	2.297	<10	Sputtering	Low temperature better	53
SiO/N	1.45–1.98	<4	RF CVD 850°C, 0.2–0.5% NO, 0.02–0.07% Silane, N_2 at 1 atm	Low loss for $n =$ 1.48–1.54	54
Si_3N_4	1.98	<0.1	Low-pressure CVD	On oxidized Si	55
$LiNbO_3$/Sapphire	2.247 TM, 2.280 TE	High	RF sputter in $Ar/O_2 = \frac{7}{3}$	Frequency doubling	56
As_2S_3/Glass	2.36	2.1 at 1.06 μm	RF sputter	0.4 dB/cm for TE_0 on thick film	57
$Ge_{29.9}Sb_{15.6}Se_{54.5}$	2.75	10	RF sputter	Anneal, 265°C, loss is 3.6 dB/cm	57
$Ge_{34.4}As_{11.1}Se_{54.5}$	2.586	20	RF sputter	Anneal improves loss	57
GaAs	3.27	1–7.4 at 10.6 μm	Platelet single crystal	Electro-optics modulator CdTe or As_2S_3 cladding layer	58
SiO_2–TiO_2	—	0.15	Flame hydrolysis deposition, then 1250°C heat	Core–cladding index ratio 1.011	59, 60
SiO_2–GeO_2	—	1.5	Same as above	Index ratio 1.009	59
Pb–silica 1:2.5	1.664	0.5	Solution coat and bake at 60°C	Adjustable by ratio	23

Note: Data for 0.6328 μm optical wavelength.

prepared by the RF sputtering technique. Films must be sputtered with oxygen between 20 and 100% in order to avoid excessive optical absorption in the waveguide due to oxygen deficiency. The waveguide loss at the He–Ne laser wavelength can be lower than 1 dB/cm. However, it was observed that the sputtered 7059 films have a different barium oxide content than the bulk material and that the refractive index of the sputtered film varies as a function of the deposition rate. At 0.6328 μm wavelength, the film refractive index could vary from 1.53 to 1.585 as the deposition rate increases. Another low-loss glass film is the barium silicate glass sputtered from a target formed by a hot-pressed mixture of barium carbonate and silica. By varying the barium oxide content from 0 to 40 wt %, the film index can be varied from 1.48 to 1.62.

Tantalum pentoxide is another popular waveguide material with film index as high as 2.08 at 0.6328 μm. When nitrogen is blended into the sputtering chamber, the resultant film index can vary from 1.85 to 2.13 depending on the N_2/O_2 ratio. Optical propagation loss of less than 1 dB/cm has been reported. Another way of changing the film index of refraction is by using a target containing SiO_2 and Ta_2O_5.

Niobium pentoxide must be sputtered at a low substrate temperature in order to keep the waveguide loss low. Further reduction of waveguide loss can be done by laser annealing. With 100% oxygen sputtering, a refractive index of 2.297 was obtained.

Other sputtered films of interest includes ZnO, Nd-doped glass, aluminum oxide, lithium niobate, and several chalcogenide glasses. The ZnO film can be deposited with (002) oriented crystallites and therefore can provide piezoelectric as well as electro-optical properties to an optical waveguide on a glass type of substrate. However, the oriented film also showes high surface scattering loss due to surface roughness unless hand polished. The Nd-doped glass exhibits low loss (0.5 dB/cm) and an optical gain of about 1 cm^{-1}. The aluminum oxide film and the lithium niobate film provides possibility for frequency doubling. But the high waveguide loss prevents it from being practical. The chalcogenide films are useful for far-IR waveguiding devices.

A rather interesting deposited optical waveguide is the silicon oxynitride film that results in the chemical vapor deposition process. Silicon oxynitride is a glassy, amorphous, stable silicon–oxygen–nitrogen polymer of variable composition. This film was deposited at 850°C in a RF-heated silica tube reactor. The gas was at 1 atm pressure, typically comprising 0.2–0.5% nitric oxide, 0.02–0.07% silane, and the remainder nitrogen. By controlling the NO-to-Silane concentration ratio, the composition ratio of silicon dioxide ($n = 1.455$) to silicon nitride ($n = 1.98$) may be adjusted. Deposition rate was about 700 Å/min. For the index range between 1.48 to 1.54, this film shows low loss. By using a low-vapor-pressure chemical vapor deposition technique on a thermally oxidized silicon wafer, a waveguide loss of <0.1 dB/cm was observed for the TE_0 mode.

Another chemical vapor deposition process with great potential for producing low-loss optical waveguides is the flame hydrolysis deposition technique similar to the OVPO and VAD processes for optical fiber fabrication. To fabricate the SiO_2–TiO_2 planar waveguide, a mixture of $SiCl_4$–$TiCl_4$ was fed into an oxyhydrogen torch. Fine glass particles synthesised by flame hydrolysis were deposited on the glass substrate on a turntable. Small amounts of BCl_3 and PCl_3 were added to the gas to lower the melting point of the glass particles. A cladding layer was deposited over the core layer to reduce surface scattering. After the deposition, the samples were heated up to 1250°C for consolidation. The measured waveguide loss was 0.15 dB/cm at 0.6328 μm optical wavelength.

Also included in Table 3.3 are a lapped GaAs platelet as the far-IR waveguide electro-optical modulator and a solution-coated low-loss glass film. The GaAs platelet is coated with CdTe (or As_2S_3) cladding layers and supports many waveguide modes. Propagation losses as low as 1–5 dB/cm have been measured on some of the modes.

Single-Crystal Epitaxial Waveguides. The previously described waveguide fabrication techniques are useful for the preparation of passive waveguides. However, a major reason for developing integrated optics technology has been the desire to have affordable miniaturized and highly efficient electro-optical devices or subsystems. For active optical power distribution functions as well as optical modulation and switching, the optical waveguide must be constructed with single-crystal materials having large electro-optical coefficients. In order to fabricate a single-crystal waveguiding layer with a higher index of refraction than the substrate material, the epitaxial crystal growth technique appears promising. During the past decade, epitaxial growth technology for the growth of semiconductive thin films has matured rapidly. It has been the major contributor for the recent successes in the laser-diode-related optoelectronics industry. However, the required optical quality in a waveguiding film for integrated optics far exceeds what is needed for laser diodes and photodetectors. This is mainly due to the longer propagation distances as well as angles and bends in a typical integrated optical device. It will continue to be a big challenge to material scientists to refine the epitaxial growth technique such that low-loss optical waveguide devices can be constructed.

In this section, optical waveguides made by epitaxial techniques will be described. Emphasis will be placed on the growth of ferroelectric electro-optical crystals. Although the epitaxial growth technique for compound semiconductor crystal materials is much more mature and is currently advancing toward integrated optoelectronics devices, it will not be the subject of this chapter, which mainly deals with planar optical waveguides and components. Readers interested in the epitaxial growth of compound semiconductive optical waveguides are referred to the chapter on laser diodes and optoelectronic devices.

Table 3.4 lists the reported dielectric waveguiding films prepared by epitaxial techniques. Most of the efforts are for the growth of ferroelectric crystals due to its large electro-optical coefficient. Also included are efforts in ZnO, an electro-optical material, and in a garnet film, a magneto-optical material.

The ZnO has been sputtered as oriented polycrystal films showing electro-optical and electromechanical properties. Early experiments using this film as optical waveguides yielded a waveguide loss greater than 20 dB/cm. Even after surface polishing, the grain boundaries due to the columnar structure as well as voids among crystallites still contributes to large amounts of scattering. Magnetron sputtering produces oriented ZnO films with improved density as well as smoothness. As-grown films on glass substrates with optical waveguide losses of 6–10 dB/cm have been reported. Recently, reduction of optical losses has been observed on sputtered ZnO films after CO_2 laser annealing. The laser annealing process is believed to induce coalescence of neighboring crystallites, thus creating a nearly single-crystal film. The waveguide loss for the fundamental mode of a three-mode waveguide was reduced to 0.01 dB/cm. Effects on the electro-optical, electromechanical, and nonlinear optical properties of laser-annealed ZnO films are yet to be determined. Epitaxial growth of ZnO has also been reported on sapphire substrates using RF sputtering and using chemical vapor deposition. The ZnO (0001) and (1-210) epitaxial films were sputter grown on (0001) and (01-12) sapphire substrates, respectively. Demonstrated optical propagation losses were 2 and 1.9 dB/cm, respectively, using a prism-coupled He–Ne laser beam. An electromechanical coupling constant of approximately $k = 0.2$ has been reported. Chemical vapor deposition (CVD) films have shown varied results. One film must be polished in order to conduct measurements yielding a 0.3-dB/cm attenuation value for the TE_0 mode. Another as-grown ZnO film gives optical propagation losses for an eight-mode waveguide from 0.7 dB/cm for the TE_0 mode to 18.3 dB/cm for the TE_7 mode in more or less a linear fashion.

A number of gallium and iron–garnet films developed originally for magnetic bubble memory devices have been tried for optical waveguiding applications. These as-grown

TABLE 3.4 Epitaxial Waveguide

Layer	Substrate	Index	Loss (dB/cm)	Fabrication Method	References
$6Bi_2O_3$:TiO_2	$B_{12}GeO_{20}$	—	—	EGM 825–930°C	61
$LiNbO_3$	$LiTaO_3$	$n_0 = 2.288$, $n_e = 2.207$	Low	EGM 1300°C cool at 20°C/h, then polish	62–64
$LiNbO_3$	Z-$LiTaO_3$	$n_0 = 2.288$, $n_e = 2.191$	5 TM, 11 TE	LPE in LiO_2–V_2O_5–Nb_2O_5 flux at 1100°C with 5:4:1 ratio, film grown at 850°C	65, 66
Ga, Fe, Garnet	Garnet	—	1–5	Grown by LPE	67
$Eu_3Ge_5O_{15}$	$Gd_3Sc_2Al_3O_{12}$	—	—	—	67
Garnet	GGG	$n = 2.247$ or $n = 2.259$	25	LPE growth for Faraday rotation device	68
ZnO	Sapphire	$n = 1.98$	2	Sputter grown on sapphire, loss depends on orientation	39, 69
ZnO	Sapphire	$n = 1.98$	0.3	CVD grown, then hand polish	70, 71

Note: Data for 0.6328 μm optical wavelength. Abbreviations: EGM, epitaxial growth by melting; LPE, liquid phase epitaxy.

garnet films are smooth, uniform, and pin-hole free. But the absorption losses depend on the impurity content in the melt. Two garnet films, Bi0.63 Tm2.3 Fe3.8 Ga1.2 Pb.02 O12 and Bi.95 Yb2.1 Fe3.8 Ga1.1 Pb.03 O12, were grown over the GGG substrate for the fabrication of a Faraday rotation device. The films had a (111) orientation and were grown by horizontal dipping. Strong TE–TM mode conversion was observed as the optical beam propagates in this waveguide. However, the measured optical loss was between 25 and 30 dB/cm. A magneto-optical switch was fabricated on $Eu_3Ge_5O_{12}$ waveguide epitaxially grown over a $Gd_3Sc_2Al_3O_{12}$ substrate. The waveguide optical loss was 1–5 dB/cm. Although garnet films make excellent magneto-optical modulators, the optical losses limits its usefulness to optical wavelengths longer than 1 μm.

Epitaxial growth by melting (EGM) has been used to make $6Bi_2O_3{:}TiO_2$ waveguides on $Bi_{12}GeO_{20}$ crystal substrate and lithium niobate waveguide on lithium tantalate crystal substrate. The growth of lithium niobate on lithium tantalate is possible because the melting temperature of lithium tantalate is higher by about 300°C than that of lithium niobate. The epitaxial film is grown from bulk melt or melting a powder or lacquer suspension on the substrate. The film grown by EGM always has a transition region that minimizes the lattice mismatch problem and produces a gradient index at the interface. For the growth of lithium niobate on lithium tantalate, lithium niobate ceramics crushed into powder were laid on the polished c plane of the substrate. Then the substrate was heated to about 1300°C to melt the powder and cooled slowly at a rate of about 20°C h^{-1}. The as-grown film had a rough surface requiring hand polish before the optical waveguiding experiments. The waveguide loss for the polished sample was reportedly too small to measure without elaborate set-up. Subsequently, an electro-optical modulator was fabricated. Film uniformity was improved by first suspending the lithium niobate powder in a lacquer and then painting onto the lithium tantalate surface before the melting process.

Another approach of epitaxial growing lithium niobate on lithium tantalate is by the liquid phase epitaxy technique using a LiO_2–V_2O_5 flux. The molten mix contains 50 mol % LiO_2, 40 mol % V_2O_5, and 10 mol % Nb_2O_5. The composition is equivalent to 20 mol % lithium niobate in the pseudobinary system. The molten mixture was heated to about 1100°C and cooled slowly to the growth temperature of about 850°C. The lithium tantalate substrate was dipped into the molten mixture for film growth. The growth rate was estimated about 0.1 μm/min. A 3-μm colorless film was grown supporting seven TE and TM modes. Waveguiding losses of 5 and 11 dB/cm were measured for the TM_0 and the TE_0 modes, respectively. Other flux systems were tried, resulting in either Li-rich or Nb-rich films.

Graded Index Optical Waveguides. Until now, all the waveguides described belong to the step index family. In this section, optical waveguides with graded index profiles will be described. Graded index waveguides are made by ion implantation, ion exchange, or diffusion processes. The ion implantation process modifies the refractive index of the substrate material by damaging the lattice structure within a certain depth range. The ion exchange process replaces cations in the substrate material with cations from an external source forming a graded impurity concentration near the substrate surface. Ion exchange can be accelerated by the application of an external electrical field. The diffusion process introduces impurities from the exposed surface into the substrate by thermal energy. The impurity concentration profiles for the unassisted ion exchange process and the diffusion process tend to follow a complimentary error function shape unless surface impurity source is depleted. Prolonged heat treatment after the depletion of surface source yields a Gaussian impurity concentration profile. The normalized waveguide mode behavior for an optical waveguide with a refractive index profile of the erfc function and of the Gaussian function has been discussed in an earlier part of this chapter. The impurity profile for an ion-implanted optical waveguide can be tailored by careful programming of the implantation energy as well as dose. The field-assisted ion exchange also has an impurity profile depending on the applied field.

In general, graded index optical waveguides exhibit very low optical scattering as well as absorption losses. The fabrication process simply modifies the substrate material by the introduction of impurities. These processes work on a great variety of substrate materials, glass, or crystalline. In most cases, the bulk physical properties of the substrate material are preserved after the index modification. This is advantageous from device design and fabrication points of view. It has been found that the most popular waveguides used in integrated optics are fabricated using either the ion exchange process or the diffusion process.

Ion Implantation Waveguide. Many waveguides have been fabricated by ion implantation. The choice of ion and implant energy mainly affects the penetration depth. Lattice structure is damaged at the penetration depth, resulting in a decrease in the refractive index. The less damaged surface layer is able to support a few guided modes due to the relatively higher refractive index on the substrate surface compared to the buried damaged layer. One of the problems associated with ion-implanted waveguides is the defects that result in scattering losses for the guided modes. Thermal annealing is useful in reducing the damage effect with improved optical propagation loss. However, thermal annealing also reduces the refractive index change produced by the implantation process. Table 3.5 summarizes the results of reported implanted optical waveguides.

Lithium ions were implanted into fused quartz through a PMMA electron-resist mask to form a waveguide. Waveguide propagation loss after annealing was about 3 dB/cm. For lithium niobate substrate, a variety of ions were implanted. The saturation value of the index change due to implantation was approximately -7 to -10% regardless of the choice of ion. Implant energy was adjusted to obtain desired penetration depth. As implanted, the optical waveguide showed very high propagation loss. Annealing at about 200°C for 30 min reduces the optical propagation loss substantially. The index change is also reduced substantially after annealing. The exact behavior of annealing depends somewhat on the choice of ion. The most discouraging finding was the reduction in the electro-optical coefficient, r_{33}, to only 60% of its original value in an ion-implanted optical waveguide on lithium niobate substrate.

Ion Exchange Waveguide. Ion exchange techniques have been used for more than a century to produce tinted glass and to improve surface mechanical properties of glasses

TABLE 3.5 Ion Implantation Waveguide

Ion Type	Substrate	Δn	Loss (dB/cm)	Fabrication Comments	References
Li	Fused quartz	0.0375	1.8–3	200 keV 10^{15} Li + 7,	72–74
		0.013	0.2	anneal, 300°C, 1 h	—
Ne	$LiNbO_3$	-0.225	—	60 keV	75, 76
Ar	$LiNbO_3$	-0.225	—	60 keV	75
He	$LiNbO_3$	-0.156–-0.25	Low	0.86–1.86 MeV to 2–4 μm below surface, anneal, 200°C, 30 min, r_{33} reduced to 60%	76, 77
H/He or B	ZnTe	—	1–4	Lattice damage	78

Note: Data for 0.6328 μm wavelength.

79]. Recently, ion exchange processes have been employed for the fabrication of low-oss optical waveguides on various glass substrates and on ferroelectric crystal substrates. Due to the ease of fabrication, the low waveguiding loss, and the compatibility with optical fibers, ion exchange optical waveguides are becoming one of the dominant waveguides of choice in integrated optics [80–82].

When a glass containing monovalent cations is placed into a molten salt containing another monovalent cation, ion exchange takes place. A generalized ion exchange reaction can be written

$$\underline{\mathrm{A}}\,(\mathrm{glass}) + \mathrm{B}\,(\mathrm{salt}) = \underline{\mathrm{B}}\,(\mathrm{glass}) + \mathrm{A}\,(\mathrm{salt}) \tag{3.46}$$

where $\underline{\mathrm{A}}$ and $\underline{\mathrm{B}}$ are the counterions in the exchanger phase and A and B are the counterions in the liquid phase. Molten salts are required because of the temperature range needed before the cations in the glass become mobile with reference to the negatively charged oxygens of the rigid, immobile silicate network. The rate at which the exchange occurs is controlled by the diffusion of the ions in the glass. In binary ion exchange, the diffusing species A and B are charged and tend, in general, to diffuse at different rates. Thus, there is a tendency of electrical charge buildup. The gradient in the electrical potential acts to slow down the faster ion and speed up the slower ion, resulting in equal and opposite fluxes for the two ions. Thus, electrical neutrality is preserved. The diffusion coefficient depends on temperature and glass composition. Below the glass transition temperature, the temperature dependence of the diffusion coefficients can be fit to the Arrhenius equation

$$D_i = D_0 \exp\left(-\frac{Q_i}{RT}\right) \tag{3.47}$$

where Q_i is the activation energy (in joules per mole), R is the gas constant 8.314 J/deg mol, and T is temperature in kelvins. For the ion exchange process without an external electrical field, the diffusion process has a solution for the concentration of the B cation in the glass,

$$N_b(x, t) = N_0 \,\mathrm{erfc}\, \frac{x}{d_0} \tag{3.48}$$

where $d_0 = 2\sqrt{Dt}$ is the effective depth of diffusion. For example, silver ion in soda–lime glass has $D_0 = 2.26 \times 10^{-6}$ and $Q_i = 8.5 \times 10^4$ J/mol. Tables 3.6 and 3.7 list the ion radius, polarizability, activation energy, and diffusion constant, for a number of commonly used cations. Figure 3.12 is a plot of the effective diffusion constant as a function of the diffusion temperature. The normalized dispersion curves are given by Fig. 3.13. The electric field of the guided modes are given by Fig. 3.8.

When an external electric field is applied, the B cation concentration becomes

$$N_b = 0.5N_0(\mathrm{erfc}(x' - r) + \exp(4rx')\,\mathrm{erfc}(x' - r)) \tag{3.49}$$

where $x' = x/d_0$ is the normalized effective depth of diffusion without an external field and $r = \mu Et/d_0$ in which μ is the electrochemical mobility in $\mathrm{m^2/V{\cdot}s}$ square meters per volt second.

Table 3.6 lists the published waveguide parameters for a number of planar waveguides. The first planar optical waveguide fabricated by this method was by ion exchange from a mixture of thallium, sodium, and potassium salts into a borosilicate glass plate. An

TABLE 3.6 Ion Exchange Waveguide Parameters

Ion	Electron Polarizability, $\lambda = D$ ($Å^3$)	Ionic Radius (Å)	Substrate	Salt	Operating Point (°C)	Decomposition Point (°C)	Index Increase	Loss (dB/cm)	References
Na^+	0.41	0.95	Glass	$NaNO_3$	307	380	—	—	83, 84, 85
Li^+	0.03	0.65	Soda lime	$LiNO_3$ or	264	600	0.01	>1	85, 86
				$(LiSO_4)/(K_2SO_4) = 4:1$	520–620	—	0.015	—	
Tl^+	5.2	1.49	Boro-silicate	$TlNO_3 + KNO_3 + NaNO_3$	206	530	0.001–0.1	<0.1	84
Cs^+	3.34	1.65	Soda lime	$CsNO_3$	520	—	0.03	>1	87, 88
			BGG21 glass	$CsNO_3 + CsCl$	435	—	0.043	—	
Ag^+	2.4	1.26	Alumino-silicate	$AgNO_3$	225–270	444	0.09–0.13	0.1–0.5	85, 89–92
Rb^+	1.98	1.49	Soda lime	$RbNO_3$	520	—	0.015	High	87
K^+	1.33	1.33	Soda lime	KNO_3	365	400	0.009	0.2	85, 93, 94
Ag^+	2.4	1.65	Glass	Silver film	—	—	0.001	—	82
	2.4	1.65	Glass	Silver film field assisted	—	—	0.025	—	82
	2.4	1.65	X-cut $LiNbO_3$	$AgNO_3$	250	444	0.12	6	95
Tl^+	5.2	1.49	$LiNbO_3$	—	—	—	—	—	96
			$LiTaO_3$	—	—	—	—	—	96
H^+	—	—	$LiNbO_3$	Benzoic Acid	110–249	—	0.2	~2	97
			Y-$LiNbO_3$	Diluted benzoic acid	—	—	0.1	~5	—

TABLE 3.7 Diffusion Coefficients of Various Cations in Glass Used for Waveguides

Cation	Glass	Temperature (°C)	D (10^{-14} m²/s)	Q (10^4 J/mol)	References
Tl^+	Borosilicate	530	20	—	84
Li^+	Soda lime	575	64	14.2	86
Ag^+	Soda lime	374	0.7	9.1	90
	Borosilicate	615	0.26	9.1	98
	Soda lime	215	0.010	—	92
	BK-7	320	0.025	9.8	91
	Soda lime	330	0.01–0.03	8.9	99
K^+	Soda lime	385	0.11	12.5	93
	BK-7	385	0.14	—	94
	Pyrex	385	0.06	—	94
Cs^+	BGG21	407	0.32	20	88
Na^+	Soda lime	371	12	16	83

Source: After Ref. 80.

external electric field was used to enhance the ion migration rate. The ion exchange occurs between the thallium ions and the sodium as well as potassium ions in the glass. Subsequently, the mixture of salts was replaced by a sodium and potassium salts mixture to form a buried waveguide by reversing the ion exchange process. The resultant multimode optical waveguide yielded a total waveguiding loss of less than 0.1 dB/cm.

Ion exchange optical waveguides were also formed on X-cut lithium niobate substrate using silver cations. Index change only occurs for the extraordinary rays with a waveguide loss of about 6 dB/cm. No waveguide was observed on Y-cut substrates even after prolonged treatment. More recently, proton ion exchange has been employed to form waveguides with index change greater than 0.12 on lithium niobate using benzoic acid ($C_7H_6O_2$) at low temperature (110–249°C). Low-loss optical waveguides (0.5 dB/cm) were observed on X-cut and Z-cut substrates. However, treatment on Y-cut samples

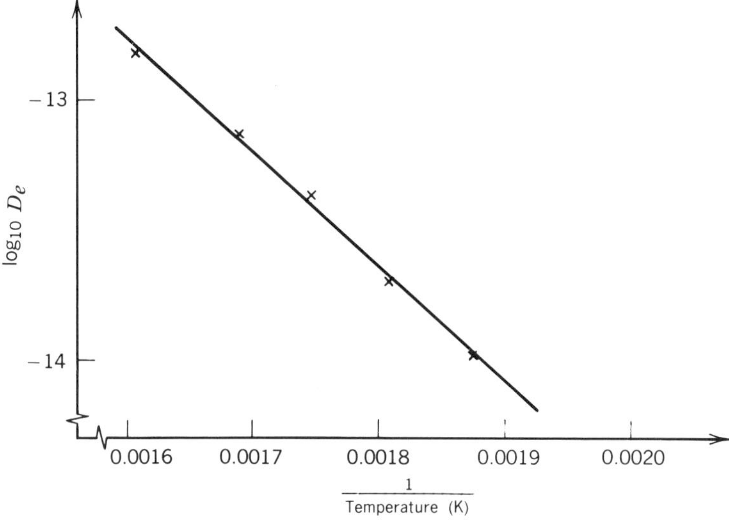

Figure 3.12 Dependence of the "effective" diffusion constant on the diffusion temperature.

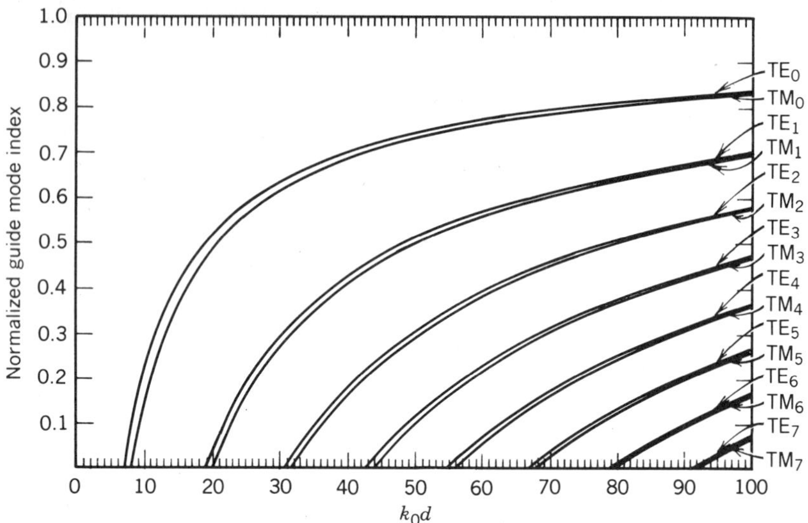

Figure 3.13 Theoretical mode dispersion curves for TE and TM modes obtained from the transcendental equation for the second-order polynomial profile.

resulted in surface damages (pits), and no waveguides were observed. Only when the benzoic acid is substantially diluted with lithium nitride, were guided modes observed on Y-cut surfaces.

Diffused Optical Waveguide. The ferroelectric crystals lithium niobate and lithium tantalate are among the best materials for electro-optical devices due to large electro-optical, electromechanical, acousto-optical, and nonlinear optical parameters. However, due to the relatively large refractive index of these materials, earlier attempts in making integrated optical devices on these substrates using deposited thin film had only limited successes. The first truly successful optical waveguide on lithium niobate and on lithium tantalate substrates were fabricated by lithium out-diffusion in a vacuum or in an oxygen environment.

Lithium niobate and lithium tantalate crystals can be grown in a slightly nonstoichiometric form with the mole content of the lithium oxide ranging from 0.48 to 0.50. It is known experimentally that for a slightly lithium-deficient crystal, the ordinary refractive index remains unchanged while the extraordinary refractive index increases approximately linearly as the lithium content decreases. For lithium niobate, the extraordinary index increases 1.63% for each percent of lithium oxide loss. The proportional constant is 0.85 for lithium tantalate crystal. Lithium out-diffusion was carried out at high temperature (850–1200°C) in vacuum or in oxygen. If the out-diffusion is done in vacuum, the crystal surface will be darkened due to the excessive loss of oxygen. Reheating the crystal at high temperature in air or in oxygen removes the discoloration. The waveguiding layer had a peak index change on the order of 0.001 and a thickness of several hundred micrometers. Such guides usually support a large number of modes and are not practical for most applications. In fact, the out-diffusion waveguides were quickly replaced by metal in-diffusion waveguides.

Interestingly, as the metal in-diffusion waveguides gained favor, the existence of uncontrolled out-diffusion of lithium oxide during the in-diffusion process became a problem. To suppress the out-diffusion, samples were packed in Li_2CO_3 or $LiNbO_3$ powder and annealed after the diffusion. In another approach, a crucible of lithium oxide was placed upstream in the gas flow during the in-diffusion process. More recently, it was found that lithium out-diffusion can be easily suppressed by wetting the incoming gas flow during diffusion.

To date, in-diffusion processes are the most commonly used in the waveguide fabrication of lithium niobate and lithium tantalate crystal substrates. The diffusion source metal, such as Ti, Nb, Mn, Co, Fe, Cu, Zn, and Mg, is deposited on the substrate by evaporation. Table 3.8 lists the reported optical waveguides fabricated by means of diffusion. The diffusion process varies slightly among research laboratories. But the diffusion temperature generally ranges between 850 and 1200°C. By far the most popular metal in-diffused waveguide is the Ti-diffused lithium niobate. Single-mode optical waveguides are easily made by diffusing 200–400 Å of titanium in a wet oxygen environment. Optical waveguiding losses under 1 dB/cm are routinely obtained. The Ti in-diffused lithium niobate waveguide has a larger index change in the extraordinary index than the ordinary index. The copper in-diffused optical waveguide provides an alternative if equal changes in these indices are desired. Another alternative is the zinc in-diffused waveguide. Because of the much larger diffusion coefficient, zinc diffusion may be carried out at lower temperatures (about 800°C) where lithium out-diffusion need not be concerned.

Here one example is given for the fabrication of a single-mode titanium-diffused waveguide on lithium niobate for operation at 0.85 μm wavelength. A titanium thickness of 475 Å is diffused for 5 h at 1000°C in wet oxygen. The wet oxygen can be prepared by feeding through a water bubbler at 95°C. The waveguide loss is sensitive to water bubbler temperature, titanium thickness, and the speed of temperature ramp. A fast temperature ramp is preferred to prevent phase changes in the lithium niobate reported in the 600–900°C range. The resultant waveguide has propagation loss of 0.5–1.5 dB/cm. When the titanium is too thick, there tends to be surface scattering. Slight hand polishing can significantly improve the waveguide loss due to surface scattering.

The lithium tantalate optical waveguide is less popular than the lithium niobate waveguides mainly due to its lower curie temperature (610°C). Unlike lithium niobate, the lithium tantalate crystals must be repoled after diffusion. However, the optical damage threshold of lithium tantalate is two orders of magnitude higher than that of lithium niobate, making it attractive for applications in the visible spectral range.

X-ray photoelectron spectroscopy has been performed to determine the state of in-diffused titanium ions in lithium niobate crystal. It was found that the titanium ions are fully ionized with no electrons in partially filled d orbitals to absorb light. Optical losses in the titanium in-diffused waveguide are mainly attributed to scattering losses.

Metal-Clad Optical Waveguide. The interest in metal-clad optical waveguides [118–125] stems from the desire to fabricate optical waveguides on ferroelectric electro-optical crystals. In the early 1970s, high-quality optical waveguides were mostly deposited organic or glass films that generally exhibit low refractive indices. In order to bring these low-loss films onto the relatively high refractive index surfaces of lithium niobates and lithium tantalates, a metal buffer layer was recommended. These waveguides have the problem of interacting with the metalized surfaces resulting in substantial waveguide losses. The waveguide losses in such waveguides had been studied using the zig-zag ray model and was found to be proportional to the square of the mode order and inversely proportional to the cube of the waveguide thickness [122]. It was also found that the loss introduced by the metalized surface has different impact on the TE modes and the TM modes. Generally, the TM modes experience greater propagation loss with the presence of metalized surfaces due to the field buildup near conducting walls and due to the existence of an additional surface plasma mode. As a result, these metal-isolated optical waveguides tend to be thick, supporting a large number of modes. The waveguiding loss could be

TABLE 3.8 Diffused Waveguide Parameters

Metal and Substrate	Diffusion E_A(eV)	Diffusion D_0(cm^2/s)	Temperature (°C)	Time (h)	Diffusion Coefficient	Metal (Å)	Waveguide Depth (μm)	Δn_0	Δn_e	Number of Modes	Loss (dB/cm)	References
Li Out, $LibO_3$	—	—	1100	23	$4.2 \times 10_{\perp}^{-9}$	—	~100	—	0.003	Many	<1	100, 101
					$1.5 \times 10_{\parallel}^{-9}$	—	~100	—	0.003	Many	<1	100, 101
Li Out, $LiTaO_3$	—	—	1150	3	$4 \times 10_{\perp}^{-9}$	—	~50	—	0.002	Many	<1	100, 101
			1400	20 min	$25 \times 10_{\perp}^{-9}$	—	~30	—	0.01	Many	<1	100, 101
V, $LiNbO_3$	—	—	950	6	—	250–500	6.2	0.0005	0.003	1 TM, 1 TE	<1	102
Ni, $LiNbO_3$	—	—	800	6	—	270	2.6–2.9	0.007	0.005	2 TE, 2 TM	<1	102
			800	6	—	500	2.8–3.1	0.01	0.006	2–3 TE, 2–3 TM	<1	114
	2.06	1.12×10^{10} μm^2/h	1000	—	2.2×10^{-10}	—	—	—	—	—	—	114
Ti, Y-	2.2	4×10^{-4}	1000	10	4.6×10^{-13}	500	2.6	—	—	—	<1	102–110
cut $LiNbO_3$			970	7	—	25–200	3	—	0.02	1	<1	102–110
			970	7	—	100–500	2.5	—	0.05	1 at 0.87 μm	<1	102–110
			960	6	—	500	$1.1_{\perp}$	0.01	0.04	4 TE, 1 TM	<1	115
							$1.6_{\parallel}$	0.006	0.025	1 TE, 5 TM	<1	115
			1000	—	$9.4 \times 10_{\parallel}^{-13}$	—	—	—	—	—	—	115
					$1.4 \times 10_{\perp}^{-12}$	—	—	—	—	—	—	115
Cu, $LiNbO_3$	1.8	4.79×10^{10} μm^2/h	—	—	—	—	—	—	—	—	—	111, 114
CuO_2, $LiTaO_3$	—	—	600	—	—	—	6	0.0075	0.0075	—	<1	112
											polish	112
Zn, $LiNbO_3$	1.6	8.15×10^7 μm^2/h	—	—	—	—	—	—	—	—	—	114
Zn, $LiTaO_3$	1.69	$5.29 \times 10_{\perp}^{-5}$	800	6	1.4×10^{-12}	Vapor	4.2	0.0027	0.0033	1 TE, 1 TM	<1	113
(Y-cut)	2.21	$3.85 \times 10_{\parallel}^{-2}$	800	6								113
MgO, $LiNbO_3$	1.4	1.7×10^{-6}	1000	—	4.5×10^{-12}	200	8.1	<0	<0	—	—	115
Co, $LiNbO_3$	1.34	1.4×10^7 μm^2/h	1000	—	1.94×10^{-10}	—	—	—	—	—	—	114
Nb_2O_5, $LiTaO_3$	—	—	—	—	—	—	—	0.0006	0.0018	—	<1	116, 117

Note: Data are for 0.62368 μm wavelength.

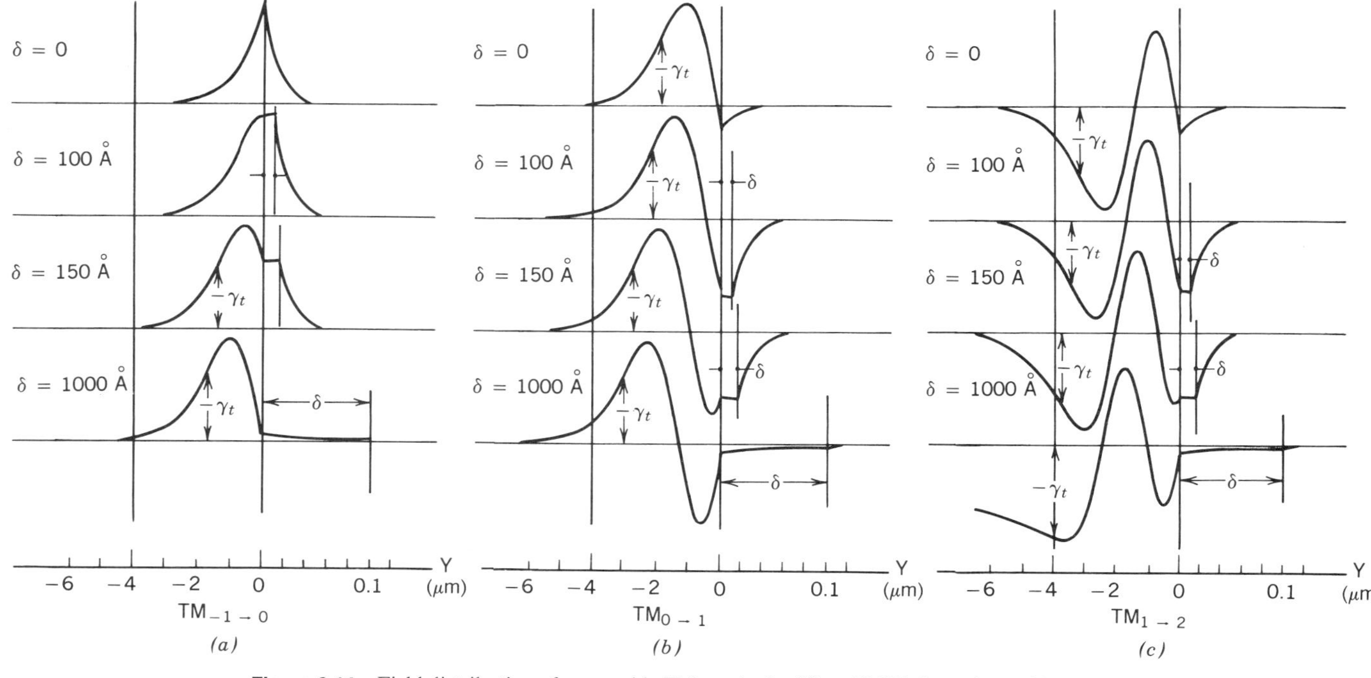

Figure 3.14 Field distribution of waveguide TM modes in diffused $LiNbO_3$ surface with the buffer SiO_2 film thickness of a parameter. With SiO_2 of more than 1000 Å thick, the field for all the three lowest-order modes become small at the metal–dielectric interface. (After Ref. 118.)

maintained low. For instance, using silver or gold as the cladding metal under a KPR photoresist waveguiding layer [123], the waveguide optical losses were reported between 0.6 and 15 dB/cm. An acousto-optical modulator was demonstrated on such a waveguide using surface acoustic waves generated by the lithium niobate substrate [126].

The advent of out-diffused and metal-diffused optical waveguides on these ferroelectric crystals basically eliminated this approach. However, the problem of guided mode interaction with the metal-cladding layer remains due to the need to have metal electrodes in active waveguide electro-optical devices. For an asymmetric dielectric waveguide with a metal–film outer coating, there is strong differential absorption between the TE_0 mode and the TM modes as well as the higher-order TE modes. It appears as if only the lower-order TE modes may be employed for the fabrication of waveguide electro-optical devices. This creates a difficult situation when working with a *C*-cut substrate that provides isotropy for guided modes propagating in all directions on the substrate surface. In order to use the large r_{33} electro-optical coefficient of lithium niobate, a TM mode will be launched that is not compatible with the metal electrodes.

To circumvent this difficulty, a dielectric buffer layer of SiO_2 film is deposited between the substrate surface and the metal electrode. The field distributions for the TM modes of a nearly three-mode diffused waveguide under an aluminum electrode are given in Fig. 3.14 with the buffer layer thickness as a parameter. As shown in Fig. 3.14, when

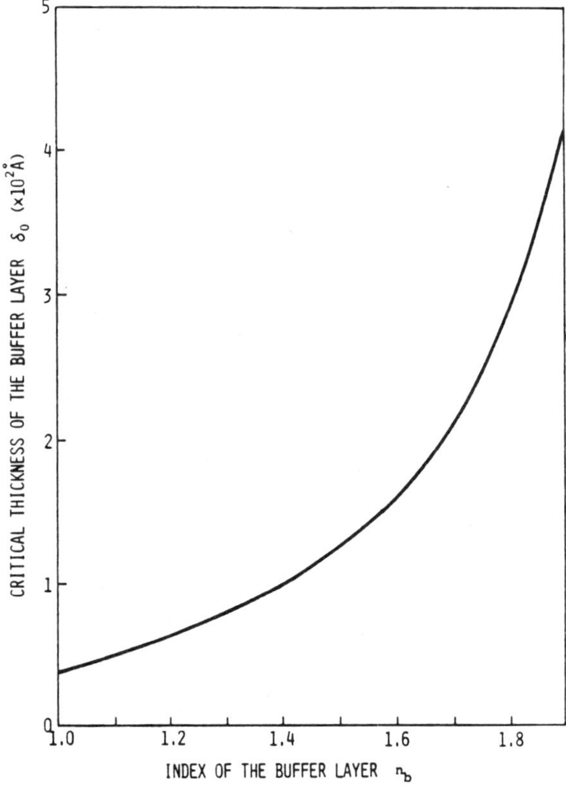

Figure 3.15 Relation between the critical thickness δ_0 and the index n_b of the buffer layer, calculated from Eq. (3.22), where aluminum is the metal, i.e., $\varepsilon_m/\varepsilon_0 = (1.2 - j7.0)^2$, $\varepsilon_{y0}/\varepsilon_0 = 2.200^2$, and $\Delta\varepsilon_y/\varepsilon_{y0} = 0.018$ for $\lambda = 0.6328$ μm. (After Ref. 118.)

there is no buffer layer (top row), the lowest-order TM mode is the highly lossy surface plasma mode, which has maximum field amplitude at the metal–substrate interface. The field distribution of the next TM mode (see Fig. 3.14*b*, top row) actually resembles the TM_0 mode in the waveguide region without the top electrode except that a negative dip occurs at the metal–substrate interface. It is expected that at the edge of the electrode area, an incident TM_0 mode is decomposed into mainly the combination of a strong TM_1 mode (top trace of Fig. 3.14*b*) and a weak plasma mode (top trace of Fig. 3.14*a*). As the buffer layer becomes thicker than a critical thickness, as shown by Fig. 3.14*a* from top row to the bottom row, the plasma mode gradually shifts its intensity into the dielectric substrate and becomes less lossy. When the buffer layer is substantially thicker than the critical thickness (bottom trace of Fig. 3.14*a*), the fundamental TM mode becomes practically the same as the TM_0 mode of the waveguide without the top electrode. There is little intensity in the metal–buffer layer interface. At the electrode edge, all the TM modes propagate through with minimum perturbation. The TM mode attenuation under the metalized area is significantly improved when the buffer layer thickness is greater than the critical thickness. Figures 3.14*b* and 3.14*c* show the transition of higher-order modes as the buffer layer thickness increases.

Figure 3.15 gives the critical thickness of the buffer layer as a function of the refractive index of the buffer layer, assuming a nearly three-mode Ti-diffused optical waveguide on lithium niobate substrate with a diffusion depth of 4 μm, a peak index change of 0.02, and an aluminum electrode. Figure 3.16 gives the attenuation constant of the three TM modes as a function of the buffer layer thickness. As expected, a lower refractive index

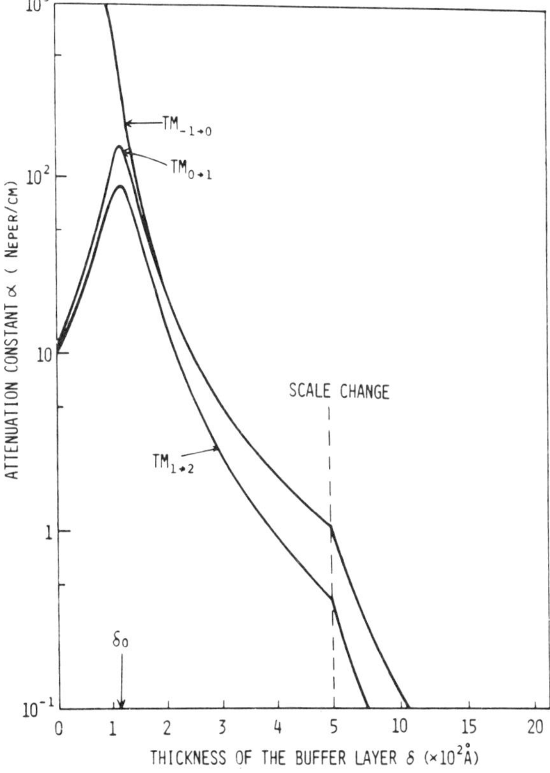

Figure 3.16 Attenuation constants α of $TM_{q\to q+1}$ modes versus δ. (After Ref. 118.)

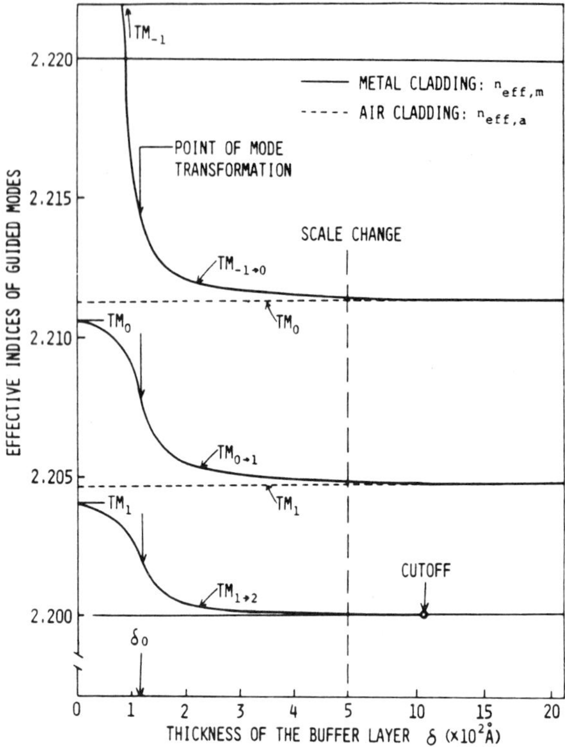

Figure 3.17 Effective indices of TM modes versus δ. The solid and dotted lines correspond to the effective indices $n_{\text{eff},m}$ of $\text{TM}_{q\to q+1}$ modes for the metal cladding and those $n_{\text{eff},a}$ of TM_q modes for the air cladding, respectively. (After Ref. 118.)

buffer layer is much more effective than one with a higher refractive index. Figure 3.17 illustrates the effective indices of the three guided modes in this example. The intersections between the dashed lines and the vertical axis indicate the guided mode indices for the three modes when nothing is coating the crystal surface. The intersections between the solid curves and the vertical axis gives the mode indices when the aluminum top electrode is applied. The rest of the solid curves illustrates how the buildup in buffer layer thickness affects the mode indices. This practice is widely employed to date in fabricating guided-wave electro-optical devices.

Free-Carrier Optical Waveguide. The presence of carrier in a semiconductor lowers the refractive index from that of the pure material. This is primarily due to the negative contribution of the free-carrier plasma to the dielectric constants. Using the free-carrier effective mass in place of the free electron mass, the change in the refractive index due to free carriers is

$$\Delta n_s = -\frac{N\lambda^2 e^2}{\varepsilon_0 n_s 8\pi^2 m^* c^2} \tag{3.50}$$

where n is the refractive index of the semiconductor, λ is the wavelength of light, e is the free carrier charge, ε_0 is the permittivity of the vacuum, and c is the velocity of light.

For example, in *n*-type GaAs with a carrier concentration of 5×10^{18} cm^{-3}, the index contribution due to the free carrier is -0.01 at an optical wavelength of 1 μm. If a high-resistivity GaAs region of more than 1 μm thick is surrounded by such doped GaAs regions, an optical waveguiding layer can be formed [127]. Light will be confined to the high-resistivity region, which is less lossy than the low-resistivity region due to lack of free-carrier absorption.

Optical waveguides based on this principle can be fabricated by an epitaxially grown high-resistivity layer over a low-resistivity substrate, by a diffused *pin* junction layer, or by ion implantation. Although these waveguides are relatively easy to fabricate, their disadvantages are the inability to independently vary the refractive index and the carrier concentration simultaneously, the inability to make waveguides thinner than 1 μm, and the free-carrier optical losses due to the cladding region. These waveguides are replaced by epitaxially grown ternary and quarternary semiconductor systems.

3.2 PASSIVE WAVEGUIDE OPTICAL COMPONENTS

The development of integrated optics requires the integration of a number of waveguide optical components into an optical device substrate. This can be done in two ways. The first approach is to construct a two-dimensional optical lens system on a planar waveguide surface that will manipulate optical wavefronts. The other approach is to construct an electro-optical system using optical channel waveguides to mimic an electronics system. In this section, the passive optical components for the first approach will be discussed.

3.2.1 Mode-launching Couplers

The first waveguide optical component to be discussed is the mode-launching couplers that served to interface guided modes with free propagating radiation modes. Mode launching can be achieved by use of a prism coupler, a grating coupler, a tapered thickness coupler, and an end-fire coupler. These couplers will be discussed next.

Prism couplers [5,128–135] and grating couplers [134–141] are mode converters capable of transferring energy between guided modes and radiation modes. In addition to being convenient devices for launching laser beams into and out of the optical waveguiding films, these devices also provide the means for investigating various properties of guided modes. Of particular importance is the prism coupler, an extremely versatile instrument for bringing light into and out of a waveguiding film. It can be applied almost anywhere on the waveguide substrate surface, requiring no special preparations. It also maps the effective refractive indices of the guided modes into propagation angles of the radiation modes. The only disadvantage is the requirement of relatively bulky mechanical clamping structures. Therefore, the prism couplers are considered a laboratory tool instead of part of an integrated optical device.

On the other hand, the grating couplers must be fabricated to designated areas of the planar waveguides with sophisticated submicrometer lithographic techniques. Once fabricated, the grating couplers cannot be removed or adjusted. The main advantage of a grating coupler is in its planar geometry, which makes it part of the guided wave device instead of just a tool. The problems of a grating coupler are the very high skill level required in fabrication and the difficulty in obtaining input coupling efficiency greater than 50%. Therefore, although grating couplers have been investigated extensively, they are seldom used in integrated optical device research activities other than for grating coupler research. Even their future prospect in packaged integrated optical devices has been shadowed by the alternative of an end-fire coupler.

The Prism Coupler. Figure 3.18 illustrates the physical mechanism of a prism coupler. When a laser beam is incident onto the base of a prism at an angle larger than the total internal reflection angle of the prism–air interface, as shown in Fig. 3.18*a*, the light will be reflected, yielding a standing wave above the prism–air interface and an evanescent wave below the prism–air interface. The evanescent field decays at the rate of $\exp(-\sqrt{n_4^2 \sin \theta^2 - n_1^2}\, kx)$ and has a phase constant $n_4 k \sin \theta$ along the z direction. In the meantime, the waveguiding mode of Fig. 3.18*a* also carries an evanescent wave above the air–guide interface with a phase constant β along the z direction. This evanescent wave decays at the rate of $\exp(-\sqrt{n_2^2 - (\beta/k)^2 - 1}\, kx)$. Figure 3.18*b* illustrates the field distribution in the air gap between the prism base and the waveguide surface in a magnified

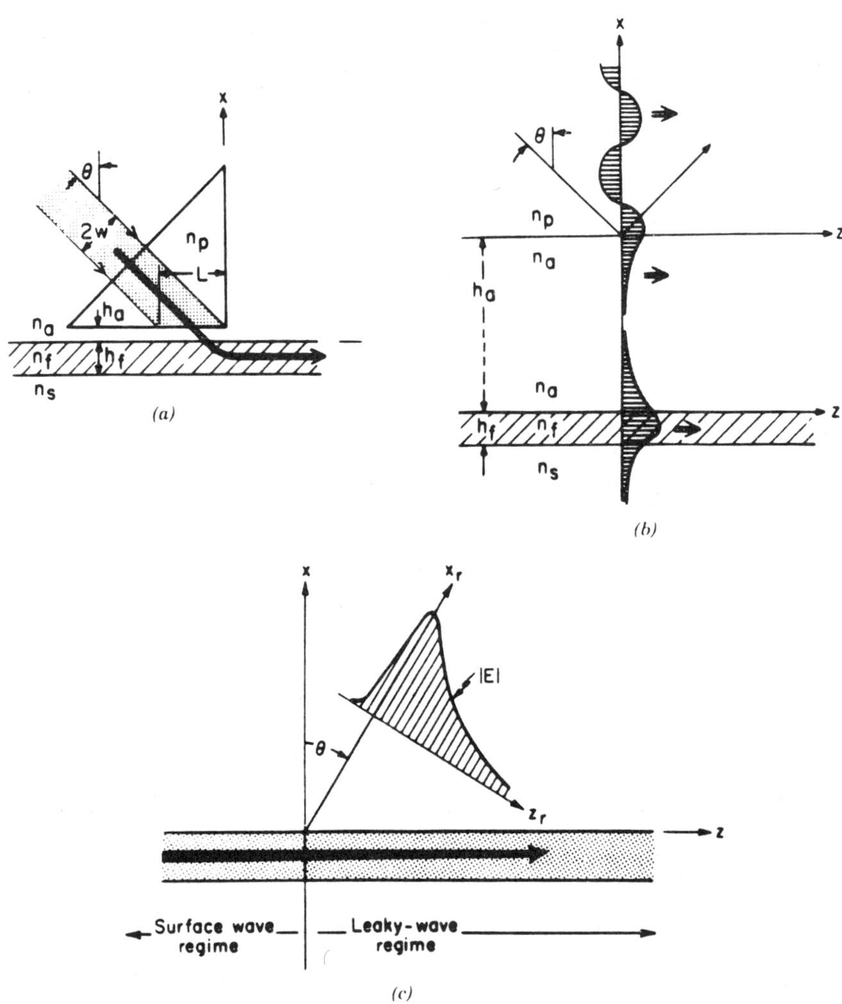

Figure 3.18 Prism coupler: (*a*) launching guided wave modes; (*b*) enlarged view of the prism–air gap–waveguide interface and field distribution; (*c*) amplitude variation in the output-coupled field. (After Ref. 135.)

scale. In these descriptions, n_4 is the refractive index of the prism, n_2 is the refractive index of the waveguide material, n_1 is the refractive index of the prism–waveguide gap, and β is the phase constant of the guided mode of interest. Now, if the gap is reduced such that the two evanescent waves may reach the other interface with appreciable field magnitude, a prism coupling condition is established. The evanescent field component at the other interface simply will leak into the other side of the system, causing the two systems to be coupled.

When the prism coupler is used as an output coupler for an optical waveguide, the guided mode will enter the waveguide area under the prism coupler with its evanescent field reaching the prism–air interface. This in turn induces a field inside the prism with a phase constant in the z direction equal to the phase constant of the guided mode. If the refractive index of the prism is higher than the effective mode index of the guided wave, then this induced field inside the prism is expected to propagate away from the prism–air interface at an angle to the normal,

$$\theta = \sin^{-1} \frac{\beta}{n_4 k} \tag{3.51}$$

Thus, the guided mode is coupled out by this prism coupler with the output angle nearly proportional to the effective mode index. When the optical waveguide supports a number of modes, the output will exhibit a fan of beams from which the waveguide mode indices can be obtained. In case of a poor-quality waveguide that scatters, each of the scattered guided modes will come out of the prism as a line in the three-dimensional space. This is the so-called *m*-lines. Figure 3.18*c* illustrates the output coupling. Since the guided wave is leaking at a constant rate as it enters the prism-covered area of constant air gap, the output beam is expected to have an exponential intensity profile due to depletion of the source. The width of the output beam profile may be adjusted by the height of the air gap and hence the coupling strength or leakage rate.

For the case of input coupling, the laser beam arrives from the inside of the prism toward the air gap, as shown in Fig. 3.18*a*. The evanescent wave now extends from the prism base to the air–guide interface and induces an optical field in the waveguide having a phase constant equal to $n_4 k \sin\theta$. This phase constant may not correspond to an allowed propagation constant of the particular waveguide of interest. However, this phase constant can be easily adjusted by varying the incidence angle of the laser beam. As long as the prism refractive index n_4 is higher than the refractive index of the waveguide, there is always an angle such that the phase constant of the induced field due to the incident laser beam is equal to the phase constant of the guided mode. This angle is again given by Eq. (3.51). The two waves are said to be phase matched when Eq. (3.51) is satisfied and the induced field becomes a guided mode. However, at any arbitrary point of the illuminated prism base of Fig. 3.18*a*, there are four wave components:

1. an incoming evanescent wave from the incident laser beam,
2. an incoming guided wave from the left side due to the induced field,
3. an outgoing guided wave toward the right side that is a combination of the induced field on this site and part of 2, and
4. an outgoing evanescent wave from the reflection of the incident laser beam and the leakage of 2.

Under the phase-matching condition, wave components in 3 will be in phase with each other and the magnitude of the guided wave will increase rapidly as the wave travels toward the z direction. In the meantime, the two components in 4 will interfere destructively to conserve the total power. An ideal condition is for the incident laser beam to have an exponentially increasing magnitude toward the z direction so as to keep up with the increasingly stronger guided wave. Complete power transfer is possible. Interestingly, this ideal condition corresponds to the reciprocal path of the output prism coupler illustrated in Fig. 3.18*c*.

In reality, the input laser beam has a Gaussian distribution instead of the ideal case of exponential distribution. In such a case, the field strength in the guided wave will eventually get to a point such that its leakage equals the power it receives from the incident beam. Optimal power launching occurs when both the illuminated area and the prism base end at that point. This is illustrated in Fig. 3.19. The optimal launch efficiency for a Gaussian laser beam and a prism coupler with constant air gap is 80.1%. In practice, one often adjusts the pressure on the prism to tune the leakage rate for best launch performance. Figure 3.20 gives the calculated maximum possible launch efficiency as a function of the prism coupler leakage rate for a Gaussian beam of width $2W_0 \cos\theta$ with the center of the laser beam offset from the prism corner by an optimal amount Z_c. Near-

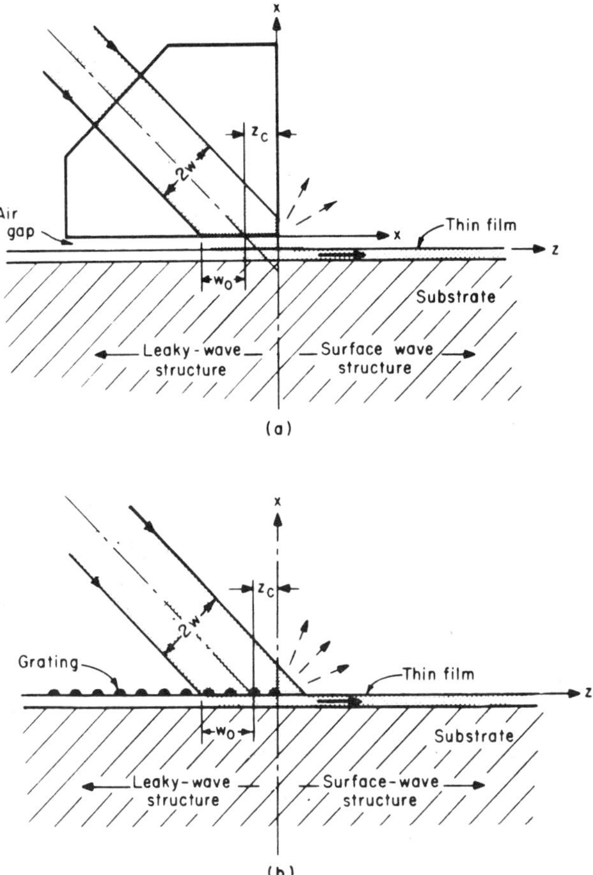

Figure 3.19 Alignment of incident beam in prism and grating input couplers for increasing coupling efficiency. Maximum efficiency achievable for Gaussian beams is 80.1%, which is obtained if $\alpha w_0 = 0.68$ and $z_c/w_0 = 0.733$. (After Ref. 134.)

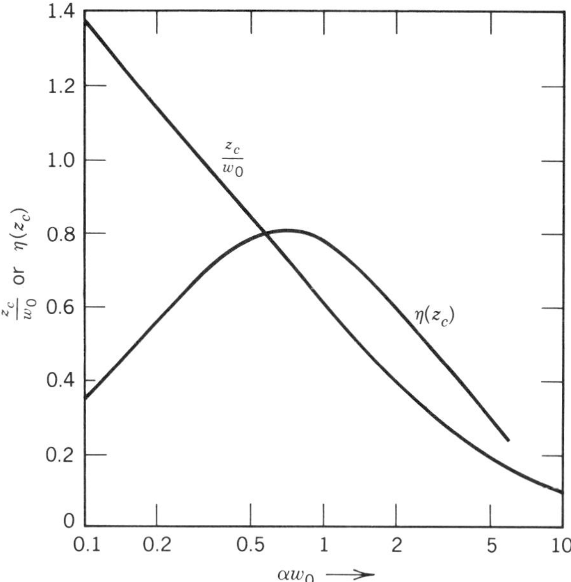

Figure 3.20 Optimal beam position offset, z_c/w_0, and maximum coupling efficiency at optimal offset, $\eta(z_c)$, as a function of normalized beam diameter, αw_0, for a Gaussian beam incident on a prism coupler of leakage rate α. (After Ref. 135.)

optimal launch performance is achievable over a reasonable range of leakage rate (or laser beam size). Figure 3.21 gives the optimal launch efficiency when the incidence angle of the laser beam is deviated from the phase-matching condition. As expected, when the laser incidence angle error equals the laser beam divergence angle, the launch efficiency will be reduced by a factor of 2.

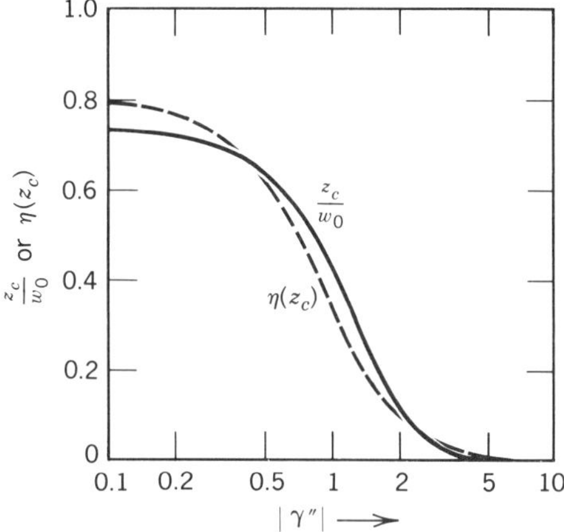

Figure 3.21 Optimal beam offset, z_c/w_0, and maximum coupling efficiency at optimal offset, $\eta(z_c)$, as a function of normalized beam incidence angle error $\gamma'' = \pi w_0 \Delta/\lambda$. The angular error Δ is normalized to the laser beam divergence angle $\lambda/\pi w_0$. (After Ref. 135.)

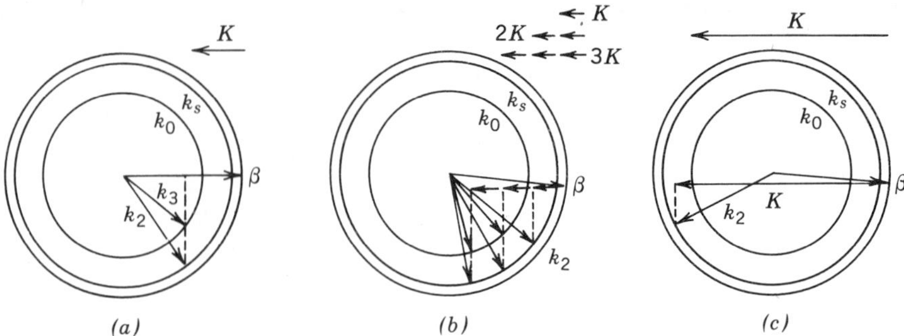

Figure 3.22 Grating coupler momentum vector diagrams showing (*a*) waveguide mode vector β is scattered by grating vector K producing radiation modes K_2 to the substrate and K_3 to the air; (*b*) a short grating vector K, intended to couple only to one radiation mode K_2 to the substrate, may have many radiation modes due to harmonic grating vectors K, $2K$, and $3K$; (*c*) a very large grating vector (periodicity much shorter than λ/n) can scatter the waveguide mode to a single backward substrate mode K_2.

The Grating Coupler. The most important part of a grating coupler design is the periodicity. The grating periodicity provides a wave vector in the z direction for momentum conservation along the z projection between the guided mode and one or more of the radiation modes. This is illustrated in Fig. 3.22. Since the refractive index of the substrate is higher than that of the air, coupling to radiation modes in the air always has a counterpart radiation mode in the substrate. For efficient mode launching, the coupler must not couple the waveguide mode to more than one radiation mode. This requires coupling between a guided mode and a substrate mode. Further complication arises due to the possibility of momentum conservation through higher-order harmonics of the grating wave vector. Therefore, only the reverse grating coupler between the guided modes and the substrate mode can have high efficiency. The problem is in the fabrication of gratings having a wave vector greater than the wave vector of the guided wave itself. The grating periodicity must be less than 0.3 μm for such grating coupler to work with the He–Ne laser light on lithium niobate waveguides. In addition, the substrate mode will stay in the substrate unless a prism is mounted to the substrate to bring the light out, as shown in Fig. 3.23. The other important parameter of a grating coupler is the corrugation as well as shape of the phase elements. Due to the very small thickness of the optical waveguide films, there is not much room for the shaping of the phase elements. Most grating couplers reported to date did not attempt to employ shaped geometries.

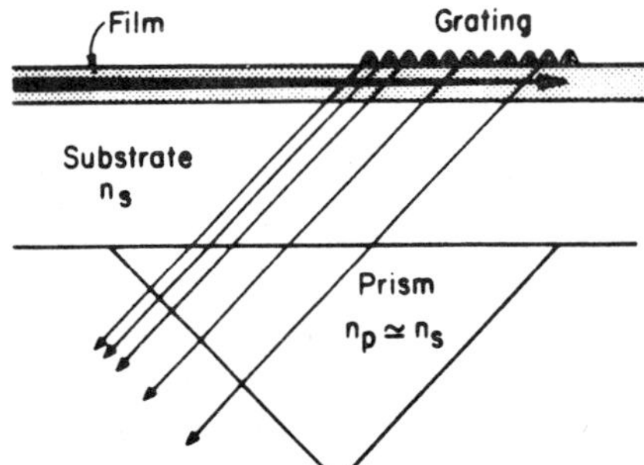

Figure 3.23 A prism mounted on the back of substrate serves to bring out the single back scattered output beam from a grating coupler of Fig. 3.22*c*.

3.2.2 Waveguide Refractive Elements

Optical components are usually made of reflective elements, refractive elements, or diffraction elements. In integrated optics, single-surface reflective elements tend to have rather limited usefulness. Although waveguides have been etched and waveguide substrates have been cleaved to form clean reflective surfaces, quality as well as reproducibility issues have jeopardized its future. Refractive elements and diffractive elements are far more compatible with integrated optics technologies in general and will be discussed in detail.

Waveguide refractive elements differ from conventional optical refractive elements in many ways. First, the index boundary for waveguide refractive elements tends to be one order of magnitude smaller than for bulk optical refractive elements. Thus, to achieve the same refractive effect, the radius of curvature of refractive boundaries must be considerably smaller than for conventional optics. Second, the material choices as well as considerations are very different. Special considerations must be given to mode conversions at the refractive boundaries and efficiency. Simplicity is a must, and the well-established multielement compound lens concept is not accepted in guided-wave optics. As a result, the waveguide optical refractive lenses used are varieties of fish-eye gradient lenses [141,142].

The most important part of refractive lens development is the development of waveguide index modification techniques. Two distinctively different techniques have been successfully developed for this purpose: the multilayer waveguide index modification technique and the geodesic curvature waveguide index modification technique. In the following sections, lenses made with these two approaches will be described in detail. Application of these waveguide index modifications to other waveguide refractive elements such as prisms is considered straightforward and will not be discussed here.

Waveguide Mode Index Modification Techniques. There are four known methods by which waveguide index modification can be accomplished. Effective index perturbation by waveguide thickness adjustment such as surface corrugation can be obtained by etching or by deposition. Volume index perturbation can be obtained by optical damage in lithium niobate or by ion exchange techniques. The effectiveness of any of these methods in producing a large index perturbation will be affected seriously by the properties of the waveguide employed. Among these techniques, the deposited thin-film overlay is by far the most versatile and thus most widely utilized method. It will be covered with more detail.

Multilayer Step Index Waveguide. Consider the four-layer structure illustrated in Fig. 3.24. It consists of a substrate having a refractive index n_4 and superstrate having a refractive index n_1, each with an infinite extent with one or two thin films between that serve as the optical waveguides or as a waveguide with an isolation layer. The planar structure considered also has infinite extent in y with optical propagation in the z direction. For wave binding, it is necessary that n_2 and/or n_3 be greater than n_1 and n_4. We will not consider the modes bound to the substrate when it is of limited extent. The modes bound to this type of structure may be either transverse electric (TE_m) or transverse magnetic (TM_m) of order m. A transverse field component bounded to both layers (overlayer and underlayer) is illustrated in Fig. 3.24, showing the zero order with the evanescent field extending into the superstrate and substrate. It is possible that the refractive index and the film thickness of these layers be such that wave binding occurs only in that region that has the highest index with an evanescent component in the other layers.

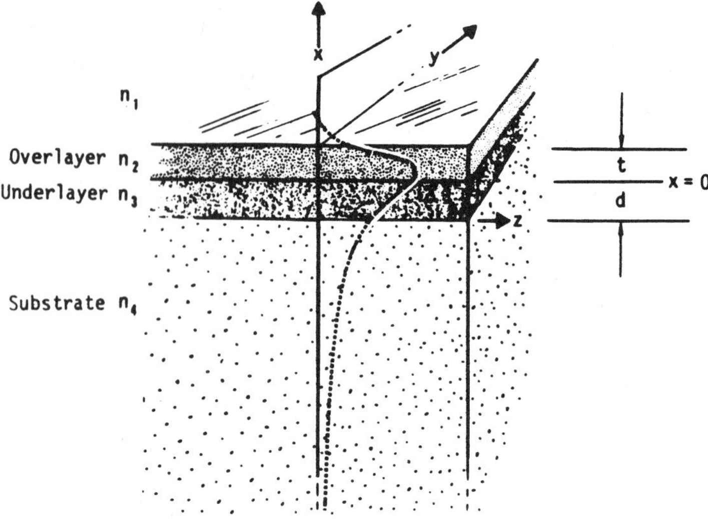

Figure 3.24 Cross section of asymmetrical multilayer planar dielectric waveguide showing zero-order mode bound in both layers with evanescent field.

The transverse electric modes (TE) are the eigensolutions of the field equation where the time variation $e^{i\omega t}$ has been suppressed,

$$\left[\frac{\partial^2}{\partial x^2} + \frac{\partial^2}{\partial z^2} + \omega^2 \varepsilon(x)\mu_0\right] E_y(x, z) = 0 \tag{3.52}$$

subject to the following boundary conditions: $E_y \to 0$ as $x \to \pm\infty$, and the tangential components of E and H are continuous across the boundaries. The z dependence will be the same in all four regions for a given mode and can be written as

$$e^{-i\beta_m z}$$

where m is the mode order. Additionally $\varepsilon(x)$ can be expressed as follows

$$\varepsilon(x) = \begin{cases} n_1^2 \varepsilon_0 & x \geqslant t \\ n_2^2 \varepsilon_0 & t > x > 0 \\ n_3^2 \varepsilon_0 & 0 > x > -d \\ n_4^2 \varepsilon_0 & -d \geqslant x \end{cases} \tag{3.53}$$

Applying the boundary conditions on E to the solutions of (3.52), the fields in the four regions can be written as

$$E_{m1} = A_m e^{-p_m(x-t)} \sin(h_m t + \phi_m) \sin \gamma_m e^{-i\beta_m z} \tag{3.54a}$$

$$E_{m2} = A_m \sin(h_m x + \phi_m) \sin \gamma_m e^{-i\beta_m z} \tag{3.54b}$$

$$E_{m3} = A_m \sin(l_m x + \gamma_m) \sin \phi_m e^{-i\beta_m z} \tag{3.54c}$$

$$E_{m4} = A_m e^{q_m(x+d)} \sin(-l_m d + \gamma_m) \sin \phi_m e^{-i\beta_m z} \tag{3.54d}$$

where A_m is a normalization constant and p_m, h_m, l_m, and q_m are the transverse components of the propagation vector in regions 1, 2, 3, and 4, respectively. The parameters p_m, h_m, l_m, q_m, and β_m are related by the following dispersion equations:

$$p_m = (\beta_m^2 - k^2 n_1^2)^{1/2} \tag{3.55a}$$

$$h_m = (k^2 n_2^2 - \beta_m^2)^{1/2} \tag{3.55b}$$

$$l_m = (k^2 n_3^2 - \beta_m^2)^{1/2} \tag{3.55c}$$

$$q_m = (\beta_m^2 - k^2 n_4^2)^{1/2} \tag{3.55d}$$

Application of the boundary conditions on H yields

$$\tan(h_m t + \phi_m) = -\frac{h_m}{p_m} \tag{3.56}$$

$$\tan \phi_m = \frac{h_m}{l_m} \tan \gamma_m \tag{3.57}$$

$$\tan(-l_m d + \gamma_m) = \frac{l_m}{q_m} \tag{3.58}$$

When combined, these equations lead to the following transcendental equation:

$$\frac{1}{l_m} \tan\left(l_m d + \tan^{-1} \frac{l_m}{q_m}\right) + \frac{1}{h_m} \tan\left(h_m t + \tan^{-1} \frac{h_m}{p_m}\right) = 0 \tag{3.59}$$

The roots of this equation are the allowed values of the propagation constants.

The transverse magnetic modes (TM) are the eigensolutions of the field equation

$$\left[\frac{\partial^2}{\partial x^2} + \frac{\partial^2}{\partial z^2} + \omega^2 \varepsilon(x) \mu_0\right] H_y(x, z) = 0 \tag{3.60}$$

The development of the TM modes is identical to that for the TE modes, and again the allowed values of the propagation constants are roots of a transcendental equation, which is

$$\frac{n_3^2}{l_m} \tan\left[l_m d + \tan^{-1}\left(\frac{n_4^2}{n_3^2} \frac{l_m}{q_m}\right)\right] + \frac{n_2^2}{h_m} \tan\left[h_m t + \tan^{-1}\left(\frac{n_1^2}{n_2^2} \frac{h_m}{p_m}\right)\right] = 0 \tag{3.61}$$

Tantalum pentoxide and Corning 7059 glass have been employed to form waveguides and lenses on thermally grown SiO_2 substrates because they exhibit low scattering loss and may be deposited reproducibly. The single-layer thin-film waveguide dispersion of Ta_2O_5 structure on Corning 7440 glass is shown in Fig. 3.25*a* for both TE and TM modes as a function of the normalized film thickness. The thin-film waveguide dispersion of a Corning 7059 glass structure on an SiO_2 substrate is shown in Fig. 3.25*b* for both TE and TM modes as a function of the normalized thickness. The effective refractive index approaches the bulk value for thick films and approaches the substrate for thin films. Each of these thin-film waveguide modes exhibits a cutoff where $n_e(m) = 1.47$.

The dispersion for a two-layer structure for only the transverse electric modes is shown in Figs. 3.26*a* and *b*, where Ta_2O_5 is employed as the *overlayer* of variable thickness T and where the *underlayer* normalized thickness is constant ($kt = 2.37\pi$), which is equivalent to a 7059 layer having a thickness of 0.75 μm and an optical wavelength of 0.63 μm.

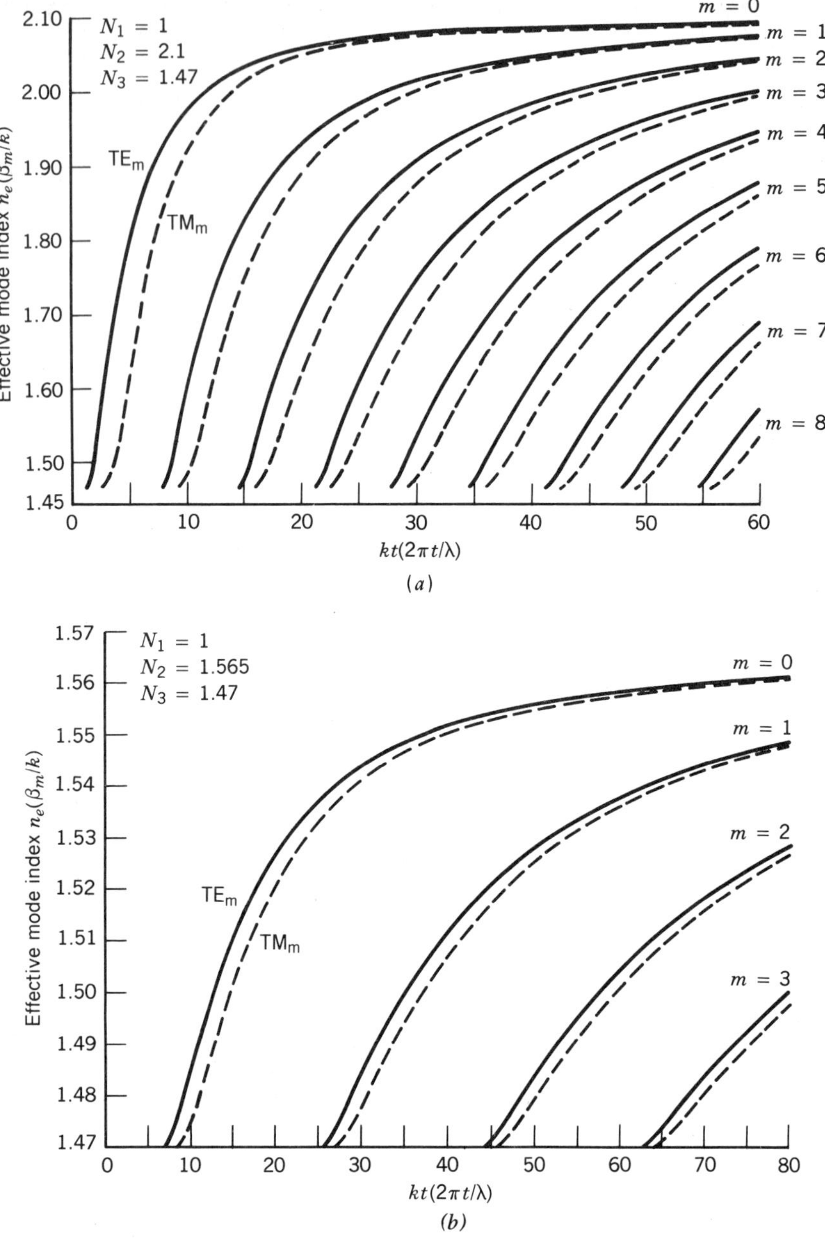

Figure 3.25 Effective refractive index versus normalized film thickness for (*a*) Ta_2O_5 film on SiO_2 substrate and (*b*) 7059 glass film on SiO_2 substrate: (——) Te modes; (– – – –) TM modes.

Each of the dispersion curves exhibits an inflection representing the transition region where wave binding occurs to only the dense overlayer and where wave binding occurs in both layers. Figure 3.26*b* is an expansion in the region where wave binding occurs in both layers showing the effect of substrate refractive index. The index modification is sufficiently large for practical waveguide refraction element applications.

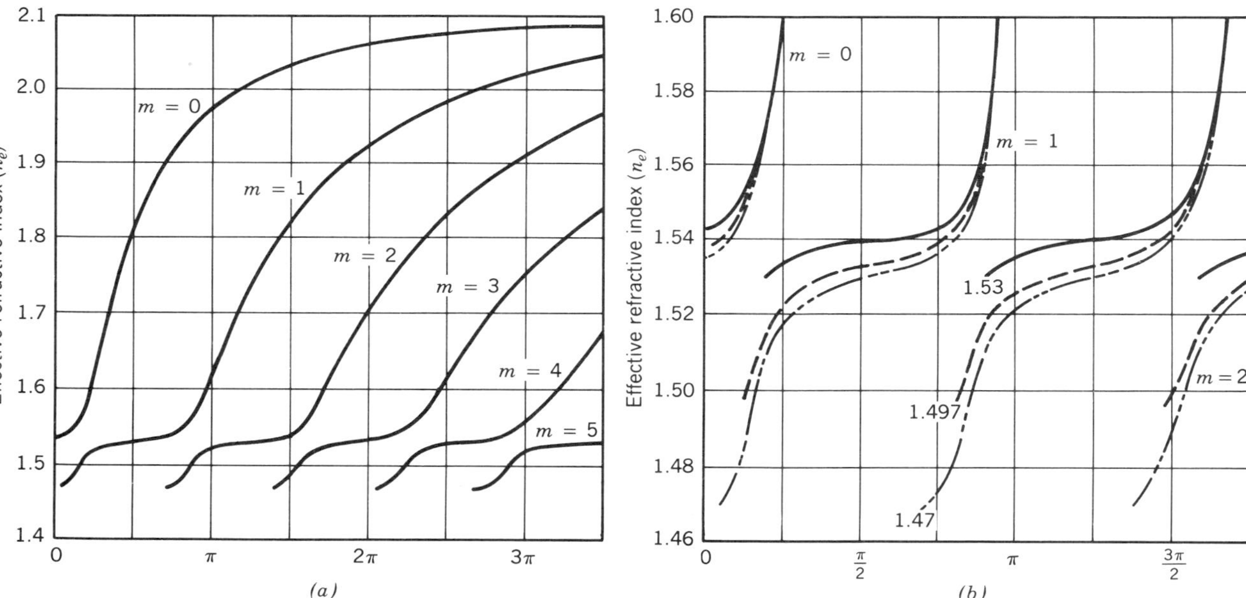

Figure 3.26 Effective refractive index of composite structure as a function of normalized thickness of the dense overlay film (lens) where the overlay film normalized thickness $kt = 2.37\pi$ for TE modes only (*a*); expanded scale showing effect of three substrates for TE modes only (*b*).

Superstrate	$n_1 = 1.0$		
Overlayer	$n_2 = 2.1$	kT	Lens
Underlayer	$n_3 = 1.565$	$kt = 2.37\pi$	($t = 0.75\ \mu$m at $\lambda = 0.63\ \mu$m) waveguide
Substrate	$n_4 = 1.47, 1.497, 1.53$		

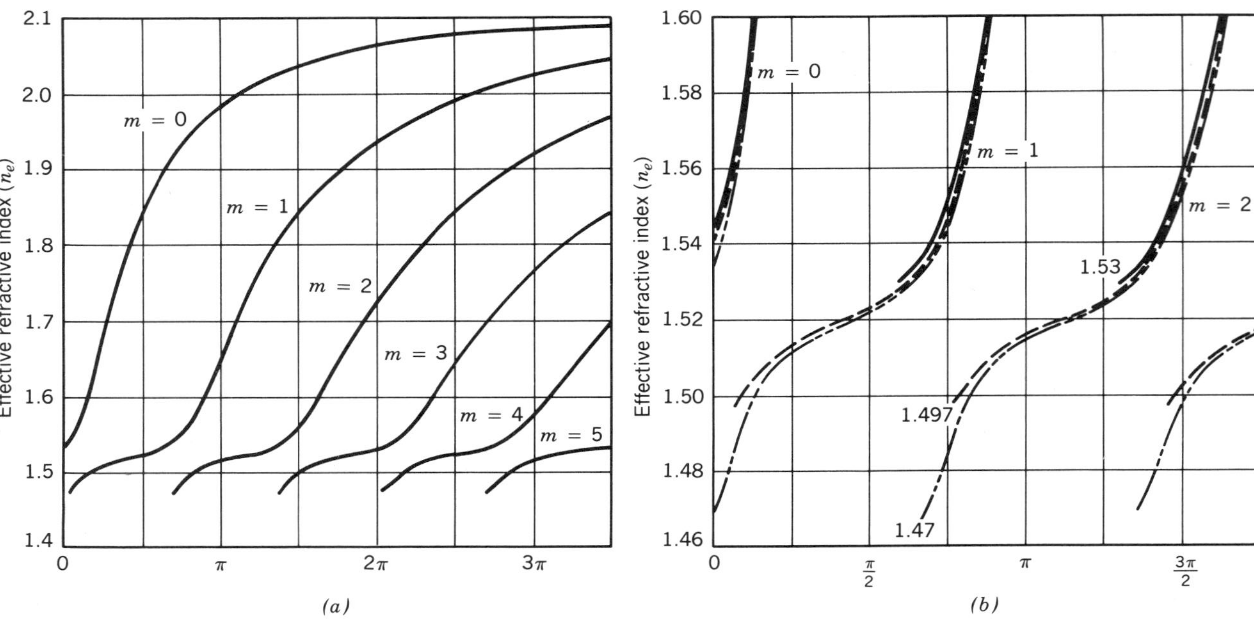

Figure 3.27 Effective refractive index of composite structure as a function of normalized thickness of the dense underlay film (lens) where the underlay film normalized thickness $kt = 2.37\pi$ for TE modes only (*a*); expanded scale showing effect of three substrates for TE modes only (*b*).

Superstrate	$n_1 = 1.0$			
Overlayer	$n_2 = 1.565$	$kt = 2.37\pi$	($t = 0.75\ \mu$m at $\lambda = 0.63\ \mu$m)	waveguide
Underlayer	$n_3 = 2.1$	kT	Lens	
Substrate	$n_4 = 1.47$			

The dispersion for a two-layer structure for only the transverse electric modes where 7059 is employed as the overlayer of fixed thickness and where the underlayer is Ta_2O_5 of variable thickness T is shown in Figs. 3.27*a* and *b*. The fixed-thickness waveguide ($kt = 2.37\pi$) is equivalent to a 7059 layer having a thickness of 0.75 μm for an optical wavelength of 0.63 μm. Although the dispersion data in Figs. 3.26 and 3.27 are similar, the waveguide index modification means by Ta_2O_5 underlayer (Fig. 3.27) causes large optical scattering at the refractive boundaries due to the large discontinuity in the waveguiding 7059 glass layer. Therefore, in practice, successful thin-film Luneburg lenses are made by Ta_2O_5 overlayer approaches only.

Overlay Film on Graded Index Waveguide. Due to the large variations possible in graded index optical waveguides, it is rather difficult to derive the dispersion characteristics for every possible graded index system here. Basically, the mathematical method is similar to the one described in the previous section. One example is given here for the case of TE modes in a graded index optical waveguide with exponential profile.

From Table 1.1, the guided mode is expected to have a field distribution described by the Bessel function of the first kind with noninteger order and with an exponential argument. The higher index overlay film will have an oscillatory field inside the overlay. By matching the fields at the boundaries, the following transcendental equation is obtained for the TE_0 mode:

$$\frac{\sqrt{n_c^2 - n_b^2}}{\sqrt{n_s^2 - n_b^2}} - \frac{\sqrt{n_c^2 - n_e^2}}{\sqrt{n_s^2 - n_b^2}} \frac{q_2 \sin(q_2 s) + p_1 \cos(q_2 s)}{p_1 \sin(q_2 s) - q_2 \cos(q_2 s)} = \frac{J_{v+1}[(4\pi d/\lambda)/\sqrt{\Delta\varepsilon}]}{J_v[(4\pi d/\lambda)\sqrt{\Delta\varepsilon}]} \tag{3.62}$$

where n_c, n_s, and n_b are refractive indices for the overlay, the surface of the graded waveguide, and the bulk of the substrate, respectively; s is the overlay thickness; and d is the effective waveguide depth. The modified effective mode index for a Ti in-diffused waveguide on lithium niobate substrate at a wavelength of 0.6328 μm is given in Fig. 3.28 as a function of overlay thickness with the overlay refractive index as a parameter. Note that an overlay thickness of more than 0.1 μm is needed to cause a mode index

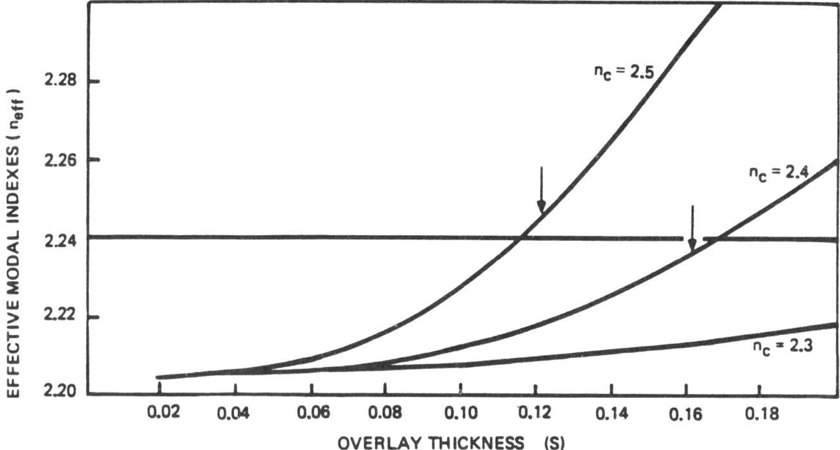

Figure 3.28 Effective modal indices of Ti:$LiNbO_3$ waveguides with a higher index overlay film. The index of the overlay varies from 2.3 to 2.5. Other waveguide parameters are $n_s = 2.24$, $n_b = 2.20$, $d = 0.5$ μm, and wavelength $= 0.6328$ μm. The arrows indicate when the second-order mode occurs.

change of only 0.03. A further increase in overlay thickness results in a multimode waveguide as well as larger mode index modification. This is generally true for index perturbation on a graded index waveguide that has a rather large effective waveguide thickness and a small index difference between the guided mode and the substrate.

In practice, mode index modification on lithium niobate poses special problems due to the difficulty in depositing a film having a higher refractive index than the surface index of metal-diffused lithium niobate [143–145]. Two kinds of films were commonly used for this purpose, namely the TiO_2 film of the rutile phase and the Nb_2O_5 film. The niobium pentoxide film can be sputtered on lithium niobate surface with low propagation loss (3 dB/cm). However, this film has limited use because its index of refraction is only slightly higher than lithium niobate. The rutile film has a considerably higher refractive index and may be deposited by RF sputtering deposition, reactive sputtering deposition from titanium target, sputtering of titanium followed by oxidation, and E-beam evaporation of titanium followed by oxidation. So far, the first three methods have yielded TiO_2 film with either too much attenuation or too low a refractive index. This is reportedly due to the incorporation of contamination in the sputtering process.

Extremely hard and durable transparent films of TiO_2 with a refractive index close to that of the pure rutile crystal have been fabricated by E-beam evaporation of a few hundred angstroms thick of Ti under very high vacuum conditions and followed by oxidation in dry oxygen at 450°C for 2–4 hours [143,144]. Experimentally, it was found that the refractive index of this film depends on the Ti deposition rate as well as on the vacuum background. For a background pressure of 6×10^{-6} Torr, the oxidized film index ranges from 2.54 at a slow deposition rate of 0.8 Å/s to 2.72 at a fast deposition rate of 35 Å/s. When the background vacuum pressure is reduced to 6×10^{-7} Torr, the oxidized film index becomes a constant value of 2.68 at all deposition rates. This overlay film process has been found compatible with the titanium-diffused waveguide on lithium niobate. However, the waveguide propagation loss in the region covered by 700 Å of this rutile film showed a substantial increase to approximately 20 dB/cm. This relatively high loss and the requirement of heating at 450°C limit the application of the oxidized film to certain diffraction elements on high-temperature metal-diffused waveguides only. Furthermore, the possibility of high waveguide loss due to incomplete oxidation also limits the maximum film thickness to under 0.1 μm, which corresponds to a waveguide propagation loss of about 40 dB/cm.

Other Index Perturbation Methods. Two other methods may be used for the creation of index perturbation [145] of optical waveguides in general: the thickness modulation method and the ion exchange or double ion exchange method. These methods can be used to produce high-quality index-modified optical waveguides. The treated waveguide area tends to have a uniform change of waveguide index that limits its application from gradient index type of waveguide optical components such as the Luneburg lenses.

The thickness modulation method is very simple. For practically any waveguide dispersion curve, the guided mode index is a function of waveguide thickness. Therefore, local adjustment of waveguide thickness will produce a corresponding change in the effective waveguide index. This method is extremely easy to implement. A two-step deposition procedure or an etch-back procedure may be employed. Often, the quality of the waveguide is preserved as long as one stays in the single-mode region at all times. However, the magnitude of index modulation is often limited due to the desire to stay in the single-mode region. For step index waveguides, the difference between the effective waveguide index and the substrate refractive index can be large. As a result, significant index perturbation may be obtained. On the other hand, very deep etched steps are required to perturb a graded index waveguide. Due to the small difference between the effective waveguide index and the substrate index, the amount of index modification tends to be very small.

Ion exchange [97] over an existing optical waveguide [146] causes further index perturbation in the volume of the waveguide. The relatively low-temperature process of ion exchange makes it compatible with high-temperature metal-diffused waveguides and many deposited waveguides. The limitation is the relatively small index perturbation that may be achieved by using this method. The one exception is proton ion exchange, which can lead to large index changes in lithium niobate. In this case, index changes greater than 0.1 have been produced.

However, due to the large thickness of graded index waveguides and the volume treatment of the ion exchange process, the proton ion exchange treated graded-index optical waveguide often exhibits multimode behavior unless the amount of dose is significantly reduced. For the purpose of index perturbation, the ion exchange process can be localized by an aluminum mask 0.2 μm thick. Figure 3.29*a* gives experimental results

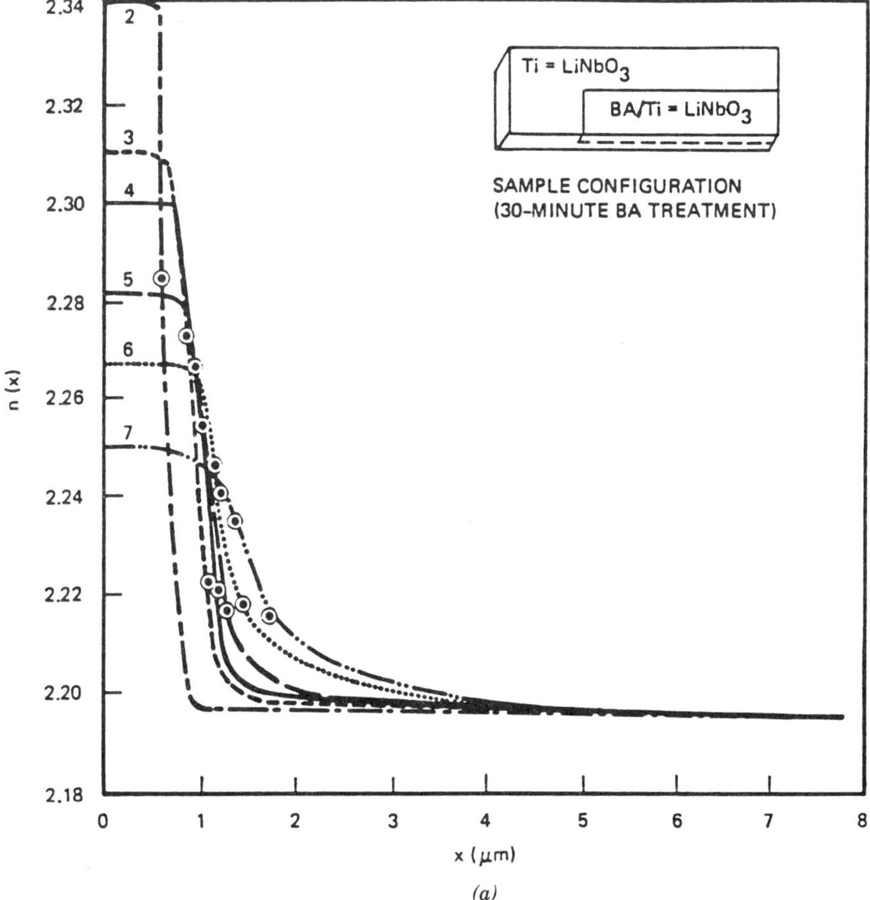

(a)

Figure 3.29 Index profiles and guided modes in a BA-treated Ti:$LiNbO_3$ waveguide with subsequent thermoannealing (*a*); with 10-min BA treatment at 0.63 μm (*b*).

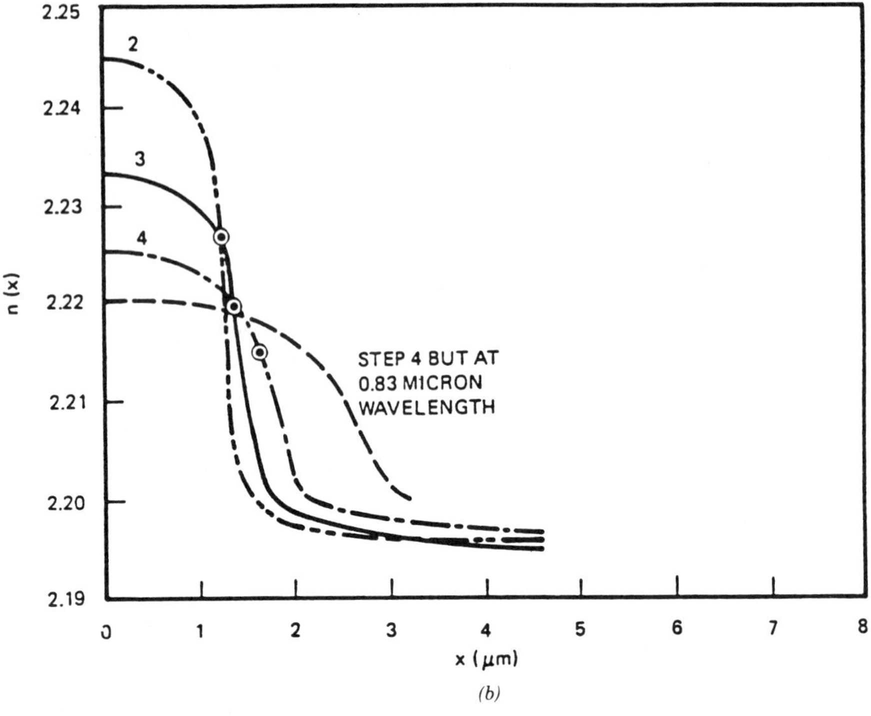

(b)

Figure 3.29 (continued)

of the refraction index profile of a single-mode titanium in-diffused waveguide on *Z*-cut lithium niobate substrate after 30 min of benzoic acid (BA) treatment with subsequent thermal annealing time as a parameter. The annealing temperature is 240°C. The treated waveguide appears to have two more modes, and the coupling of the titanium waveguide mode and the newer proton ion exchange waveguide modes is very poor. The titanium mode appears unperturbed. Curves 3, 4, 5, 6, and 7 correspond to annealing times of 1, 2, 4, 8, and 16 h, respectively. The annealing process drives the proton ions deeper into the waveguide with reduced surface index value. In the meantime, the titanium mode power becomes more and more coupled to the proton ion exchange waveguide modes that have a higher effective refractive index. After 16 h of annealing, 58% of the titanium waveguide mode power moves into the first proton ion exchange waveguide mode, and the waveguide mode index perturbation is only about 0.03 (from 2.205 for the titanium mode to 2.233 for the ion exchange waveguide mode). Another set of data is given in Fig. 3.29*b*, where the benzoic acid treatment time is reduced to 10 min. At the He–Ne laser wavelength, there are nearly two titanium modes and one ion exchange mode. Curve 2 is the as-perturbed waveguide index profile. Curves 3 and 4 are index profiles after 2 and 8 h of annealing at 240°C. After 8 h of annealing, the conversion efficiency of the titanium waveguide mode to the ion exchange waveguide mode is 74%. Of great interest are the results obtained from the same waveguide at the laser diode wavelength of 0.85 μm. At the longer wavelength, the titanium mode supports only one mode, and over 90% of the titanium mode power is converted to the ion exchange mode when the sample has been annealed for 8 h. However, the index change was only about 0.01.

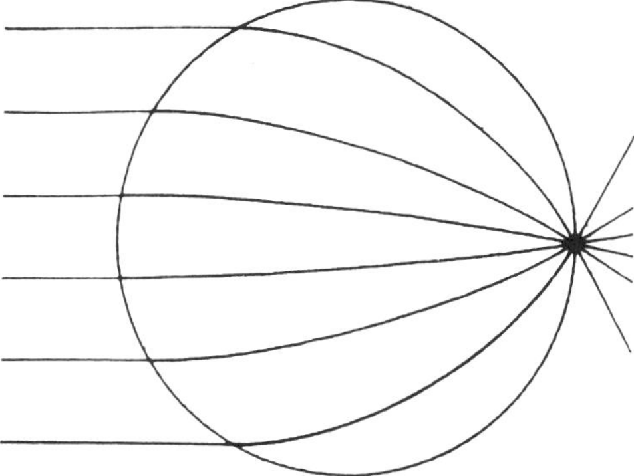

Figure 3.30 Ray trajectory through Luneburg lens where f-number = 0.5, which is equivalent to Morgan's $s = 1$.

Thin-Film Waveguide Luneburg Lens

Waveguide Luneburg Lens Theory. Optical lenses with simple form suffer from spherical aberrations particularly when the lenses are thick and the numerical aperture is large. Unfortunately, the desire to develop miniaturized integrated optical devices requires the development of large-numerical-aperture waveguide lenses using thick lenses made of the small index perturbation available on optical waveguides. In order to overcome this serious problem, graded index optical lenses were investigated. It was found that the Luneburg lens can provide large-numerical-aperture imaging without spherical aberration while only a small index perturbation is needed.

The "classical" Luneburg lens [147] is an inhomogeneous positive refractor with radial symmetry that images a plane wavefront to a hemispherical spot located on the opposite boundary of the refractor. Ray traces through the classical Luneburg lens are illustrated in Fig. 3.30. Interest in the Luneburg lens arises because of its 4π field of view and because it is free of all aberrations except field curvature. The cross section of a dielectric waveguide Luneburg lens is illustrated in Fig. 3.31, where a circular overlay film is employed with a prescribed radial thickness function. For integrated optics, its simplicity of fabrication in a thin-film structure is an important attribute. The circular symmetry is of particular interest where the substrate area is limited because folding of the optical axis may use the same lens for several purposes. The waveguide Luneburg lens has a further advantage that all refractive index variations are continuous, smooth, slow functions with respect to the optical wavelength. Mode conversion in the lens is

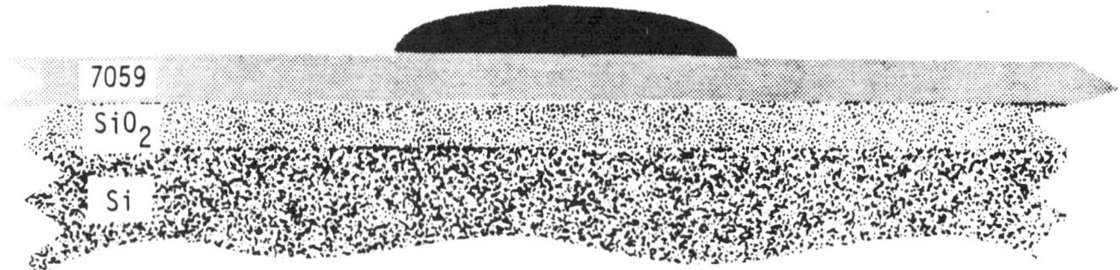

Figure 3.31 Cross section of a waveguide Luneburg lens using a dense Ta_2O_5 overlay lens on a Corning 7059 waveguide on a thermally oxidized silicon substrate.

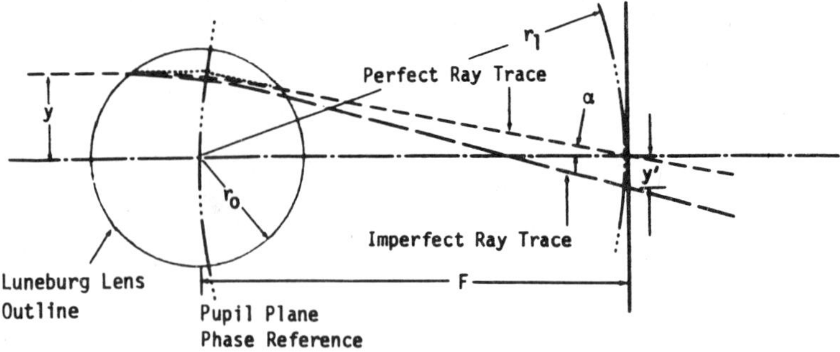

Figure 3.32 Generalized Luneburg lens and ray trace definitions.

suppressed by the graded thickness profile and by symmetry. For a classical Luneburg lens the refractive index $n(r, m)$ is

$$n(r, m) = n_e(m)\sqrt{2 - \left(\frac{r}{r_0}\right)^2} \qquad r \leqslant r_0 \tag{3.63}$$

where n_e is the refractive index of the medium surrounding the lens, r_0 is the lens radius, and r is the radial coordinate.

The classical Luneburg lens has been generalized by Morgan [148]. The generalized Luneburg lenses may have incidence rays originated from a point at a finite distance from the lens, and the focused image point is not restricted to the edge of the lens. Figure 3.32 illustrates a generalized Luneburg lens with a focal plane curvature of r_1. A perfect ray is shown to arrive at the focal point of the focal plane. An imperfect ray is shown to intercept the focal plane with a ray intercept error of y'.

The radial refractive index profile for some generalized Luneburg lenses can be obtained from a paraxial approximation when the f-number is over 1.5. The normalized refractive index profile for a focal length of n is given by

$$\frac{n(r, m)}{n_e(m)} = \exp \frac{\sqrt{n_e(m)^2 r_0^2 - r^2 n^2(r, m)}}{\pi n_e(m) r_1}. \tag{3.64}$$

Since Eq. (3.64) is dependent on the waveguide mode index, a multimode waveguide will have a number of focal planes corresponding to the number of waveguide modes. Only one of these modes will experience perfect focusing for a given index profile. From Eq. (3.64), the normalized refractive index profiles for a number of generalized Luneburg lenses with different f-numbers can be calculated and are given in Fig. 3.33.

For the general Luneburg lenses without the restriction that the f-number be greater than 1.5, the radially symmetric index profile is the solution of the following integral equations:

$$n = \exp[\omega(\rho, s)] \tag{3.65}$$

where $\rho = rn$ and

$$\omega(\rho, s) = \frac{1}{\pi} \int_\rho^1 \frac{\arcsin(x/s)}{(x^2 - \rho^2)^{1/2}}\, dx \tag{3.66}$$

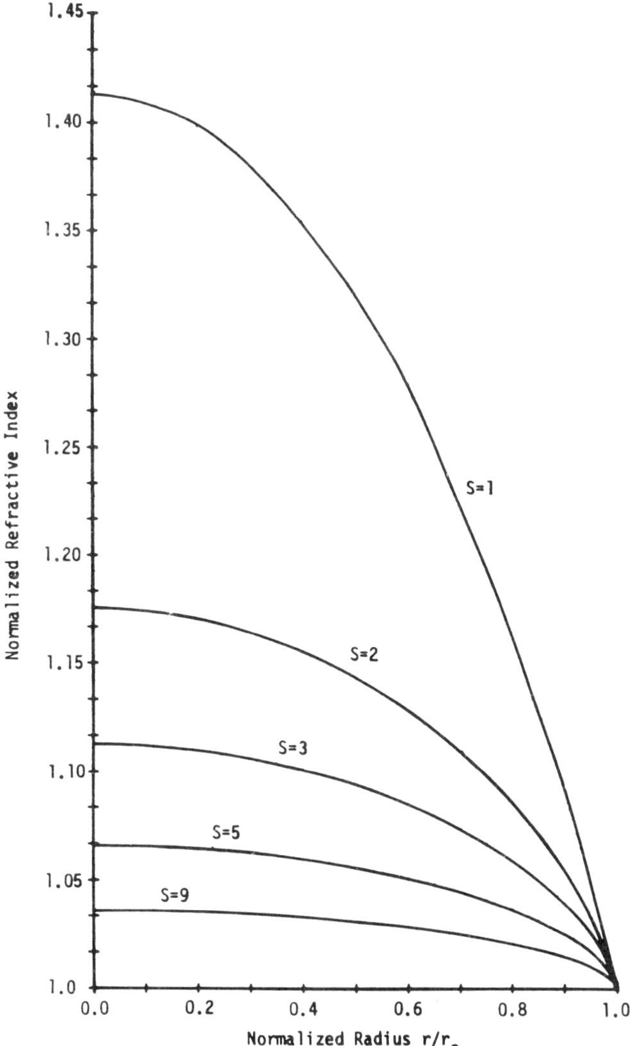

Figure 3.33 Normalized refractive index profile for generalized Luneburg lenses for various f-numbers $= s/2$. (After Ref. 150.)

Southwell [149] has been able to describe the solution of these integral equations by a power series expansion of only five terms, which is given by Eqs. (3.67). The coefficients for this power series can be determined by least mean squares fit to numerically generated index values from the preceding integral equations. The errors due to the power series approximation is reportedly as small as 10^{-6}. Table 3.9 lists the coefficients for several generalized Luneburg lenses with different normalized focus values. Table 3.10 gives numerical values for the index profile of four generalized Luneburg lenses:

$$\omega(\rho, s) = p_1(1-\rho)^{1/2} + p_2(1-\rho)^{3/2} + p_3(1-\rho)^{5/2} + p_4(1-\rho)^{7/2} + p_5(1-\rho)^{9/2} \tag{3.67a}$$

TABLE 3.9 Coefficients for Four Examples of Generalized Luneburg Lenses

	$s = 2$	$s = 3$	$s = 5$	$s = 9$
p_1	0.235687835	0.152976535	0.0906399959	0.0501194645
p_2	$-7.47500358 \times 10^{-2}$	$-4.22494061 \times 10^{-2}$	$-2.34300068 \times 10^{-2}$	$-1.26356394 \times 10^{-2}$
p_3	$6.72894476 \times 10^{-3}$	$-1.93175172 \times 10^{-3}$	$-2.51100017 \times 10^{-3}$	$-1.63934418 \times 10^{-3}$
p_4	$-5.14447054 \times 10^{-3}$	$-8.25244897 \times 10^{-4}$	$-1.49192458 \times 10^{-4}$	$-4.07947091 \times 10^{-5}$
p_5	$-9.89299661 \times 10^{-4}$	$-1.19124255 \times 10^{-3}$	$-7.44793717 \times 10^{-4}$	$-4.11582122 \times 10^{-4}$

with

$$p_1 = \frac{\sqrt{2}}{\pi} \arcsin \frac{1}{s} \tag{3.67b}$$

and

$$p_5 = \omega(0, s) - p_1 - p_2 - p_3 - p_4 \tag{3.67c}$$

TABLE 3.10 Normalized Refractive Index Profiles for Generalized Luneburg Lenses

r	$s = 2$	$s = 3$	$s = 5$	$s = 9$
0	1.175311212	1.112688200	1.065884534	1.036025859
0.05	1.174999407	1.112507930	1.065788087	1.035976411
0.10	1.174071705	1.111969546	1.065500332	1.035828873
0.15	1.172521576	1.111069129	1.065018948	1.035581965
0.20	1.170340993	1.109801290	1.064340603	1.035233817
0.25	1.167519780	1.108158811	1.063460759	1.034781859
0.30	1.164044879	1.106132212	1.062373426	1.034222692
0.35	1.159899514	1.103709214	1.061070836	1.033551908
0.40	1.155062156	1.100874045	1.059543028	1.032763858
0.45	1.149505212	1.097606509	1.057777265	1.031851324
0.50	1.143193243	1.093880684	1.055757219	1.030805062
0.55	1.136080437	1.089663045	1.053461790	1.029613132
0.60	1.128106836	1.084909633	1.050863316	1.028259883
0.65	1.119192434	1.079561599	1.047924746	1.026724355
0.70	1.109227400	1.073537799	1.044594938	1.024977603
0.75	1.098054844	1.066721643	1.040800259	1.022977928
0.80	1.085437716	1.058935566	1.036428193	1.020661529
0.85	1.070987513	1.049884988	1.031290974	1.017921692
0.90	1.053981685	1.039009952	1.025027636	1.014552365
0.91	1.050159236	1.036524794	1.023580201	1.013768729
0.92	1.046155078	1.033902395	1.022045349	1.012935510
0.93	1.041942017	1.031119755	1.020407545	1.012043672
0.94	1.037484456	1.028146025	1.018645667	1.011080865
0.95	1.032734249	1.024938133	1.016729693	1.010029408
0.96	1.027623448	1.021432662	1.014614324	1.008862349
0.97	1.022050589	1.017529080	1.012225136	1.007534739
0.98	1.015851602	1.013049431	1.009422596	1.005960167
0.99	1.008726930	1.007614146	1.005874179	1.003921693
1.00	1	1	1	1

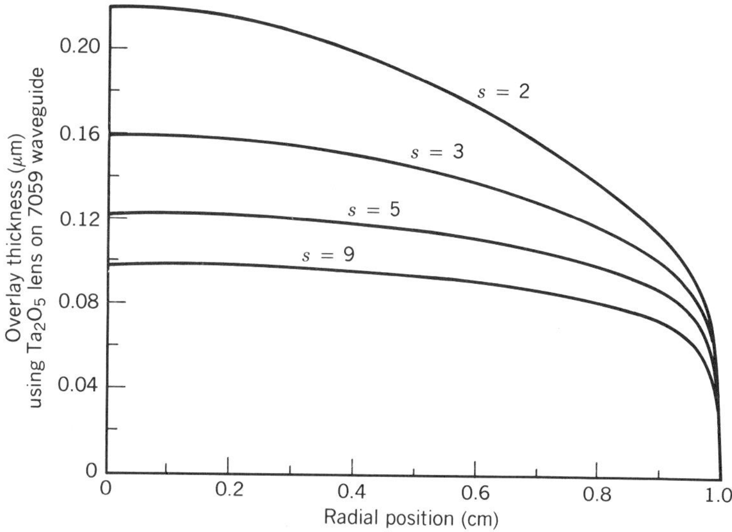

Figure 3.34 Waveguide overlay thickness profile for generalized Luneburg lenses using Ta_2O_5 on Corning 7059 ($d = 1.0665$ μm) on SiO_2, which is tabulated in Table 3.11. (After Ref. 142.)

Once the lens index profile is known, the waveguide mode index of interest must be tailored precisely using index modification techniques. When a high index overlay film is employed for this purpose, the overlay film thickness must be calculated using Eqs. (3.59) and (3.61). For the case of TE modes, Eq. (3.59) is rewritten as

$$ht = \begin{cases} \tan^{-1} \dfrac{h[1-(l/q)\tan(ld)] + (ph/l)[l/q + \tan(ld)]}{(h^2/l)[l/q + \tan(ld)] - p[1-(l/q)\tan(ld)]} & n < n_3 \\[2ex] \tan^{-1} \dfrac{h[(q+l)e^{2ld} + q - l] + (ph/l)[(q+l)e^{2ld} - q + l]}{(h^2/l)[(q+l)e^{2ld} - q + l] - p[(q+l)e^{2ld} + q - l]} & n \geqslant n_3 \end{cases} \tag{3.68}$$

for the TE modes. If the parameters p, h, l, and q are replaced by p/n_1^2, h/n_2^2, l/n_3^2, and q/n_4^2, Eq. (3.68) can be employed for TM modes.

Figure 3.34 gives the calculated Ta_2O_5 overlay film thickness for generalized Luneburg lenses on a 7059 glass waveguide on oxidized silicon substrate. The 7059 glass waveguide has a refractive index of 1.565 and uniform thickness of 1.0665 μm, and the silicon dioxide layer has a refractive index of 1.47. The optical wavelength is assumed to be 0.9 μm. Numerical data are given in Table 3.11. A ray tracing program has been developed for gradient index optical lenses and has been used to evaluate the Luneburg lens with the index profile given in Tables 3.10 and 3.11. Figure 3.35 gives the ray tracing results for a Luneburg lens with $s = 2$, showing good potential performances. For a given thin-film Luneburg lens thickness profile, the focal length may be varied slightly when a thickness error occurs. Figure 3.36 gives the focus sensitivity of a Ta_2O_5 thin-film Luneburg lens over a 7059 glass waveguide using computer simulation programs.

Fabrication of Thin-Film Luneburg Lens. One method of forming thin-film waveguide lenses is by shadowed deposition through an appropriate mask. The RF sputter process has been employed throughout for this purpose. The cross section of the RF sputter

TABLE 3.11 Overlay thickness (μm) for Optical waveguide Luneburg Lenses

r (cm)	$s = 2$	$s = 3$	$s = 5$	$s = 9$
0	0.2191	0.1591	0.1223	0.0979
0.05	0.2187	0.1590	0.1222	0.0979
0.10	0.2177	0.1585	0.1220	0.0978
0.15	0.2160	0.1578	0.1216	0.0975
0.20	0.2135	0.1567	0.1211	0.0972
0.25	0.2105	0.1554	0.1204	0.0968
0.30	0.2068	0.1537	0.1196	0.0963
0.35	0.2024	0.1518	0.1186	0.0957
0.40	0.1975	0.1495	0.1174	0.0949
0.45	0.1920	0.1469	0.1160	0.0940
0.50	0.1860	0.1440	0.1141	0.0930
0.55	0.1794	0.1406	0.1126	0.0918
0.60	0.1722	0.1370	0.1105	0.0905
0.65	0.1646	0.1328	0.1081	0.0888
0.70	0.1563	0.1282	0.1054	0.0869
0.75	0.1472	0.1229	0.1022	0.0846
0.80	0.1374	0.1169	0.0983	0.0817
0.85	0.1262	0.1097	0.0935	0.0708
0.90	0.1130	0.1006	0.0870	0.0727
0.91	0.1099	0.0984	0.0853	0.0713
0.92	0.1067	0.0960	0.0835	0.0697
0.93	0.1031	0.0933	0.0814	0.0679
0.94	0.0993	0.0903	0.0790	0.0657
0.95	0.0949	0.0869	0.0762	0.0632
0.96	0.0898	0.0827	0.0728	0.0601
0.97	0.0835	0.0774	0.0682	0.0559
0.98	0.0749	0.0699	0.0616	0.0499
0.99	0.0597	0.0562	0.0496	0.0395
1.00	0	0	0	0

Note: The waveguide parameters are $n_1 = 1$, $n_2 = 2.1$, $n_3 = 1.565$, $n_4 = 1.47$, $d = 1.0665$ μm, and $\lambda = 0.9$ μm.

deposition sample is depicted in Fig. 3.37*a*. Argon ions are accelerated toward the Ta or Nb cathode target where bombardment releases molecules that are oxidized during their flight until collision. In the sputtering environment employed, the mean-free path length is greater than the thickness of the mask, and thus, effects of collision and diffusion are avoided. A Lambertian distribution has been assumed and verified. Thus, the molecular beam arrives at the mask entrance with a predictable angular momentum distribution $A(\theta, v)$. Where the mask produces a shadow, integration over the particle velocity range provides an angular distribution function $A_d(\theta)$. The distribution function $A_d(\theta)$ was experimentally measured with a knife-edge mask confirming the Lambertian source. Based upon this knowledge, the effect of a shadow mask upon sputtering has been predicted with the aid of a computer [150,151].

The shadow mask computation is relatively simple when the mask has circular symmetry. The coordinates employed to compute the distribution are illustrated in Fig. 3.37*b*. For an arbitrary point on the substrate, P_0, the thickness distribution becomes

$$T(x_0, y_0) = \iint_S \frac{A(\theta)\cos\theta}{R^2}\, dx\, dy \tag{3.69}$$

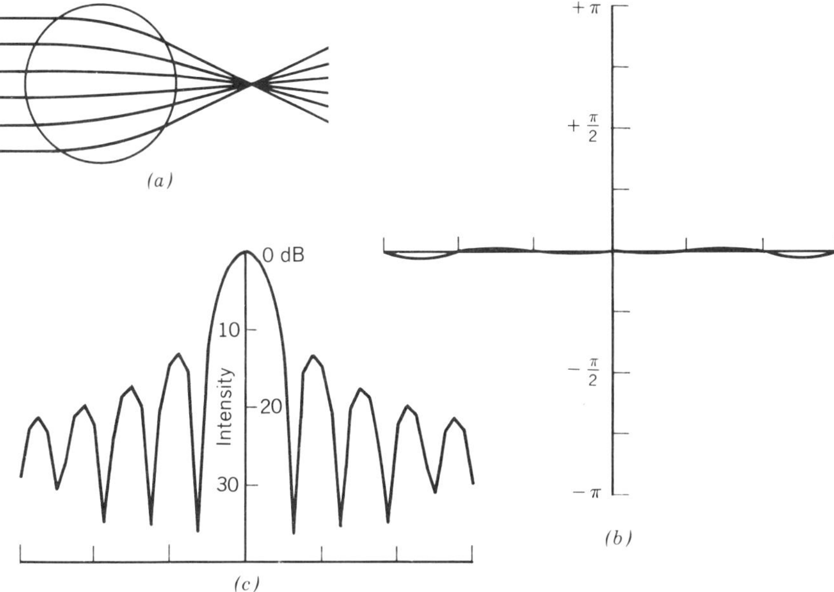

Figure 3.35 (*a*) Ray traces of an $s = 2$ generalized Luneburg lens using the index profile derived from Eq. (3.67), (*b*) the wavefront phase error in the pupil plane, and (*c*) the corresponding intensity diffraction pattern using a logarithmic scale in the image plane. (After Ref. 142.)

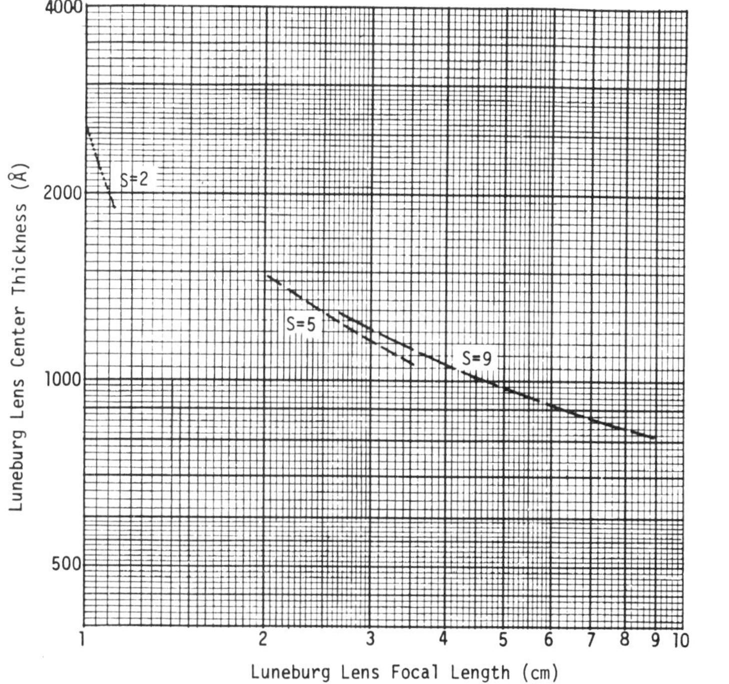

Figure 3.36 Calculated waveguide Luneburg lens focal length for $s = 2$, $s = 5$, and $s = 9$ perfect profiles where a fractional thickness error has been introduced. (After Ref. 142.)

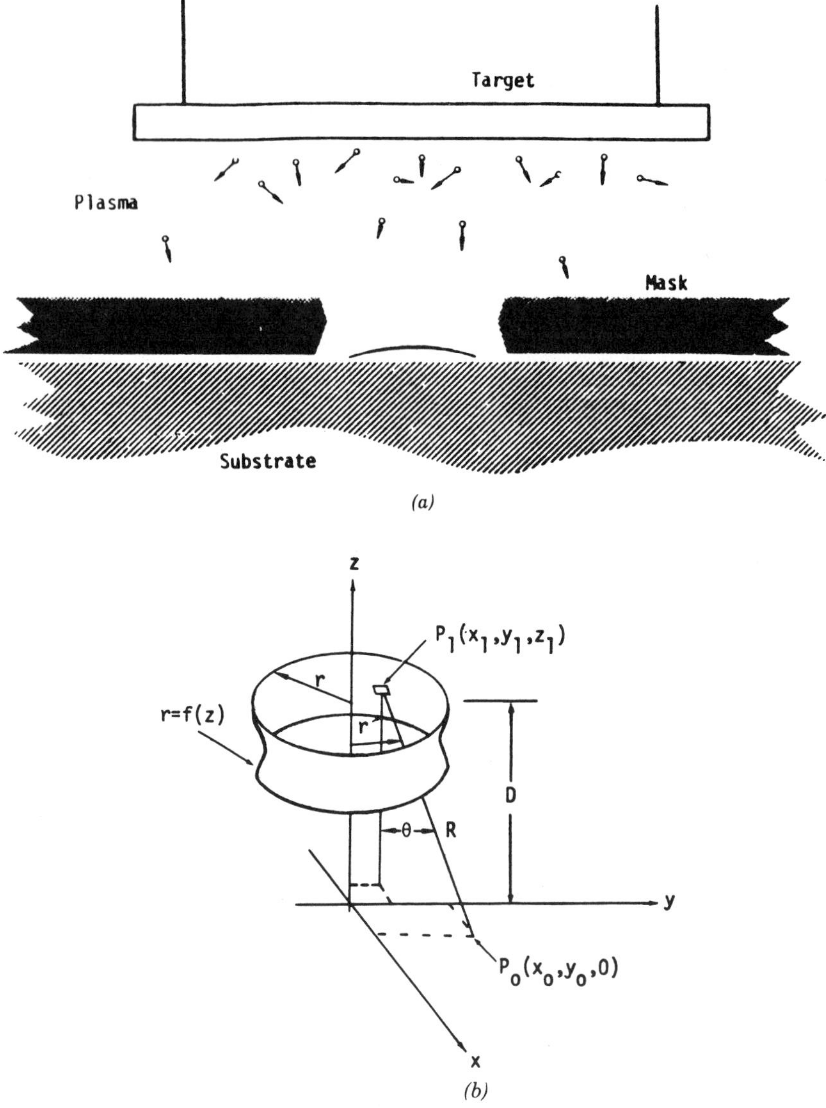

Figure 3.37 (*a*) Cross section of RF sputter deposition chamber used to profile Luneburg lenses and taper waveguide films. (After Ref. 152.) (*b*) Coordinate system for Luneburg lens mask edge synthesis. (After Ref. 152.)

where

$$\theta = \cos^{-1} \frac{D}{\sqrt{(x_1 - x_0)^2 + (y_1 - y_0)^2}} \qquad R = \sqrt{(x_1 - x_0)^2 + (y_1 - y_0)^2 + z_1^2}$$

$dx\,dy$ is an infinitesimally small area of the mask entrance window, and S is the portion of the entrance window that is visible from point P. The distance D in Fig. 3.37*b* extends from the plane containing the top of the mask to the substrate surface.

The determination of the effective window function, S, for a mask contour of a general cylinder is also illustrated by Fig. 3.37. The edge contour is conveniently described by $r = f(z)$. The line of sight between points P_0 and P_1 is given by the solution of

$$\begin{aligned} z_1 y + (y_0 - y_1)z - z_1 y_0 &= 0 \\ z_1 x + (x_0 - x_1)z - z_1 x_0 &= 0 \end{aligned} \tag{3.70}$$

which defines two planes each parallel to the x- or y-axis while containing the line of sight P_0P_1. At any point on this line, the distance to the axis of rotation of the mask is therefore

$$r' = \left[\left(\frac{(x_0 - x_1)z}{z_1} - x_0\right)^2 + \left(\frac{(y_0 - y_1)z}{z_1} - y_0\right)^2\right]^{1/2} \tag{3.71}$$

If the line of sight is not interrupted by the edge of the shadow mask, the following function must not have a real solution:

$$(r')^2 - r^2 = \left(\frac{(x_0 - x_1)z}{z_1} - x_0\right)^2 + \left(\frac{(y_0 - y_1)z}{z_1} - y_0\right)^2 - f^2(z) = 0$$

For $f(z)$ described by quadratic functions, an analytic result may be easily obtained. For $f(z)$ of higher-order functions, numerical evaluation must be employed. When the preceding function becomes greater than or equal to zero, it means that interference with the line of sight is occurring. Thus, the deposition distribution function becomes

$$T(x_0, y_0) = \iint_W \frac{A(\theta)\cos\theta G}{R^2}\,dx\,dy \tag{3.72}$$

where W is the entrance window and G is a logic function that is 1 (or zero) when the line of sight is clear (or interrupted).

With a computer model a conical mask can be computed for a given deposition thickness profile. A cross section of the conical mask and its descriptive parameters are depicted in Fig. 3.38*a*. The effect of changing the conical angle is depicted in Fig. 3.38*b*. The effect of the mask spacing above the substrate is depicted in Fig. 3.38*c*, and the effect of mask aperture diameter is depicted in Fig. 3.38*d*. The accuracy of these computations has been confirmed by a stylus probe, including measurements of the slope and steps for a number of shadow masks.

From these variational effects, depicted in Fig. 3.39, a first-iteration approximation to a Luneburg lens thickness profile was developed using two conical sections, as illustrated by the cross section in Fig. 3.39. The resulting calculated thickness profile is indicated together with an experimentally measured segment near the edge of the deposition. Using the preceding computer simulation programs, the thickness profile is converted to a refractive index profile with ray traces through the lens calculated to find the ray intercept errors at the focus. Figure 3.40 depicts the calculated ray intercepts for the thickness profile of Fig. 3.39 derived from the conical mask. The insert illustrates the ray traces. The experimental points concerning the calculations have been taken from the measured ray traces photographed in Fig. 3.41 using a waveguide Luneburg lens formed with mask SK-1.

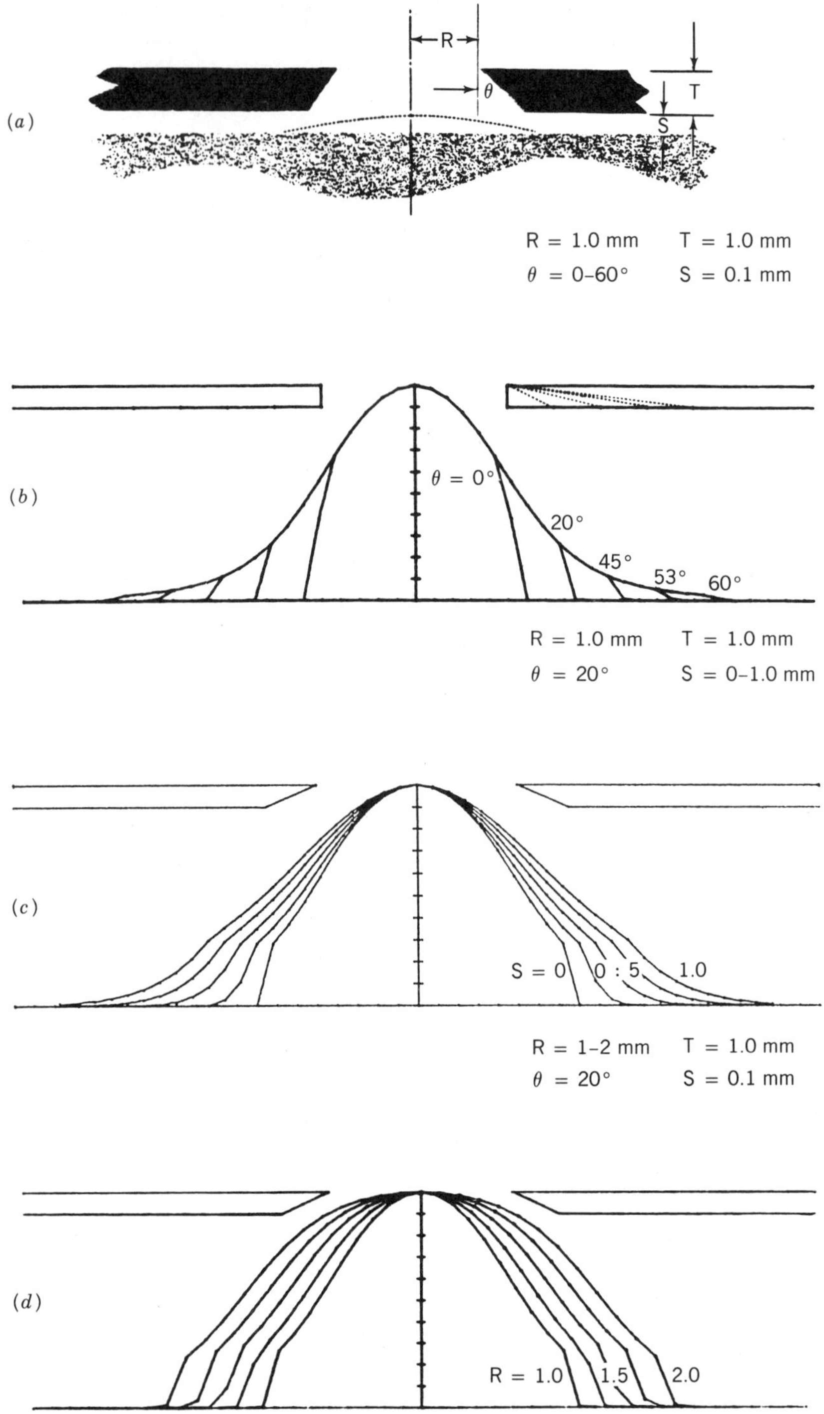

(a)
R
θ
T
S
R = 1.0 mm T = 1.0 mm
θ = 0–60° S = 0.1 mm
(b)
θ = 0°
20°
45°
53°
60°
R = 1.0 mm T = 1.0 mm
θ = 20° S = 0–1.0 mm
(c)
S = 0
0 : 5
1.0
R = 1–2 mm T = 1.0 mm
θ = 20° S = 0.1 mm
(d)
R = 1.0
1.5
2.0

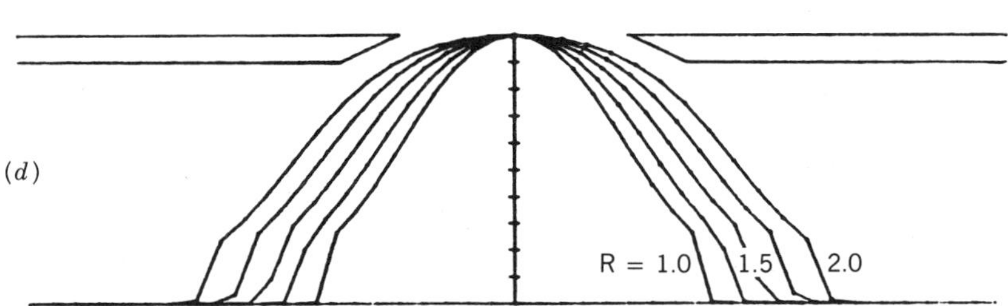

Figure 3.38 Cross section of mask (*a*) and calculated thin-film deposition profiles for variable cone angle (*b*), variable mask spacing above substrate (*c*), and variable mask aperture diameter (*d*).

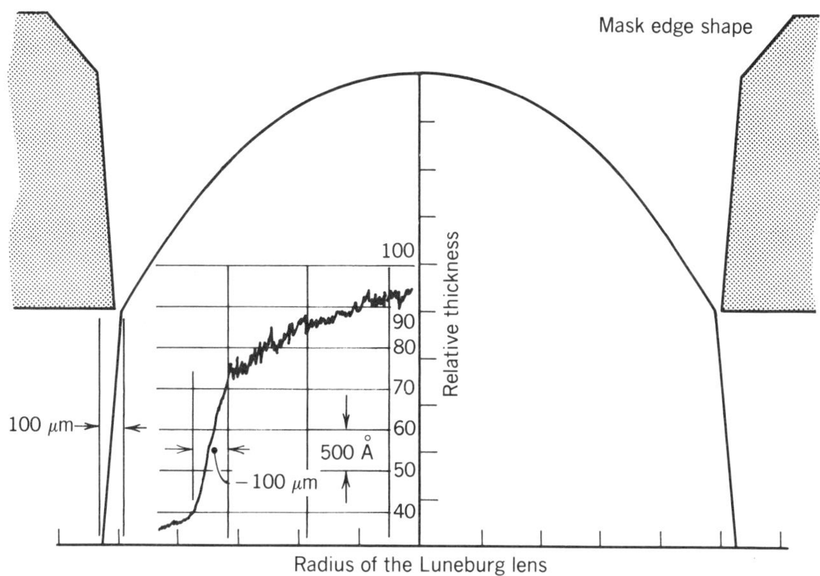

Figure 3.39 Computed Luneburg lens thickness profile for SK-1 mask edge shape indicated with a confirming measured stylus thickness profile. (After Ref. 151.)

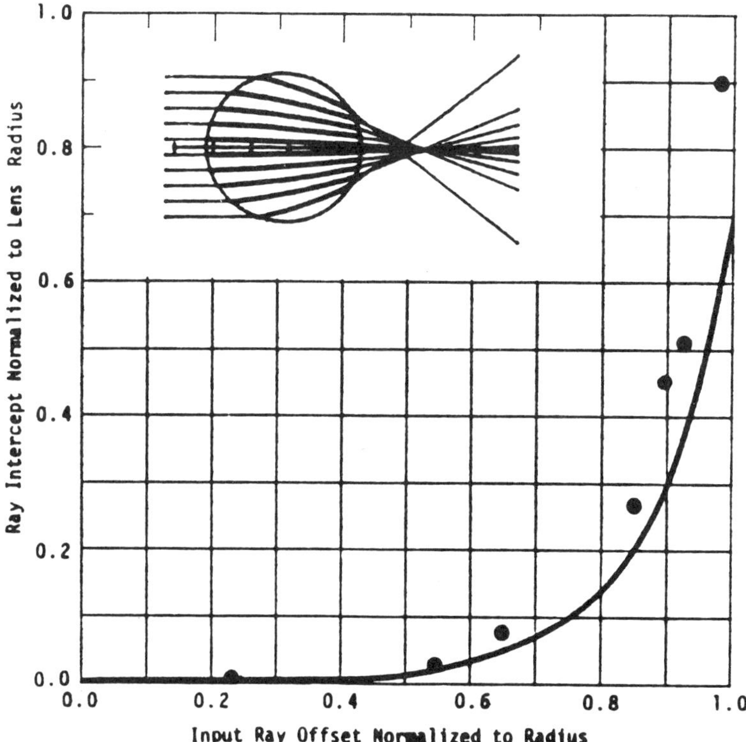

Figure 3.40 Calculated and measured ray intercept for waveguide Luneburg lens using mask SK-1. (After Ref. 151.)

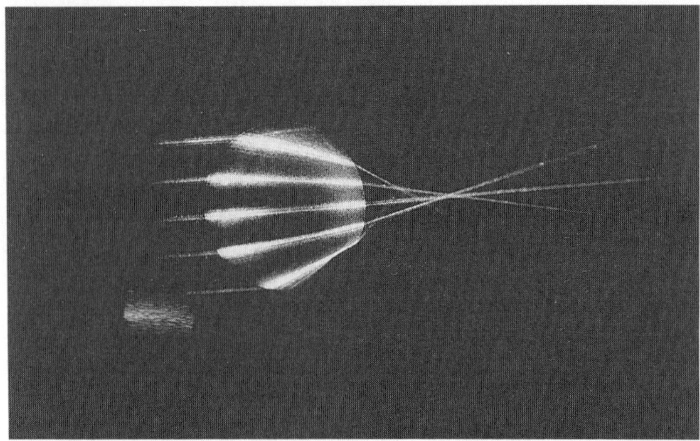

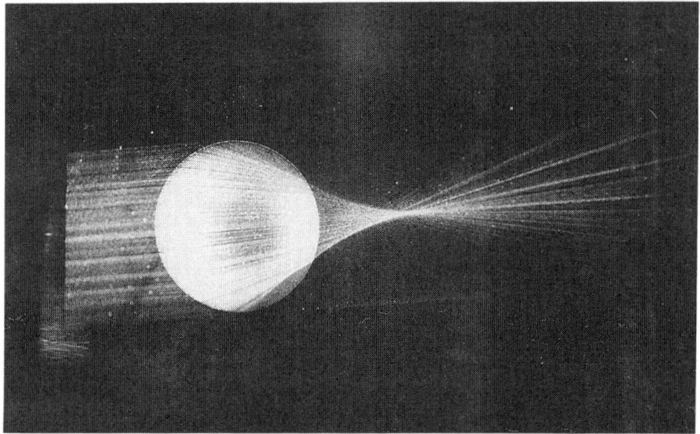

Figure 3.41 Measured ray traces for waveguide Luneburg lens formed with mask SK-1. (After Ref. 151.)

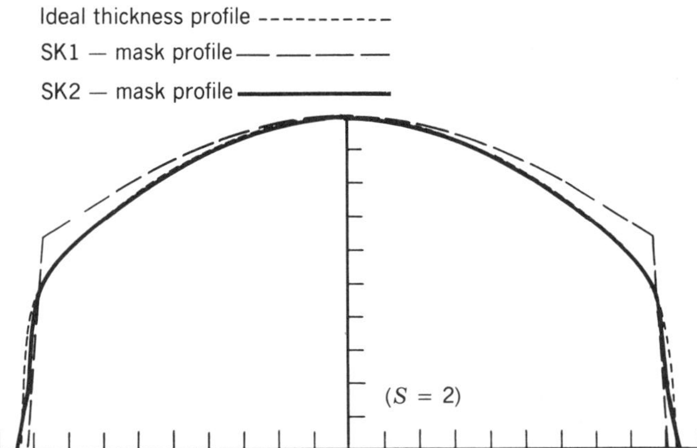

Figure 3.42 Sputtered deposition profiles for waveguide Luneburg lens using computer-generated masks. (After Ref. 152.)

A second mask (SK-2) to more closely approximate the ideal lens thickness profile has been developed that employs 12 conical segments to shape the mask edge. The resulting computed thickness profile is depicted in Fig. 3.42 and compared with the profile generated from mask SK-1 and the Luneburg lens ideal thickness profile. A much closer approximation has been achieved. The computed enlarged cross section of the improved mask SK-2 is depicted in Fig. 3.43. Its thickness is 6 mm and its diameter is 8 mm. This computer-generated profile was transferred directly for fabrication of the mask. Figures 3.44*a* and *b* show the measured focal distribution for thin-film Luneburg lenses made by the 2-section mask SK-1 and the 12-section mask SK-2, respectively. The trace of Fig. 3.44*b* indicates diffraction-limited performance.

Geodesic Lenses. Geodesic [153] waveguide lenses have been known and utilized for many years in microwave applications. In waveguide optics form, a shallow spherical depression is drilled into the substrate before the fabrication of a uniform waveguiding layer. As a guided optical ray enters the depression region, its actual path will follow a geodesic over the curved depression as ascertained by Fermat's principle. Thus, for a spherical depression, this is along an arc of a great circle. When this optical path is projected to the plane of the waveguide, the incident rays exhibit a change in direction at the entrance and exit of the waveguide depression region. The behavior is somewhat analogous to a ray plot of a meridional fan through a conventional lens wherein an abrupt change in the direction of the rays occurs at the lens boundaries. However, although a spherical optical depression is easy to prepare, spherical geodesic lenses behave rather poorly in terms of diffraction-limited performances. One exception is the case of a hemisphere in which all rays from one edge come to perfect focus on the opposite edge. This is the geodesic analog of the Maxwell "fisheye" lens. Unfortunately, the near 90° bend for waveguides at the hemispherical edges is not practical for optical waveguides as it resembles a perfect exit termination.

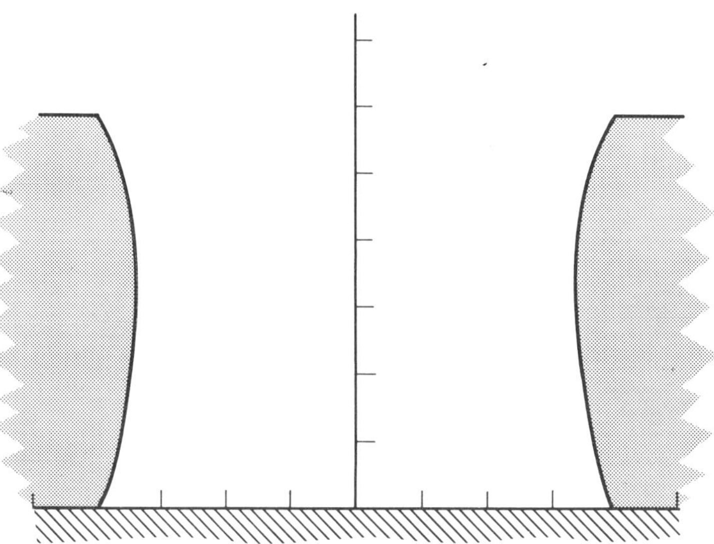

Figure 3.43 Computed cross section of SK-2 mask employed for optical waveguide Luneburg lenses which yield diffraction-limited results (mask thickness 6 mm, diameter 8 mm). (After Ref. 152.)

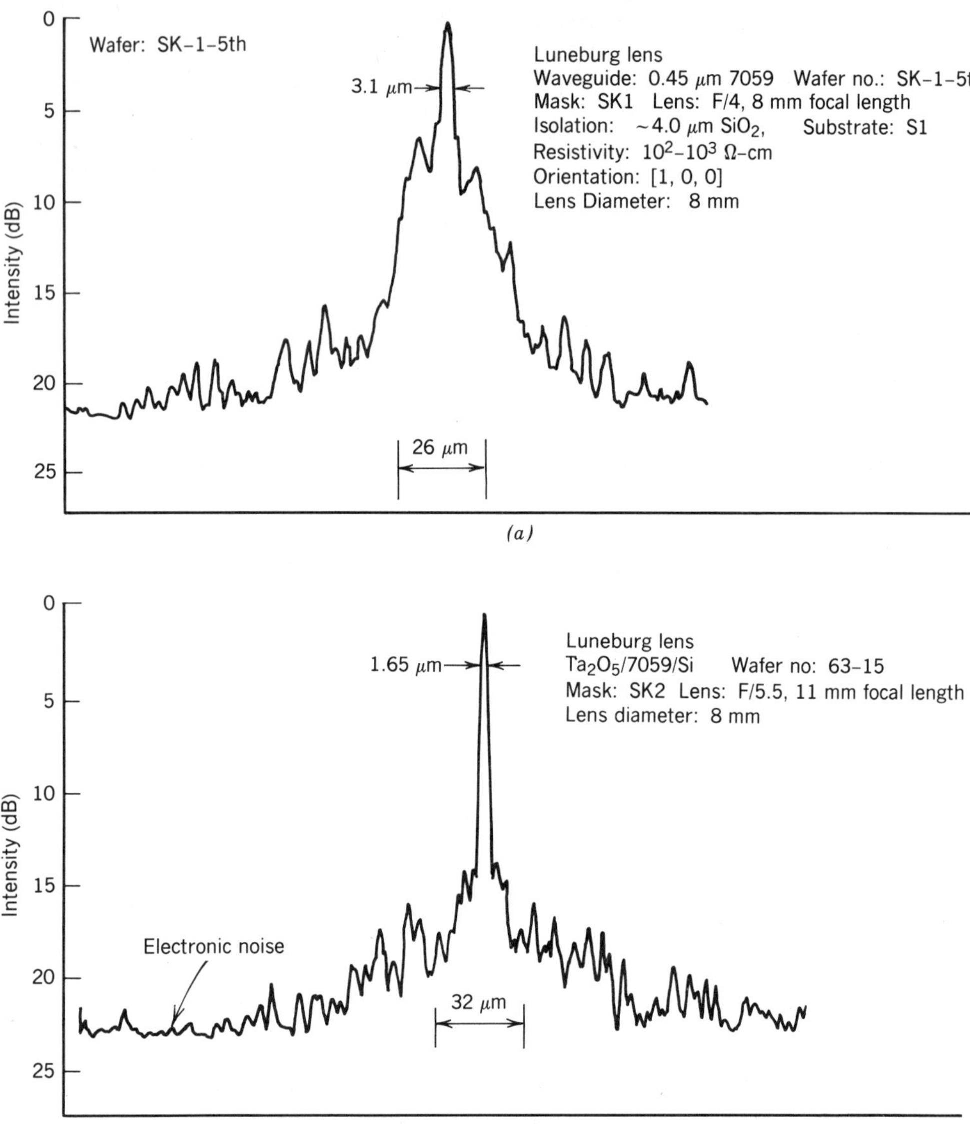

Figure 3.44 Measured diffraction pattern: (*a*) Lens using mask SK-1 and Tropel F/3 reimaging lens: lens, F/4, focal length 8 mm; wafer, SK-1-5th. (*b*) Waveguide Luneburg lens fabricated using mask SK-2 using a cleaved exit face at the focus and Zeiss X40, F/.67 microscope objective showing diffraction-limited performance: lens, F/5.5, focal length 11 mm (expanded-scale insert).

Various approaches have been investigated for the correction of spherical aberration in a spherical geodesic lens [154–161]. A conical transition region [159] and a toroidal transition region [157] have been recommended near the boundary of the spherical geodesic depression. When properly designed, the added geodesic transition region serves to not only improve the mode conversion loss at the lens edges but also compensate for part of the spherical aberration in a spherical geodesic lens. Modification of the waveguide parameters on the two sides of the geodesic boundary can also improve lens performance. Another approach calls for the deposition of a phase front correction overlay thin film at one edge of the geodesic lens. This thin film is to have uniform thickness but a carefully designed area shape. A more elegant approach is to generate an aspherical geodesic surface in analogy of the thin-film Luneburg lens. This aspherical geodesic lens is called a generalized Rinehart–Luneberg lens [153]. The issue is in the fabrication of such aspherical geodesic surfaces.

Spherical Geodesic Waveguide Lenses. Consider a spherical depression with radius of curvature R; the radius of the depression region at the lens contour intersection with the substrate surface, R_c; and the polar angle of the lens, θ, where $R_c = R\sin\theta$. The paraxial focus of the spherical geodesic lens can be determined by trigonometry. The normalized focal length, which is defined as the distance between the focus and the center of the depression measured on the substrate plane normalized to the depression radius R_c, is simply

$$\frac{f}{R_c} = \frac{1}{2(1 - \cos\theta)} \tag{3.73}$$

The normalized paraxial focal length for a spherical geodesic lens as a function of the spherical geodesic half angle is given in Fig. 3.45. Deeper geodesic depression is required for shorter focal lengths. The ray tracing of a spherical geodesic lens is given in Fig.

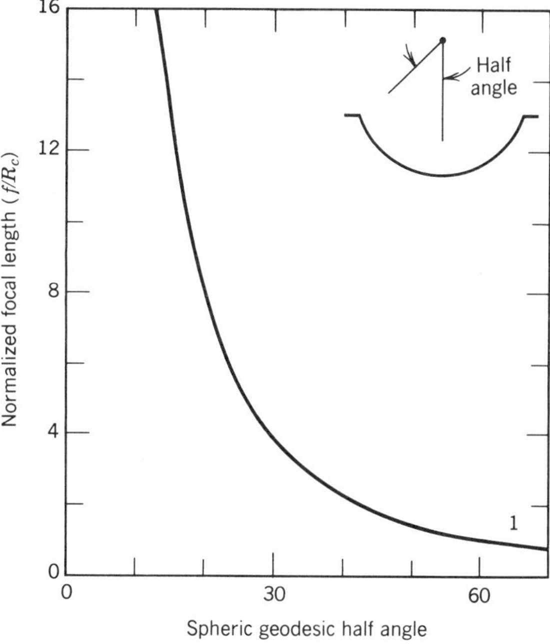

Figure 3.45 Spheric geodesic lens normalized focal length as function of half-angular sector.

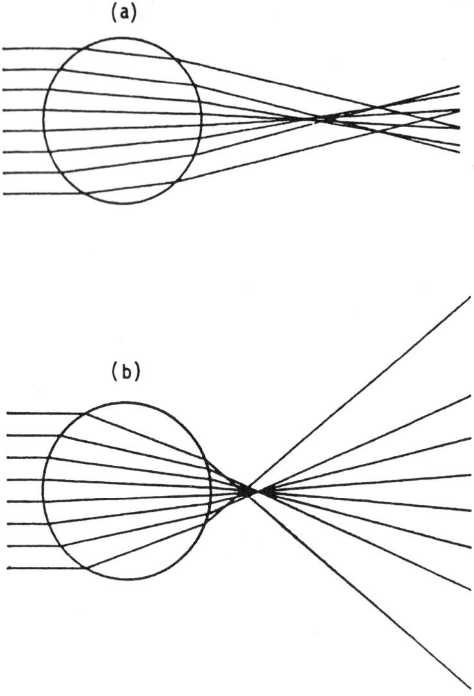

Geodesic Lens Parameters

	(*a*)	(*b*)
Angular radius	39.4°	39.4°
Geodesic radius of curvature	1.575	1.618
Toric rim radius	0	0.1027
Lens radius in waveguide plane	1.0000	1.0634
Angle α	0.6876	0.6660
Paraxial focal length	2.211	1.597

Figure 3.46 Computed ray traces through spheric geodesic lens having (*a*) sharp edge and (*b*) toroidal lip for parameters listed having half-angular width 39.4°.

3.46*a* in which the poor quality of focus is obvious. The rather large longitudinal aberration is displayed in Fig. 3.47 as the curve with $A = 0$.

Transition to the edge of the spherical geodesic lens has been studied for its effect on lens aberrations using ray tracing techniques. Figure 3.48 defines the radius of curvature for a rounded toroidal edge. The ray traces through a spherical geodesic lens with a normalized (relative to R_c) toroidal curvature of 0.1 are given in Fig. 3.46*b*, and the longitudinal aberration is plotted in Fig. 3.47 showing remarkable improvement.

The spherical geodesic lens can also be modified by simply making the optical waveguides on the two sides of the geodesic edge different [155]. Figure 3.49*a* shows the improvement in longitudinal aberration when the effective waveguide index in the geodesic depression region is made slightly higher than the waveguide index in the planar

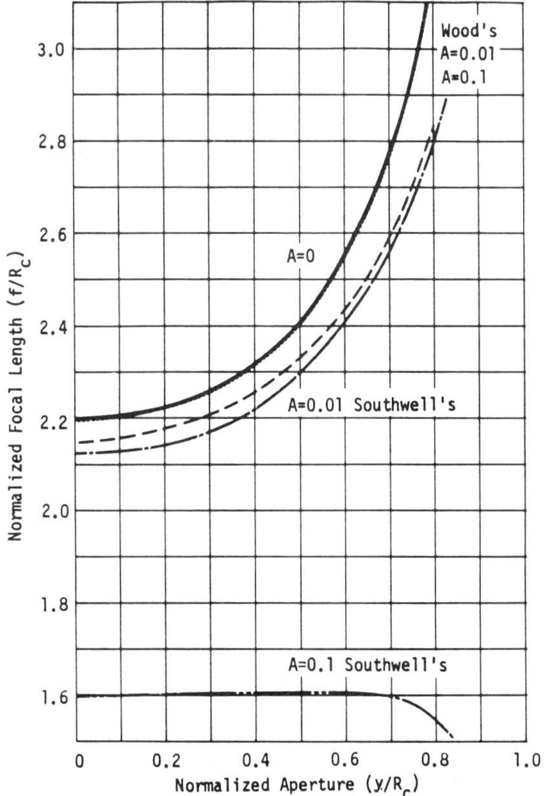

Figure 3.47 Normalized focal length L of the Geodesic lens having a spherical depression half angle $\theta_0 = 39.4$ with different amounts of edge rounding. The quantities L, Y, and A are in units of waveguide lens radius prior to edge rounding. Also shown are the results of Wood's approximation [24]. Note the improved focusing properties of the lens with edge rounding $A = 0.1$.

region. Experimental data are given in Fig. 3.49*b*. Figure 3.50 gives the optimal index ratio for waveguides inside and outside of the geodesic boundary. The lower trace in Fig. 3.50 compares the normalized focal length for an optimized spherical geodesic lens and a regular spherical geodesic lens. As expected, the higher index waveguide in the geodesic region increases the focusing power slightly.

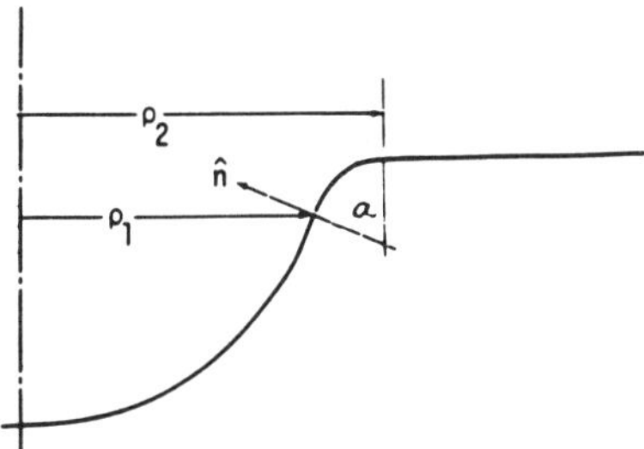

Figure 3.48 Toroidal edge rounding. Note that the slope of the geodesic curve must match the slope of the toroidal region at the point of intersection.

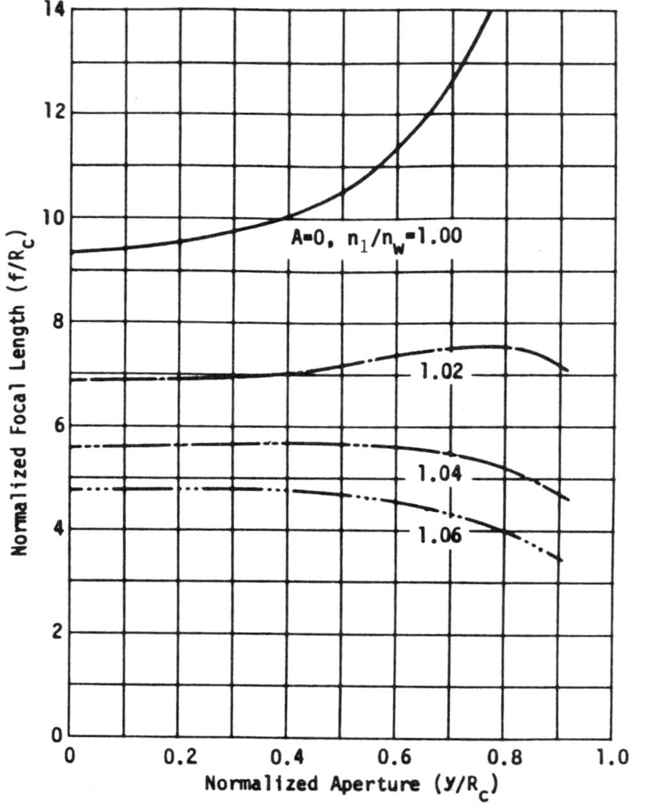

(a)

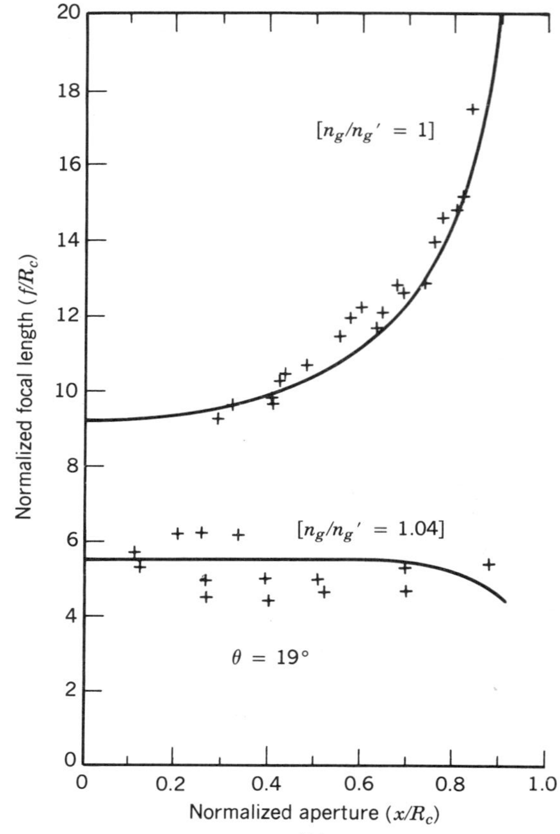

(b)

Figure 3.49 (a) Normalized focal length f/R_c versus normalized distance y/R_c of the incident ray from the axis. The parameter n_1/n_w is the ratio of the mode indices inside and outside of the depression. (After Ref. 155.) (*b*) Normalized focal length f/R_c versus normalized distance x/R_c of the incident beam from the axis for a spherical depression before and after compensation of the aberrations. Points are experimental; curves are theoretical result ($\theta = 19°$, $R_c = 2.5$ mm, $\lambda = 0.633$ μm). (After Ref. 155.)

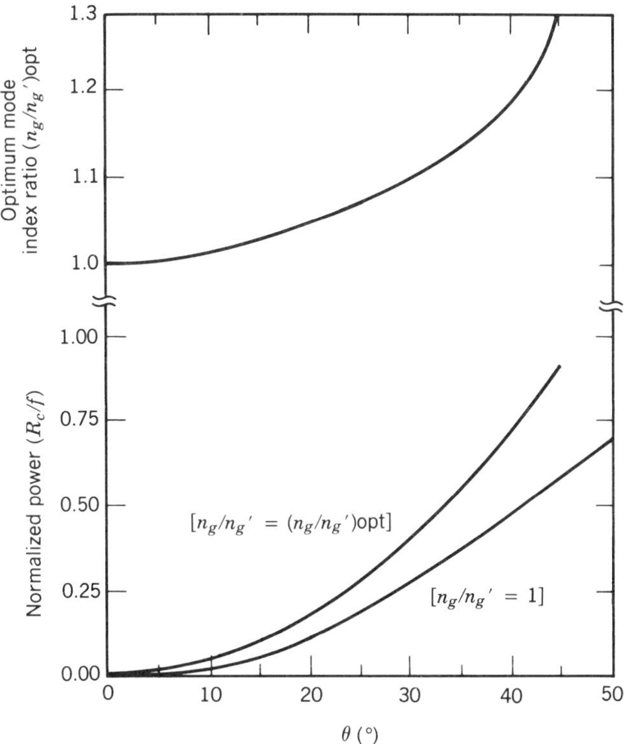

Figure 3.50 Ratio of the mode indices n_g/n_g' inside and outside of the depression for best compensation of the aberration and normalized power R_c/f for an uncompensated $(n_g/n_g' = 1)$ depression lens and a depression lens with optimum compensation plotted versus the angle θ. (After Ref. 155.)

The Generalized Rinehart–Luneburg Geodesic Lens. The modification on the waveguide effective index due to surface curvature can be utilized to generate the effective index profile of a generalized Luneburg lens [153,159,162]. This can be accomplished by letting the optical path length of a ray path over the assumed aspherical geodesic surface be equal to the optical path length of a corresponding ray through the gradient index thin-film Luneburg lens. Since the thin-film Luneburg lens has no surface depression while the geodesic lens has a constant refractive index, the Fermat principle leads to

$$\rho = rn(r) \tag{3.74a}$$

and

$$\frac{ds}{dr} = n(r) \tag{3.74b}$$

where ρ is the radius for an arbitrary point of the geodesic depression projected onto the waveguide plan, $n(r)$ is the refractive index of the Luneburg lens normalized to the waveguide index, and ds is the derivative along the profile of the geodesic curve. The definition of these variables for the geodesic lens are given in Fig. 3.51. From Eq. (3.74a),

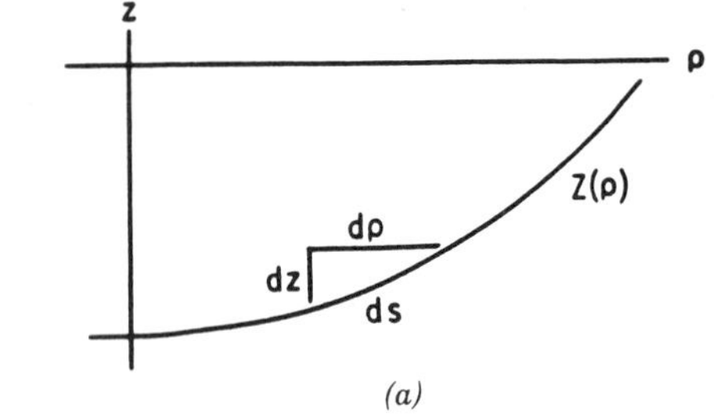

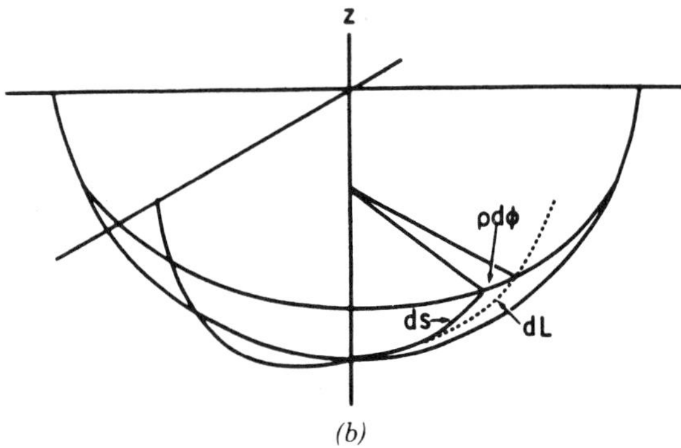

Figure 3.51 (*a*) Profile of waveguide depression showing the generating curve $Z(\rho)$ for the surface of revolution. The geodesic lens is represented by rotating $Z(\rho)$ about the z-axis. (*b*) Waveguide depression showing orthogonal surface infinitesimals ds and $\rho\, d\phi$. Distances between neighboring points on the surface are given by $dL = [ds^2 + \rho^2\, d\phi^2]^{1/2}$. The angle θ is between dL and ds, so that $\sin\theta = \rho\, d\phi/dL$.

Eq. (3.74b), and Fig. 3.51*a*, the geodesic generating function $Z(\rho)$ can be obtained as

$$Z = -Z_0 + \int_0^{\rho} \left[\frac{\rho}{n}\frac{dn}{d\rho}\left(\frac{\rho}{n}\frac{dn}{d\rho} - 2 \right) \right]^{1/2} d\rho \tag{3.75}$$

Equation (3.75) may be evaluated numerically. Table 3.12 gives four sets of the aspherical geodesic lens depression depth values for four normalized focal lengths. From these values, the cross-sectional profiles of these four ideal aspherical geodesic lenses (also called generalized Rinehart–Luneburg lenses) are shown in Fig. 3.52.

Fabrication of Geodesic Lenses. A spherical geodesic lens can be fabricated using universal curve generators similar to the making of spherical lenses. Correction of the spherical geodesic lenses may be done by overlay thin-film deposition on selected areas. A toroidal lip or a conical lip can also be fabricated using conventional polishing techniques. Care must be taken in the depth control [163] because of the sensitivity of focal length to the depression depth, $(df/f)/(dz/z)$, which has been found to be approximately equal to 1.9.

TABLE 3.12 Aspheric Geodesic Lens Depression Depth Values

ρ	$L = 1$	$L = 2$	$L = 3$	$L = 5$
0.0	−0.63261854	−0.41516450	−0.32577595	−0.24632306
0.05	−0.63173418	−0.41362671	−0.32497713	−0.24581597
0.1	−0.62907524	−0.41049277	−0.32302122	−0.24442130
0.15	−0.62462402	−0.40601189	−0.31993946	−0.24213778
0.2	−0.61835026	−0.40021762	−0.31572078	−0.23895513
0.25	−0.61021006	−0.39309853	−0.31034511	−0.23485890
0.3	−0.60014406	−0.38462124	−0.30378519	−0.22983014
0.35	−0.58807494	−0.37473657	−0.29600621	−0.22384467
0.4	−0.57390386	−0.36338075	−0.28696460	−0.21687209
0.45	−0.55750528	−0.35047422	−0.27660619	−0.20887433
0.5	−0.53871973	−0.33591890	−0.26486356	−0.19980375
0.55	−0.51734302	−0.31959338	−0.25165224	−0.18960030
0.6	−0.49311002	−0.30134550	−0.23686495	−0.17818739
0.65	−0.46566912	−0.28098068	−0.22036284	−0.16546551
0.7	−0.43453994	−0.25824265	−0.20196111	−0.15130186
0.75	−0.39903824	−0.23277983	−0.18140406	−0.13551237
0.8	−0.35813042	−0.20408137	−0.15831775	−0.11782775
0.85	−0.31011230	−0.17133889	−0.13210832	−0.09782079
0.9	−0.25174672	−0.13308526	−0.10169755	−0.07471869
0.91	−0.23830090	−0.12454115	−0.09494112	−0.0696055
0.92	−0.2240644	−0.11561444	−0.08789797	−0.06428411
0.93	−0.20889917	−0.10624774	−0.08052637	−0.05872504
0.94	−0.19261701	−0.09636395	−0.07277039	−0.05288894
0.95	−0.17494911	−0.08585453	−0.06455161	−0.04672091
0.96	−0.15548637	−0.07455703	−0.05575321	−0.04013959
0.97	−0.13354614	−0.06220543	−0.04618507	−0.03301353
0.98	−0.10780997	−0.04829728	−0.03549105	−0.0250989
0.99	−0.07490968	−0.03158515	−0.02279784	−0.01580685
0.999	−0.02283366	−0.00837304	−0.00572737	−0.00370575
0.9999	−0.00712019	−0.00245407	−0.00162569	−0.00100000
1	0	0	0	0

Aspherical geodesic lenses are more difficult to fabricate. The two leading techniques are the ultrasonic impact grinder [160] approach and the diamond machining approach. These techniques provide the proper surface contour, and the substrate is subsequently polished and covered by a waveguide. It is time consuming but relatively straightforward.

Figure 3.53 shows experimental data on the normalized focal length for a Rinehart–Luneburg geodesic lens fabricated by an ultrasonic impact grinder on lithium niobate substrate. The focal spot intensity distribution is given in Fig. 3.54 indicating a diffraction-limited spot size.

3.2.3 Waveguide Diffraction Elements

In the section on multilayer step index waveguides, the principles of waveguide index modification using overlay films were described. Modification of the waveguide effective refractive index tends to be small, causing difficulties in fabricating waveguide refractive elements. This is particularly true for graded index waveguides. However, large variations in refractive indices are not required in most diffractive optical elements. In general, the

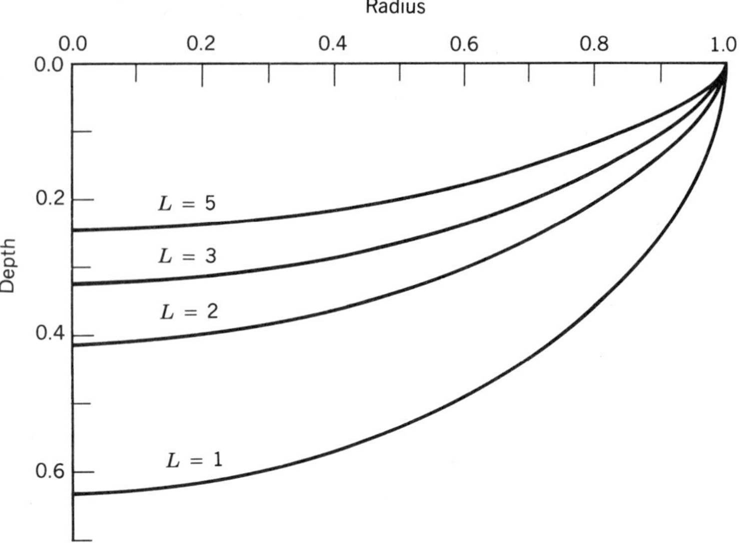

Figure 3.52 Cross section of four designs of generalized Rinehart–Luneburg geodesic lenses. Focal length L is normalized to geodesic edge radius R_c.

maximum optical path length difference in diffractive elements is on the order of micrometers instead of on the order of millimeters for refractive optical elements. On the other hand, the proper operation of diffraction elements requires the use of a highly monochromatic light source, which is generally true in waveguide integrated optics. Therefore, there seems to be a natural compatibility between guided-wave optics and diffraction optical components.

Various types of waveguide optical diffraction elements [164–184] have been reported. Some have been experimentally demonstrated. Some have only been analyzed and published in the open literature. These include lenses, mirror/beam splitters, channel-to-planar waveguide expanders, spatial light modulators, and input/output optical couplers. In this section, attention will be given to lenses and to a lesser extent to mirrors/beam splitters. The waveguide input/output couplers has been discussed before. Channel waveguides and modulators will be discussed in Chapters 4–6. The discussion on waveguide diffraction elements will begin with basic design and fabrication considerations for simple waveguide optical gratings.

Basics of Waveguide Optical Gratings. Waveguide optical diffraction elements are volume gratings of different types. In most cases, these two-dimensional volume gratings have very similar behavior as conventional volume gratings. However, due to the uniqueness of planar optical waveguides, there are several special features. First, gratings in optical waveguides can be made in such a way that the grating phase vector is colinear to the optical phase vector, an arrangement more like an optical interference filter than ordinary gratings. Second, these volume gratings can easily be accessed and prepared from the waveguide surface. This feature allows the use of sophisticated planar fabrication technologies such as photo- or E-beam lithography in making grating devices. Third, guided modes can only be manipulated gracefully or it will leak into the substrate. This

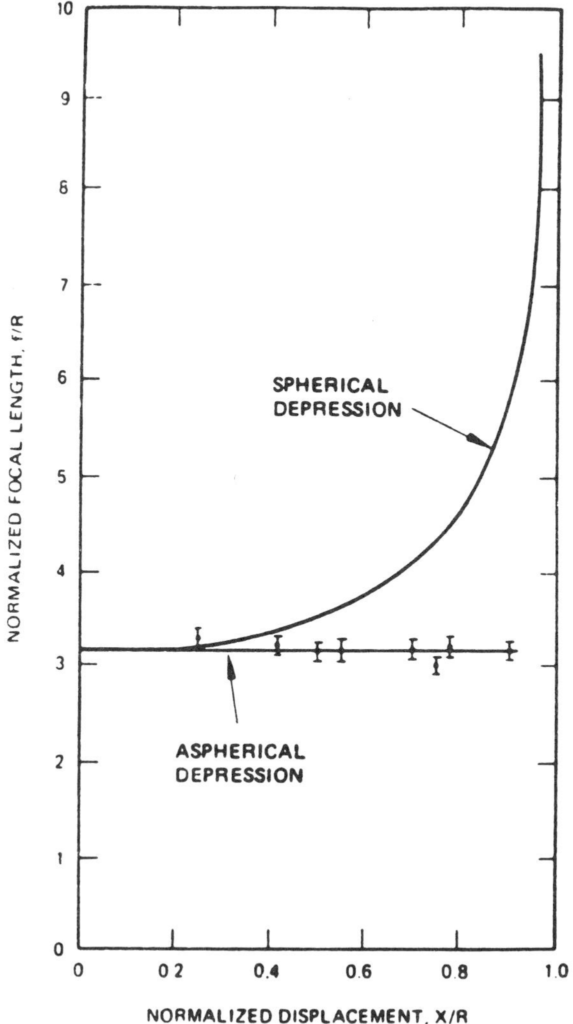

Figure 3.53 Measured focal distance versus lateral displacement. (After Ref. 160.)

last feature is very important since it severely limits the design and application of diffraction elements on many integrated optical waveguides [181].

Consider the phase-matching diagram shown in Fig. 3.55. Consider also a grating structure that has a constant periodicity Λ. The phase-matching condition for efficient volume interaction between the incident guided-wave beam and the diffracted guided-wave beam is satisfied for a given transverse section when the periodicity Λ of the grating grooves satisfies the condition $K = 2n_{\text{eff}}k_0 \sin(\theta/2)$, where $K = 2\pi/\Lambda$ and θ is the total angle of deflection. However, if the projection of the K vector on the direction of the incident beam, $K \sin(\theta/2)$, is large enough such that $K \sin(\theta/2) \geqslant (n_{\text{eff}} - n_s)k_0$, then the incident guided-wave beam will also be coupled to substrate modes that have a propagation wave number equal to or smaller than $n_s k_0$. This may happen despite the fact that the phase-matching condition to the substrate mode may be violated in the direction

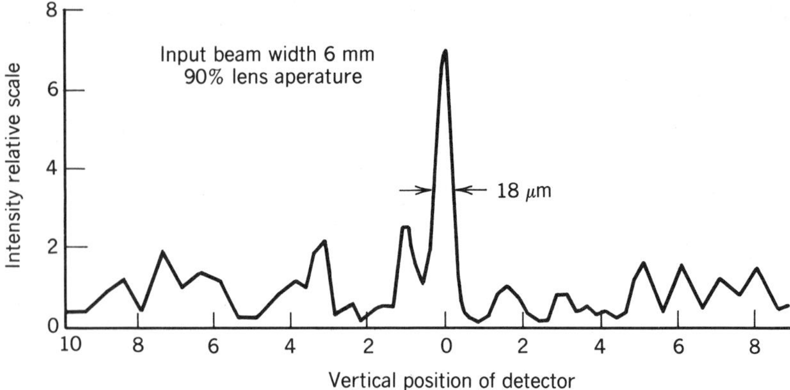

Figure 3.54 Intensity distribution at focal plane for a He–He input beam with 6 mm width. (After Ref. 160.)

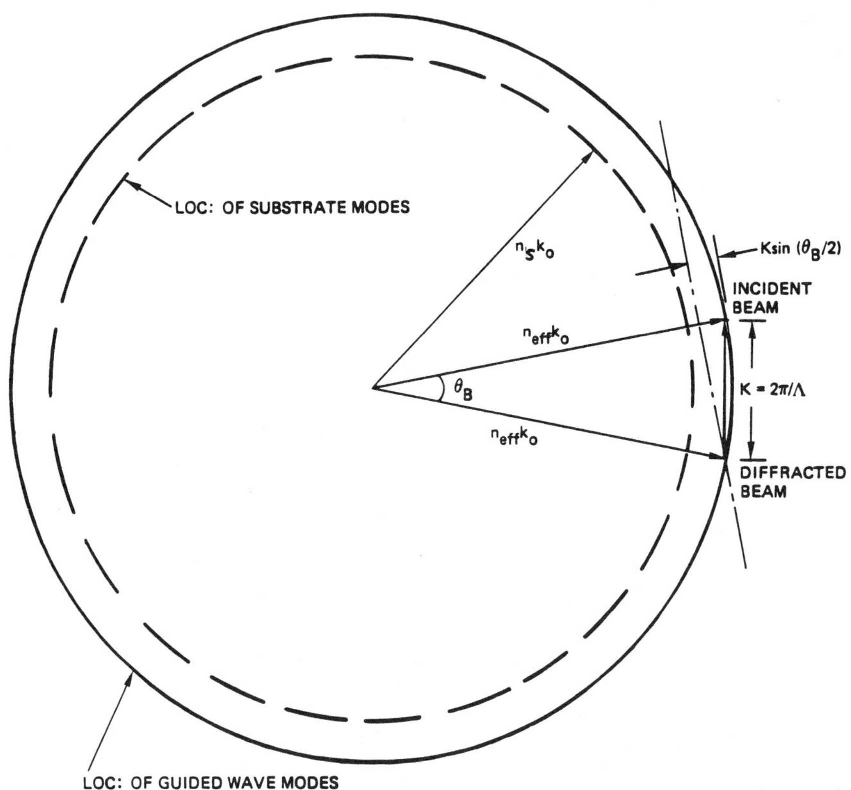

Figure 3.55 Phase-matching diagram for the incident guided-wave beam, the diffracted guided-wave beam, and the substrate modes.

perpendicular to the direction of the incident beam because each transverse section is very narrow, creating only a very broadband phase-matching condition in the perpendicular direction. Thus, the condition for avoiding the excitation of substrate mode is the maximum grating diffraction angle

$$\theta_M \leqslant \sqrt{\frac{2(n_{\text{eff}} - n_s)}{n_{\text{eff}}}} \tag{3.76}$$

However, it is believed that this effect will be dependent on optical arrangements and optical beam size. For collimating lenses, this condition implies a limitation on how small an f-number may be used without significant reduction of the diffraction efficiency due to coupling into the substrate modes. For step index waveguides, the maximum diffraction angle can be tens of degrees. However, for graded index waveguides such as metal-diffused lithium niobate, the maximum angle can be less than 10°.

Waveguide Diffraction Lenses. Waveguide diffraction lenses can be made as either a Fresnel lens [165,166] or a chirp grating lens [164]. The Fresnel lens is simply a waveguide optics analogy of the Fresnel zone lens. The basic principle of a zone lens will not be repeated here. Figures 3.56*a* and *b* illustrate a digital zone lens and an analog zone lens, respectively, on planar waveguide. The location of each zone boundary is given by, for the *m*th zone,

$$S_m = \sqrt{\frac{m\lambda f}{n_{\text{eff}}}} \tag{3.77}$$

where λ/n_{eff} is the optical wavelength in the waveguide and f is the focal length of the zone lens. The digital zone plate has a number of phase shifters, and the output optical phase front usually decomposes into a number of cylindrical components each focusing to a focus. There are convergent as well as divergent wavefronts. If this zone lens is employed to focus a laser beam, the efficiency will be poor. However, most of the unwanted focal spots may be eliminated when the analog zone lens in Fig. 3.56*b* is employed and when the maximum phase shift is 2π. In both cases, the zone plate has its smallest features at the two ends of the optical aperture. The smallest feature line width can be derived from the equation

$$\begin{aligned} d &= S_m - S_{m-1} \\ &\cong \frac{F\lambda}{n_{\text{eff}}}, \end{aligned} \tag{3.78}$$

which equals the diffraction-limited focal spot of this zone lens. That is, the ideal focal spot size equals the smallest feature size of a waveguide zone lens.

So far, all Fresnel lenses have been made on waveguides that have low index substrates. Their performances are listed by Table 3.13.

Chang and Ashley [186] obtained Fresnel zone lenses on BaO glass waveguides ($n_f = 1.55$, $n_5 = 1.512$). The phase shift zone pads are obtained by depositing high-index CeO ($n^1 = 2$) on the waveguide. The typical width of the lenses are from 25 to 50 μm, and 23% diffraction efficiency, 3 μm spot size, $F = 5$, 15° angular field of view, and 18 dB of signal-to-noise ratio have been obtained experimentally.

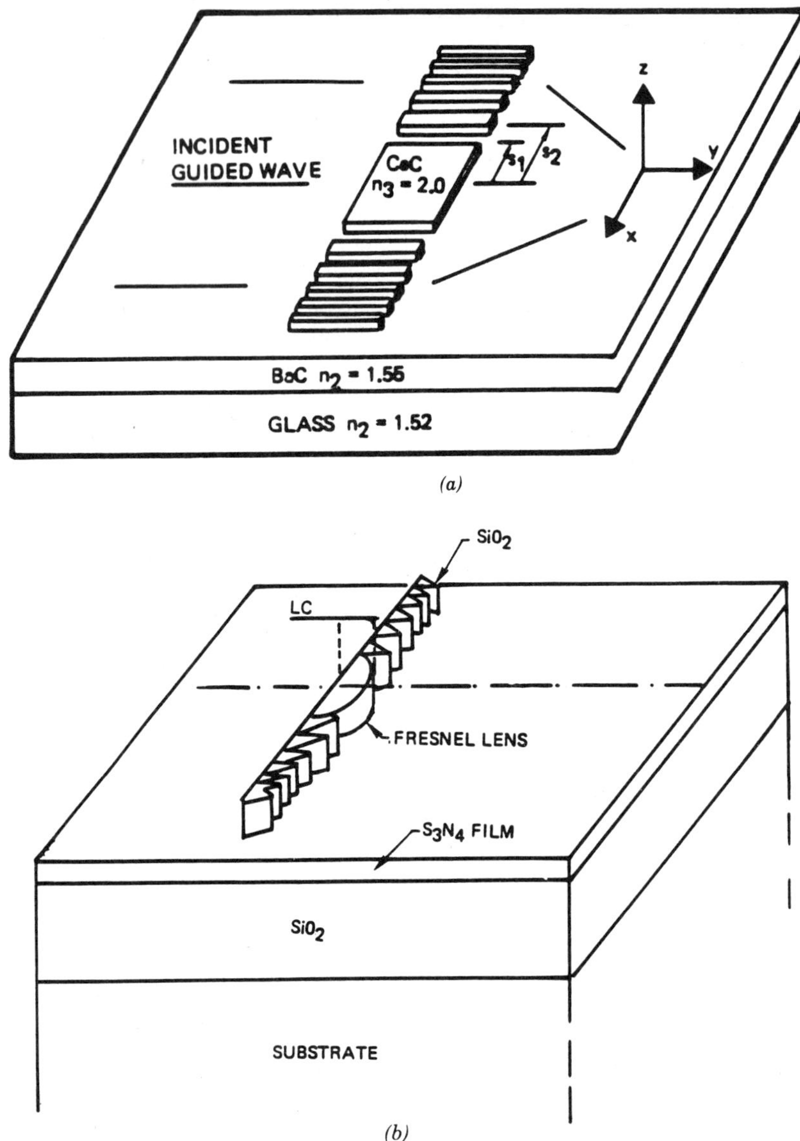

Figure 3.56 (*a*) Diagram of a surface-deposited Fresnel zone lens in a thin-film waveguide. (After Ref. 166.) (*b*) Schematic diagram of waveguide analog Fresnel lens. (After Ref. 185.)

The side lobe was limited to -12 dB of the main lobe because of the truncation of the incident beam by the finite aperture size of the lens.

Valette et al. [185] made graded index Fresnel lenses on Si_3N_4/SiO_2 waveguides where the phase shift is obtained by deposition of an SiO_2 film with a prescribed pattern, as shown in Fig. 3.56*b*. The width of the lens is 20 μm. The focused spot size is 3.6 μm, and the efficiency is 60–70%. The signal-to-noise ratio for a Gaussian incident beam was -27 dB at 150 μm off-axis ($\cong 1°$).

TABLE 3.13 Experimental Results on Grating Lenses for Guided-Wave Optics

Lens Type	Waveguide/ Substrate Material	Lens Fabrication Technique	Grating Period Λ (μm)	Groove Depth or ΔN	Focal Length f (mm)	Grating Interaction Length (μm)	Focal Spot Size (μm)	f-Number	Acceptance Angle (degrees)	Signal-to-Noise Ratio (dB)	Diffraction Efficiency (%)	Reference
Analog Fresnel lens	Si_3N_4 film on Si–SiO_2	SiO_2 over layer pattern etched	—	4000 Å	8.5	20	3.6	—	10	27	60–70	185
Graded index Fresnel lens	As_2S_3 film on Si–SiO_2	Direct electron beam writing	—	0.05	5.0	20	10.0	5	—	20	48	172
Digital Fresnel lens	Si_3N_4 film on Si–SiO_2	SiO_2 over layer pattern etched	—	4000 Å	10.2	10	4–5	8.5	—	8	19	165
Digital Fresnel lens	BaO layer on glass	CeO over layer pattern etched	—	300 Å	4.0	25–100	3.4	2.5–5	15	18	23	186
Chirp grating lens	Corning 7059 glass on Si–SiO_2	Photolithography and plasma etch	3.4–6.8	400 Å	32.0	150	6.1	16	—	−15	90	164
Chirp grating lens	Corning 7059 glass on pyrex	E-Beam lithography and chemical etch	3–100	3000 Å	20.0	3 mm	—	—	—	—	—	167

Handa et al. [172] have fabricated graded index Fresnel lenses (F/5, 1 mm aperture) in amorphous As_2S_3 waveguides on SiO_2 substrates. The phase shift pattern was written directly into the As_2S_3 waveguide by direct electron beam exposure of the As_2S_3. The maximum index change in As_2S_3 that can be obtained by this method is $\Delta n \sim 0.06$. By keeping the electron beam current and the electron beam scanning speed fixed, the number of the scanning repetition on a line was varied to give the appropriate dose distribution for obtaining the desired index variation. Nearly diffraction-limited focusing characteristics (10 μm spot size) and an efficiency of 48% have been experimentally obtained.

Although these results are impressive, Fresnel zone lenses have a number of problems.

1. Unless a graded analog zone lens is employed, the large number of focal orders means low efficiency as well as high background noise.
2. With graded analog zone lenses, the high efficiency demands a maximum phase shift equal to 2π; otherwise efficiency as well as noise background will deteriorate.
3. They demand a large index change in the phase elements so that the phase elements can be made sufficiently short to maintain a wide acceptance angle.

Another waveguide diffraction lens is the chirp grating lens [164] of Fig. 3.57*b*. The device is a thick grating operating in the Bragg diffraction regime. The grating has a variable diffraction angle over its aperture due to variable grating periodicity. A key feature is the off-axis focusing behavior that separates the focused beam from other grating diffraction orders. As a result, it provides low background noise and insensitivity to fabrication variations and can be quite efficient. The disadvantages of chirp grating lenses when compared with zone lenses are the need of a smaller line feature width for a given focal spot size, smaller numerical aperture, and narrower field of view. A brief discussion of the theory for chirp grating lenses follows.

To understand the operation of a grating lens, one can start with a simple uniform grating. The angle of deviation between an arbitrary grating diffraction order m and the incident optical beam can be obtained from

$$\theta_d = \sin^{-1}\left(\frac{m\lambda}{\eta\Lambda} - \sin\theta_i\right) \tag{3.79}$$

$$\theta_m = \theta_i + \theta_d \approx \frac{m\lambda}{\eta\Lambda} \tag{3.80}$$

where λ is the optical wavelength, n is the index of refraction, Λ is grating periodicity, and angles θ_i and θ_d are given in Fig. 3.57*b*. The error in this approximation is less than 1% when the angle is less than 15°. If the grating spatial frequency $p(x) = 1/\Lambda(x)$ has a nearly linear chirp, the planar optical wave can be diffracted into a converging cone similar to the focusing properties of a thin lens. In practice, the grating spatial frequencies $p(x)$ may be adjusted to compensate the phase front error. To focus a collimated light beam as shown in Fig. 3.57*b*, the following must be satisfied ($\theta_i = \text{const}$):

$$\theta_d(x) = \tan^{-1}\frac{x}{f} \tag{3.81a}$$

$$\theta_m(x) = \theta_i + \tan^{-1}\frac{x}{f} \tag{3.81b}$$

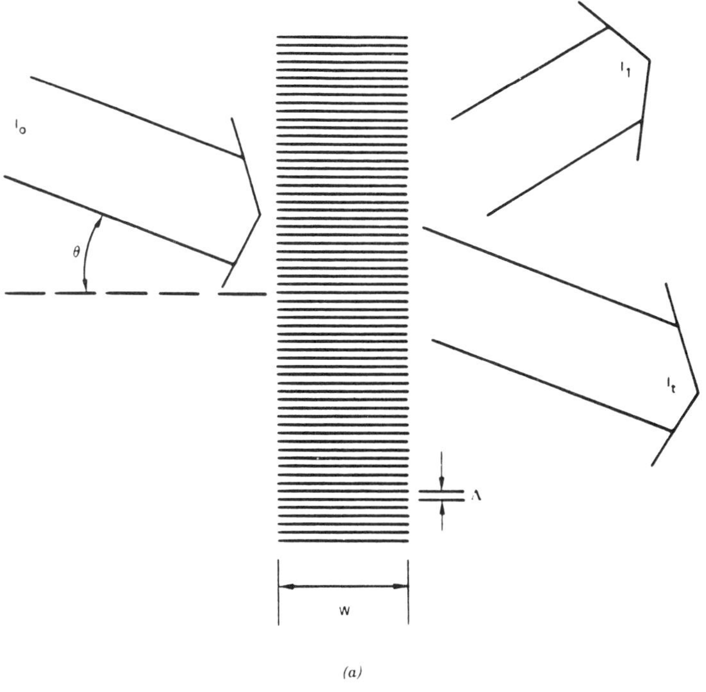

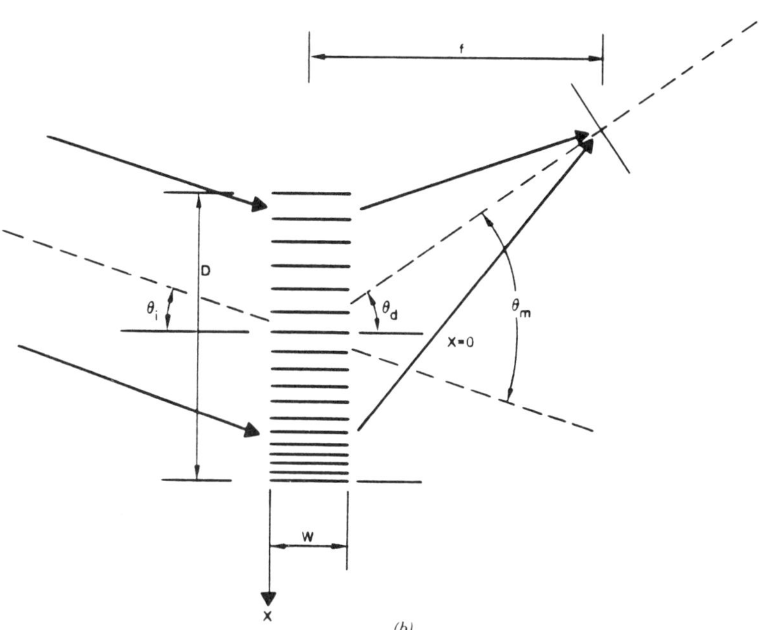

Figure 3.57 (*a*) Uniform grating and (*b*) chirp grating lens.

The range of x is related to the lens aperture and the f-number. Equation (3.79) can be rewritten for grating periodicity as

$$\Lambda(x) = \frac{m\lambda}{2n\sin[\frac{1}{2}\theta_m(x)]\cos[\frac{1}{2}(\theta_i - \theta_d(x)]} \tag{3.82}$$

which in small-angle approximation is linear to x as shown in the following:

$$p(x) = \frac{n}{m\lambda}\left(\theta_i + \frac{x}{f}\right) \tag{3.83}$$

Figure 3.58 gives the grating periodicity computed from Eq. (3.82). The similarity between Eq. (3.82) and an optical interferogram allows the use of a holographic recording process in making this grating lens pattern. However, in such a case, the grating periodicity will be correct only if the recording wavelength in the recording medium is the same as the reconstruction wavelength in the optical waveguide. The recording must be performed in an index matching fluid of proper refractive index.

The lens properties of a chirp grating can be determined by the grating diffraction angles at the extremeties of the optical aperture, and the results are

$$f\text{-number} = \frac{n\Lambda_{\max}\Lambda_{\min}}{\lambda(\Lambda_{\max} - \Lambda_{\min})} \tag{3.84}$$

$$\text{Focal spot } \delta \cong \frac{F\lambda}{n} = \frac{\Lambda_{\max}\Lambda_{\min}}{(\Lambda_{\max} - \Lambda_{\min})} \tag{3.85}$$

$$\text{Focal length } f = \frac{W}{2} + \frac{D}{\tan(\theta_d)_{\max} - \tan(\theta_d)_{\min}} \tag{3.86}$$

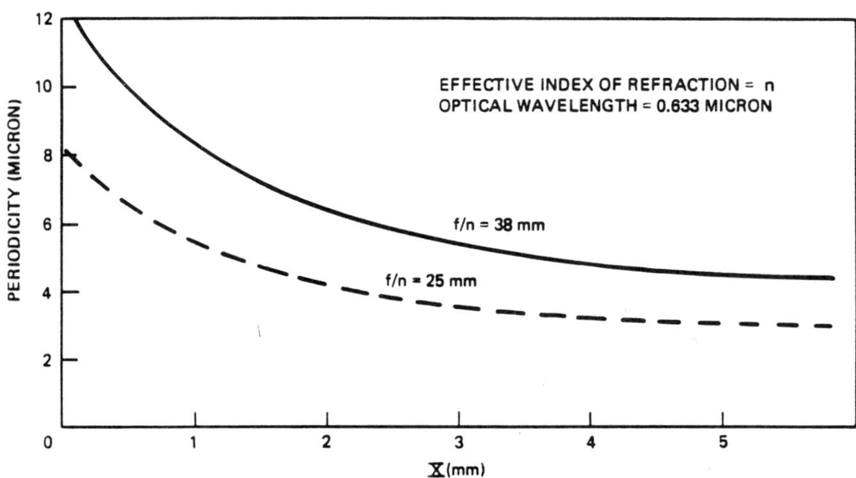

Figure 3.58 Distribution of grating periodicity for two examples of chirp grating lenses. Focal lengths f/n are 28 and 18.6 mm at 0.85 μm optical wavelength. (After Ref. 187.)

where D is the optical aperture of the grating lens. The focal length can also be described by the chirp rate as

$$f \approx \frac{n/\lambda}{dp/dx} \tag{3.87}$$

In many practical chirp grating lens designs, the special case of $\Lambda_{max} = 2\Lambda_{min}$ corresponding to the octave spatial bandwidth is employed. Thus, Eqs. (3.84) and (3.85) can be simplified to

$$f\text{-number} = \frac{n\Lambda_{max}}{\lambda} \tag{3.88}$$

$$\text{Focal spot} = \Lambda_{max} \tag{3.89}$$

Since the optical properties of the chirp grating lens depends only on line geometry, it is very insensitive to fabrication process variations and it is highly predictable.

From small-angle approximations, the focal loci of the chirp grating lens can be determined as the loci of the constant-output beam convergence angle shown in Fig. 3.59. Note that the focal loci is slightly tilted due to the off-axis nature of a chirp grating lens. However, this tilt is small since for all practical applications, all angles in a grating waveguide lens must be small to avoid substrate mode conversion. For a grating lens of 10° bend on titanium-diffused lithium niobate waveguides, the smallest focal spot size is approximately 2 μm at the He–Ne laser wavelength.

A very important feature of the chirp grating lens is the separation of the higher-order diffractions from the optical axis of interest. It can be shown using simple geometry that the clear angular field (clear from spurious responses) for the case of $\Lambda_{max} = 2\Lambda_{min}$

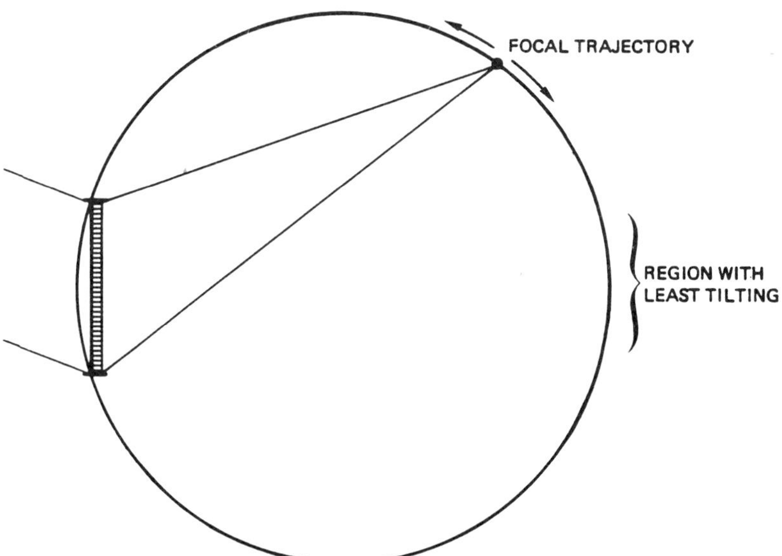

Figure 3.59 Focal plane (trajectory) of a diffraction lens (Fresnel as well as chirp grating).

is related to other lens parameters as

$$\text{Clear angular field} = \frac{\lambda}{n\Lambda_{\max}} = \frac{1}{f\text{-number}} \tag{3.90}$$

$$\text{Clear spatial field} = D \tag{3.91}$$

$$\text{Resolution elements} = \frac{D}{\Lambda_{\max}} \quad \text{(pixels)} \tag{3.92}$$

The spatial resolution of such a lens is simply the size of the lens aperture divide by the maximum line width.

Practical waveguide grating lenses have long interaction phase element lengths due to the small index perturbation achievable on optical waveguides. These devices tend to operate in the Bragg diffraction regime like thick gratings. As is known in conventional optics, thick gratings can have very high diffraction efficiency, approaching 100% over a restricted field of view. Therefore, the application of chirp grating lenses must be accompanied with careful design to ensure satisfaction of the Bragg incidence angle condition at the center of the chirp grating. Thus,

$$\theta_i = \sin^{-1} \frac{\lambda}{2n\Lambda_0} \tag{3.93}$$

where

$$\Lambda_0 = \frac{2\Lambda_{\max}\Lambda_{\min}}{\Lambda_{\max} + \Lambda_{\min}} \tag{3.94}$$

Due to the variable grating periodicity in a chirp grating lens, residual angle mismatch exists over the optical aperture:

$$\begin{aligned} \Delta(x) &= \theta_B(x) - \theta_i \\ &\cong \frac{\lambda}{2n}\left[\frac{1}{\Lambda(x)} - \frac{1}{\Lambda_0}\right] \end{aligned} \tag{3.95}$$

Note that the maximum angle mismatch equals $1/4F$. The efficiency of a thick grating with finite angular mismatch is given by

$$\frac{I_d}{I_0} = \frac{\sin^2 \pi\sqrt{(n'W/2\lambda)^2 + (W\Delta/\Lambda)^2}}{1 + (2\lambda\Delta/n'\Lambda)^2} \tag{3.96}$$

where W is the grating thickness (interaction length), n' is the magnitude of index perturbation, and Δ is the angular mismatch, which is a function of x in a chirp grating lens. Figure 3.60 gives the diffraction efficiency of a thick grating as a function of normalized angular mismatch and as a function of normalized optical index modulation over the length of a phase element. It is observed that maximum efficiency occurs when the accumulated index perturbation (i.e., the phase perturbation) $n'W$ equals one optical wavelength and that angular mismatch as high as $0.4\Lambda/W$ may be tolerated. These observations will be extremely valuable for the design of a chirp grating lens with parallel straight-line phase elements.

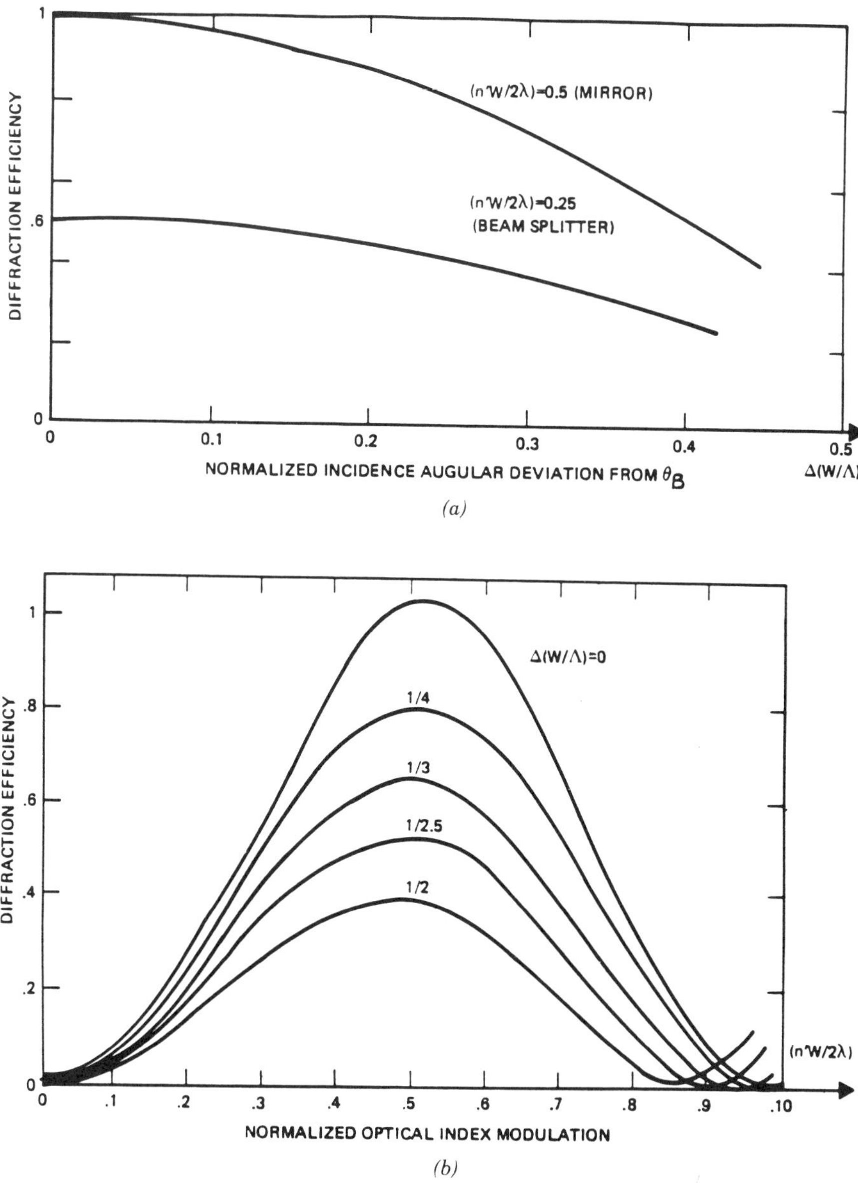

Figure 3.60 Thick grating diffraction efficiency as function of (*a*) incident angle deviation and (*b*) phase perturbation.

However, since the chirp grating lens can be produced by advanced lithographic means, the phase elements do not have to be parallel straight lines. When the phase elements are rotated to completely satisfy the Bragg incidence condition over the chirp grating lens aperture, the only angular mismatch will be due to the finite angular spread of the incoming spatially modulated optical wavefront. Figure 3.61 illustrates the phase element design using this concept. From previous discussions on chirp grating lens

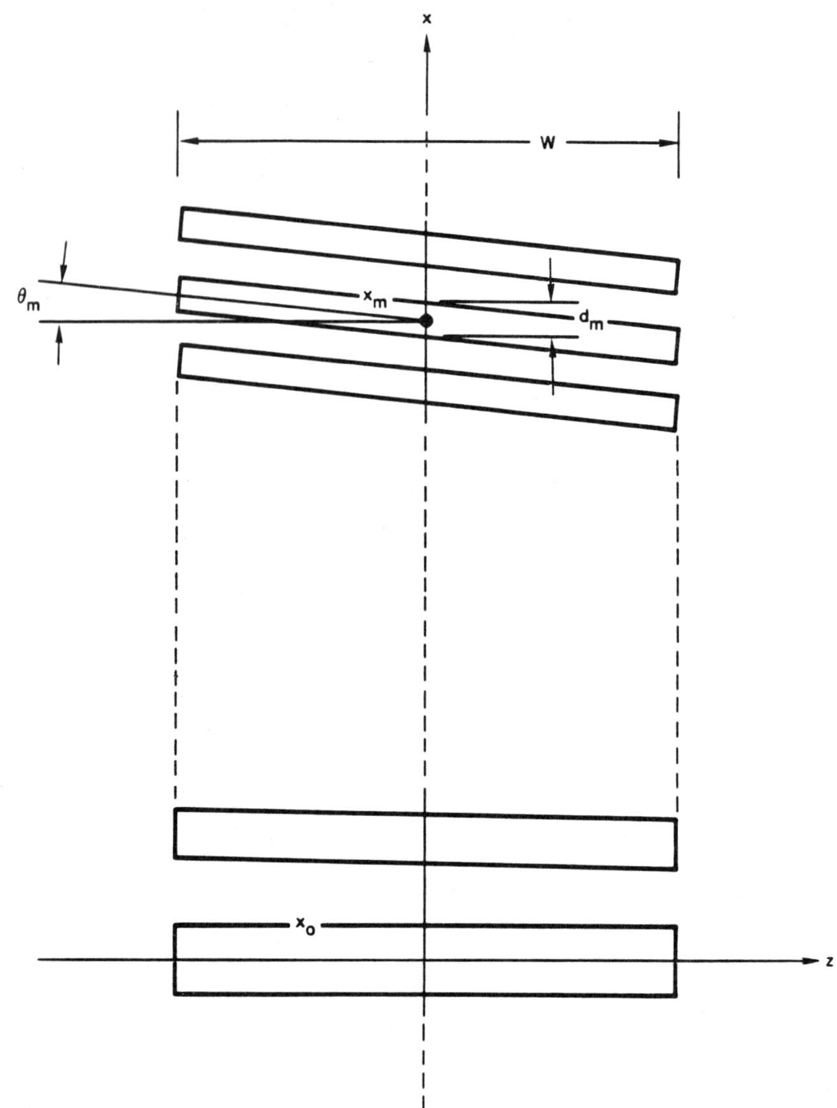

Figure 3.61 Rotation of chirp grating lines to satisfy Bragg conditions over the entire aperture of a chirp grating lens.

properties, it can be shown that a chirp grating lens with rotated phase elements only has to accommodate half as much angular mismatch as a chirp grating lens with parallel phase elements. Consequently, for a constant normalized mismatch, the grating thickness can be made longer by a factor of 2 using rotated phase elements.

Finally, the field of view of waveguide diffraction elements is limited by the small refractive index difference between the guided mode and the substrate. Figure 3.62 is the momentum vector diagram illustrating the diffraction of an optical wave vector by a grating wave vector K. The triangle formed by the dashed lines indicates guided-mode-

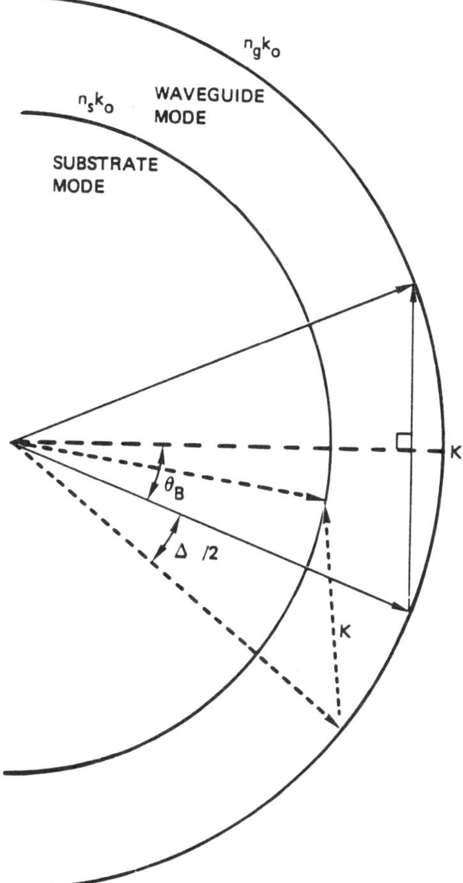

Figure 3.62 Momentum vector diagram

to-substrate-mode conversion by the same device when the optical incoming wave vector is tilted off-axis, as is required to provide angular field of view. This circumstance may be analyzed by simple geometry, which yields

$$2nk_0 K \sin\left(\theta_B + \frac{\Delta'}{2}\right) = (n^2 - n_s^2)k_0^2 + K^2 \tag{3.97}$$

which can be reduced to

$$\theta_B \Delta' < \frac{1}{2[1 - (n_s/n)^2]} \tag{3.98}$$

The constant Bragg angle and mismatch angle product can be very small for a graded index waveguide. For example, this product is about 0.0027 rad^2 for a titanium-diffused single-mode waveguide on lithium niobate. This corresponds to a field of view of about

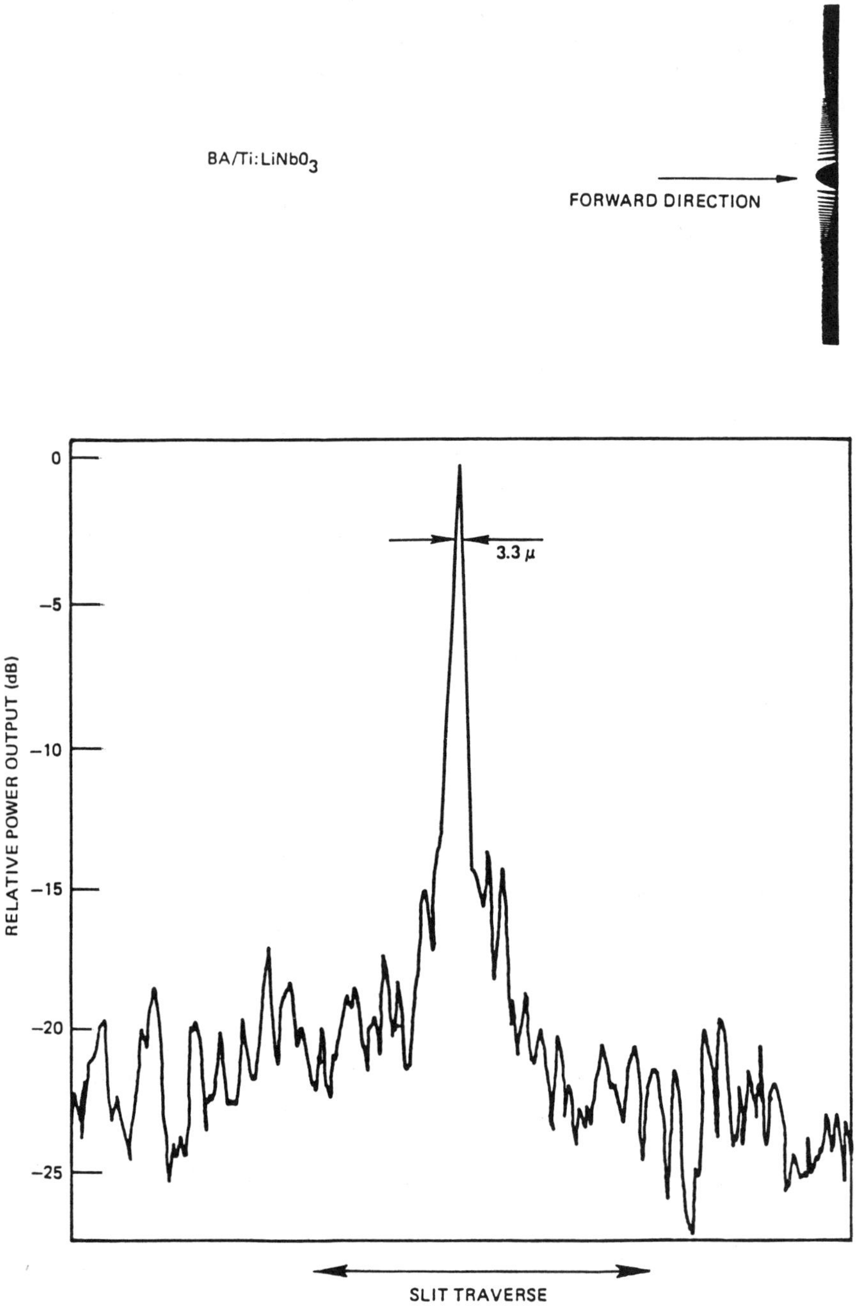

Figure 3.63 Focal spot pattern of the shaped-zone Fresnel lens at 0.83 μm wavelength. (After Ref. 145.)

3.5°. Although small, this field of view can provide over 400 pixels over an optical aperture of 2 mm. Much better values are achievable on shallower waveguides.

Table 3.13 includes the performance of several chirp grating lenses on glass substrates. The relatively large index step between the glass waveguide and substrate is the major factor for the high performance of these waveguide diffraction lenses. Recent experiments on titanium-diffused waveguides on lithium niobate verified the expected difficulties due to the small index difference between the graded index waveguide and the substrate. However, a diffraction-limited focal spot was routinely obtained. With due carefulness, diffraction efficiencies of 30 and 40% were obtained on titanium-diffused lithium niobate waveguides using benzoic acid treatment and a rutile overlay film, respectively, as index pertubation methods. The focal intensity profile of a shaped-zone (analog) Fresnel lens on lithium niobate is given in Fig. 3.63. Figure 3.64 is the focal distribution of a chirp grating lens on lithium niobate with 75% diffraction efficiency.

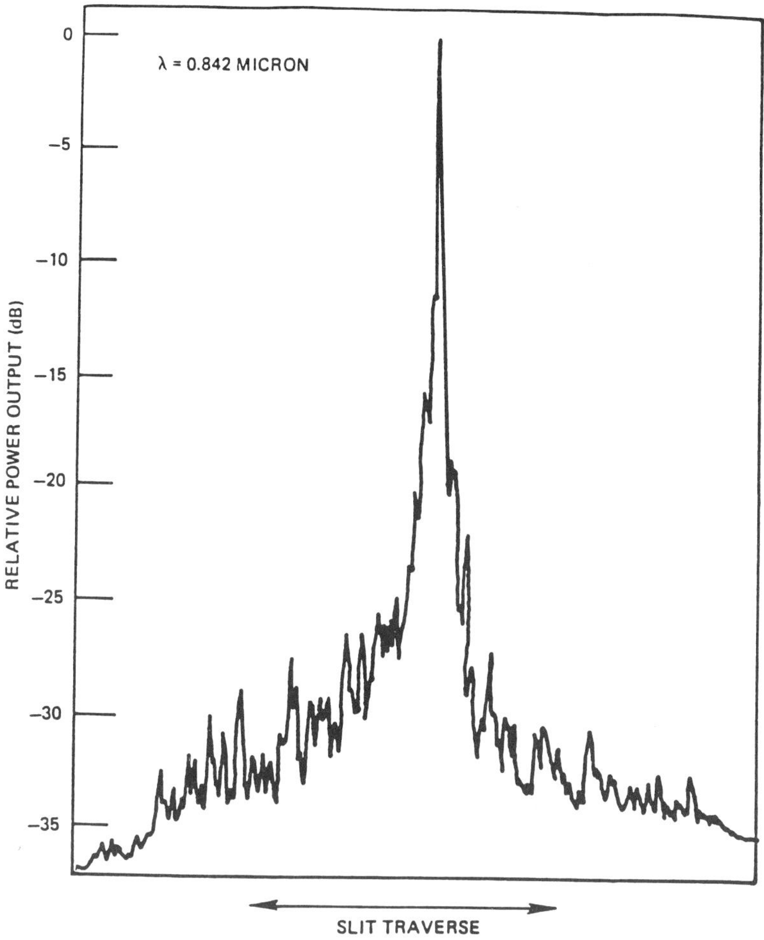

Figure 3.64 Chirp grating lens focal spot: efficiency 75%. (After Ref. 145.)

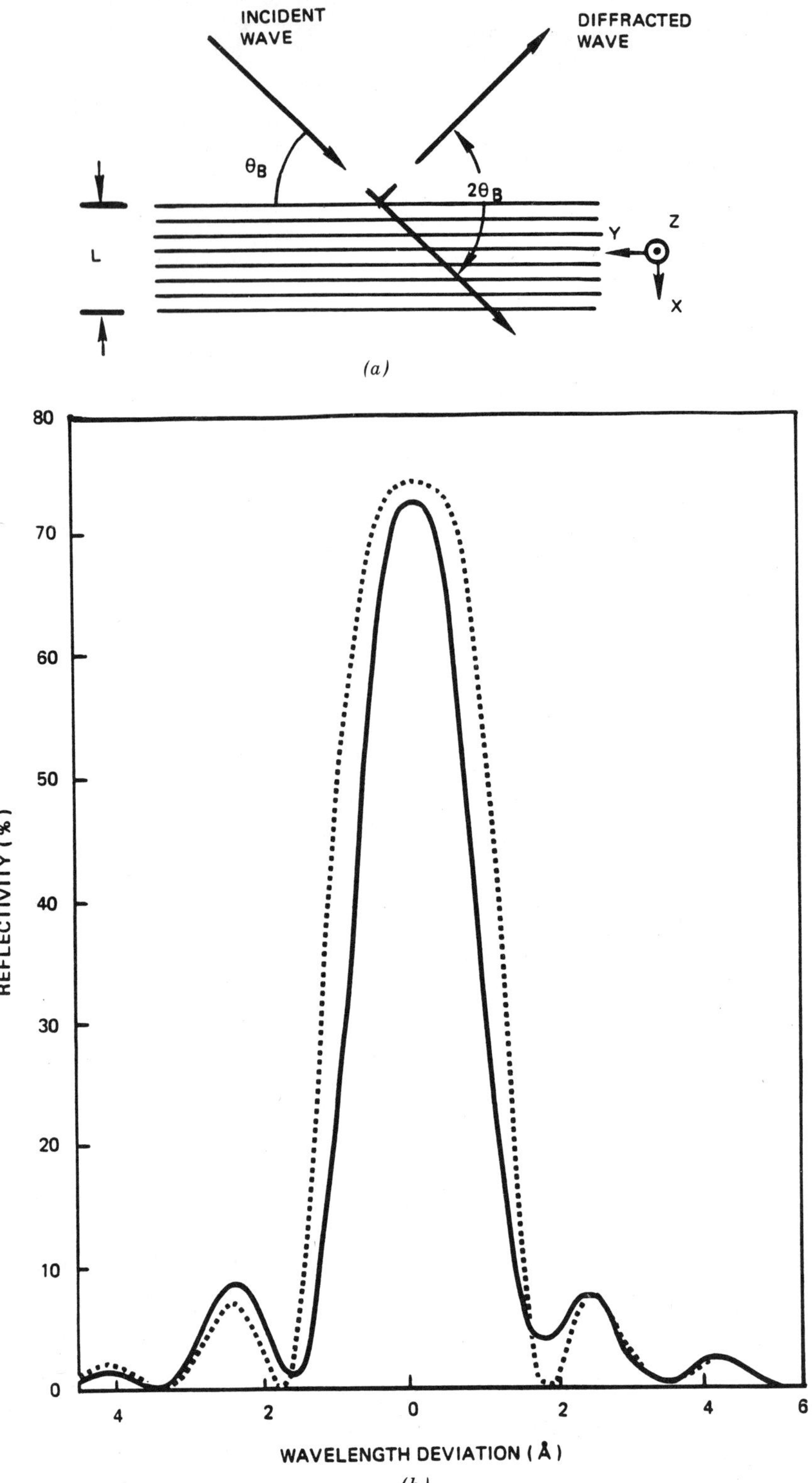
INCIDENT
WAVE
DIFFRACTED
WAVE
θB
2θB
L
Y
Z
X
(a)
80
70
60
50
40
30
20
10
0
REFLECTIVITY (%)
4
2
0
2
4
6
WAVELENGTH DEVIATION (Å)
(b)

Waveguide Diffraction Mirrors and Filters. For waveguide diffraction lenses, the optical wavefront and the grating phase front are nearly perpendicular to each other. In this section, the waveguide optical components with the optical wavefront parallel to the grating wavefront are discussed. This parallel arrangement produces waveguide mirrors with spectral response strongly influenced by grating periodicity similar to multilayer interference coatings in conventional optics.

Figure 3.65 illustrates the general condition between the grating and the optical guided mode. The reflectivity from each boundary of the grating lines is very weak. Thus, the response of the grating is simply the algebraic sum of the reflections,

$$r_T = \sum_{n=1}^{N} \left[-r + re^{j\phi} \right] B_n e^{j2n\phi} \tag{3.99}$$

where r_T is total reflection, N is number of lines, r is reflectivity at the edge of each line, $\phi = k(\Lambda/2)$ is the propagation delay through each line, and a uniform grating periodicity of Λ is assumed with each grating line weighted by B_n. The response of the reflector is simply $|r_T|^2$. For instance, if B_n is constant,

$$r_T = \frac{rB_n \sin(N\phi)}{\sin[(\phi/2) + 90^\circ]} \tag{3.100}$$

which has a peak value of $N_r B_n$ and a wavelength bandwidth of $\Delta\lambda/\lambda \approx (2N)^{-1}$ centered at $\lambda = \Lambda$ for a 90° reflector of projected periodicity Λ. For control of the passband shape, one simply introduces various weighting coefficients B_n that are generally the Fourier transform of the desired filter passband shape (transversal filter concept). A transmission line model is readily available to take into account multiple reflections within the grating lines. The filter response will be distorted in the case of strong reflection efficiency.

A waveguide grating mirror can be employed as a mirror, a beam splitter, a beam expander, and wavelength filter banks. Table 3.14 lists a number of published grating mirrorlike devices. Because the reflectivity is the algebraic sum of the reflection from a large number of grating lines, small index perturbation can still provide very high reflection efficiency. However, the spectral bandwidth will be narrower and the optical alignment will be more critical for grating reflectors made of lower index perturbation values. Chirp grating has been made to reduce the wavelength selectivity as well as to increase the acceptance angle of a grating reflector with the penalty of reduced efficiency. Figure 3.66 illustrates the use of a long grating as a beam expander. Figure 3.67 illustrates the use of a grating bank to achieve a controlled beam splitter array. Figures 3.68*a* and *b* illustrate the use of a variable-periodicity grating reflector as a waveguide optical filter bank, which can be useful for wavelength division multiplexed optical communications. Grating reflectors with curved grating lines can serve as focusing reflectors on optical waveguides. Figure 3.69 illustrates one of the most celebrated use of grating reflectors in waveguide optics, the distributed Bragg reflector (DBR) diode laser. The DBR laser and a similar distributed feedback (DFB) diode laser use the wavelength selectivity of

Figure 3.65 (*a*) Schematic diagram of grating waveguide beam splitter. (After Ref. 172.) (*b*) Reflectivity versus wavelength deviation from the Bragg condition for thin-film grating filter. The solid line is measured and the dotted line is calculated. (After Ref. 180.)

TABLE 3.14 Experimental Results of Wide-Angle In-Plane Bragg Deflection Integrated Optics Devices

Waveguide Device	Waveguide/Substrate Material	Grating Fabrication Technique	Grating Type	Grating Spacing Λ (μm)	Number of Grooves, N	Depth or ΔN	Bragg Angle (degrees)	Wavelength λ (Å)	Diffraction Efficiency (%)	Reference
Beam expander	$Cs^+ \leftrightarrow Na^+$ ex. in soda lime glass	Holographic and ion beam mill	Uniform	0.6	—	1000–3000 Å	70	6328	16 (total)	171
Beam deflector splitter	$Ag^+ \leftrightarrow Na^+$ ex. in soda lime glass	Photographic mask inhibits ion exchange	Uniform	3.0	—	2.6×10^{-4}	4	6328	50–70	188
Beam deflector and WDM	As_2S_3–SiO_2–In_2O_3–glass	Electron beam lithography	Uniform Chirped	0.5 ~0.5	400 1000	5×10^{-3} 3×10^{-3}	28 —	1.153 μm 1.153 μm	~100 ~50	172
Beam deflector splitter	PMMA/fused quartz	Electron beam lithography	Uniform	2.0 0.8	100 375	8.4×10^{-4} 8.4×10^{-4}	6.2 15.7	6328 6328	5 15	174
WDM	Corning 7059 glass on glass substrate	Holographic and ion beam mill.	Chirped	0.298–0.336	—	400–500 Å	50	6030–6300	~50 high IL	177
WDM	Corning 7059 glass on SiO_2	Holographic and AZ1350 grating	Chirped	0.42–0.46	—	700 Å	~60	6471	60–90	178
Beam deflector and WDM	Three-layer glass	Holographic ion beam etch	Uniform	0.237	—	120 Å	39.2	5700	—	170
Beam deflector	$Ag^+ \leftrightarrow Na^+$ ex. in glass	Holographic and etch	Uniform	0.29	—	—	45–75	6328	—	175
Multiple beam splitter	As_2S_3–SiO_2–In_2O_3–glass	Electron beam lithography	Chirped Uniform	0.5 0.5	500 40–200	7×10^{-3} 7×10^{-3}	29 29	1.153 μm 1.153 μm	90–100 30–100	173
Beam deflector	Corning 7059 glass on SiO_2	Holographic and ion beam mill	Chirped	~0.4	—	—	—	8450	—	176

Abbreviations: WDM, wave division multiplexer; IL, insertion loss.

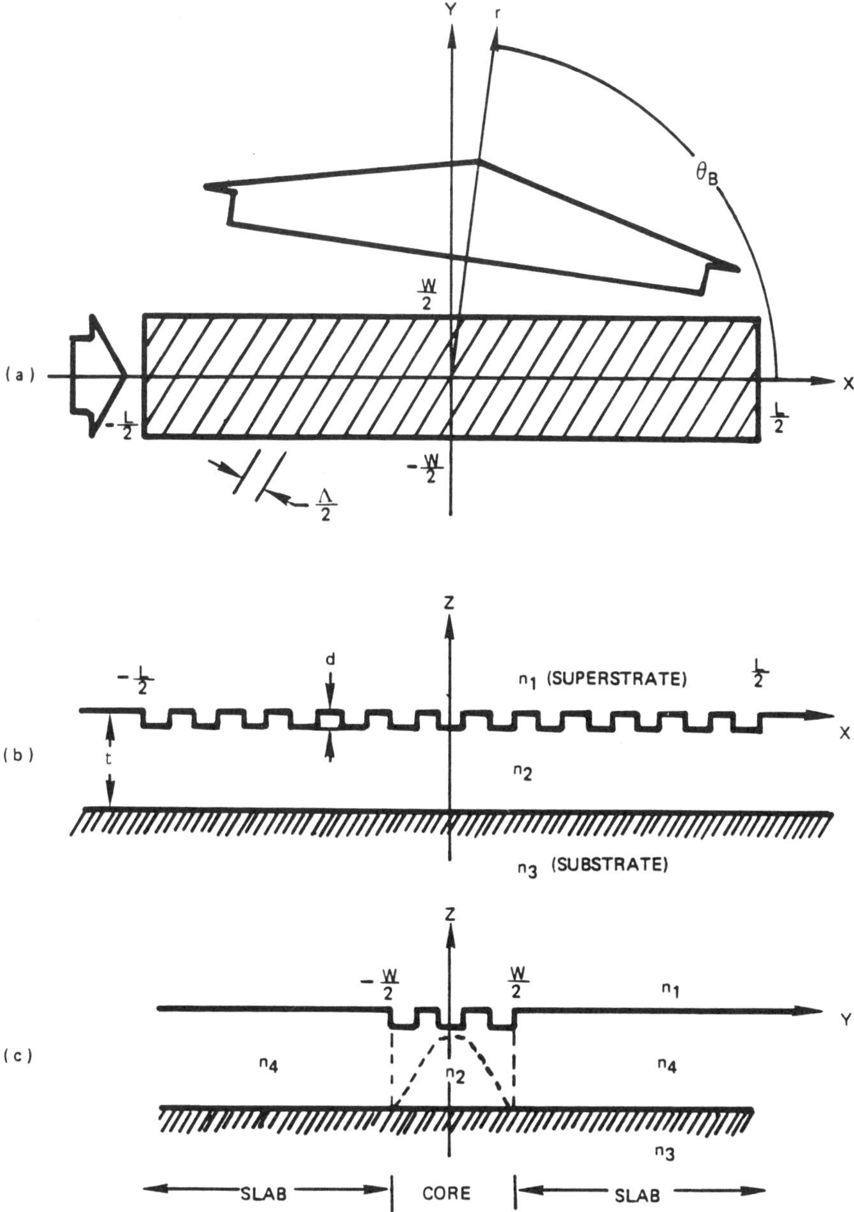

Figure 3.66 (*a*) Schematic diagram of the beam deflection and expansion process. (*b*) Side view of the channel waveguide and grating. (*c*) Cross-sectional view of the channel waveguide and grating. (After Ref. 169.)

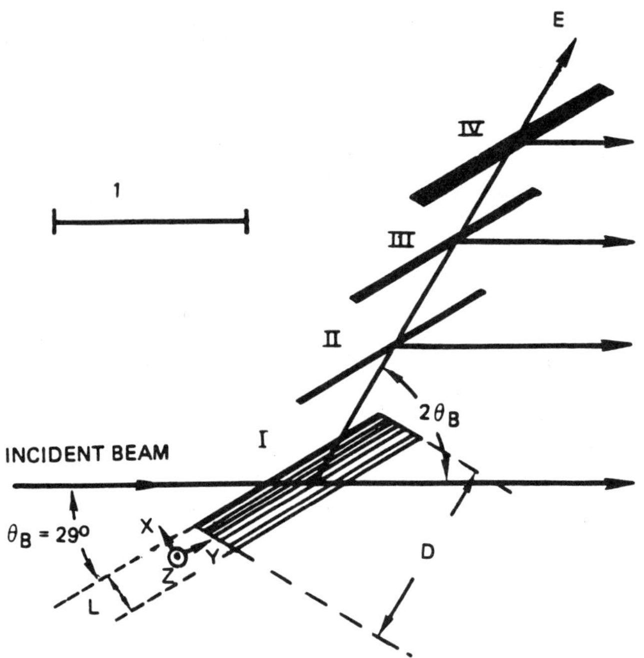

Figure 3.67 Schematic diagram of a multiple-grating waveguide beam splitter. (After Ref. 173.)

grating reflectors as laser wavelength control devices, resulting in more stable laser performances.

In general, grating reflectors require much smaller line widths than grating lenses. The fabrication of a uniform large-area submicrometer line pattern has been difficult. Imperfections cause scattering and nonuniform responses. Large-angle reflection from a grating reflector is highly polarization dependent, and there is strong TE–TM mode coupling upon such reflections.

3.2.4 Waveguide Optical Harmonic Generator

Second-order nonlinear optical harmonic generation has been observed in waveguides on lithium niobate substrates, in GaAs waveguides, and in polycrystalline ZnS thin-film waveguides deposited on BK-7 glass substrate [190–198]. Second-order nonlinear effects in randomly oriented crystalline films exist because there will be a nonvanishing component after the volume average. Recently emphasis is shifted to waveguides made of nonlinear organic materials either in the form of crystalline films or electric-field-oriented polymer films [193,195,196].

The theory of nonlinear integrated optics has been treated by Kuhn [199], Anderson et al. [200], and Conwell [201]. More recently, attention has been given to third-order nonlinear integrated optics [202] for potential applications in all-optical-signal processing. Optical waveguides are attractive for efficient nonlinear optical interactions due to (1) the high power density possible with moderate total power due to beam confinement,

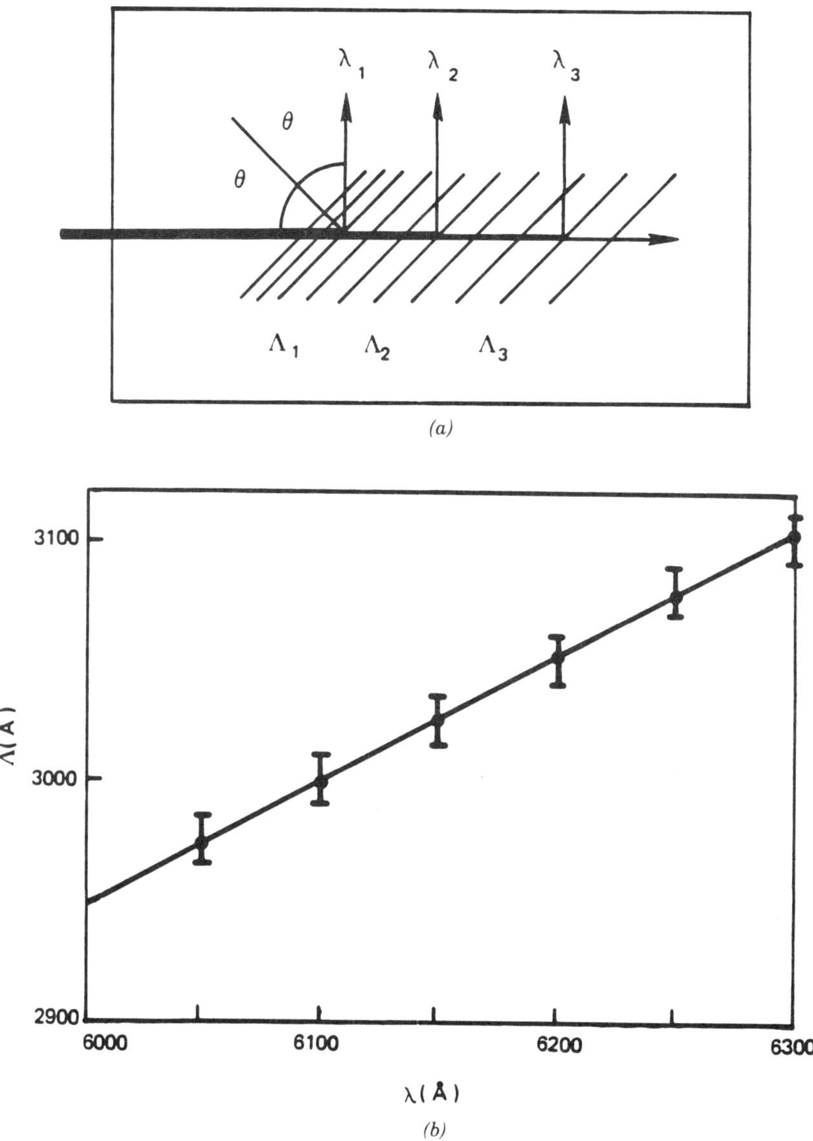

Figure 3.68 (*a*) Chirped grating waveguide optical demultiplexer. (After Ref. 178.) (*b*) Wavelength versus grating periodicity Λ in the reflecting region, where Λ is a function of distance along the grating. (After Ref. 177.)

(2) the diffractionless propagation for long interaction length, and (3) the possibility of phase matching without birefringence using waveguide mode dispersion relations. However, most recent devices tend to use channel waveguides, which provide better field confinement than planar waveguides. Therefore, no detailed treatment will be given here.

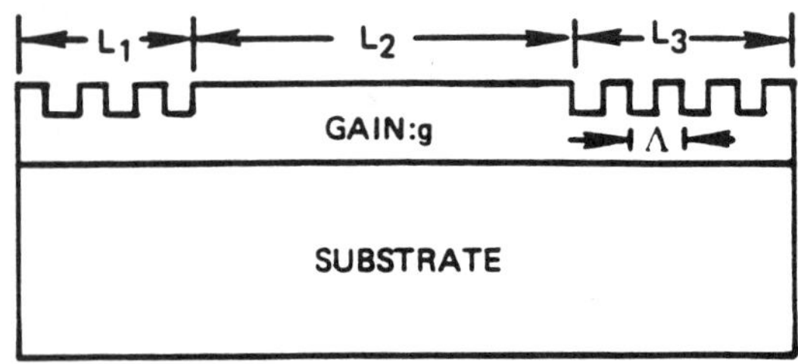

Figure 3.69 Schematic diagram of a DBR laser. The feedback is supplied by Bragg reflectors outside the gain region of the laser. (After Ref. 189.)

3.3 WAVEGUIDE MEASUREMENT TECHNIQUES AND APPLICATIONS

3.3.1 Waveguide Parameter Measurement

When conducting waveguide parameter measurements, very often it is necessary to couple the light into and out of the optical waveguide for observation. The most commonly employed technique is the prism coupler (discussed in an earlier section). A prism with index of refraction greater than the waveguide must be used. Typically, rutile prisms are used for experiments with lithium niobate substrates and dense flint glass prisms are used with glass waveguides. The prism is pressed to the substrate surface to reduce the air gap between the prism base and the substrate surface. Sometimes, a drop of water may be used to make this step easier if contamination of the waveguide surface by water is not a concern. A laser beam is directed toward the 90° edge of the prism, and the incidence angle as well as the prism–substrate air gap are adjusted until a streak of light appears either on the waveguide surface of at the exit end of the waveguide/substrate. For substrate modes, the intensity is not sensitive to small changes in incidence angle. For waveguide modes, the intensity flashes as the laser beam incidence angle is adjusted slightly. If a pair of prism couplers are used, the output beam from the second prism can be observed for evidence of waveguiding. Note that an output beam from the second prism does not mean the successful launching of a guided-wave mode. Substrate modes can also be coupled out by the second prism in discrete angles when the substrate mode bounces to the base of the output prism coupler. In most cases, the guided mode exhibits more in-plane scattering than the substrate modes, and the output beam from a guided mode may spread out as a number of nearly evenly spaced *m*-lines. The number of lines and the evenness in line spacing depends on the number of modes supported and the specific index profile of the waveguide being tested.

Sometimes, tests are conducted by end-fire coupling through cleaved waveguide edges or polished waveguide edges [203]. In this case, light is focused to the sharp edges by a high-power microscope objective lens. Success of the launching depends on the quality of the waveguide edges.

In the following sections, the commonly used method in characterizing waveguide impurity profiles, waveguide index profiles, optical propagation losses, and waveguide device performances are described.

Impurity Profile Measurement. The measurement of an impurity profile under the surface of substrate materials is a useful step in characterizing a newly formed waveguide. The most popular method is the use of an X-ray microanalyzer with an electron micro-

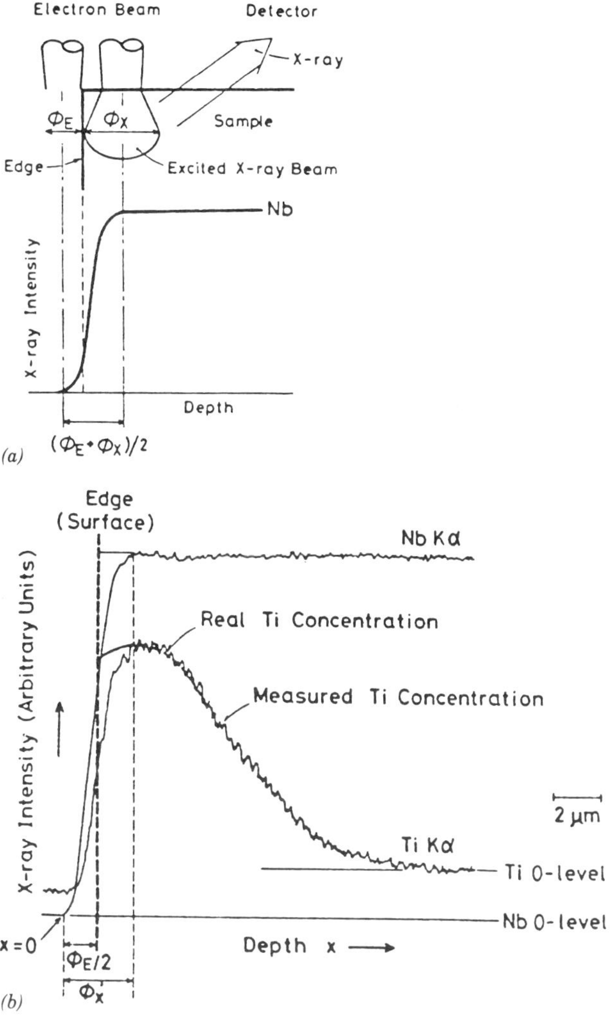

Figure 3.70 Illustrated edge correction: (*a*) transient region of measured Nb X-ray intensity for a sample with a steplike Nb concentration: (*b*) relation between real and measured Ti concentration in the $LiNbO_3$ sample diffused at 1000°C for 100 h with 1400-Å-thick and 30-μm-wide strip Ti film. (After Ref. 204.)

probe, as illustrated by Fig. 3.70. The substrate with optical waveguide is cleaved or edge polished and is viewed by a scanning electron microscope. The presence of heavy atoms is detected by surface emission of characteristic X-rays. Measuring the intensity of emmited X-rays as the electron beam is focused to different locations yields a plot of impurity concentrations. The limitation of this method is the inability to measure lighter atoms and the lack of accuracy when concentration of the interest species is low. Resolution is limited by the beam size, beam energy, and penetration. The relative statistical counting uncertainty was believed to be better than 1% with absolute accuracy only ±20%. An improved method has been reported by fixing the electron beam while moving

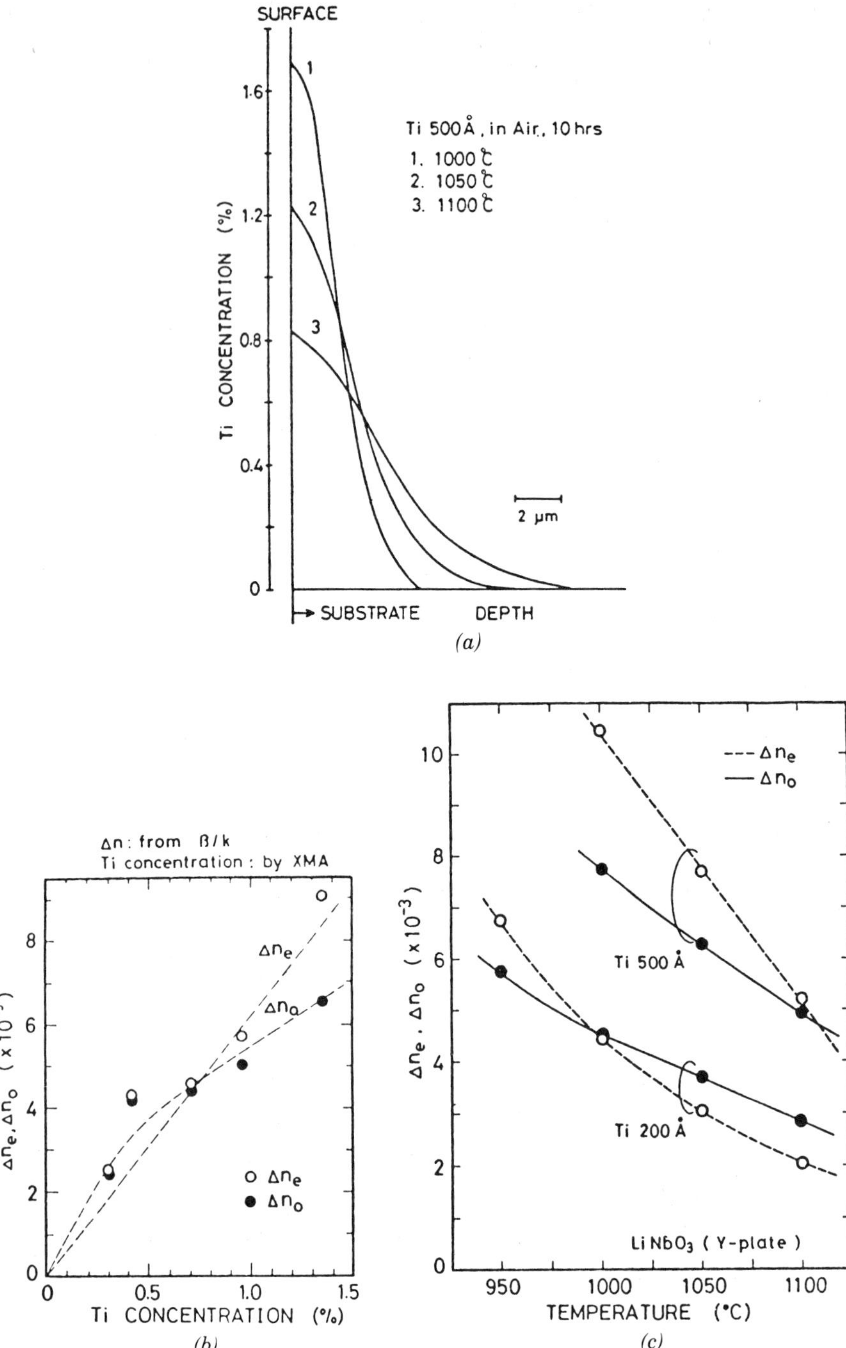

Figure 3.71 (*a*) Diffusion profiles for Ti in $LiNbO_3$ optical waveguides for different diffusion temperatures. (*b*) Refractive index changes in $LiNbO_3$ optical waveguides as a function of Ti concentration. Refractive index changes are determined independently by using a prism-coupling method. Ti concentrations are measured by an XMA. (*c*) Maximum refractive index changes of $LiNbO_3$ optical waveguides obtained by the proposed method. Thicknesses of the Ti films are given. (After Ref. 204.)

the sample at a constant speed [204]. Spatial resolution better than 0.2 μm has been achieved. However, the measured data need to be corrected for the edge effect, as illustrated in Fig. 3.70, due to the finite size of the probe beam. Figure 3.71*a* gives the measured titanium concentration profile in lithium niobate substrate. Figures 3.71*b* and *c* give the relation between titanium concentration and index changes and the relation between maximum index changes and diffusion temperature.

Secondary ion mass spectroscopy (SIMS) and Rutherford alpha particle backscattering are two other methods [205] that have been employed for the measurement of impurity profiles. The SIMS method uses an imaging mass spectrometer. A large area of the sample surface is sputtered away, and the resulting atomic and molecular species are analyzed. It exhibits poor absolute concentration accuracy ($\pm 100\%$) due to uncertainty in component sputtering rates, but it has excellent depth resolution (about 0.01 μm). The Rutherford backscattering method analyzes the alpha particle scattered from nuclei of the sample to determine the energy loss with respect to the incident beam, giving simultaneous information on concentration and depth of the nuclei present. Depth resolution is about 0.01 μm, and absolute concentration accuracy can be as good as 10%. These methods are sometimes used together with the electron microprobe to achieve accuracy in all respects.

Refractive Index Profile Measurement. The refractive index profile of an optical waveguide can be measured by direct viewing of the sample cross section in an optical interferometer or by waveguide mode index measurements and curve fitted by a WKB numerical program. For a step index waveguide, the thickness is measured either by a scanning electron microscope or a high-power microscope at its cross section or by optical thin-film measurement instruments such as an ellipsometer or a spectrameter or simply an etch-and-probe mechanical method. The ellipsometer and the optical spectrameter data also give refractive index values of the thin film as well as the substrate. These measured values are cross checked by correlating effective guided-wave mode indices with predicted values from theory. On the other hand, measurement of a graded index profile is much more involved.

Direct observation of the optical index profile in an optical interferometer setup is conceptually straightforward. The only critical requirement is that sample edges must be polished to perfect sharpness. This method has been very useful, particularly for diffused graded index waveguides of large thickness. However, cutting the sample into thin slices and polishing is a time-consuming process. It is mostly used to provide reference points for other measurements.

By far the most popular method is the use of a prism coupler and the WKB numerical process to determine the refractive index profile of a graded index film. Using a prism coupler, the mode indices of a waveguiding film can be easily determined from the coupling angles of each waveguide modes using the formula

$$N = n_p \sin\left(A + \sin^{-1}\frac{\sin i}{n_p}\right) \tag{3.101}$$

However, there is no data on effective mode depth. The WKB method is used in a numerical program iteratively until an assumed index profile satisfies the measured mode indices. For a single-mode waveguide, the single-mode index value is insufficient to generate both the film index profile and the thickness. One needs to assume a general index profile shape and an effective depth or a surface index value. For a graded index waveguide of known shape containing more than two modes, the index profile can be

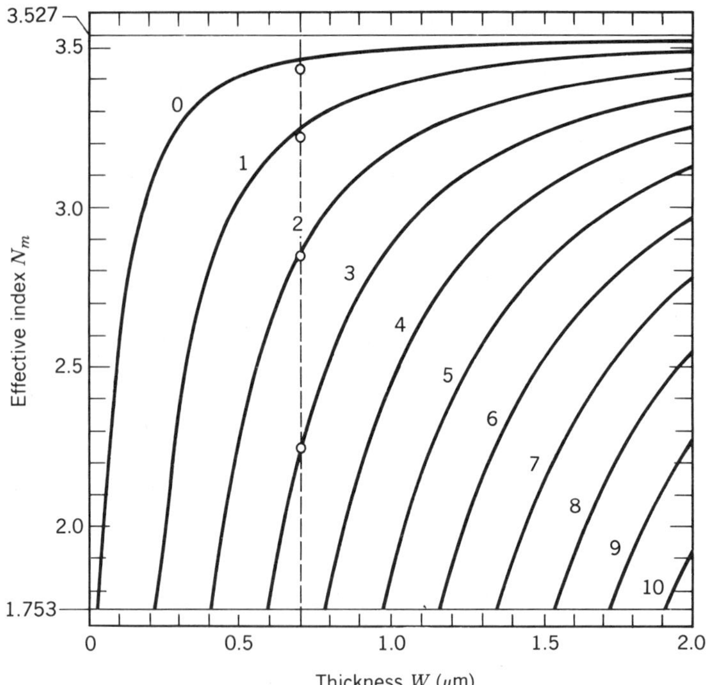

Figure 3.72 Waveguide thickness value can be obtained by fitting the measured mode index values to the dispersion diagram.

determined by the least mean-square fit as illustrated in Fig. 3.72. If the index profile is unknown, often a deeply diffused waveguide supporting many modes is fabricated to provide a large number of mode indices for the computation of the index profile. One may assume a surface index slightly higher than the highest-mode index value. Using Eq. (3.45) and assuming a piecewise linear profile, the index profile can be computed. Then, the surface index may be adjusted and more iteration taken to refine the results.

Optical Loss Measurement. Optical loss measurement requires the launching of a waveguide mode by whatever means. A highly lossy film can have its loss factor estimated by simply observing the length of the surface streak. Assuming the average dynamic response of the human eye is in the range of 37 dB, the waveguide loss can be estimated as 37 dB divided by the observed streak length. Less lossy films must be measured more carefully.

One of the most commonly used methods is taking output intensity readings from a prism output coupler that is allowed to slide on the surface of the waveguide. This is because a prism output waveguide coupler can be made to couple out nearly 100% of the waveguide mode even when the air gap between the prism base and the substrate surface varies. Plotting the output power relative to the prism location on a log scale should form a data distribution that can be approximated by a straight line in the least root-mean-square sense. The slope of this line is the waveguide propagation loss. Figure 3.73 is an example of such waveguide propagation loss measurement for a titanium-diffused waveguide on lithium niobate substrate after several steps of surface polishing.

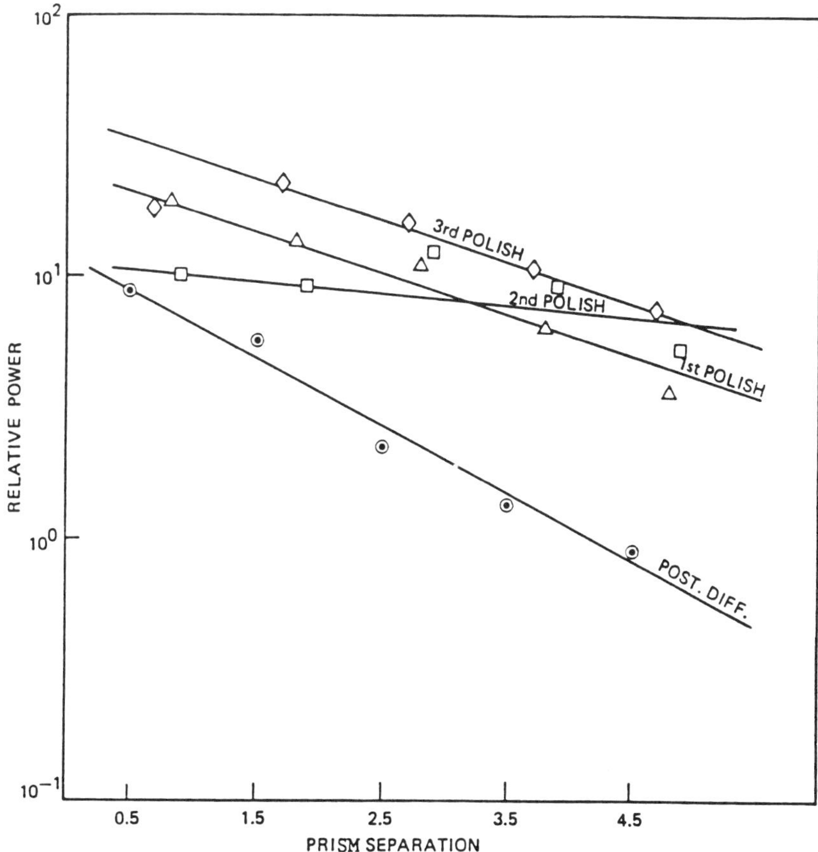

Figure 3.73 Waveguide propagation loss data for a Ti:$LiNbO_3$ waveguide with 500 Å starting titanium and 5 h wet oxygen diffusion at 1000°C after several polish steps. (After Ref. 145.)

The data tend to be well behaved, and accuracy on the order of a fraction of a decibel per centimeter can be obtained.

Automated waveguide loss measurement can be made by scanning a fiber-optical pick-up probe over the waveguide following the streak of scattered light by the guided mode. Its accuracy is particularly good when a high-quality waveguide is fabricated on an optically absorptive substrate such as oxidized silicon (see Fig. 3.74) propagation loss of 0.2 dB/cm can be measured. It is particularly interesting when waveguide thin-film components are involved that may prevent the sliding of a prism coupler. Figure 3.75 is the surface scattering trace of a glass waveguide with a thin-film Luneburg lens on top of the waveguide. Note that the laser beam is coming from the right on the trace. The scattering of the waveguide mode at lens edges is obvious. The slopes of straight lines I and II are optical losses in the lens region and in the glass waveguide region, respectively. Gap III denotes the insertion loss of the Luneburg lens including mode conversion losses at the two edges and the propagation loss within the lens. However, the problem of the surface probe is the ambiguity to distinguish the cases of normal

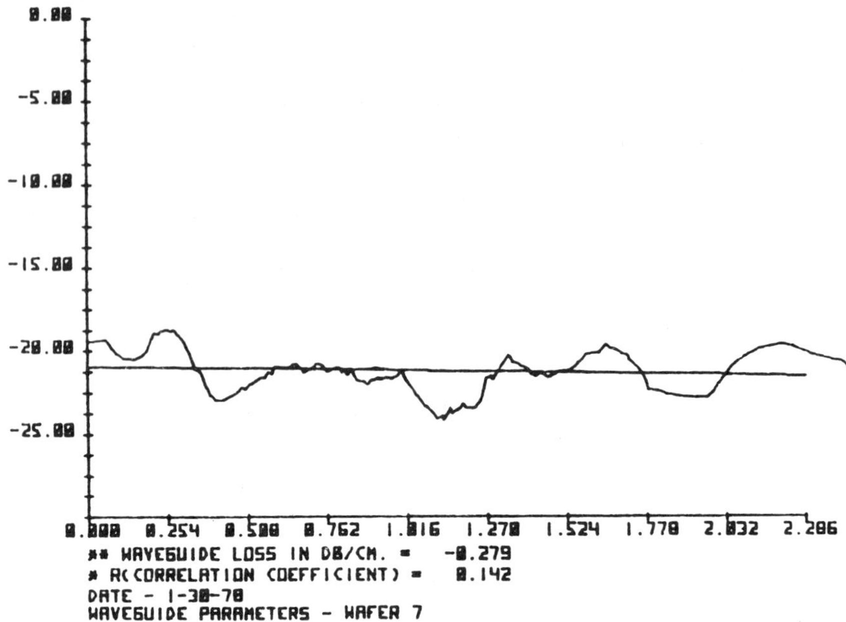

Figure 3.74 Longitudinal scan of scattered radiation from Corning 7059 waveguide on silicon water. (After Ref. 142.)

scattering with high guided-wave intensity and large surface scattering with moderate guided-wave intensity. An example is given in Fig. 3.76 in which the slope may be interpreted in many ways.

For measurement of extremely low-loss waveguides, a pyroelectric probe that will slide on the surface of the waveguide has been developed (see Fig. 3.77). Measurement accuracy of 0.01–0.05 dB/cm has been reported.

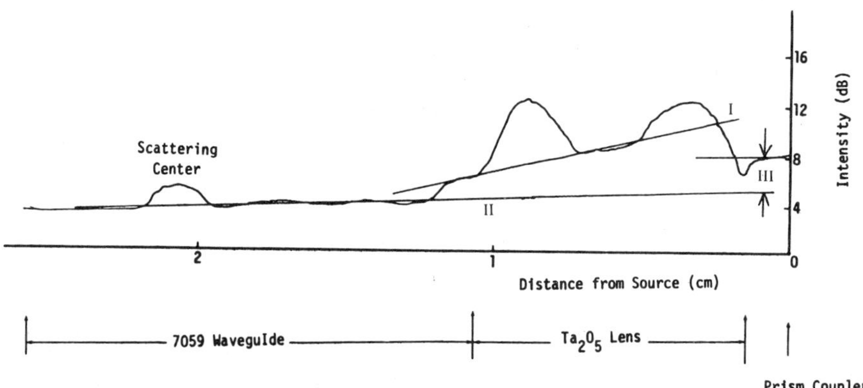

Figure 3.75 Waveguide scattering trace across Ta_2O_5 thin-film Luneburg lens and subsequent Corning 7059 waveguide. (After Ref. 142.)

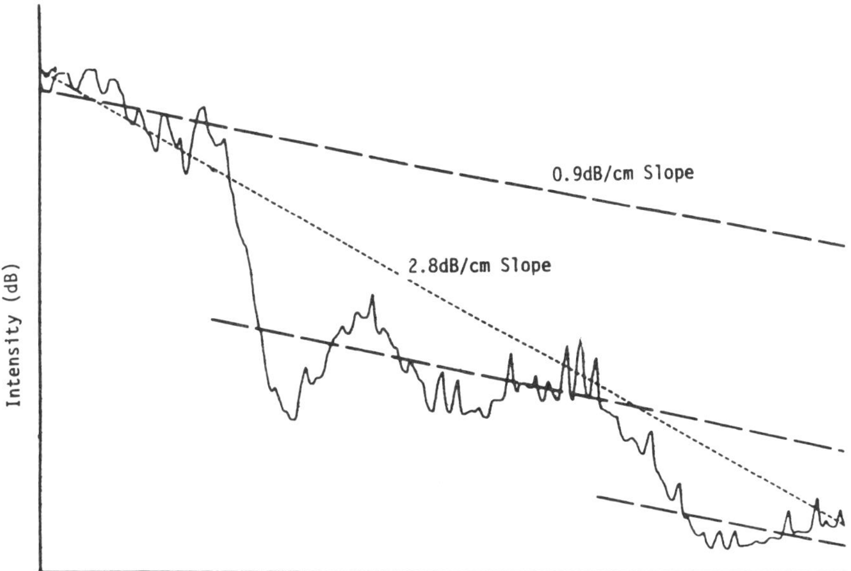

Figure 3.76 Measured optical waveguide scattering for Corning 7059 waveguide on a thermally oxidized silicon substrate. Two sharp steps are due to scattering loss and absorption in silicon substrate. Waveguide effective loss is 2.8 dB/cm. Reduced slope lines represent 0.9 dB/cm loss. (After Ref. 142.)

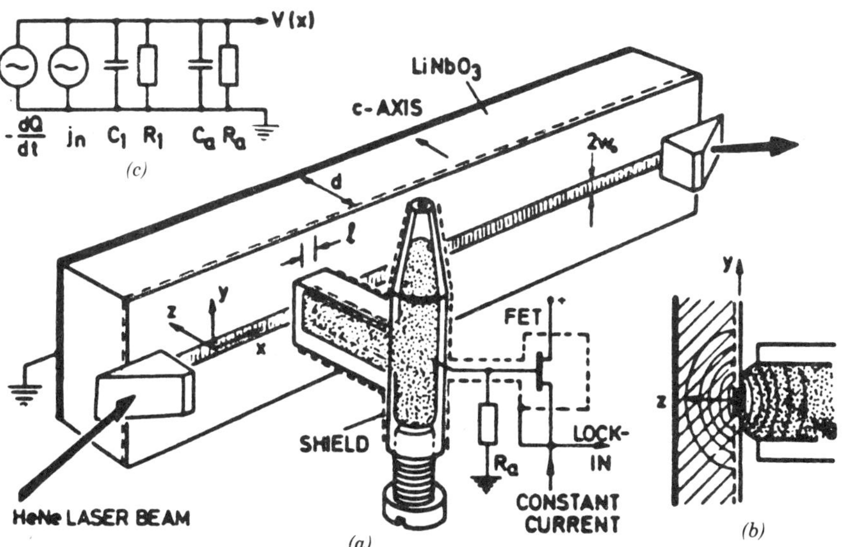

Figure 3.77 (*a*) Schematic arrangement of the Hg electrode; (*b*) cross section of contact region; (*c*) equivalent circuit. (After Ref. 206.)

Waveguide Optical Component Measurement. To measure waveguide optical component performances such as focal spots or in-plane scattering, an output coupler must be used to bring the guided mode out for evaluation (see Fig. 3.78). However, care must be taken to consider the effect of reimaging optical quality as the point spread function of the measurement system will convolve with the beam being evaluated. This is particularly critical for the measurement of diffraction-limited focal spots of a waveguide lens. Figure 3.79 shows the results of two thin-film Luneburg lenses measured by two reimaging optics of different quality. The much better performance SK-2 lens is observed only with the better reimaging optics.

3.3.2 Planar Optical Waveguide Applications

This chapter is summarized by several examples of planar waveguide applications. The planar waveguide is suitable for the construction of advanced electro-optical signal-processing subsystems as long as there is no need of a two-dimensional spatial light modulator. Many high-speed signal processors fall into this category. Examples are acousto-optical RF spectrum analyzers, and acousto-optical correlators [126, 207–213]. The major components are planar optical waveguides, waveguide optical lenses, and surface acousto-optical modulators [208]. The coupling of light from the laser diode to the photodetectors can be done in several ways including end-fire coupling. An intensive effort has been dedicated by many research organizations over the world on this subject during the past decade. The device is made on lithium niobate substrate that has good surface acoustic wave generation as well as surface acousto-optical interaction capabilities. Most approaches use a geodesic lens fabricated by either impact grinding [212] or diamond machining [209]. Thin-film Luneburg lenses made of niobium pentoxide and chirp grating lenses have also been used in certain cases. Figure 3.80 shows one of the designs. It promises much improved manufacturability and compact device size.

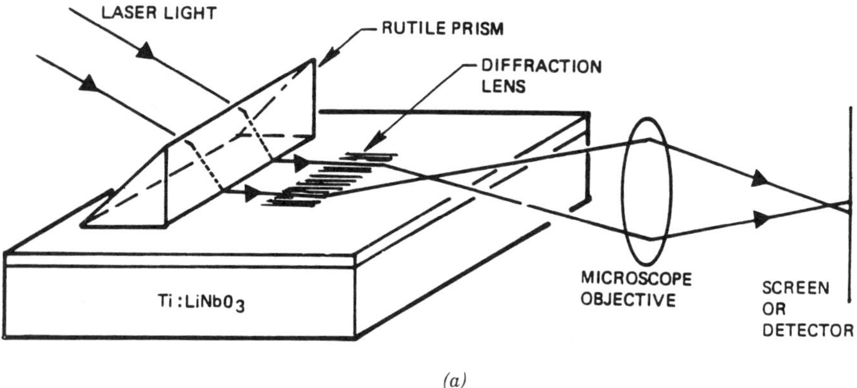

(*a*)

Figure 3.78 (*a*) Experimental setup for lens evaluation using a polished sharp edge for output coupling. (*b*) Apparatus for evaluation of waveguide geodesic lens diffraction patterns using a prism output coupler.

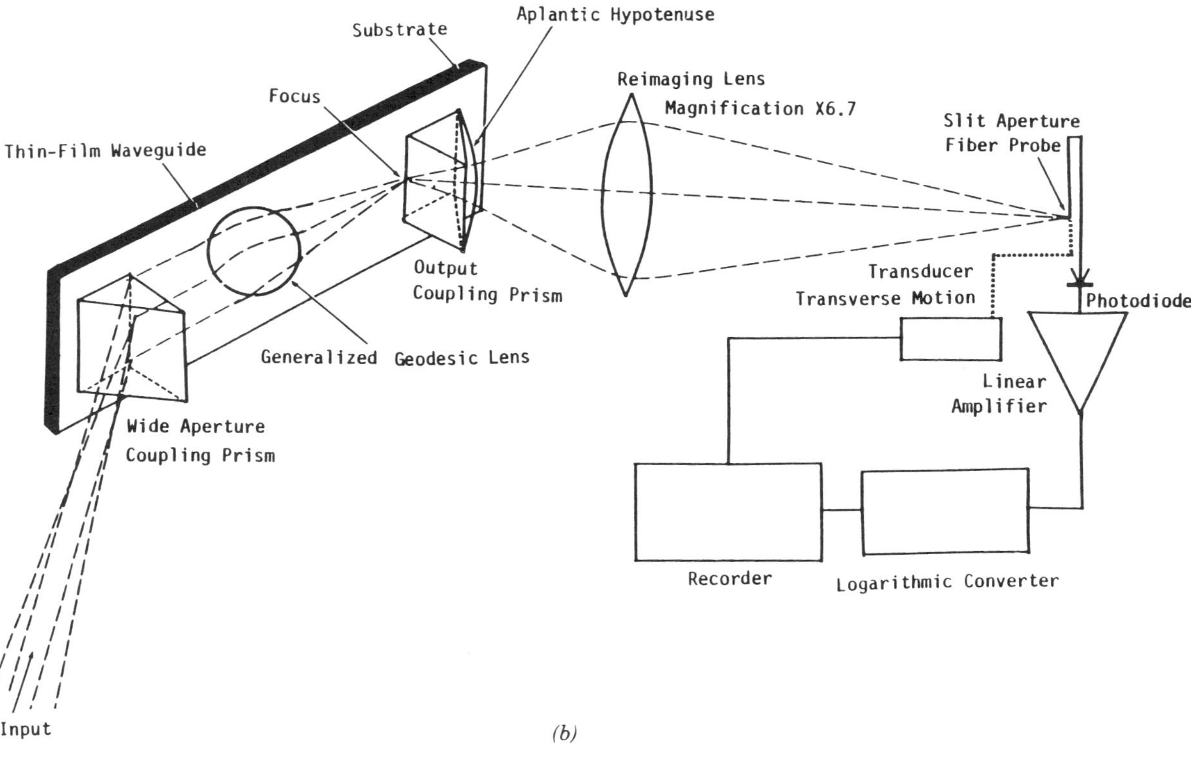

(b)

Figure 3.78 (continued)

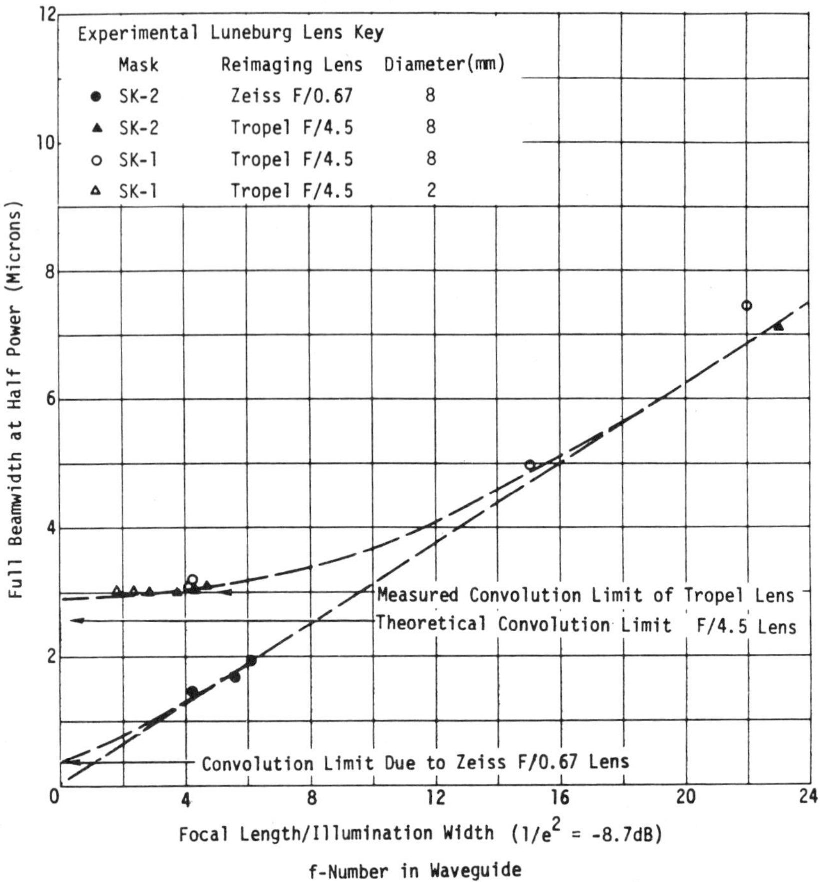

Figure 3.79 Measured diffraction pattern beamwidth (3 dB) as a function of the f-number in the waveguide, limited by convolution with the reimaging lens showing experimental results. (After Ref. 152.)

Another planar waveguide device that has been investigated extensively is the wavelength division multiplexer made of waveguide optical filter banks [170–178, 210]. In this case the major candidate substrate material is an oxidized silicon wafer on which photodetector arrays can be fabricated. The optical waveguide is formed by a deposited thin film, and grating reflectors are employed to separate the various optical wavelength components.

Optical waveguide principles have been applied for the characterization of thin-film materials with great success [214–217]. The procedure is simply the measurement of the optical mode indices of the dielectric film of interest with a subsequent fitting to modal dispersion curves. Normally, a pair of prism couplers are used, and the coupling angles for the guided modes are observed. A modified approach uses a single prism coupler for both input and output coupling and a convergent laser beam, as shown in Fig. 3.81. This method eliminates the rotation and observation process, but it does not rely on the actual propagation of the guided modes for appreciable distances. Therefore, it can operate irrespective of the polarization, absorption, or scattering of the guided light. The mode

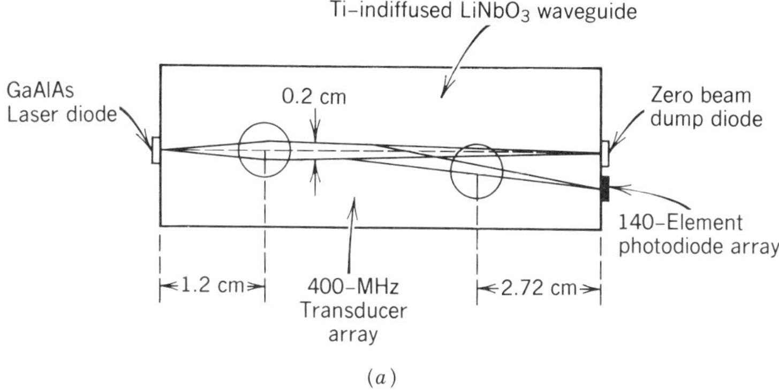

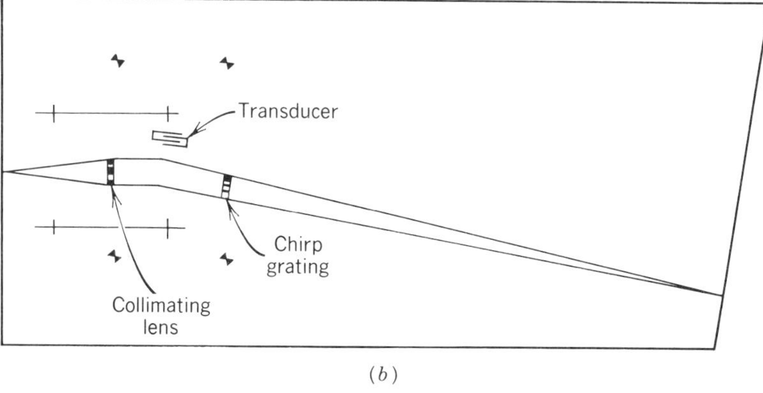

Figure 3.80 Integrated optical RF spectrum analyzer on $LiNbO_3$ substrate: (*a*) design using geodesic lenses; (*b*) design using diffraction grating lenses. (After Ref. 145.)

indices appear as dark lines in the reflected beams, indicating the trapping of laser power by the existence of guided modes. The mode indices are fit to normalized dispersion curves by minimizing the least mean-square error for the determination of film thickness and index. Accuracy better than 0.3% is possible.

Guided-wave optics is becoming increasingly mature. Planar waveguide technology has been used as an investigative tool for certain materials in addition to its intended application with integrated optical devices for signal processing and optical communication. Novel phenomena and applications are discovered each year. Complete signal-processing subsystems such as the RF spectrum analyzer have been demonstrated successfully. Future developments will certainly result in more implementation of guided-wave technology in the form of integrated optical devices.

REFERENCES

1. D. B. Anderson and R. R. August, *Proc. IEEE*, Vol. 54, p. 657, 1966.
2. A. Yariv and R. C. C. Leite, *Appl. Phys. Lett.*, Vol. 2, p. 55, 1963.
3. W. L. Bond, B. G. Cohen, R. C. C. Leite, and A. Yariv, *Appl. Phys. Lett.*, Vol. 2, p. 57, 1963.

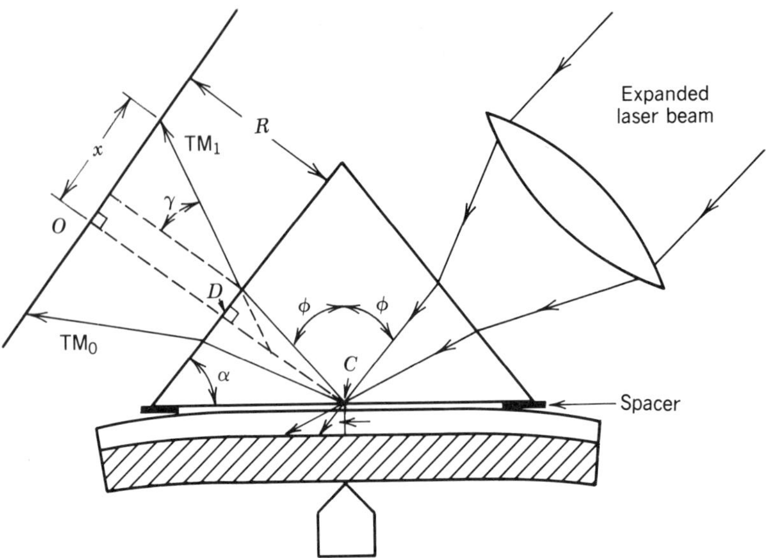

Figure 3.81 Schematic of prism film arrangement showing expanded collimated laser beam focused onto the prism base and the illuminated area on the screen at distance R from the prism face. A ray incident on prism base at ϕ is coupled into guide at angle θ; the "missing ray" makes angle γ with normal to prism exit face and appears as a dark line at distance x from the position, O, of an undeviated ray. (After Ref. 217.)

4. D. F. Nelson and F. K. Reinhardt, *Appl. Phys. Lett.*, Vol. 5, p. 148, 1964.
5. H. Osterberg and L. W. Smith, *J. Opt. Soc. Amer.*, Vol. 54, p. 1073, 1964.
6. R. Shubert and J. H. Harris, *IEEE Trans. Microwave Theory Tech.*, Vol. 16, p. 1048, 1968.
7. D. B. Anderson "An Integrated Circuit Approach to Optical Waveguide," *IEEE Microelectron. Symp. Dig.*, St. Louis, MO, 1968.
8. S. E. Miller, *Bell Syst. Tech. J.*, Vol. 48, p. 2059, 1970.
9. P. K. Tien, *Appl. Opt.*, Vol. 10, p. 2395, 1971.
10. S. J. Maurer and L. B. Felson, *Pro. IEEE*, Vol. 55, p. 1718, 1967.
11. H. K. V. Lotsh, *Optik*, Vol. 27, p. 239, 1968.
12. H. Kogelnik and V. Ramaswamy, *Appl. Opt.*, Vol. 13, p. 1857, 1974.
13. D. Marcuse, *Theory of Dielectric Optical Waveguides*, Academic, New York, 1974.
14. J. McKenna, *Bell Syst. Tech. J.*, Vol. 46, p. 1491, 1967.
15. H. Kogelnik, "Theory of Dielectric Waveguides," in T. Tamir, Ed., *Integrated Optics*, Vol. 7, Topics in Applied Physics, Springer-Verlag, New York, Heidelberg, and Berlin, 1975, Chap. 2.
16. J. R. Carruthers, I. P. Kaminov, and L. W. Stulz, *Appl. Opt.*, Vol. 13, p. 2333, 1974.
17. E. M. Conwell, *Appl. Phys. Lett.*, Vol. 23, p. 328, 1973.
18. R. D. Standley and V. Ramaswamy, *Appl. Phys. Lett.*, Vol. 25, p. 711, 1974.
19. A. K. Ghatak, E. Khular, and K. Thyagarajan, *IEEE J. Quantum Electron.*, Vol. QE-14, No. 6, p. 389, 1978.
20. J. P. Gordon, *Bell Syst. Tech. J.*, Vol. 45, p. 34, 1966.
21. L. B. Felson and N. Marcuvitz, *Radiation and Scattering of Waves*, Prentice-Hall, Englewood Cliffs, NJ, 1973.

22. G. B. Hocker and W. K. Burns, *IEEE J. Quantum Electron.*, Vol. QE-11, No. 6, p. 270, 1975.
23. R. Ulrich and H. P. Weber, *Appl. Opt.*, Vol. 11, No. 2, p. 428, 1972.
24. D. B. Ostrowsky and A. Jacques, *Appl. Phys. Lett.*, Vol. 18, p. 556, 1971.
25. P. K. Tien, G. Smolinsky, and R. J. Martin, *Appl. Opt.*, Vol. 11, p. 637, 1972.
26. M. J. Vasile and G. Smolinsky, *J. Electrochem. Soc.*, Vol. 119, p. 451, 1972.
27. P. K. Tien, R. J. Martin, and G. Smolinsky, *Appl. Opt.*, Vol. 12, p. 1909, 1973.
28. T. P. Sosnowski and H. P. Weber, *Appl. Phys. Lett.*, Vol. 21, p. 310, 1972.
29. J. P. Kaminow, H. P. Weber, and E. A. Chandross, *Appl. Phys. Lett.*, Vol. 18, p. 497, 1971.
30. E. A. Chandross, C. A. Pryde, W. J. Tomlinson, and H. P. Weber, *Appl. Phys. Lett.*, Vol. 24, No. 2, p. 72, 1974.
31. R. Selvaraj, H. T. Lin, and J. F. McDonald, *IEEE J. Lightwave Technol.*. Vol. 6, p. 1034, 1988.
32. J. Nishigawa and A. Otsuka, *Appl. Phys. Lett.*, Vol. 21, p. 48, 1972.
33. C. Hu and J. R. Whinnery, *J. Opt. Soc. Amer.*, Vol. 64, p. 1424, 1974.
34. J. P. Sheridan, J. M. Schmur, and T. G. Giallorenzi, *Appl. Phys. Lett.*, Vol. 22, p. 56, 1973.
35. Y. A. Bykovskii, A. V. Makovkin, V. L. Smirnov, and V. N. Sorokovikov, *Sov. J. Quantum Electron.*, Vol. 5, p. 1361, 1976.
36. D. P. Gra Russo and J. H. Harris, *Appl. Opt.*, Vol. 10, p. 2786, 1971.
37. L. N. Polky and J. H. Harris, *Appl. Phys. Lett.*, Vol. 21, p. 307, 1972.
38. P. K. Tien et al., *Appl. Phys. Lett.*, Vol. 14, p. 291, 1969.
39. E. L. Paradis and A. J. Shuskus, *Thin Solid Films*, Vol. 38, p. 131, 1976.
40. S. Dutta, H. E. Jackson, J. T. Boyd, F. S. Hickernell, and R. L. Davis, *Appl. Phys. Lett.*, Vol. 39, p. 206, 1981.
41. R. Th. Kersten and W. Rauscher, *Opt. Commun.*, Vol. 13, p. 189, 1975.
42. J. E. Goel and R. D. Standley, *Bell Syst. Tech. J.*, Vol. 48, p. 3445, 1969.
43. P. D. Davidse and L. I. Maissel, *J. Appl. Phys.*, Vol. 37, p. 574, 1966.
44. T. Nishimura, Y. Murayana, K. Dota, and H. Matsumaru, *Dig. 3rd Symp. Deposition of Thin Films by Sputtering*, Rochester, 1969, p. 96.
45. C. W. Pitt, *Optiss Lett.*, Vol. 9, p. 401, 1973.
46. J. E. Goell, *Appl. Opt.*, Vol. 12, p. 737, 1973.
47. H. Yajima, S. Kawase, and Y. Sekimoto, *Appl. Phys. Lett.*, Vol. 21, p. 407, 1972.
48. B. Chen and C. L. Tang, *Appl. Phys. Lett.*, Vol. 28, p. 435, 1976.
49. B. Chen, C. L. Tang, and J. M. Telle, *Appl. Phys. Lett.*, Vol. 25, p. 495, 1974.
50. D. H. Hemster, J. D. Cuthbert, R. J. Martin, and P. K. Tien, *Appl. Opt.*, Vol. 10, p. 1037, 1971.
51. H. Terui and M. Kobayashi, *Appl. Phys. Lett.*, Vol. 32, p. 666, 1978.
52. S. J. Ingrey, W. D. Westwood, Y. C. Cheng, and J. Wei, *Appl. Opt.*, Vol. 14, p. 2194, 1975.
53. S. J. Ingrey and W. D. Westwood, *Appl. Opt.*, Vol. 15, p. 607, 1976.
54. M. J. Rand and R. D. Standley, *Appl. Opt.*, Vol. 11, p. 2482, 1972.
55. W. Stutius and W. Streifer, *Appl. Opt.*, Vol. 16, p. 3218, 1977.
56. G. H. Hewig and K. Jain, *J. Appl. Phys.*, Vol. 54, p. 57, 1983.
57. A. K. Watts, M. deWit, and W. C. Holton, *Appl. Opt.*, Vol. 13, p. 2329, 1974.
58. J. F. Lotspeich, *Appl. Opt.*, Vol. 13, p. 2529, 1974.
59. M. Kawachi, M. Yasu, and T. Edahiro, *Electron. Lett.*, Vol. 19, p. 583, 1983.
60. M. Kawachi, M. Yasu, and M. Kobayashi, *Japan J. Appl. Phys.*, Vol. 22, p. 1932, 1983.
61. A. A. Ballman, H. Brown, P. K. Tien, and R. J. Martin, *J. Crystal Growth*, Vol. 20, p. 251, 1973.
62. S. Miyazawa, *Appl. Phys. Lett.*, Vol. 23, p. 198, 1973.
63. S. Fukunishi, N. Uchinda, S. Miyazawa, and J. Noda, *Appl. Phys. Lett.*, Vol. 24, p. 424, 1974.

64. P. K. Tien, S. Riva-Sanseverino, R. J. Martin, A. A. Ballman, and H. Bro, *Appl. Phys. Lett.*, Vol. 24, p. 503, 1974.
65. S. Miyazawa, S. Fushimi, and S. Kondo, *Appl. Phys. Lett.*, Vol. 26, p. 8, 1975.
66. P. K. Tien and A. A. Ballman, *J. Vac. Sci. Tech.*, Vol. 12, p. 892, 1975.
67. P. K. Tien, R. J. Martin, S. L. Blank, S. H. Wemple, and L. J. Varnerin, *Appl. Phys. Lett.*, Vol. 21, p. 207, 1972.
68. P. G. Van Engen, *J. Appl. Phys. Lett.*, Vol. 49, p. 4660, 1978.
69. N. Chubachi, *Proc. IEEE*, Vol. 64, p. 772, 1976.
70. D. J. Chanin, J. M. Hammer, and M. T. Duffy, *Appl. Opt.*, Vol. 14, p. 923, 1975.
71. T. Shiosaki, S. Ohnishi, Y. Hirokawa, and A. Kawabata, *Appl. Phys. Lett.*, Vol. 35, p. 406, 1978.
72. J. E. Goell, R. D. Standley, W. M. Gibson, and J. W. Rodgers, *Appl. Phys. Lett.*, Vol. 21, p. 72, 1972.
73. A. P. Webb and P. D. Townsend, *J. Phys. D, Appl. Phys.*, Vol. 9, p. 1343, 1976.
74. R. D. Standley, W. M. Gibson, and J. W. Rodgers, *Appl. Opt.*, Vol. 11, p. 1313, 1972.
75. D. T. Y. Wei, W. W. Lee, and L. R. Bloom, *Appl. Phys. Lett.*, Vol. 25, p. 329, 1974.
76. G. L. Destefanis, J. P. Gailliard, E. L. Ligeon, and S. Valette, *J. Appl. Phys.* Vol. 50, p. 7898, 1979.
77. G. L. Destefanis, P. D. Townsend, and J. P. Gailliard, *Appl. Phys. Lett.*, Vol. 32, p. 29, 1978.
78. S. Valette, G. Labrunie, J-C Deutsch, and J. Lizet, *Appl. Opt.*, Vol. 16, p. 1289, 1977.
79. R. F. Bartholomew and H. M. Garfinkel, *Glass: Science and Tech*, Vol. 5, Academic, New York, 1980, Chap. 6.
80. R. V. Ramaswamy and R. Srivastava, *J. Lightwave Tech.*, Vol. 6, p. 984, 1988.
81. C. A. Millar and R. H. Hutchins, *J. Phys. D, Appl. Phys.*, Vol. 11, p. 1567, 1978.
82. T. Findakly, *Opt. Eng.*, Vol. 24, p. 244, 1985.
83. R. H. Doremus, *J. Phys. Chem.*, Vol. 68, p. 2212, 1964.
84. T. Izawa and H. Nakagome, *Appl. Phys. Lett.*, Vol. 21, p. 584, 1972.
85. T. G. Giallorenzi, E. J. West, R. Kirk, R. Ginther, and R. A. Andrews, *Appl. Opt.*, Vol. 12, p. 1240, 1973.
86. G. H. Chartier, P. Jaussaud, A. D. de Oliveira, and O. Parriaux, *Electron. Lett.*, Vol. 13, p. 763, 1977.
87. V. Neuman, O. Parriaux, and L. M. Walpita, *Electron. Lett.*, Vol. 15, p. 704, 1979.
88. L. Ross, H. J. Lilienhof, H. Holscher, H. F. Schlaak, and A. Brandenburg, Topical Meeting Integrated and Guidewave Optics, Atlanta, GA, 1986, pp. 25–26.
89. G. Stewart, C. A. Millar, P. J. R. Laybourn, C. D. W. Wilkinson, and R. M. DeLaRue, *IEEE J. Quantum Electron.*, Vol. QE-13, p. 192, 1977.
90. G. Stewart and P. J. R. Laybourn, *IEEE J. Quantum Electron.*, Vol. QE-14, p. 930, 1978.
91. R. G. Eguchi, E. A. Maunders, and I. K. Naik, *Proc. SPIE*, Vol. 408, p. 21, 1983.
92. R. G. Walker, C. D. W. Wilkinson, and J. A. H. Wilkinson, *Appl. Opt.*, Vol. 22, p. 1923, 1983.
93. G. L. Yip and J. Albert, *Opt. Lett.*, Vol. 10, p. 151, 1985.
94. J. E. Gortych and D. G. Hall, *IEEE J. Quantum Electron.*, Vol. QE-22, p. 892, 1986.
95. M. Shah, *Appl. Phys. Lett.*, Vol. 26, p. 652, 1975.
96. J. Jackel, *Appl. Phys. Lett.*, Vol. 37, p. 739, 1980.
97. J. L. Jackel, C. E. Rice, and J. J. Vaselka, Jr., IEEE/OSA Topical Meeting in Integrated and Guidedwave Optics, Pacific Grove, CA, 1982.
98. H. J. Lilienhof, E. Vogas, D. Ritter, and B. Pantschew, *IEEE J. Quantum Electron.*, Vol. QE-18, p. 1877, 1982.
99. R. V. Ramaswamy and S. I. Najafi, *IEEE J. Quantum Electron.*, Vol. QE-22, p. 883, 1986.
100. I. P. Kaminov and J. R. Carruthers, *Appl. Phys. Lett.*, Vol. 22, p. 326, 1973.
101. J. R. Carruthers, I. P. Kaminov, and L. W. Stulz, *Appl. Opt.*, Vol. 13, p. 233, 1974.

102. R. V. Schmidt and I. P. Kaminov, *Appl. Phys. Lett.*, Vol. 25, p. 458, 1974.
103. G. J. Griffith and R. J. Esdail, *IEEE J. Quantum Electron.*, Vol. QE-20, p. 149, 1984.
104. O. Eknoyan, A. S. Greenblatt, W. K. Burns, and C. H. Bulmer, *Appl. Opt.*, Vol. 25, p. 737, 1986.
105. R. J. Esdaile, *Appl. Phys. Lett.*, Vol. 33, p. 733, 1978.
106. T. R. Ranganath and S. Wang, *Appl. Phys. Lett.*, Vol. 30, p. 376, 1977.
107. J. L. Jackel, *J. Opt. Commun.*, Vol. 3, p. 82, 1982.
108. T. P. Pearsall, S. Chiang, and R. V. Schmidt, *J. Appl. Phys.*, Vol. 47, p. 4794, 1976.
109. A. M. Glass, J. P. Kaminow, A. A. Ballman, and D. H. Olson, *Appl. Opt.*, Vol. 19, p. 276, 1980.
110. M. Minakata, S. Saita, M. Shibata, and S. Miyazawa, *J. Appl. Phys.*, Vol. 49, p. 4677, 1978.
111. J. Noda, T. Saku, and N. Uchida, *Appl. Phys. Lett.*, Vol. 25, p. 308, 1974.
112. Y. Okamura, S. Yamamoto, and T. Makimoto, *Appl. Phys. Lett.*, Vol. 32, p. 161, 1978.
113. O. Eknoyan, D. W. Yoon, and H. F. Taylor, *Appl. Phys. Lett.*, Vol. 51, p. 384, 1987.
114. G. D. Boyd, R. V. Schmidt, and F. D. Storz, *J. Appl. Phys.*, Vol. 48, p. 2880, 1977.
115. J. Noda, M. Fukuma, and S. Saito, *J. Appl. Phys.*, Vol. 49, p. 3150, 1978.
116. M. Minakata, J. Noda, and N. Uchida, *Appl. Phys. Lett.*, Vol. 26, p. 395, 1975.
117. V. Ramaswamy and R. D. Standley, *Appl. Phys. Lett.*, Vol. 26, p. 10, 1975.
118. M. Masuda and J. Koyama, *Appl. Opt.*, Vol. 16, p. 2994, 1977.
119. T. Findakly and C. L. Chen, *Appl. Opt.*, Vol. 17, p. 469, 1978.
120. Y. Suematsu, M. Hakuta, K. Furuya, K. Chiba, and R. Hasumi, *Appl. Phys. Lett.*, Vol. 21, p. 291, 1972.
121. A. Reisinger, *Appl. Opt.*, Vol. 12, p. 1015, 1973.
122. A. Reisinger, *Appl. Phys. Lett.*, Vol. 23, p. 237, 1973.
123. I. P. Kaminov, W. L. Mammel, and H. P. Weber, *Appl. Opt.*, Vol. 13, p. 396, 1974.
124. E. M. Garmire and H. Stoll, *IEEE J. Quantum Electron.*, Vol. QE-8, p. 763, 1972.
125. H. F. Taylor, W. E. Martin, D. B. Hall, and V. N. Smiley, *Appl. Phys. Lett.*, Vol. 21, p. 95, 197.
126. M. C. Hamilton, D. A. Wille, and W. J. Miceli, *IEEE 1976 Ultrason. Symp. Proc.*, Annapolis, MD, 1976.
127. D. Hall, A. Yariv, and E. Garmire, *Opt. Commun.*, Vol. 1, p. 403, 1970.
128. R. Shubert and J. H. Harris, *IEEE Trans. Microwave Theory Tech.*, Vol. MTT-16, p. 1048, 1968.
129. D. Marcuse and E. A. J. Marcatili, *Bell Syst. Tech. J.*, Vol. 50, p. 43, 1971.
130. J. H. Harris and R. Shubert, URSI (International Radio Scientific Union) Spring Meeting, Washington DC, 1969, p. 71.
131. P. K. Tien, R. Ulrich, and R. J. Martin, *Appl. Phys. Lett.*, Vol. 14, p. 291, 1969.
132. F. Zernike and J. E. Midwinter, *IEEE J. Quantum Electron.*, Vol. QE-6, p. 577, 1970.
133. P. K. Tien and R. Ulrich, *J. Opt. Soc. Amer.*, Vol. 60, p. 1325, 1970.
134. T. Tamir and H. L. Bertoni, *J. Opt. Soc. Amer.*, Vol. 61, p. 1397, 1971.
135. T. Tamir, "Beam and Waveguide Couplers," in T. Tamir, Ed., *Integrated Optics*, Vol. 7, Topics in Applied Physics, Springer-Verlag, New York, Heidelberg, and Berlin, 1975, Chap. 3.
136. M. L. Dakss, L. Kuhn, P. F. Heidrich, and B. A. Scott, *Appl. Phys. Lett.*, Vol. 16, p. 523, 1970.
137. H. Kogelnik and T. P. Sosnowski, *Bell Syst. Tech. J.*, Vol. 49, p. 1602, 1970.
138. P. K. Tien, *Appl. Opt.*, Vol. 10, p. 2395, 1971.
139. S. T. Peng and T. Tamir, *Opt. Commun.*, Vol. 11, p. 405, 1974.
140. S. T. Peng, T. Tamir, and H. L. Bertoni, *IEEE Trans. Microwave Theory Tech.*, Vol. MTT-23, p. 123, 1975.
141. D. B. Anderson, R. L. Davis, J. T. Boyd, and R. R. August, *IEEE J. Quantum Electron.*, Vol. QE-13, p. 275, 1977.

142. D. B. Anderson, R. R. August, and S. K. Yao, "Waveguide Optics for Coherent Optical Processing," Final Technical Report, Rockwell International, AFAL-TR-78-83, June 1978.
143. J. M. Delavaux, S. Forouhar, W. S. C. Chang, and R. X. Lu, "Experimental Fabrication and Evaluation of Diffraction Lenses in Planar Optical Waveguides," IEEE/OSA Conference on Lasers and Electro-Optics Phoenix, AZ, April 1982.
144. S. Forouhan, C. Warren, R. X. Lu, and W. S. C. Chang, "Tech. for the Fabri. of High Index Overlay Films on $LiNbO_3$," SPIE Technical Symposium, Arlington, VA, April 1983.
145. S. K. Yao and L. Jostad, "Optical Waveguide Diffraction Elements Final Technical Report," TRW, AFWAL-TR-85-1061, May 1985.
146. C. Warren, S. Forouhan, W. S. C. Chang, and S. K. Yao, *Appl. Phys. Lett.*, Vol. 43, p. 424, 1983.
147. R. K. Luneburg, *The Mathematical Theory of Optics*, University of California Press, Berkeley, 1954.
148. S. P. Morgan, *J. Appl. Phys.*, Vol. 29, p. 1358, 1958.
149. W. H. Southwell, *J. Opt. Soc. Amer.*, Vol. 67, p. 1010, 1977.
150. S. K. Yao, *J. Appl. Phys.*, Vol. 50, p. 3390, 1979.
151. S. K. Yao and D. B. Anderson, *Appl. Phys. Lett.*, Vol. 33, p. 307, 1978.
152. S. K. Yao, D. B. Anderson, R. R. August, B. R. Youmans, and C. M. Oania, *Appl. Opt.*, Vol. 18, p. 4067, 1979.
153. R. F. Rinehart, *J. Appl. Phys.*, Vol. 19, p. 860, 1948.
154. G. C. Righini, V. Russo, S. Sottini, and G. T. diFrancia, *Appl. Opt.*, Vol. 12, p. 1477, 1973.
155. E. Spiller and J. S. Harper, *Appl. Opt.*, Vol. 13, p. 2105, 1974.
156. C. M. Verber, D. W. Vahey, and V. E. Wood, *Appl. Phys. Lett.*, Vol. 28, p. 514, 1976.
157. V. E. Wood, *Appl. Opt.*, Vol. 15, p. 2817, 1976.
158. K. S. Kunz, *J. Appl. Phys.*, Vol. 25, p. 642, 1954.
159. G. C. Righini, V. Russo, and S. Sottini, *IEEE J. Quantum Electron.*, Vol. QE-15, p. 1, 1979.
160. B. Chen, E. Marom, and R. J. Morrison, *Appl. Phys. Lett.*, Vol. 33, p. 511, 1978.
161. G. E. Betts and G. E. Merx, *Appl. Opt.*, Vol. 17, p. 3969, 1978.
162. W. H. Southwell, *J. Opt. Soc. Amer.*, Vol. 67, p. 1293, 1977.
163. S. Sottini, V. Russo, and G. C. Righini, *IEEE Trans. Circ. Syst.*, Vol. CAS-26, p. 1036, 1979.
164. S. K. Yao and D. E. Thompson, *Appl. Phys. Lett.*, Vol. 33, p. 635, 1978.
165. P. Mottier and S. Valette, *Appl. Opt.*, Vol. 20, p. 1630, 1981.
166. P. R. Ashley and W. S. C. Chang, *Appl. Phys. Lett.*, Vol. 33, p. 491, 1978.
167. G. Hatakoshi and S. Tamaka, *Opt. Lett.*, Vol. 2, p. 142, 1978.
168. P. K. Tien, *Opt. Lett.*, Vol. 1, p. 64, 1977.
169. H. M. Stoll, *Appl. Opt.*, Vol. 17, p. 2562, 1978.
170. K. Wagatsuma, H. Sakaki, and S. Saito, *IEEE J. Quantum Electron.*, Vol. QE-15, p. 632, 1979.
171. V. Neuman, C. W. Pitt, and L. M. Walpita, *Electron. Lett.*, Vol. 17, p. 165, 1981.
172. Y. Handa, T. Suhara, H. Nishihara, and J. Koyama, *Appl. Opt.*, Vol. 9, p. 2842, 1980.
173. Y. Handa, T. Suhara, H. Nishihara, and J. Koyama, *Opt. Lett.*, Vol. 5, p. 309, 1980.
174. H. Kotani, M. Kawabe, and S. Namba, *Jap. J. Appl. Phys.*, Vol. 18, p. 279, 1979.
175. N. Gremillet, G. Thomin, and J. Marcou, *Opt. Commun.*, Vol. 32, p. 69, 1980.
176. D. A. Bryan and J. K. Powers, *Opt. Lett.*, Vol. 5, p. 407, 1980.
177. A. C. Livanos, A. Katzir, A. Yariv, and C. S. Hong, *Appl. Phys. Lett.*, Vol. 30, p. 519, 1977.
178. T. Fukuzawa and M. Nakamura, *Opt. Lett.*, Vol. 4, p. 343, 1979.
179. A. Yariv and M. Nakamura, *IEEE J. Quantum Electron.*, Vol. QE-13, p. 233, 1977.
180. D. C. Flanders, H. Kogelnik, R. V. Schmidt, and C. V. Shank, *Appl. Phys. Lett.*, Vol. 24, p. 194, 1974.

181. S. K. Yao, *SPIE Proc.*, Vol. 269, 1981.
182. M. Stockman and W. Beinvogl, *Wave Electron.*, Vol. 4, p. 221, 1983.
183. S. Forouhan, R. X. Lu, W. S. C. Chang, R. L. Davis, and S. K. Yao, *Appl. Opt.*, Vol. 22, p. 3128, 1983.
184. S. Wang, *IEEE J. Quantum Electron.*, Vol. QE-13, p. 176, 1977.
185. S. Valette, A. Morgue, and P. Mottier, *Electron. Lett.*, Vol. 18, p. 13, 1982.
186. W. S. C. Chang and P. R. Ashley, *IEEE J. Quantum Electron.*, Vol. QE-16, p. 744, 1980.
187. S. K. Yao and E. H. Young, *SPIE Proc.*, Vol. 90, p. 23, 1976.
188. E. Y. B. Pun and A. Yi-Yan, *Appl. Phys. Lett.*, Vol. 38, p. 673, 1981.
189. H. W. Yen, W. Ng, I. Samid, and A. Yariv, *Opt. Commun.*, Vol. 17, p. 213, 1976.
190. D. B. Anderson and J. T. Boyd, *Appl. Phys. Lett.*, Vol. 19, p. 266, 1971.
191. W. Sohler and H. Suche, *SPIE Proc.*, Vol. 408, p. 163, 1983.
192. R. Regener and W. Sohler, *J. Opt. Soc. Amer. B*, Vol. 5, p. 267, 1988.
193. G. H. Hewig and K. Jain, *Opt. Commun.*, Vol. 47, p. 347, 1983.
194. H. Ito, N. Uesugi, and H. Inaba, *Appl. Phys. Lett.*, Vol. 25, p. 385, 1974.
195. K. Sasaki, T. Kinoshita, and N. Karasawa, *Appl. Phys. Lett.*, Vol. 45, p. 333, 1984.
196. K. D. Singer, J. E. Sohn, and S. J. Lalama, *Appl. Phys. Lett.*, Vol. 49, p. 248, 1986.
197. P. K. Tien, R. Ulrich, and R. J. Martin, *Appl. Phys. Lett.*, Vol. 17, p. 447, 1970.
198. S. Zemon, R. R. Alfano, S. L. Shapiro, and E. Conwell, *Appl. Phys. Lett.*, Vol. 21, p. 327, 1972.
199. L. Kuhn, *IEEE J. Quantum Electron.*, Vol. QE-5, p. 383, 1969.
200. D. B. Anderson, J. T. Boyd, and J. D. McMullen, *Proc. Symp. Submilliter Waves, MRI Series*, Vol. 20, p. 19, 1971.
201. E. M. Conwell, *IEEE J. Quantum Electron.*, Vol. QE-9, p. 867, 1973.
202. G. I. Stegeman, E. M. Wright, N. Finlayson, R. Zanoni, and C. T. Seaton, *IEEE J. Lightwave Tech.*, Vol. 6, p. 953, 1988.
203. J. T. Boyd and D. B. Anderson, *Opt. Commun.*, Vol. 13, p. 353, 1975.
204. M. Minakata, S. Saito, M. Shibata, and S. Miyazawa, *J. Appl. Phys.*, Vol. 49, p. 4677, 1978.
205. M. Johnson and C. W. Pitt, *Opt. Commun.*, Vol. 23, p. 121, 1977.
206. K. H. Haegele and R. Ulrich, *Opt. Lett.*, Vol. 4, p. 60, 1979.
207. D. B. Anderson, R. L. Davis, J. T. Boyd, and R. R. August, *IEEE J. Quantum Electron.*, Vol. QE-13, p. 268, 1977.
208. C. S. Tsai, Le T. Nguyen, S. K. Yao, and M. A. Alhaider, *Appl. Phys. Lett.*, Vol. 26, p. 140, 1975.
C. J. Li, C. S. Tsai, and C. C. Lee, *IEEE J. Quantum Electron.*, Vol. QE-22, p. 868, 1986.
D. A. Wille and M. C. Hamilton, *Appl. Phys. Lett.*, Vol. 24, p. 159, 1974.
209. D. Mergerian, E. C. Malarkey, R. P. Pautienus, J. C. Bradley, G. E. Marx, L. D. Hutcheson, and A. L. Kellner, *Appl. Opt.*, Vol. 19, p. 3033, 1980.
210. J. D. Spear-Zino, R. R. Rice, J. K. Powers, D. A. Bryan, D. G. Hall, E. A. Dalke, and W. R. Reed, *Proc. SPIE*, Vol. 239, p. 293, 1980.
211. C. M. Verber, R. P. Kenan, and J. R. Busch, *J. Lightwave Tech.*, Vol. LT-1, p. 256, 1983.
212. T. R. Joseph and B. U. Chen, Technical Digest of the Conference on Integrated and Guided wave Optics, Paper ME-2, Incline Village, NV, 1980.
213. C. S. Tsai, *IEEE Trans. Circuits Systems*, Vol. CAS-26, p. 1072, 1979.
214. M. J. Sun, *Appl. Phys. Lett.*, Vol. 33, p. 291, 1978.
215. M. Olivier and J-C Peuzin, *Appl. Phys. Lett.*, Vol. 32, p. 385, 1978.
216. J. D. Swalen, R. Santo, M. Tacke, and J. Fischer, *IBM J. Res. Dev.*, p. 168, March 1977.
217. J. S. Wei and W. D. Westwood, *Appl. Phys. Lett.*, Vol. 32, p. 819, 1978.

4

OPTICAL MODULATION: ELECTRO-OPTICAL DEVICES

S. Thaniyavarn

Boeing High Technology Center
Seattle, Washington

4.1 INTRODUCTION

In this chapter, we describe electro-optical devices whose optical transmission characteristics can be controlled by an applied electric field. In general, there are a variety of electro-optical effects such as the linear electro-optical (Pockels) effect, the quadratic electro-optical (Kerr) effect, and electro absorption. The most common and widely used electro-optical effect is the linear electro-optical effect.

The devices whose operation are based on this linear electro-optical effect have been receiving increasingly more attention due to their application to the rapid growth of low-loss, high-bandwidth single-mode fiber-optical systems. The technology to fabricate low-loss single-mode waveguide structures on electro-optical material such as lithium niobate and III–V semiconductors has made it possible to realize highly efficient, voltage-controlled optical devices. These devices are used as building blocks to perform various fundamental voltage-controlled optical signal-processing functions such as modulation, switching, and polarization control. Waveguides offer not only compact devices that are readily compatible with optical fibers but also the convenience of being able to fabricate many interconnecting devices on the same substrate. Several devices can be integrated together to form optical circuits to perform complex functions such as a switch array, an analog-to-digital converter, and various kinds of optical sensors.

This chapter is focused on the linear electro-optical devices based on single-mode waveguide structures. We begin the chapter with a discussion of the electro-optical effect. The basic principles of operation of fundamental electro-optical modulators are then reviewed.

There are many electro-optical materials such as lithium niobate, lithium tantalate, KTP, various III–V semiconductors, and organic polymers that have been used to fabricate electro-optical devices. Each material offers some advantages and disadvantages. The two material systems that have been receiving the most attention are lithium niobate and III–V semiconductor. At present, lithium niobate device technology is the most mature technology. This does not by any means imply that the III–V semiconductor electro-optical devices are less promising. However, for clarity, practical examples of devices based on lithium niobate waveguide structures will be used to illustrate the operation of electro-optical devices.

4.2 ELECTRO-OPTICAL EFFECT

The most basic signal processing of a light wave is to modulate (or alter) one of the characteristics (such as amplitude, phase, polarization, or propagation direction) of the light wave by an input signal. Information (such as in the form of an electrical signal) can be transferred onto an optical wave.

In certain types of material, the application of an electric field to the medium results in a change in the optical properties of the medium, which in turn changes the characteristics of light waves propagating through that medium.

In particular, for an electro-optical crystal such as lithium niobate, it is possible to change the index of refraction of the crystal by applying an electric field to the crystal. The change in the refractive index induced by the electric field in this case is proportional to the strength of the electric field. This is referred to as the linear electro-optical (Pockels) effect [1–2]. It provides a very convenient and efficient means of realizing various types of optical devices whose optical transfer characteristics can be controlled by an electrical signal.

The most widely used optical modulators are based on this linear electro-optical effect. The way in which the induced refractive index change results in intensity, phase, or polarization modulation depends upon the device configuration, the crystal orientation, and the direction of the applied electric field with respect to the optical polarization.

The basic operation of the electro-optical interaction can be best described by the use of an index ellipsoid [2]. The change in the refractive index due to the linear electro-optical effect is represented by the change in the components of the optical indicatrix matrix. The electro-optical effect is described in terms of third-rank tensors, r_{ij}, which express the linear dependence of the coefficients of the index ellipsoid on the applied electric field.

4.2.1 Index Ellipsoid

In general, the indices of refraction of an optical medium can be written in the form of a 3 × 3 indicatrix matrix ($\mathbf{n}$), which is related to the dielectric tensor ($\boldsymbol{\varepsilon}$) by the relation

$$\boldsymbol{\varepsilon} = \varepsilon_0 \cdot \left(\frac{1}{\mathbf{n}^2}\right)^{-1} \tag{4.1}$$

In the principal coordinate system,

$$\boldsymbol{\varepsilon} = \varepsilon_0 \begin{bmatrix} \frac{1}{n_{11}^2} & 0 & 0 \\ 0 & \frac{1}{n_{22}^2} & 0 \\ 0 & 0 & \frac{1}{n_{33}^2} \end{bmatrix}^{-1} = \varepsilon_0 \begin{bmatrix} n_{11}^2 & 0 & \\ 0 & n_{22}^2 & 0 \\ 0 & 0 & n_{33}^2 \end{bmatrix} \tag{4.2}$$

where 11, 22, and 33 correspond to x, y, and z, respectively.

The properties of the propagation of optical wave through the medium can be fully determined from the index ellipsoid surface [2].

In the principal coordinate system, the equation of the index ellipsoid in the absence of an applied electric field is expressed as [2]

$$\frac{x^2}{n_x^2} + \frac{y^2}{n_y^2} + \frac{z^2}{n_z^2} = 1 \tag{4.3}$$

where the directions x, y, and z are the principal dielectric axes of the crystal with the corresponding indices of refraction n_x, n_y, and n_z.

For a birefringent crystal, the z-axis is normally chosen to be along the optical C-axis of the crystal. The two refractive indices are then $n_x = n_y = n_o$ (ordinary index) and $n_z = n_e$ (extraordinary index).

For the electro-optical crystal, the application of an applied electric field results in the change in both the size and orientation of the index ellipsoid. The effect of the electric field on the propagation of the optical field can be analyzed most conveniently in terms of changes in the constants, $1/n^2$, of the index ellipsoid.

In the presence of an electric field, the general equation of the index ellipsoid is modified to

$$x^2\left(\frac{1}{n^2}\right)_1 + y^2\left(\frac{1}{n^2}\right)_2 + z^2\left(\frac{1}{n^2}\right)_3 + 2yz\left(\frac{1}{n^2}\right)_4 + 2zx\left(\frac{1}{n^2}\right)_5 + 2xy\left(\frac{1}{n^2}\right)_6 = 1 \quad (4.4)$$

where the subscripts 1, 2, 3, 4, 5, and 6 correspond to 11, 22, 33, 23, 31, and 12; the subscripts 1, 2, and 3 are associated with x, y, and z, respectively.

If x, y, and z are chosen to be along the principal dielectric axes, then in the absence of an external applied electric field, the equation of the index ellipsoid is reduced to Eq. (4.3). Therefore, for $\mathbf{E} \equiv 0$, $n_1 = n_x$, $n_2 = n_y$, and $n_3 = n_z$,

$$\left(\frac{1}{n^2}\right)_4 = \left(\frac{1}{n^2}\right)_5 = \left(\frac{1}{n^2}\right)_6 = 0 \quad (4.5)$$

The electro-optically induced change in the indices due to an electric field $\mathbf{E}$ ($\mathbf{E} = E_1, E_2, E_3$) is given by

$$\Delta\left(\frac{1}{n^2}\right)_i = \sum_{j=1}^{3} r_{ij}E_j \quad (4.6)$$

where E_j is the electric field component in the jth direction and the electro-optical coefficient r_{ij} is the ij(th) components of the electro-optical tensor.

The electro-optical relation can be expressed in matrix form as

$$\begin{vmatrix} \Delta\left(\frac{1}{n^2}\right)_1 \\ \Delta\left(\frac{1}{n^2}\right)_2 \\ \Delta\left(\frac{1}{n^2}\right)_3 \\ \Delta\left(\frac{1}{n^2}\right)_4 \\ \Delta\left(\frac{1}{n^2}\right)_5 \\ \Delta\left(\frac{1}{n^2}\right)_6 \end{vmatrix} = \begin{vmatrix} r_{11} & r_{12} & r_{13} \\ r_{21} & r_{22} & r_{23} \\ r_{31} & r_{32} & r_{33} \\ r_{41} & r_{42} & r_{43} \\ r_{51} & r_{52} & r_{53} \\ r_{61} & r_{62} & r_{63} \end{vmatrix} \begin{vmatrix} E_1 \\ E_2 \\ E_3 \end{vmatrix} \quad (4.7)$$

The linear electro-optical effect occurs only in the crystals that lack a center of inversion. Tables of electro-optical coefficients for various material can be found in Refs. 1 and 2. The way in which an applied electric field induces changes in the properties of the propagating optical wave depends on the polarization state of the optical wave, the direction of the electric field, and the electro-optical coefficients.

4.2.2 Lithium Niobate

In order to best describe electro-optical devices, we present practical examples of the devices fabricated on the electro-optical lithium niobate. Lithium niobate ($LiNbO_3$) is the most widely used material for fabricating electro-optical devices based on single-mode waveguide structures. Lithium niobate is a birefringent crystal. It has a large electro-optical coefficient. Most importantly, techniques for fabricating very low loss optical waveguides have reached the level of a mature technology. The waveguide size can be made to closely match that of a typical single-mode fiber.

The most popular method of fabricating waveguides on lithium niobate is by titanium diffusion [3]. Diffusing titanium into $LiNbO_3$ results in an increase in the refractive index just below the surface that is large enough to produce optical waveguiding. The dimensions of the optical waveguides can be controlled by choosing the proper initial titanium strip width, film thickness, and diffusion environment. Typically, single-mode waveguides for operation in the visible to near-infrared (IR) wavelengths ($\sim$0.6–1.6 μm) are formed by diffusion of a titanium strip of $\sim$2–8 μm in width and $\sim$0.02–0.1 μm thick at approximately 950–1050°C for 4–10 h in a wet argon or oxygen atmosphere. The resulting waveguides have a very low loss of typically only a few tenths of a decibel per centimeter.

The dimensions of the optical waveguides are normally designed to support a single spatial mode. The mode size is generally controlled to closely match that of the single-mode fibers for efficient optical coupling between the waveguide and the single-mode fiber. The waveguide device geometry can be formed by patterning the titanium film using typical photolithographic techniques. After the diffusion, the end surfaces are polished to optical quality. Coplanar electrode structures are then added to the waveguide circuits. Optimal electrode placement with respect to the waveguides are chosen to maximize the electro-optical interaction.

The refractive indices of the $LiNbO_3$ can be written as

$$n_x = n_o \qquad n_y = n_o \qquad n_z = n_e \tag{4.8}$$

where n_o (the ordinary index) is 2.29 and n_e (the extraordinary index) is 2.2.

$LiNbO_3$ belongs to the $3m$ crystallographic group. The electro-optical coefficients are in the form

$$\begin{vmatrix} 0 & -r_{22} & r_{13} \\ 0 & r_{22} & r_{13} \\ 0 & 0 & r_{33} \\ 0 & r_{51} & 0 \\ r_{51} & 0 & 0 \\ -r_{22} & 0 & 0 \end{vmatrix} \tag{4.9}$$

where

$$\begin{aligned} r_{61} &= -r_{22} \qquad & r_{42} &= r_{51} \\ r_{12} &= -r_{22} \qquad & r_{23} &= r_{13} \end{aligned} \tag{4.10}$$

and

$$\begin{aligned} r_{13} &= 8.6 \text{ pm/V} \qquad r_{33} = 30.8 \text{ pm/V} \\ r_{22} &= 3.4 \text{ pm/V} \qquad r_{51} = 28.0 \text{ pm/V} \end{aligned} \tag{4.11}$$

The equation of the index ellipsoid in the presence of electric field **E** (**E** $= E_1, E_2, E_3$) for lithium niobate is given by

$$\begin{aligned} & x^2\left(\frac{1}{n_o^2} - r_{22}E_2 + r_{13}E_3\right)_1 + 2yz(r_{51}E_2)_4 + y^2\left(\frac{1}{n_o^2} + r_{22}E_2 + r_{13}E_3\right)_2 \\ & \quad + 2zx(r_{51}E_1)_5 + z^2\left(\frac{1}{n_e^2} + r_{33}E_3\right)_3 - 2xy(r_{22}E_1)_6 = 1 \end{aligned} \tag{4.12}$$

The perturbation in the diagonal terms of the indicatrix matrix results in the phase change for an incident optical wave polarized along these principal axes:

$$\begin{aligned} \Delta\left(\frac{1}{n^2}\right)_1 &= -r_{22}E_2 + r_{13}E_3 \\ \Delta\left(\frac{1}{n^2}\right)_2 &= r_{22}E_2 + r_{13}E_3 \\ \Delta\left(\frac{1}{n^2}\right)_3 &= r_{33}E_3 \end{aligned} \tag{4.13}$$

For small perturbation, that is, $rE \ll 1/n^2$; the electro-optically induced changes in the diagonal terms of the indices (Δn's) are

$$\begin{aligned} n_{11} &= n_x = n_o + \Delta n_1 \rightarrow n_o - \tfrac{1}{2}n_o^3(r_{13}E_3 - r_{22}E_2) \\ n_{22} &= n_y = n_o + \Delta n_2 \rightarrow n_o - \tfrac{1}{2}n_o^3(r_{13}E_3 + r_{22}E_2) \\ n_{33} &= n_z = n_e + \Delta n_3 \rightarrow n_e - \tfrac{1}{2}n_e^3(r_{33}E_3) \end{aligned} \tag{4.14}$$

For $LiNbO_3$, since r_{33} is the largest coefficient, most of the devices are designed to utilize this interaction by applying the field along the z direction.

In particular, when the electric field is only in the z direction (E_3), only the diagonal terms of the indices are affected. The equation of the index ellipsoid reduces to

$$x^2\left(\frac{1}{n_o^2} + r_{13}E_3\right)_1 + y^2\left(\frac{1}{n_o^2} + r_{13}E_3\right)_2 + z^2\left(\frac{1}{n_e^2} + r_{33}E_3\right)_3 = 1 \tag{4.15}$$

For small perturbation, that is, $rE \ll 1/n^2$,

$$\begin{aligned} n_{11} &= n_x = n_o + \Delta n_1 \rightarrow n_o - \tfrac{1}{2}(n_o^3 r_{13}E_3) \\ n_{22} &= n_y = n_o + \Delta n_2 \rightarrow n_o - \tfrac{1}{2}(n_o^3 r_{13}E_3) \\ n_{33} &= n_z = n_e + \Delta n_3 \rightarrow n_e - \tfrac{1}{2}(n_e^3 r_{33}E_3) \end{aligned} \tag{4.16}$$

The off-diagonal elements of the indicatrix matrix represents electro-optically induced mixing between the orthogonal polarization components:

$$\Delta\left(\frac{1}{n^2}\right)_4 = r_{51}E_2 \qquad \Delta\left(\frac{1}{n^2}\right)_5 = r_{51}E_1 \qquad \Delta\left(\frac{1}{n^2}\right)_6 = r_{22}E_1 \tag{4.17}$$

Perturbation in the off-diagonal terms of the indices results in coupling between the two orthogonal polarization states of the incident optical wave polarized along these principal axes. This effect is utilized in the operation of polarization modulation.

4.3 ELECTRO-OPTICAL DEVICES

There are a variety of electro-optical devices that can be realized in waveguide forms [4]. Such waveguides offer not only devices that are readily compatible with optical fibers but also the convenience of being able to fabricate many interconnecting devices on the same substrate. Furthermore, electro-optical devices in waveguide structures are much more efficient than the free-space bulk optical structure counterpart. In free-space propagation, the beam diameter of the optical wave cannot be made arbitrarily small for a long propagation distance due to diffraction. In contrast, the optical wave is tightly confined to a very small transverse dimension as it propagates along the waveguide. As a result, electrode structures having a very small electrode spacing comparable to the waveguide width can be placed over or alongside the optical waveguide. Thus, a much higher electric field can be created inside the waveguide with a much smaller applied voltage resulting in a far more efficient and practical device than is possible with the free-space bulk optical version.

Electro-optical waveguide devices provide compact, voltage-tunable components that can alter the characteristics of light waves such as phase, amplitude, polarization, and direction of propagation. The most fundamental voltage-controlled devices are optical phase shifters, amplitude modulators, spatial switches, and polarization converters. In many cases, the same device can perform many functions. For example, an electro-optical directional coupler can function as either a modulator or a switch, depending on the strength of the interaction between the optical waves and the controlling electrical signal as well as on the arrangement of input and output ports.

The primary function of the modulator is to impress information on a light wave by temporally varying one of its properties. The switch either turns the light on and off or shifts the direction of output light from one port to another. Many of the same factors must be considered in designing and evaluating both modulators and switches.

4.3.1 Phase Modulator

The simplest form of electro-optical waveguide device is the optical phase modulator. The basic device configurations are shown in Figs. 4.1*a*,*b*. The device consists of a single-mode waveguide and a pair of coplanar electrodes placed above or alongside the waveguide.

The electrode configuration in Fig. 4.1*a* provides a maximum vertical electric field, while the electrode configuration in Fig. 4.1*b* provides the maximum horizontal electric field inside the optical waveguide. The choice between the two electrode configurations depends on the crystal cut and the waveguide orientation and is chosen to maximize the electro-optical interaction.

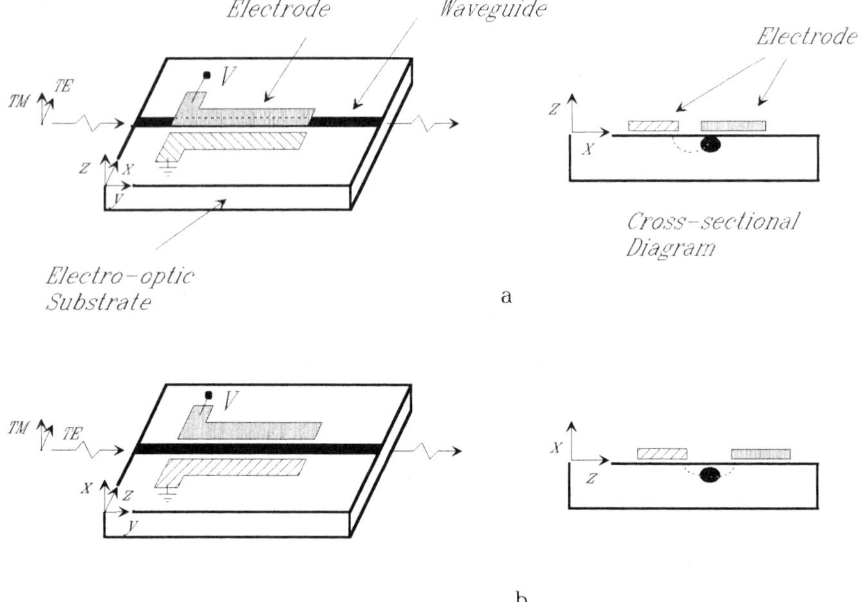

Figure 4.1 Schematics of electro-optical waveguide phase modulators.

When a modulation voltage is applied to the electrode pair, an electric field is created inside the waveguide that in turn changes the waveguide index via the electro-optical effect. The electro-optically induced index change results in the optical phase modulation of the optical wave propagating along the waveguide.

For a Ti:$LiNbO_3$ device, a typical waveguide size is fabricated to closely match that of the single-mode fiber, that is, $\sim$3–8 μm wide. The gap between the two electrodes is typically 5–10 μm wide. The waveguide propagation direction is normally oriented perpendicular to the z-axis of the crystal. In order to make use of the largest electro-optical coefficient (r_{33}), the electrode is designed to maximize the electric field inside the waveguide along the z-axis.

For a Z-cut substrate, the electrode is placed directly over the waveguide to create a predominately vertical (E_3) electric field within the waveguide (Fig. 4.1a). In order to eliminate optical propagation loss due to the metallic electrode loading effect, a thin dielectric buffer layer such as silicon dioxide is included to optically isolate the waveguide from the metallic electrode.

The electro-optically induced index changes due to the vertical electric field (E_3) for the vertical optical polarization (TM mode) and the horizontal optical polarization (TE mode) are, for the phase modulator in Fig. 4.1a,

$$n_{33} = n_z = n_e + \Delta n_e = n_e - \tfrac{1}{2}(n_e^3 r_{33} E_3) \quad \text{(TM mode)} \tag{4.18}$$

$$n_{11} = n_x = n_o + \Delta n_o = n_o - \tfrac{1}{2}(n_o^3 r_{13} E_3) \quad \text{(TE mode)} \tag{4.19}$$

For an X-cut (or Y-cut) substrate, the electrode is placed symmetrically along both sides of the waveguide to create a predominately horizontal (E_3) field inside the waveguide (Fig. 4.1b).

The electro-optically induced index changes due to the horizontal electric field (E_3) for the vertical optical polarization (TM mode) and the horizontal optical polarization (TE mode) are, for the phase modulator in Fig. 4.1*b*,

$$n_{11} = n_x = n_o + \Delta n_o = n_o - \tfrac{1}{2}(n_o^3 r_{13} E_3) \quad \text{(TM mode)} \tag{4.20}$$

$$n_{33} = n_z = n_e + \Delta n_e = n_e - \tfrac{1}{2}(n_e^3 r_{33} E_3) \quad \text{(TE mode)} \tag{4.21}$$

In order to achieve the most efficient phase modulation, the largest r_{33} electro-optical coefficient should be utilized. This means that the extraordinary wave that polarized along the z-direction should be utilized.

The effective change in the extraordinary index n_e is linearly proportional to the applied voltage V and can be written as

$$\Delta n_e = -\tfrac{1}{2}\left(\frac{n_e^3 r_{33} V}{d}\right)\alpha \tag{4.22}$$

where d is the electrode gap and α is the overlapping integral between the applied electric and optical fields ($0 \leqslant \alpha \leqslant 1$). Since the direction and strength of the applied electric field are nonuniform with respect to the optical field, the efficiency of the phase modulator depends on this overlapping integral of the electric field and the optical field. Theoretical calculation of the overlapping integral can be found in Refs. 5 and 6.

The corresponding change in the propagation constant ($\Delta\beta$) of the optical wave at wavelength λ is then

$$\Delta\beta = \frac{2\pi}{\lambda}\,\Delta n_e \tag{4.23}$$

The total electro-optically induced phase change ($\Delta\phi$) over an interaction length L is then

$$\Delta\phi = \Delta\beta\, L = \frac{2\pi L}{\lambda}\,\Delta n_e \tag{4.24}$$

As an example, a typical lithium niobate phase modulator with an electrode length of 1 cm for operation at $\sim 1\ \mu$m wavelength, the voltage V_π required to induce a phase change of 180° is of an order of a few volts.

4.3.2 Mach–Zehnder Interferometer

The most common electro-optical waveguide amplitude modulator is based on the Y-junction Mach–Zehnder Interferometer structure [7]. The device is illustrated in Fig. 4.2. It consists of a single-mode waveguide that branches out at a symmetric Y-junction into two single-mode waveguide arms. The two waveguide arms in the electrode region are far enough apart that there is no appreciable optical field interaction between them. Electrodes of a certain length L are provided as shown so that voltage can be applied to electro-optically induce a phase change to one or both of the waveguide paths. The electrodes can be arranged in a push–pull configuration to maximize the electro-optically induced phase difference between the paths. The two waveguide branches are then merged into one single-mode waveguide at another Y-junction.

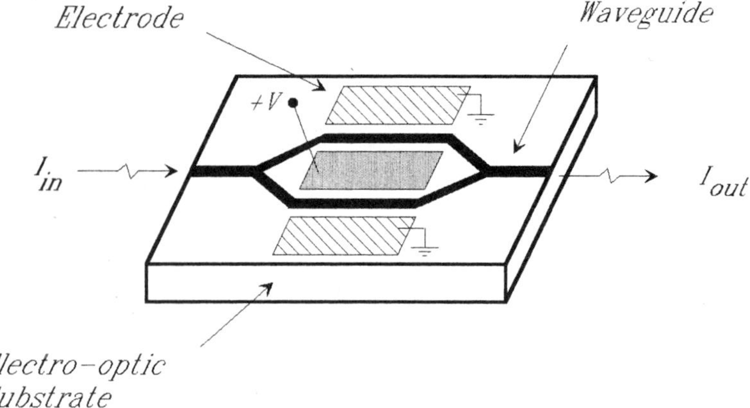

Figure 4.2 The Y-junction Mach–Zehnder interferometric waveguide modulator.

Light entering the input waveguide is first divided equally into the two arms of the interferometer, where the two optical waves propagate independently. If the optical path lengths of the two arms are identical, the two optical waves arrive at the output Y-junction in phase and interfere "constructively," producing a symmetric optical field profile that closely matches the lowest first-order mode profile of the output waveguide, and the light is coupled into the output waveguide.

However, if there is a 180° phase difference between the two optical path lengths, the two optical waves arrive at the output Y-junction with opposite phase and interfere "destructively," producing an antisymmetric field profile that closely matches the second-order mode profile. Since the output single-mode waveguide does not support the second-order antisymmetric mode, the optical power is not confined inside the output waveguide. The optical wave is lost by radiating into the substrate.

Using this interferometric structure, an electro-optically induced phase modulation can be converted to an amplitude modulation. The optical transmission of the interferometer is a sinusoidal function of the relative phase shifts between the two optical paths. The normalized optical power at the output waveguide (I_{out}) as a function of the electro-optically induced propagation mismatch ($\Delta\beta$) between the two waveguide arms is [7]

$$\begin{aligned} I_{\text{out}} &= \tfrac{1}{2}[1 + \cos(\Delta\phi)] \\ &= \cos^2(\tfrac{1}{2}\Delta\phi) \\ &= \cos^2(\tfrac{1}{2}\Delta\beta L) \\ &= \cos^2\left[(\tfrac{1}{2}\pi)\left(\frac{\Delta\beta L}{\pi}\right)\right] \end{aligned} \tag{4.25}$$

To facilitate the comparison of various kinds of modulator designs, the modulation characteristic is best expressed in terms of the optical transmission as a function of the normalized electro-optically induced propagation constant mismatch in the unit of $\Delta\beta L/\pi$. This propagation mismatch $\Delta\beta L/\pi$ is linearly proportional the applied voltage V.

The optical transmission characteristic I_{out} versus $\Delta\beta L/\pi$ of the interferometric modulator is shown in Fig. 4.3. The condition for a complete switching on and off of light

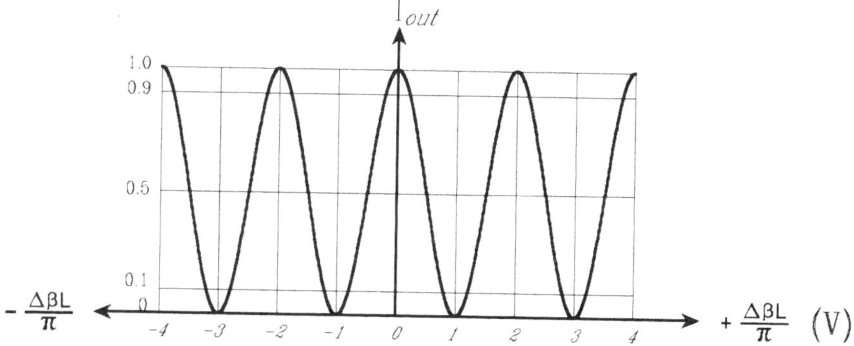

Figure 4.3 The I_{out} versus $\Delta\beta\, L/\pi$ characteristics of the Mach–Zehnder interferometer.

at the output waveguide (i.e., the switching condition) is given by

$$\frac{\Delta\beta\, L}{\pi} = 1 \tag{4.26}$$

For a typical operation as a linear modulator, a phase bias of $\frac{1}{2}\pi$ between the two arms must be introduced. This means a dc voltage bias is required. Alternatively, an intrinsic phase bias can be built into the device structure. This can be accomplished by making the device with a physical path length difference between the two interferometric arms by using different waveguide widths or different waveguide indices.

Another configuration of a similar device is the balanced bridge modulator [8,9]. The device is illustrated in Fig. 4.4. The device has two input and two output waveguides. Two 3-dB (50:50 power splitter) directional couplers are used instead of the symmetric Y-junctions. Similarly, electro-optical phase modulation of the two optical waves in the central section results in an amplitude modulation at each of the output waveguides. Depending on the relative phases of the two waves at the output coupler, the light will interfere, causing the optical power to transfer back and forth between the upper and the lower waveguides. Thus, this device can be used as a 2×2 directional switch as well.

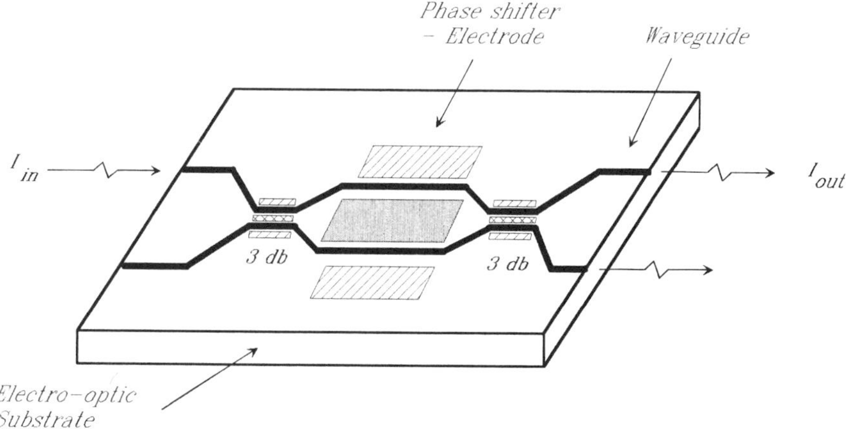

Figure 4.4 Schematic of the 2×2 balanced bridge modulator.

4.3.3 Directional Coupler

Another fundamental waveguide structure is the directional coupler. An optical waveguide directional coupler consists of two single-mode waveguides that are placed close to each other so that light propagating in one of the waveguides experiences the presence of the other guide. In this "interaction" region, optical power can be transferred back and forth between the two waveguides via the coupling of the evanescent fields [10].

The directional coupler can be characterized by the interaction length L, the coupling length l, and the mismatch $\Delta\beta$ between the propagation constants of the two waveguides. The coupling length is defined as the minimum length required in order to obtain a complete transfer of light from one guide to the other in the absence of the propagation mismatch $\Delta\beta$ between the two waveguides. The coupling length is related to the coupling constant κ by the relation $l = \pi/2\kappa$.

Operation of a directional coupler can be analyzed fully by the coupled-mode formulation [10]. The optical transmission characteristic depends on the normalized interaction length L/l and the propagation constant mismatch $\Delta\beta$ between the two coupling waveguides.

Electro-optical modulation and switching of the light between the two waveguides can be achieved by introducing electrodes into the coupler interaction region. When voltage is applied, the electric field created inside the waveguide coupler induces a change in the waveguide indices via the electro-optical effect that results in a change in the propagation constants of the two waveguides. By arranging electrodes with respect to the waveguides in a push–pull configuration, the electro-optically induced change in the propagation constants of each waveguide are of opposite polarity. The induced $\Delta\beta$ phase mismatch results in the modulation of the transfer characteristics of the directional coupler. The modulation or switching characteristics of the device depend not only on the waveguide structure, interaction length, and applied voltage but also on the electrode configuration as well. The interaction between these various design factors will be illustrated by considering both the 2×2 and 1×2 directional coupler waveguide in conjunction with single and multiple alternating multisection electrode configurations.

Uniform 2 × 2 Directional Coupler Modulator. The most common electro-optical directional coupler waveguide device is the 2×2 directional coupler switch. The device is illustrated in Fig. 4.5. The waveguides are all single-mode. The waveguide pairs in the input and output regions are far enough apart so that there is no optical field coupling between them. In the center region where the two waveguides are close enough to be optically coupled, a uniform electrode section is placed typically in a push–pull config-

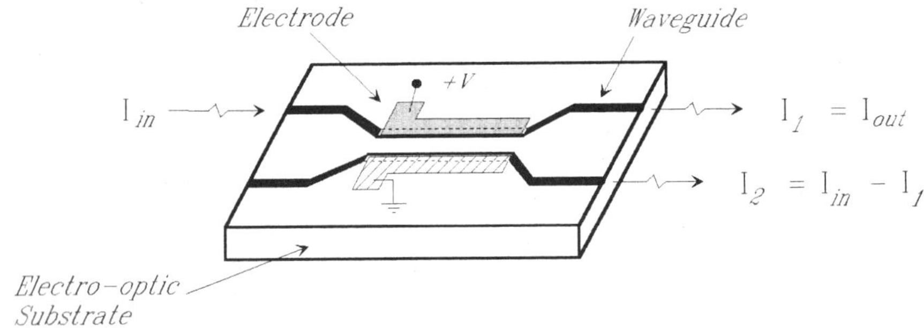

Figure 4.5 The 2×2 directional coupler waveguide switch.

uration. When a voltage is applied, the electric field inside the two coupling waveguides induces an equal but opposite change in the propagation constants ($\Delta\beta$) of the two waveguide paths, which in turn results in a modulation of the optical transfer characteristics.

Using the coupled-mode theory, the optical transfer characteristics of the device can be analyzed as follows.

Let $R(z)$ and $S(z)$ represent the normalized complex amplitude of the optical waves in the two coupled waveguides at a propagation distance z. The output optical fields $R(L)$ and $S(L)$ are related to the input $R(0)$ and $S(0)$ via a coupled-wave transfer matrix as follows [11]:

$$\begin{bmatrix} R(L) \\ S(L) \end{bmatrix} = \begin{bmatrix} A & -jB \\ -jB^* & A^* \end{bmatrix} \begin{bmatrix} R(0) \\ S(0) \end{bmatrix} \tag{4.27}$$

where A, B are the complex transferred matrix elements given by

$$\begin{aligned} A &= \cos(L\sqrt{\kappa^2 + \delta^2}) + j\frac{\delta}{\sqrt{\kappa^2 + \delta^2}}\sin(L\sqrt{\kappa^2 + \delta^2}) \\ B &= \frac{\kappa}{\sqrt{\kappa^2 + \delta^2}}\sin(L\sqrt{\kappa^2 + \delta^2}) \end{aligned} \tag{4.28}$$

where

$$\kappa = \frac{\pi}{2l} \qquad \delta = \frac{\Delta\beta}{2}$$

The terms A, B can be rewritten in terms of the normalized interaction length L/l and the normalized propagation mismatch $\Delta\beta\, L/\pi$ as follows:

$$\begin{aligned} A &= \cos(\theta) + j\frac{(\Delta\beta\, L/\pi)\sin(\theta)}{\sqrt{(L/l)^2 + (\Delta\beta\, L/\pi)^2}} \\ B &= \frac{(L/l)\sin(\theta)}{\sqrt{(L/l)^2 + (\Delta\beta\, L/\pi)^2}} \end{aligned} \tag{4.29}$$

where

$$\theta = \tfrac{1}{2}\pi\sqrt{\left(\frac{L}{l}\right)^2 + \left(\frac{\Delta\beta\, L}{\pi}\right)^2}$$

Therefore, the optical output power from the upper waveguide ($|R(L)|^2$) and lower waveguide ($|S(L)|^2$) can be expressed as functions of the normalized coupler length (L/l) and the normalized applied voltage in the unit of the electro-optically induced propagation constant mismatch ($\Delta\beta\, L/\pi$).

When light is launched only into the upper waveguide, that is, $R(0) = 1$ and $S(0) = 0$, the optical output (I_{out}) from the upper waveguide becomes

$$I_{\text{out}} = |R(L)|^2 = 1 - |S(L)|^2 = 1 - \frac{(L/l)^2 \sin^2\theta}{(L/l)^2 + (\Delta\beta\, L/\pi)^2} \tag{4.30}$$

For a switch operation, the switch is said to be in the "straight-through" state when all the optical power exits from the same waveguide as the input, that is, $R(L) = 1$ and $S(L) = 0$, and in a "crossover" state if all the optical power is transferred over to the other waveguide, that is, $R(L) = 0$ and $S(L) = 1$.

In its normal operation, the interaction length L of the coupling region is made to be exactly one coupling length l. A plot of the output optical power (I_{out}) as a function of the electro-optically induced propagation constant mismatch ($\Delta\beta\, L/\pi$) for this case is shown in Fig. 4.6.

With no voltage applied, all the light is in the crossover state. The voltage required to switch the light completely from the crossover state back to the straight-through state is expressed by the switching condition

$$\frac{\Delta\beta\, L}{\pi} = \sqrt{3} \tag{4.31}$$

In general, for a directional coupler device with an arbitrary interaction length L/l, it is not always possible to switch the light completely between the two outputs. The generalized modulation characteristics of the devices with an arbitrary coupling length can be described most conveniently by the switching/modulation diagrams [11].

By using Eq. (4.30), a diagram can be constructed to represent the contour map of the two switching states as a function of the coupling length L/l and the propagation constant mismatch $\Delta\beta\, L/\pi$. We will discuss the use of the switching diagram in more detail in a later section.

Symmetric 1 × 2 Directional Coupler Modulator. The symmetric 1 × 2 directional coupler modulator is illustrated in Fig. 4.7. The device consists of only one input single-mode waveguide that branches out into a pair of symmetric single-mode waveguide directional couplers of length L and a uniform electrode section [12].

Because of the symmetry of the device configuration, equal optical power with exactly the same phase is launched from the input waveguide into the waveguide directional coupler pairs. The optical power at each output waveguide is automatically set at the 3-dB half-power point at zero voltage. Thus, unlike the conventional Mach–Zehnder interferometer or the 2 × 2 directional coupler, the symmetric 1 × 2 directional coupler modulator is a self-biased structure, requiring no dc voltage bias.

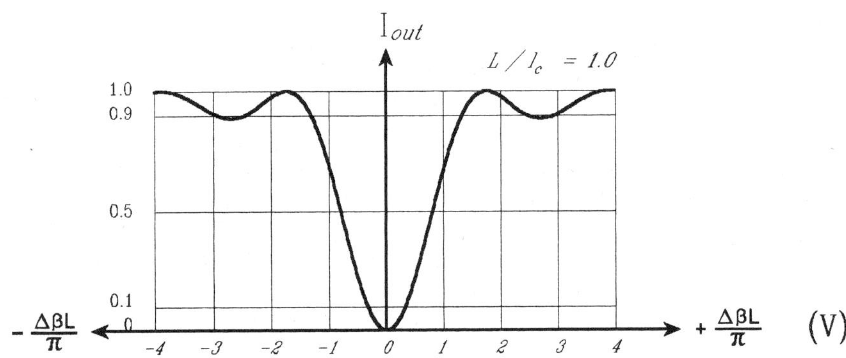

Figure 4.6 The I_{out} versus $\Delta\beta\, L/\pi$ characteristics of the 2 × 2 uniform directional coupler modulator.

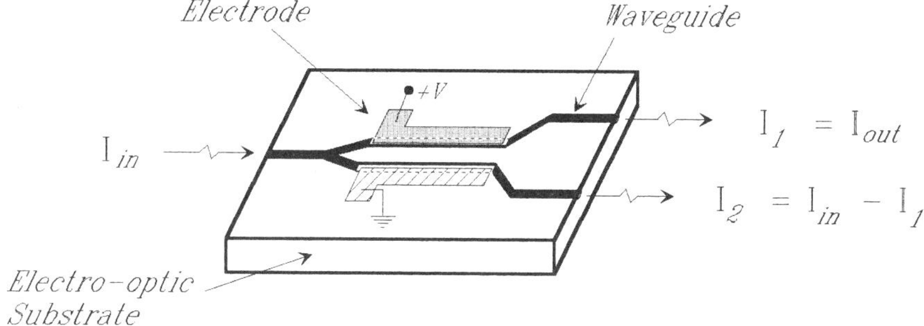

Figure 4.7 The symmetric 1 × 2 directional coupler waveguide modulator.

Using the same coupled-mode analysis as in the previous section, the output power in the upper waveguide is given by

$$
\begin{aligned}
I_{\text{out}} &= |R(L)|^2 = 1 - |S(L)|^2 \\
&= \frac{1}{2}\left[1 - \frac{2(L/l)(\Delta\beta\, L/\pi)}{(L/l)^2 + (\Delta\beta\, L/\pi)^2}\sin^2(\theta)\right]
\end{aligned}
\tag{4.32}
$$

For normal operation as a linear modulator or a 1 × 2 optical switch, the device is designed so that the interaction length $L/l = 1/\sqrt{2} \approx 0.7$.

The plot of the output (I_{out}) as a function of the electro-optically induced propagation constant mismatch ($\Delta\beta\, L/\pi$) for this case is shown in Fig. 4.8. The light can be switched between the two outputs by applying a voltage corresponding to a $\Delta\beta\, L/\pi$ of $\pm 1/\sqrt{2}$.

Thus, the switching condition for switching the light between the two outputs is

$$
\frac{\Delta\beta\, L}{\pi} = \sqrt{2} \tag{4.33}
$$

Notice that the switching condition between the two switch states for the 1 × 2 directional coupler is more efficient than that of the conventional 2 × 2 directional coupler

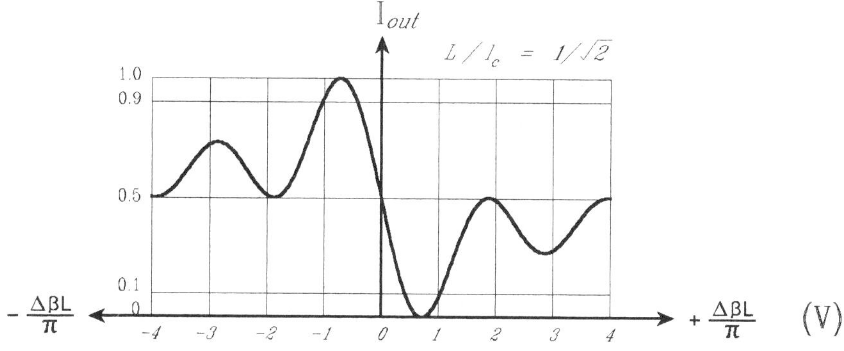

Figure 4.8 The I_{out} versus $\Delta\beta\, L/\pi$ characteristics of the symmetric 1 × 2 directional coupler modulator.

($\Delta\beta L/\pi = \sqrt{3}$). Modulation characteristics of the device for other coupling lengths will also be examined using the modulation diagram in Section 4.4.

Stepped $\Delta\beta$ Phase Reversal Directional Coupler Switch. In general, the interaction length L of a directional coupler switch with uniform electrode design has to be controlled with a very tight tolerance to be exactly one coupling length in order to obtain complete switching. This stringent requirement can be greatly relaxed with the use of multiple sections of electrode having alternating polarities. The device is known as a stepped $\Delta\beta$ phase reversal directional coupler [11,13].

The schematic diagram of the simplest device of this type is illustrated in Fig. 4.9. The device consists of a 2×2 directional coupler waveguide structure. The electrode consists of two equal-length sections in which voltages of equal magnitude but opposite in polarity are applied to the alternating section. The length of each electrode section is equal to half the interaction length ($d_1 = d_2 = \frac{1}{2}L$); and the electro-optically induced phase mismatch in the first section is the reverse of that in the second section, that is, $\Delta\beta_1 = -\Delta\beta_2 = \Delta\beta$. Detail of the device operation is theoretically described fully in Ref. 12. With this electrode configuration, it is always possible to achieve complete switching as long as the interaction length L is controlled to be in the range 1–3 of the coupling length l using only voltage adjustment.

The plot of the output optical power (I_{out}) as a function of the electro-optically induced propagation constant mismatch ($\Delta\beta L/\pi$) for the case $L/l = 1.5$, 2, 2.5 is shown in Fig. 4.10. Complete modulation characteristics using modulation diagrams will also be deferred until Section 4.4.

4.3.4 Polarization Modulator

The basic polarization modulator is the TE–TM mode converter. Electro-optical polarization conversion is accomplished by use of an off-diagonal element of the electro-optical tensor. This results in coupling between the two normally uncoupled orthogonal TE and TM polarizations. In order to convert between the two modes efficiently, the phase-matching condition between the two polarization modes must also be satisfied.

The device consists of a straight channel waveguide that supports one TE mode and one TM polarization mode. Depending on the crystal cut and the waveguide orientation, the electrode configuration is designed so that the two modes can be properly phase

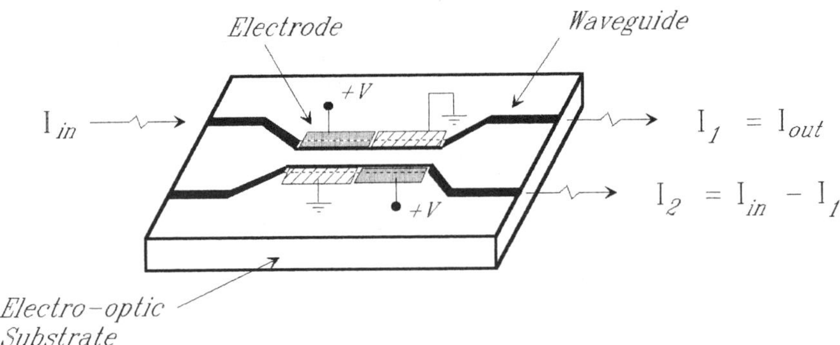

Figure 4.9 The 2×2 directional coupler waveguide switch with two-section stepped $\Delta\beta$ phase reversal electrodes.

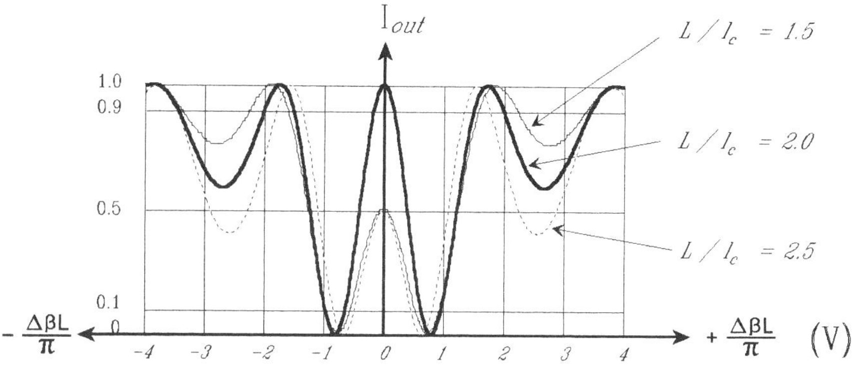

Figure 4.10 The I_{out} versus $\Delta\beta L/\pi$ characteristics of the stepped $\Delta\beta$ phase reversal directional coupler switch.

matched and the electro-optical interaction is via the off-diagonal terms to induce coupling between the two polarization modes.

In lithium niobate, there are two basic polarization converter designs. Depending on crystal orientation and waveguide orientation, a polarization conversion with a narrow or a broad wavelength response can be constructed.

Narrow-Band TE–TM Converter. The $LiNbO_3$ waveguide narrowband TE–TM mode converter is illustrated in Fig. 4.11. The device consists of a straight y-propagation waveguide on an X-cut crystal and a periodic interdigitated electrode structure. The waveguide supports one TM mode polarized along the x direction and one TE mode polarized along the z direction [14].

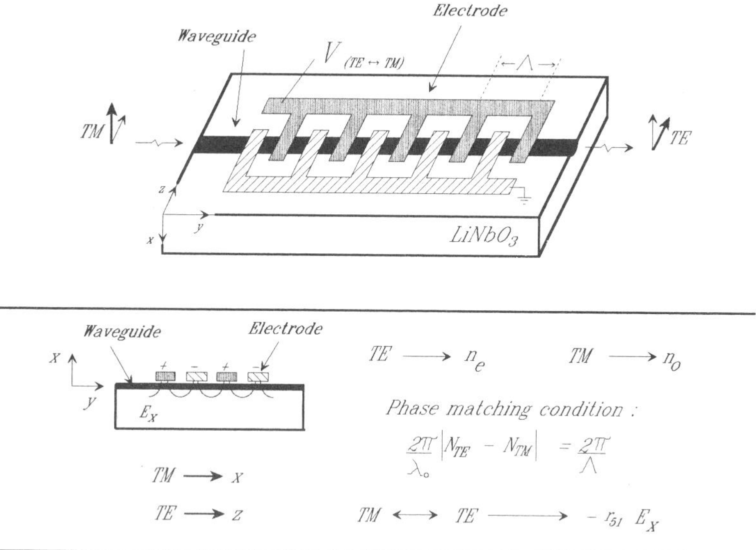

Figure 4.11 The $LiNbO_3$ waveguide narrow-band TE–TM converter.

From the fifth term in the index ellipsoid equation (4.12), in order to induce the coupling between these two modes, the r_{51} electro-optical coefficient and the applied electric field E_1 directed along the x direction must be utilized.

Using the coupled-mode theory analysis, the electro-optically induced TE–TM mode coupling (κ) is given by

$$\kappa = \frac{\pi}{2\lambda}(n^3 r_{51})E_1 \tag{4.34}$$

Because of the large birefringence property of the $LiNbO_3$ crystal, the TM and TE modes sense different crystal indices. Due to the large difference between the effective TM and TE mode indices ($N_{TM} \approx n_o$ and $N_{TE} \approx n_e$), the two polarization modes are not phase matched. In order to achieve the phase-matching condition, a periodic electrode structure, as shown in Fig. 4.11, is required to provide periodic interaction.

Let Λ be the period of the electrode. Efficient phase matching via periodic coupling is achieved only for the wavelength λ_0 that satisfies the phase-matching condition:

$$\Lambda|N_{TM} - N_{TE}| = \lambda_0 \tag{4.35}$$

Once the phase-matching condition is obtained, the electro-optically induced TE–TM (or vice versa) conversion efficiency (η) is given by

$$\eta = \sin^2(\kappa L) \tag{4.36}$$

where L is the length of the electrode.

The switching condition for a complete polarization mode conversion is given by

$$\frac{\kappa L}{\pi} = \frac{1}{2} \tag{4.37}$$

Polarization conversion in this case will be effective only in a narrow range of wavelengths determined by the period of the electrode that satisfies the phase-matching condition equation (4.35). Thus the device is used as a narrow-band wavelength filter.

In general, by adding electro-optical TE–TM phase shifters to the TE–TM polarization converter, an arbitrary polarization transformer can be constructed [14].

Broadband TE–TM Converter. The $LiNbO_3$ waveguide broadband TE–TM converter is illustrated in Fig. 4.12. The device consists of a straight z-propagation waveguide on an X-cut crystal [15]. The electrode structure consists of three separated electrode sections symmetrically located with respect to the waveguide. The waveguide supports one TM mode polarized along the x direction and one TE mode polarized along the y direction. One of the outer electrodes is grounded and two voltages V_1 and V_2 can be applied to the center and the other electrode, respectively. This makes it possible to adjust the field strength of both the vertical E_1 and the horizontal E_2 electric fields inside the waveguide.

From the last term in the index ellipsoid equation (4.12), in order to induce coupling between these two polarization modes, the r_{61} $(= -r_{22})$ electro-optical coefficient and the applied electric field E_1 directed along the x direction must be utilized.

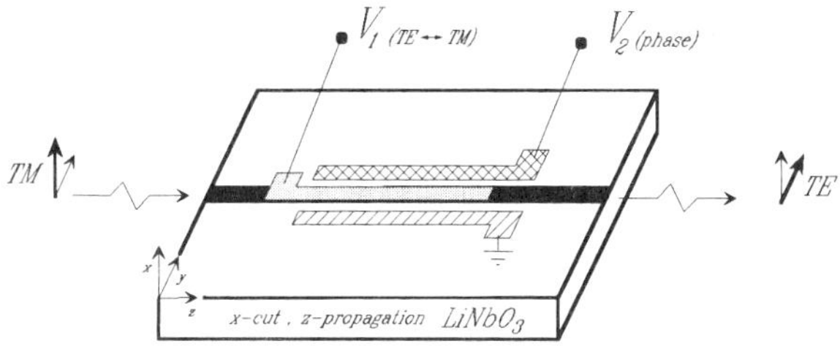

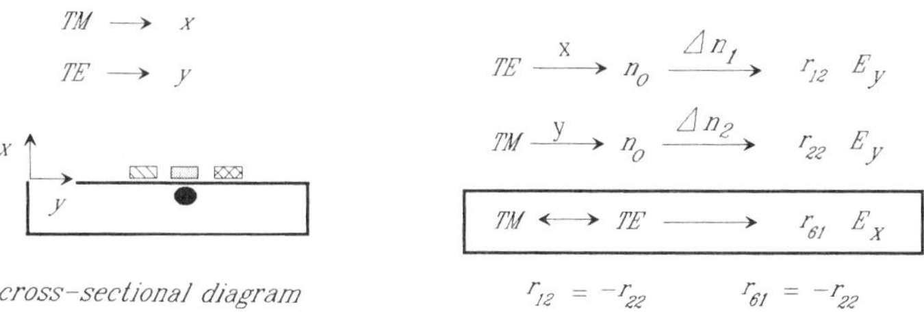

Figure 4.12 The $LiNbO_3$ waveguide broadband TE–TM converter.

Using the coupled-mode theory analysis, the electro-optically induced TE–TM mode coupling (κ) is given by

$$\kappa = \frac{\pi}{2\lambda}(n^3 r_{61})E_1 \tag{4.38}$$

Since the waveguide propagation direction is along the optic C-axis, the TM and TE modes sense the same ordinary crystal index. Thus the effective TM and TE mode indices are already almost equal, that is, $N_{\text{TM}} \approx N_{\text{TE}} \approx n_o$. Therefore, the two modes are already nearly phase matched. The only contribution to the phase mismatch comes from the small modal birefringence due to the difference in waveguide boundary conditions sensed by the two modes. This small modal mismatch can be completely eliminated electro-optically as follows.

From Eq. (4.12), the horizontal field E_2 provides electro-optical phase shifts Δn_1 via the $r_{12}(= -r_{22})$ and Δn_2 via the r_{22} coefficients for the TM and the TE modes, respectively. Since the phase shifts are of opposite signs, complete phase matching of the mode indices $N_{\text{TM}} = N_{\text{TE}}$ can be achieved by adjusting the strength of the appropriate horizontal field.

Since absolute phase matching can be achieved, extremely high polarization conversion efficiency can be achieved for a very wide range of wavelength of operation.

4.3.5 Other Electro-optical Devices

In addition to basic phase amplitude and polarization modulation, frequency modulation of light waves is also possible. The simplest electro-optical frequency modulation technique is based on *serrodyne* techniques [16]. By applying an appropriate sawtooth drive signal to the electro-optical phase modulator, a single frequency-shifted signal can be generated. The frequency shift is directly proportional to the rate of change of the applied voltage. In practice, the performance of the frequency shifter is limited because an ideal sawtooth waveform with instantaneous falltime cannot be generated. Alternatively, a double-sideband frequency shifter can be generated by sinusoidally driving a Mach–Zehnder interferometer [17]. By properly driving a combination of Mach–Zehnder interferometer circuits, a single-sideband frequency shifter can also be generated [18]. Yet another approach is to utilize the polarization TE–TM mode converter structure [19].

Various types of optical multiplexers such as wavelength and polarization multiplexers can be fabricated using basic electro-optical device structures. For example, a wavelength division multiplexer with a periodic wavelength filter response can be realized using a zero-gap directional coupler (or a two-mode-interference) structure [20]. A TE–TM polarization splitter–combiner can be realized by using an asymmetrically loaded 2×2 directional coupler waveguide switch [21].

The devices discussed represent only a few examples of many types of voltage-controlled electro-optical devices that can be realized. Most of these devices are based on the combination or modification of the basic phase, amplitude, and polarization modulator structures, as illustrated in the case of frequency modulators. There are many excellent review articles describing many other types of electro-optical devices. Interested readers are referred to the references.

4.4 SWITCHING/MODULATION DIAGRAMS

A directional coupler is by far one of the most versatile type of waveguiding elements. The optical transfer characteristics of this device depend not only on the waveguide and electrode structures but also on the length of the coupling region and the applied voltage. Perhaps the simplest and most convenient method of describing the characteristics of directional coupler switches and modulators is through the use of switching/modulation diagrams.

The switching diagram is a plot describing graphically the locations of the two switch states, namely, the straight-through state and the crossover state of an electro-optical directional coupler switch as a function of the normalized interaction length L/l and the applied voltage in units of $\Delta\beta L/\pi$, (i.e., the normalized electro-optically induced propagation mismatch) [11]. In many cases, a directional coupler device can be operated as a modulator as well as an optical switch. By enhancing the switching diagram to include contour maps of various levels of normalized optical power output, the modulation depth characteristics of a modulator and the crosstalk characteristics of the switch can be derived [22].

The modulation diagram can be used to analyze the behaviors of both conventional and nonideal or nonconventional device structures. In particular, modulation diagrams for conventional devices, including the uniform 2×2 directional coupler, the symmetric 1×2 directional coupler, and a stepped two-section $\Delta\beta$ phase reversal directional coupler are presented. For a nonideal device, a modulation diagram of a Mach–Zehnder interferometer is constructed to describe the effect on the modulation depth due to a finite optical coupling between the two interferometer waveguide arms. Finally, modulation diagrams are constructed to describe nonconventional switch structures, namely, a 1×2 $\Delta\beta$ phase reversal digital switch design.

4.4.1 Modulation Diagrams

The modulation characteristics of a directional coupler can be calculated using the coupled-mode theory analysis. As has been described in detail for the case of the 2×2

directional coupler with uniform electrode, the normalized optical output power can be related to the input optical wave via a transfer matrix characterizing the device. The modulation diagram is the contour plot of the normalized optical output power (I_{out}) for various power levels as a function of both the normalized interaction length L/l and the applied voltage in the normalized electro-optically induced phase mismatch ($\Delta\beta\, L/\pi$) unit. For clarity, the modulation diagram includes only the contour maps of output power levels of significant interest (i.e., >99.9, 99.0–99.9, 90.0–99.0, 1.0–10.0, 0.1–1.0, and <0.1%). The diagram describing modulation depth can also be used to examine the crosstalk behavior of the device operating as an optical switch as well. The power levels mentioned in the preceding correspond to the crosstalk levels of a switch in the range of less than −30, −20 to −30, and −10 to −20 dB, respectively.

For a device structure with two outputs, such as the 2 × 2 directional coupler switch, the ⊜ represents the contour map of the straight-through switch state with better than −30 dB crosstalk. Similarly, the ⊗ represents the crossover switch states with better than −30 dB crosstalk. In the case of an interferometer with only one output, the ⊜ represents the on state with >99.9% of the light at the output waveguide. The ⊗ represents the off state with <0.1% of the light at the output.

4.4.2 Uniform 2 × 2 Directional Coupler

Using Eq. (4.30), which describes the normalized optical output power as a function of L/l and $\Delta\beta\, L/\pi$, the modulation diagram is constructed and is shown in Fig. 4.13. The solid regions surrounded by the hatched and cross-hatched regions correspond to the straight-through (⊜) and crossover (⊗) switch states with better than −30 dB crosstalk respectively. The hatched regions surrounding the ⊜ region correspond to the straight-through state with a cross-talk level between −10 and −20 dB. Similarly, the cross-

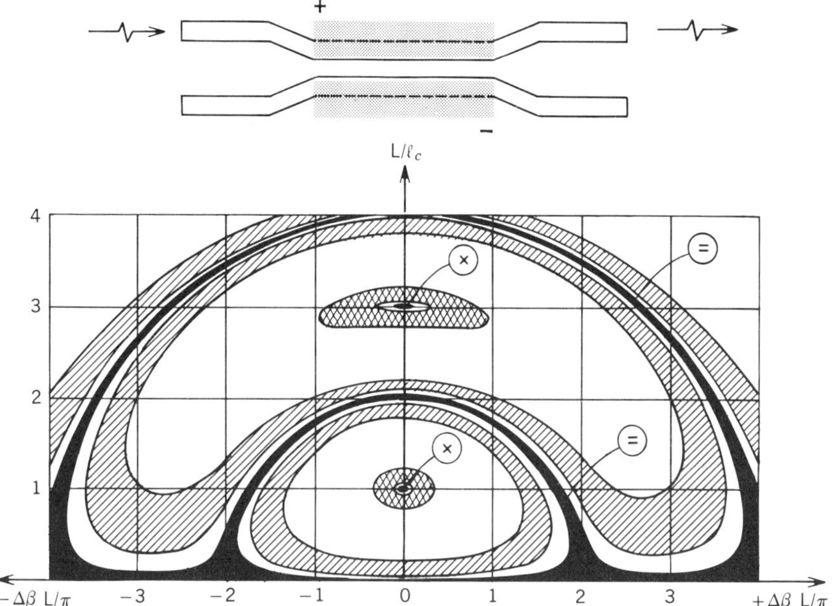

Figure 4.13 Modulation diagram of the uniform 2 × 2 directional coupler

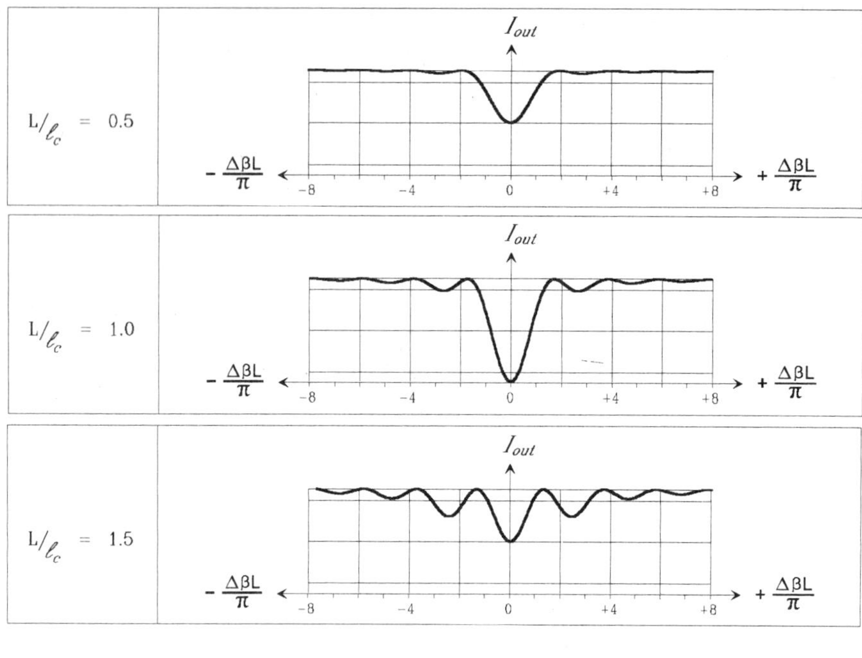

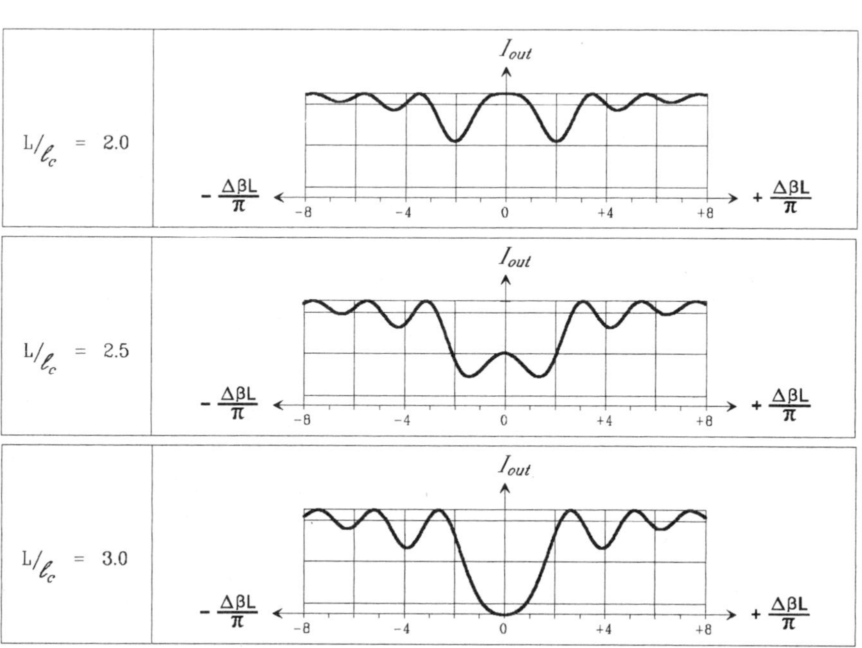

Figure 4.14 The I_{out} versus $\Delta\beta\, L/\pi$ modulation characteristics of the uniform 2×2 directional coupler.

hatched regions surrounding the ⊗ region correspond to the crossover state with a crosstalk level between −10 and −20 dB.

Using this modulation diagram, it can be readily seen that it is not always possible to switch the light into the crossover (⊗) state. To operate the device as a 2 × 2 switch, the device must be in the cross-state at no applied voltage, that is, $L/l = 1, 3, \ldots$. It can also be deduced that in order to achieve maximum modulation depth or minimum crosstalk level and lowest switching voltage, the coupler should be designed so that the normalized interaction length $L/l = 1$. Figure 4.14 shows the I_{out} versus $\Delta\beta L/\pi$ modulation characteristics for various interaction lengths, namely $L/l = 0.5$, 1.0, 1.5, 2.0, 2.5, 3.0.

4.4.3 Symmetric 1 × 2 Directional Coupler

The modulation diagram of the 1 × 2 directional coupler is shown in Fig. 4.15. The diagram shows that the light can be electro-optically switched completely between the two states if the normalized interaction length L is designed to be about 0.7, 2.1, and 3.5. Figure 4.16 shows I_{out} versus $\Delta\beta L/\pi$ modulation characteristics for the devices with various interaction lengths, namely, 0.7, 2.1, 3.5, 0.5, 1.5, and 3.0. For linear modulation and lowest-voltage operation, the appropriate interaction length $L/l = \sqrt{2}/2 \approx 0.7$.

4.4.4 $\Delta\beta$ Phase Reversal 2 × 2 Directional Coupler

The modulation diagram of the 2 × 2 directional coupler with two-section stepped $\Delta\beta$ phase reversal electrodes is shown in Fig. 4.17. The diagram shows that it is always possible to achieve complete switching between the two output states with low crosstalk as long as the normalized interaction length L/l is designed to be in the range 1–3. Figure

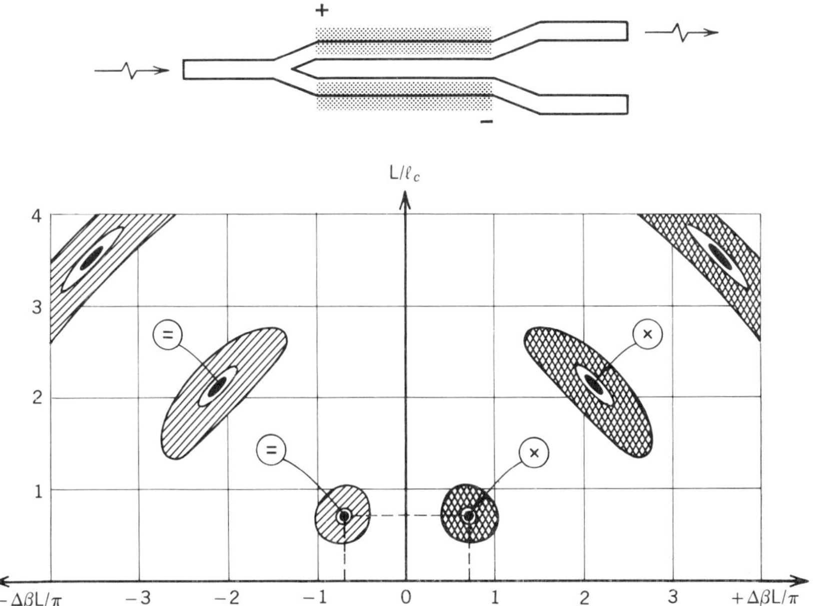

Figure 4.15 Modulation diagram of the symmetric 1 × 2 directional coupler.

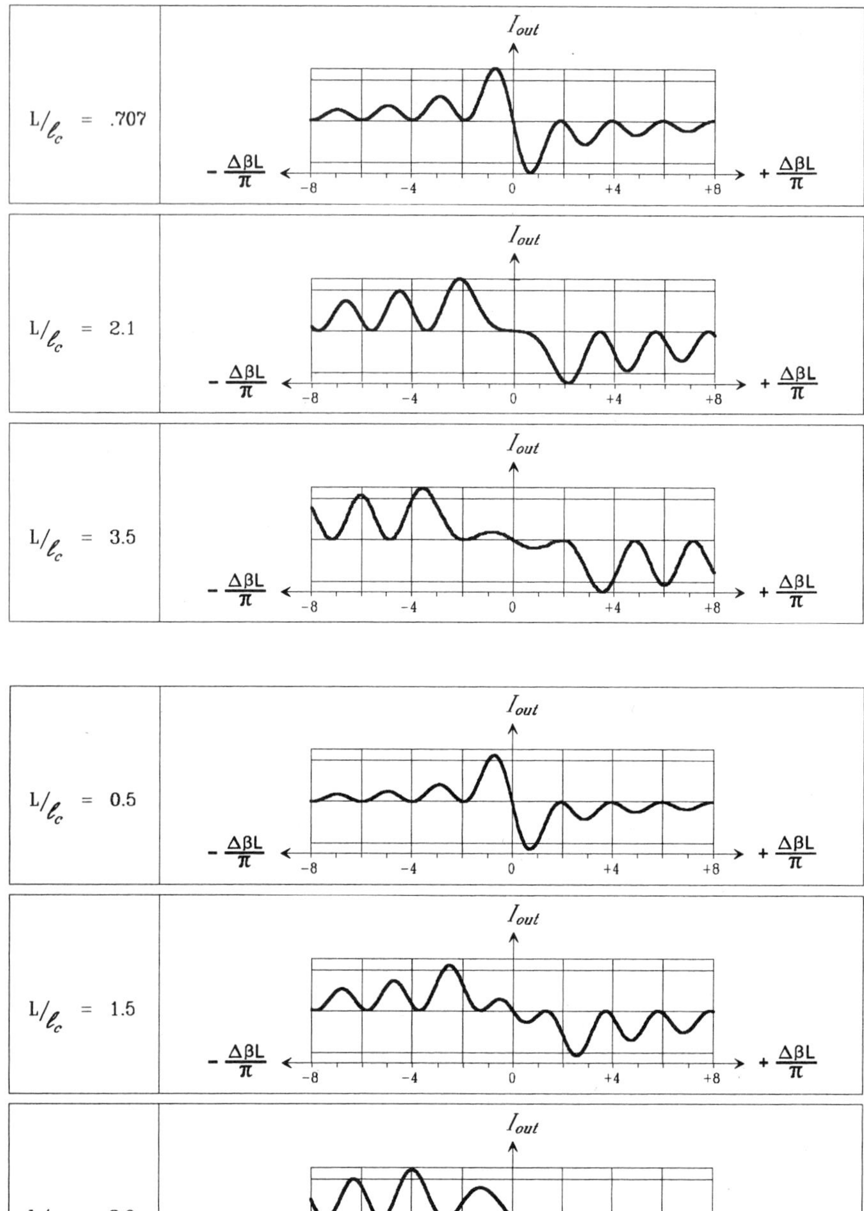

Figure 4.16 The I_{out} versus $\Delta\beta\, L/\pi$ modulation characteristics of the symmetric 1×2 directional coupler.

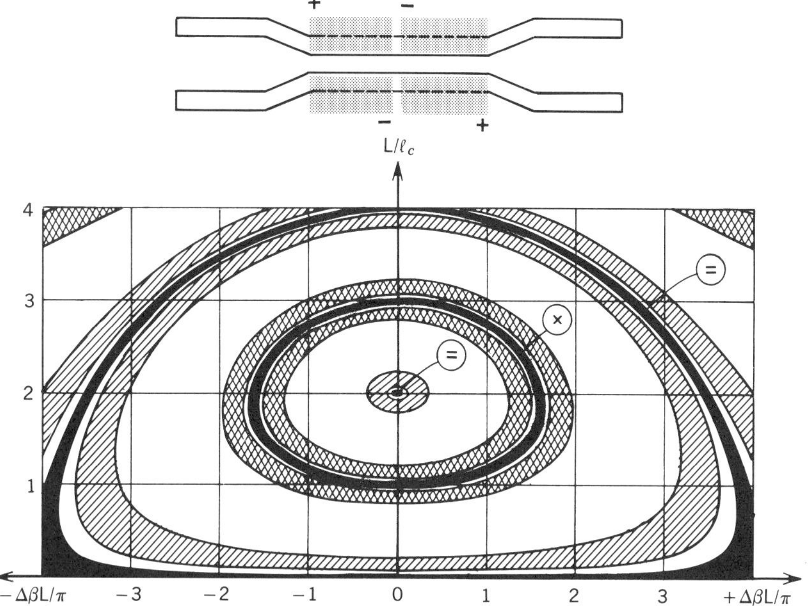

Figure 4.17 Modulation diagram of the 2 × 2 directional coupler with two-section stepped $\Delta\beta$ phase reversal electrodes.

4.18 shows I_{out} versus $\Delta\beta\, L/\pi$ modulation characteristics for devices with various interaction lengths, namely, $L/l = 0.5, 1.0, 1.5, 2.0, 2.5, 3.0$.

The modulation diagram in Fig. 4.17 is constructed for an ideal device in which the electrode sections are exactly equal. However, if the device structure is not ideal, the modulation characteristics of the device are affected. The effect on the modulation characteristics caused by various nonideality in the device structure can be easily analyzed by constructing an appropriate modulation diagram that includes the nonideal condition. In particular, modulation diagrams for various cases, such as unequal electrode-sectional length and a finite passive optical coupling section outside the interaction region, have been constructed to examine the degradation of the crosstalk characteristics of the $\Delta\beta$ phase reversal directional coupler switch. Details of these modulation diagrams are described in Ref. 23.

The next two examples demonstrate the use of modulation diagrams to examine the effect of nonideal device structure as well as nonconventional device design.

4.4.5 Mach–Zehnder Interferometer (with Coupled Waveguide Arms)

A conventional Mach–Zehnder interferometer has a sinusoidal type of modulation response. For an ideal interferometer, the two interferometer waveguide arms must be far enough apart so that there is no significant optical coupling between them. If there is a finite coupling between the two waveguide arms, the modulation depth of the Mach–Zehnder is degraded. The modulation diagram of the interferometer with finite coupling between the two arms is shown in Fig. 4.19. The diagram shows that at zero coupling ($L/l = 0$; i.e., an ideal interferometer) the light can be switched on (⊜) and off (⊗) periodically, as expected. However, when there is significant coupling, the ⊗ states can no longer be reached, that is, poor modulation depth. Figure 4.20 shows I_{out} versus $\Delta\beta\, L/$

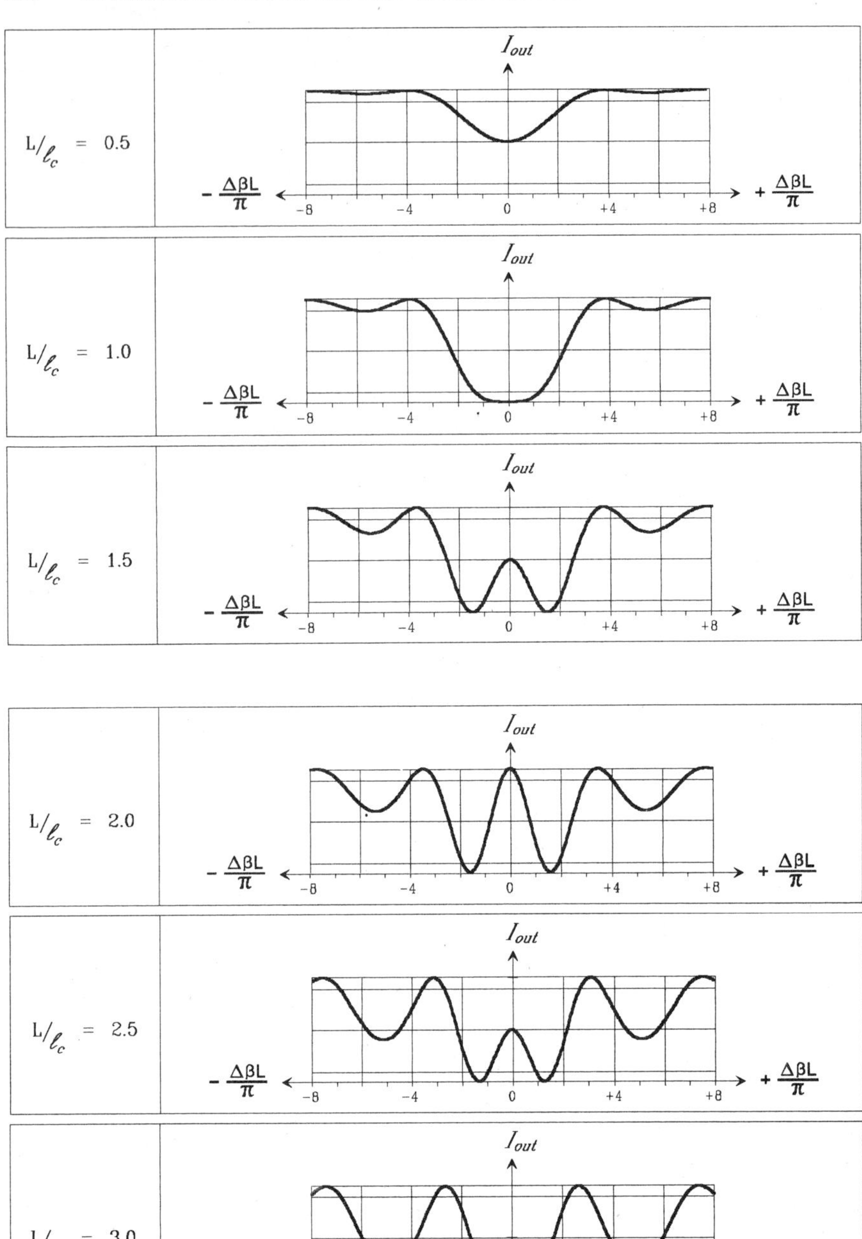

Figure 4.18 The I_{out} versus $\Delta\beta L/\pi$ modulation characteristics of the 2×2 $\Delta\beta$ phase reversal directional coupler.

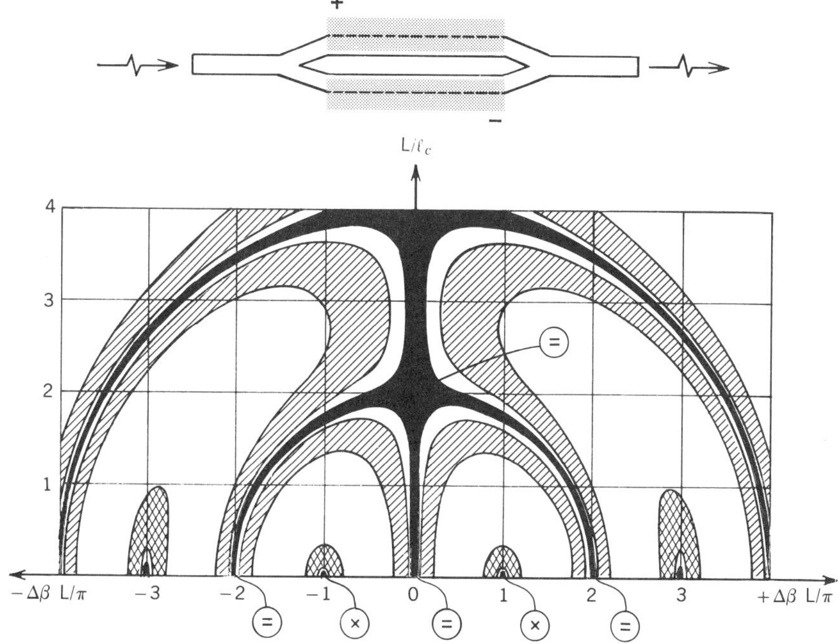

Figure 4.19 Modulation diagram of the interferometer with finite optical coupling between the two arms.

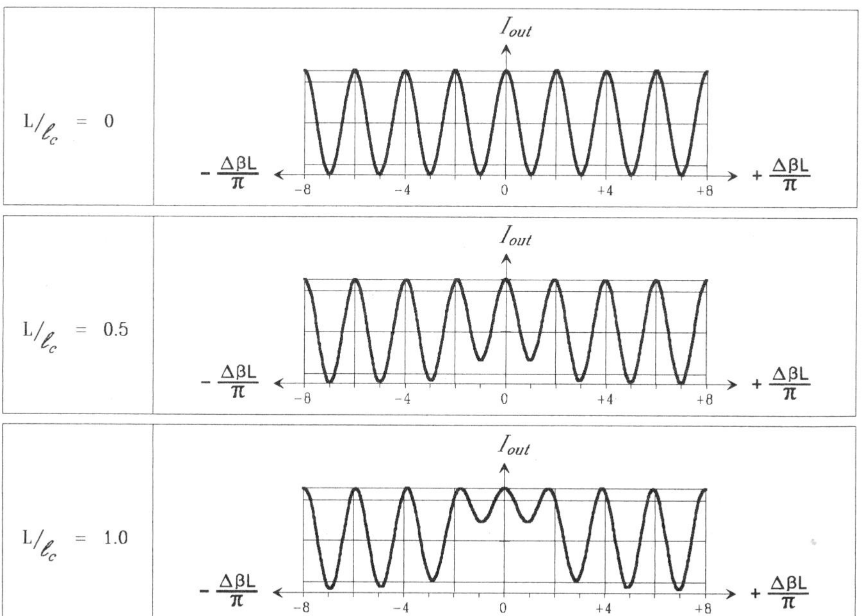

Figure 4.20 The I_{out} versus $\Delta\beta\, L/\pi$ modulation characteristics of the nonideal interferometer for various coupling strengths.

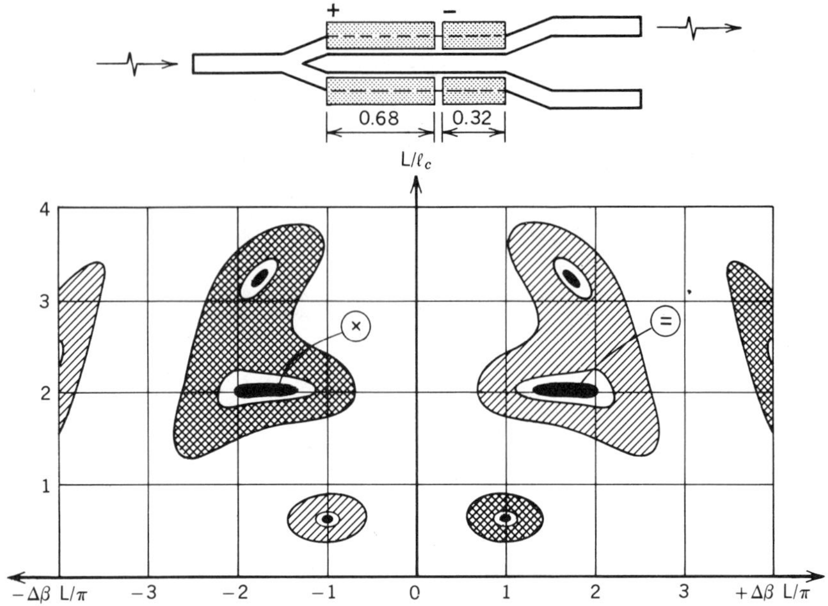

Figure 4.21 Modulation diagram of the $\Delta\beta$ phase reversal 1 × 2 "digital" switch using two sections of unequal length.

π modulation characteristics for the interferometer with various optical coupling between the two arms, namely, $L/l = 0.0, 0.5, 1.0$. Modulation depth degrades as the coupling between the two interferometric arms grows stronger.

4.4.6 $\Delta\beta$ Phase Reversal 1 × 2 Digital Switch Design

A modulation diagram can also be extremely useful to examine new device designs. As an example, a modulation diagram (shown in Fig. 4.21) is constructed for a 1 × 2 directional coupler waveguide structure design (Fig. 4.22) having an asymmetric $\Delta\beta$ phase reversal electrode of specific sectional length $L_1/L = 0.68$ and $L_2/L = 0.32$ [24]. The diagram shows that the device has an enlarged region of the two switch states when the interaction length is properly designed to be ≈ 2 coupling length. The I_{out} versus

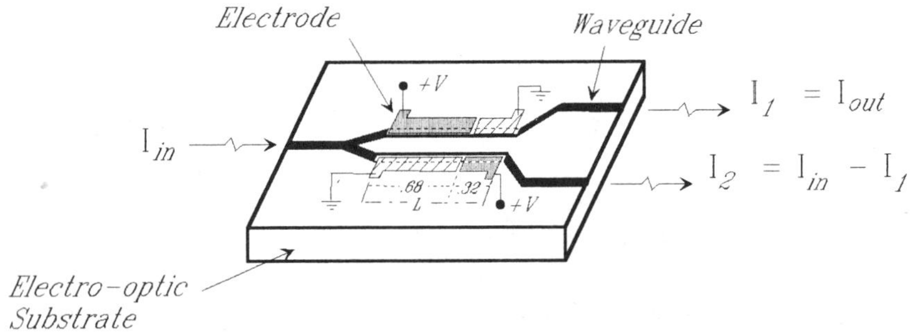

Figure 4.22 Schematic diagram of the $\Delta\beta$ phase reversal 1 × 2 directional coupler "digital" switch.

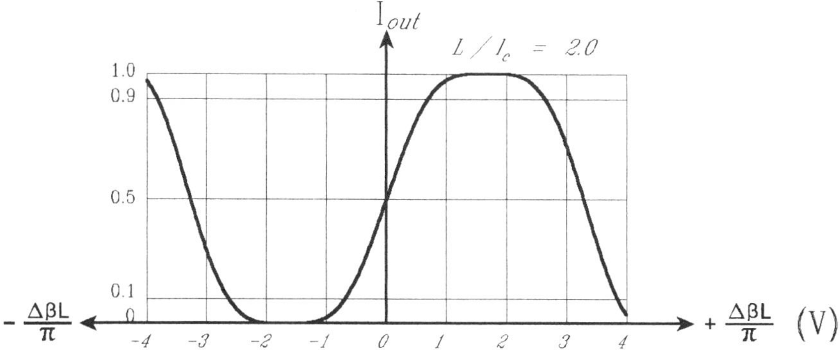

Figure 4.23 The I_{out} versus $\Delta\beta L/\pi$ modulation characteristics of the 1 × 2 "digital" switch.

$\Delta\beta L/\pi$ modulation characteristics in this case is shown in Fig. 4.23. The switch has a flat voltage response of that of a "digital" switch. This means that precise control of the switching voltage is not required since switch states with very low levels of crosstalk can be maintained for a large range of applied voltages.

4.4.7 Discussion

In general, switching/modulation diagrams can be constructed to examine any directional-coupler-based electro-optical devices. The basic characteristics of a particular device structure can first be calculated using the coupled-mode theory analysis. As has been described in some detail for the case of the 2 × 2 directional coupler with uniform electrode, the optical transfer matrix characterizing the device can be deduced. The normalized optical output power can then be related to the input optical wave via this transfer matrix. The switching or modulation diagrams can then be constructed to analyze the device characteristics in more graphical terms. One can construct diagrams to examine various nonideality in the device structures. The diagrams can also be used as a tool in the design and analysis of new device structures to achieve the desired modulation characteristics.

4.5 HIGH-SPEED MODULATION

Electro-optical effects such as in lithium niobate and III–V semiconductor exhibit extremely fast (subpicosecond scale) response to an applied modulating electric field. This makes it possible to realize very high speed devices. Electro-optical modulators with bandwidth well in excess of 10 GHz have already been used as key components in wide-band fiber-optical systems.

The modulation speed of the modulator depends on the electrode geometry and the transit time of the optical wave through the device. The electrodes can be designed to behave like lumped circuit elements or electrical transmission lines.

As a lumped circuit element, the electrode can be considered as a capacitor. The modulator response time is then limited by the RC time constant, where C is the electrode capacitance and R is the load resistance normally chosen to be $R = 50\ \Omega$ to match the standard microwave driver impedance. The response time can also be limited by the optical transit time through the device. This occurs when the transit time is comparable to the period of the modulating electric field. The transit time must therefore be kept small compared to the modulation period.

Alternatively, the electrode can be designed as a microwave transmission line, thus eliminating the RC response time limitation. In this case, the electrical modulating signal is launched as a traveling wave into the transmission line and allows it to propagate codirectionally with the propagating optical wave inside the waveguide. This results in a much more efficient modulator, having a larger bandwidth. In order to obtain an efficient wide-band response, the transmission line electrode structure must be designed to properly match to the microwave input connector and must be terminated in a matched impedance. In general, the propagation speed of the optical wave is not the same as the electrical signal propagating down the transmission line. The electrical bandwidth of the traveling-wave modulator is ultimately limited by the mismatch in the transit time between the two propagating waves [25]. The bandwidth (Δf) for a traveling-wave modulator is given by [2,19]

$$\Delta f \approx \frac{c/2L}{n - n_{\mathrm{RF}}} \tag{4.39}$$

where L is the electrode transmission line length, c is the speed of light in free space, n is the optical index, and n_{RF} is the effective RF index.

Experimental traveling-wave Ti:$LiNbO_3$ waveguide modulators with a bandwidth in excess of 10 GHz are now commonplace [26]. Advance designs such as an aperiodic phase reversal traveling-wave structure with a Barker code can extend the modulator response well into the millimeter-wave regime [27].

Furthermore, if wide-bandwidth response is not required, periodic phase reversal traveling-wave electrode structures or resonant electrode structures can be used to achieve highly efficient high-frequency modulators with narrow-band response [28].

4.6 SUMMARY

We have discussed voltage-controlled optical waveguide devices based on the linear electro-optical effect. Only the most common and fundamental electro-optical modulators have been presented. Many excellent review articles describe various types of electro-optical devices [29–35]. Interested readers will find them in the references.

REFERENCES

1. I. P. Kaminow and E. H. Turner, "Linear Electrooptical Materials," in *Handbook of Lasers*, Chemical Rubber Co., Cleveland, OH, 1971.
2. A. Yariv, *Quantum Electronics*, Wiley, New York, 1975; A. Yariv and P. Yeh, *Optical Waves in Crystals*, Wiley, New York, 1984.
3. R. V. Schmidt and I. P. Kaminow, "Metal Diffused Optical Waveguides in $LiNbO_3$," *Appl. Phys. Lett.*, Vol. 25, pp. 458–460, 1974; M. Fukuma and J. Noda, "Optical Properties of Ti-diffused $LiNbO_3$ Strip Waveguides," *J. Appl. Phys.*, Vol. 19, pp. 4677–4682, 1978.
4. I. P. Kaminow, "An Introduction to Electrooptic Devices," Academic, New York, 1974.
5. D. Marcuse, "Optimal Electrode Design for Integrated Optics Modulators," *IEEE J. Quantum Electron.*, Vol. QE-18, pp. 393–398, 1982.
6. O. G. Ramer, "Integrated Optic Electrooptic Modulator Electrode Analysis," *IEEE J. Quantum Electron.*, Vol. QE-18, pp. 386–392, 1982.

7. F. J. Leonberger, "Applications of Guided-Wave Interferometers," *Laser Focus*, pp. 125–129, March 1982; C. H. Bulmer and W. K. Burns, "Linear Interferometric Modulators in Ti:$LiNbO_3$," *IEEE J. Lightwave Tech.*, Vol. LT-2, pp. 512–521, 1984.
8. V. Ramaswamy, M. D. Divino, and R. D. Standley, "A Balanced Bridge Modulator Switch," *Appl. Phys. Lett.*, Vol. 32, pp. 644–646, 1978.
9. M. Minakata, "Efficient $LiNbO_3$ Balanced Bridge Modulator/Switch with Ion-Etched Slot," *Appl. Phys. Lett.*, Vol. 35, pp. 40–42, 1979.
10. S. E. Miller, "Coupled-Wave Theory and Waveguide Applications," *Bell Syst. Tech. J.*, Vol. 33, pp. 661–719, 1954; A. Yariv, "Coupled-Mode Theory for Guided-Wave Optics," *IEEE J. Quantum Electron.*, Vol. QE-9, pp. 919–934, 1973.
11. H. Kogelnik and R. V. Schmidt, "Switched Directional Couplers with Alternating $\Delta\beta$," *IEEE J. Quantum Electron.*, Vol. QE-12, pp. 396–401, 1976.
12. S. Thaniyavarn, "Modified 1×2 Directional Coupler Waveguide Modulator," *Elect. Lett.*, Vol. 22, p. 941, 1986.
13. R. V. Schmidt and H. Kogelnik, "Electrooptically Switched Coupler with Stepped $\Delta\beta$ Reversal Using Ti-diffused $LiNbO_3$ Waveguides," *Appl. Phys. Lett.*, Vol. 2, pp. 503–505, 1976.
14. R. C. Alferness, "Electrooptic Guided-Wave Device for General Polarization Transformations," *IEEE J. Quantum Electron.*, Vol. QE-17, pp. 965–969, 1981.
15. S. Thaniyavarn, "Wavelength-Independent, Optical-Damage-Immune $LiNbO_3$ TE–TM Mode Converter," *Opt. Lett.*, Vol. 11, pp. 39–41, 1986.
16. K. K. Wong and R. M. De La Rue, "Electrooptic-Waveguide Frequency Translator in $LiNbO_3$ Fabricated by Proton Exchange," *Opt. Lett.*, Vol. 7, pp. 546–548, 1982; L. Thylen, P. Sjoberg, and G. E. Lindqvist, "Electro-optical Serrodyne Frequency Translator for $\lambda = 1.3\ \mu m$," *IEEE Proc.*, Vol. 132, pp. 119–121, 1985.
17. W. A. Stallard, T. G. Hodgkinson, K. R. Preston, and R. C. Booth, "Novel $LiNbO_3$ Integrated Optic Component for Coherent Optical Heterodyne Detection," *Elec. Lett.*, Vol. 21, pp. 1077–1078, 1985.
18. M. Izutsu, S. Shikama, and T. Sueta, "Integrated Optical SSB Modulator/Frequency Shifter," *IEEE J. Quantum Electron.*, Vol. QE-17, pp. 2225–2227, 1981.
19. F. Heismann and R. Ulrich, "Integrated-Optical Frequency Translator with Stripe Waveguide," *Appl. Phys. Lett.*, Vol. 45, pp. 490–492, 1984.
20. A. Neyer, "Integrated-Optical Multi-channel Wavelength Multiplexer for Monomode Systems," *Elec. Lett.*, Vol. 20, pp. 744–746, 1984.
21. O. Mikami, "$LiNbO_3$ Coupled-Waveguided TE/TM Mode Splitter," *Appl. Phys. Lett.*, Vol. 36, pp. 491–493, 1980.
22. S. Thaniyavarn, "Modulation Diagrams of Coupled-Waveguide Interferometers and Directional Couplers," *SPIE Proc.*, Vol. 835, pp. 197–204, 1986.
23. S. Thaniyavarn, "Crosstalk Characteristics of $\Delta\beta$ Phase Reversal Directional Coupler Switches," *SPIE Proc.*, Vol. 578, pp. 192–198, 1985.
24. S. Thaniyavarn, "A Synthesized Digital Switch Using a 1×2 Directional Coupler with Asymmetric $\Delta\beta$ Phase Reversal Electrode," Integrated and Guided-Wave Optics 1988 Technical Digest series, Vol. 5, pp. TUC6-1 to TUC6-4, Optical Society of America, Washington DC, 1988.
25. M. Izutsu, Y. Yamane, and T. Sueta, "Broad-Band Traveling-Wave Modulator Using $LiNbO_3$ Optical Waveguide," *IEEE J. Quantum Electron.*, Vol. QE-13, pp. 287–290, 1977; K. Kubota, J. Noda, and O. Mikami, "Traveling-Wave Optical Modulator Using Directional Coupler Ti:$LiNbO_3$ Waveguide," *IEEE J. Quantum Electron.*, Vol. QE-16, pp. 754–760, 1981; S. K. Korotky and R. C. Alferness, "Time- and Frequency-Domain Response of Directional-Coupler Traveling-Wave Optical Modulators," *J. Lightwave Technol.*, Vol. LT-1, pp. 244–251, 1983.
26. C. M. Gee, G. D. Thurmond, and H. W. Yen, "17-GHz Bandwidth Electrooptic Modulator," *Appl. Phys. Lett.*, Vol. 43, pp. 998–1000, 1983; S. Korotky, G. Eisenstein, R. Tucjer, J. Veselka, and G. Raybon," Optical Intensity Modulation to 40 GHz Using a Waveguide Electro-Optic Switch," *Appl. Phys. Lett.*, Vol. 50, pp. 1631–1633, 1987.

27. D. W. Dolfi, M. Nazarathy, and R. L. Jungerman, "Wide-Bandwidth 13-Bit Baker Code $LiNbO_3$ Modulator with Low Drive Voltage," *Tech. Dig. Opt. Fiber Commun. Conf., OFC '88*, Paper ThA2, 1988.

28. R. C. Alferness, S. K. Korotky, and E. Marcatili, "Velocity-matching Techniques for Integrated Optic Traveling-Wave Switch/Modulators," *IEEE J. Quantum Electron.*, Vol. QE-20, pp. 301–310, 1984; L. Thylen and A. Djupsjobacka, "Bandpass Response Traveling-Wave Modulator with Transit Time Difference Compensation Scheme," *IEEE J. Lightwave Technol.*, Vol. LT-3, pp. 47–51, 1985; M. Izutsu, H. Murakami, and T. Sueta, "Standing-Wave Structure Optical Waveguide Modulator Using a Resonant Electrode at 10 GHz," in *Tech. Dig. Opt. Fiber Commun. Conf., OFC/IOOC '87*, Vol. TuQ 32, January 1987.

29. I. P. Kaminow, "Optical Waveguide Modulators," *IEEE Trans. Microwave Theory Tech.*, Vol. MTT-23, pp. 57–70, 1975.

30. R. C. Alferness, "Guided Wave Devices for Optical Communication," *IEEE J. Quantum Electron.*, Vol. QE-17, pp. 946–959, 1981.

31. R. C. Alferness, "Waveguide Electrooptic Modulators," *IEEE J. Trans. Microwave Theory Tech.*, Vol. MTT-30, pp. 1121–1137, 1982.

32. T. Sueta and M. Izutsu, "High Speed Guided-Wave Optical Modulators," *J. Opt. Commun.*, Vol. 3, pp. 52–58, 1982.

33. E. Voges and A. Neyer, "Integrated-Optic Devices on $LiNbO_3$ for Optical Communication," *IEEE J. Lightwave Technol.*, Vol. LT-5, pp. 1229–1238, 1987.

34. S. K. Korotky and R. C. Alferness, "Waveguide Electrooptic Devices for Optical Fiber Communication," in *Optical Fiber Telecommunications*, II, Academic Press, Boston. 1988, Chap. 11.

35. R. Syms, "Advances in Channel Waveguide Lithium Niobate Integrated-Optics," *Opt. Quantum Electron.*, Vol. 20, pp. 189–213, 1988.

5

OPTICAL MODULATION: ACOUSTO-OPTICAL DEVICES

Chen S. Tsai

Department of Electrical and Computer Engineering and Institute for Surface and Interface Science
University of California
Irvine, California

5.1 INTRODUCTION

Acousto-optics broadly refers to the interactions between optical (light) waves and acoustic (sound) waves. In engineering, however, it is now common to refer to Acousto-optics more narrowly as influences of the latter upon the former, influences that have been successfully utilized to construct various types of devices of both scientific and technological importance. Significant influences are possible under certain situations as the refractive index gratings created by acoustic waves will cause diffraction or refraction of an incident light wave. Acousto-optics is further branched into two subareas: bulk-wave and guided-wave acousto-optics. In the former [1–9] both light and sound propagate as unguided (unconfined) columns of waves inside a medium. A great many studies in the area of bulk-wave acousto-optics since the 1960s have resulted in various types of bulk acousto-optical devices and subsystems. In the latter [10–16] both light and sound waves are confined to a small depth in a suitable solid substrate. This subarea has been a subject of intensive study since the early 1970s as an outgrowth of guided-wave optics science and technology [17–26] and surface acoustic wave device technology [27–31], which had undergone intensive research and development a few years earlier. These latest studies on guided-wave acousto-optics have generated many fruitful results [15,16]. For example, the resulting wide-band planar guided-wave acousto-optical (AO) Bragg modulators and deflectors are now widely used in the development and realization of micro-optical modules for real-time processing of radar signals, for example, integrated optic RF spectrum analyzers. Serious attempts to realize a variety of other integrated AO device modules with applications to optical communications, computing, and signal processing are also being made [32–34]. This chapter presents a brief review of bulk-wave acousto-optics and a detailed treatment of guided-wave acousto-optics with emphasis on the principles of AO Bragg diffraction, the resulting wide-band modulators and deflectors, and applications.

In Section 5.2 the geometry, working principles, and device characteristics of a basic bulk-wave AO Bragg modulator together with its applications are reviewed briefly. It suffices to present a brief review on this subarea as a number of excellent review papers and chapters have been written [1–9]. The basic configuration and mechanisms for planar guided-wave AO Bragg diffraction and the resulting diffraction efficiency and frequency response are then analyzed in detail using the coupled-mode technique (Section 5.3). For convenience, the analysis is carried out for the simple but basic case involving a single-surface acoustic wave (SAW) in a $LiNbO_3$ waveguide. The salient differences between guided-wave AO Bragg diffraction and bulk-wave AO Bragg diffraction are discussed. A comparison of the three major types of AO materials that have been studied is also given in Section 5.3. The key parameters of the resulting AO Bragg modulators and deflectors or Bragg cells and their inherent limitations are then identified and discussed (Section 5.4). Extension of the coupled-mode technique to analysis of AO Bragg diffraction from multiple SAWs, namely, multiple tilted SAWs and phased SAWs, is briefly discussed in Section 5.5. Subsequently, a number of SAW transducer configurations for realization of wide-band Bragg cells are described and compared. Also presented in Section 5.5 are the design, fabrication, testing, and measured performances of wide-band AO Bragg cells in Y-cut $LiNbO_3$ and Z-cut GaAs substrates. Some of the potential applications of wide-band guided-wave AO Bragg cells and modules in optical communications, computing, and RF signal processing are described in Section 5.6.

5.2 BULK-WAVE ACOUSTO-OPTICAL BRAGG DIFFRACTION AND APPLICATIONS

As an introduction to this chapter, a brief review of the geometry, working principles, and device characteristics of a basic bulk-wave AO Bragg modulator, as shown in Fig. 5.1, together with its applications now follows. A detailed treatment on application will be given in Section 5.5 since practically all applications of bulk-wave AO devices can be carried over to their guided-wave counterparts. A basic bulk-wave AO Bragg modulator or cell consists of an acoustic cell of sufficiently large aperture and a laser source, both operating at suitable wavelengths. The acoustic cell is in turn made of a column of isotropic or anisotropic material that possesses desirable acoustic, optical, and AO properties. The acoustic waves are generated by applying an RF signal (at a frequency ranging

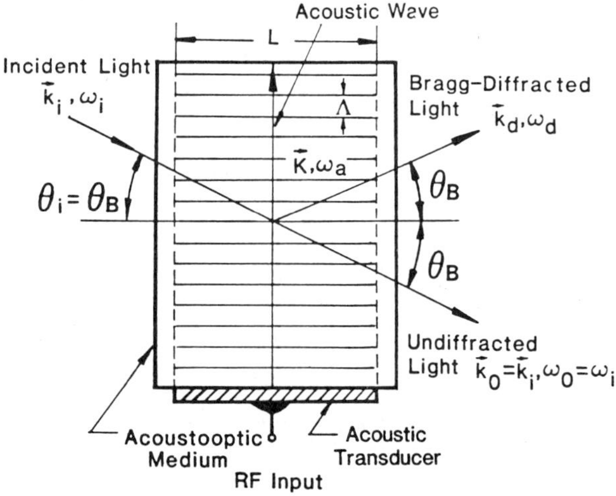

Figure 5.1 Basic bulk-wave acousto-optical Bragg modulator.

from tens to thousands of megahertz) upon a planar piezoelectric transducer that is bonded on one finely polished end face of the column [31]. A moving optical index grating is induced by the acoustic waves via an elasto-optical effect. We shall now consider isotropic and anisotropic AO Bragg diffraction separately.

5.2.1 Isotropic Acousto-optical Bragg Diffraction

For simplicity we shall first consider the case of an isotropic material such as fused quartz in which the refractive index of a light wave is independent of its direction of propagation or polarization. When a light wave is incident upon an acoustic cell at the Bragg angle θ_B as defined in Eq. (5.1), a portion of the light is diffracted mainly into one side at an angle that is twice the Bragg angle from the undiffracted (incident) light. The intensity of the diffracted light increases linearly with the power of the acoustic or RF driving signal before saturation is reached:

$$\sin \theta_B = \frac{\lambda_0}{2n\Lambda} = \frac{\lambda_0}{2nV_a} f_a \tag{5.1}$$

In Eq. (5.1), n and λ_0 designate, respectively, the refractive index of the medium and the wavelength of the light in free space; Λ, f_a, and V_a are, respectively, the wavelength, frequency, and velocity of the acoustic wave.

It has been established that the conservation of both frequency (or energy) and the wave vector (or momentum) as expressed in Eqs. (5.2) and (5.3) is satisfied in AO Bragg diffraction:

$$\omega_d = \omega_i \pm \Omega$$

That is,

$$f_d = f_i \pm f_a \tag{5.2}$$

and

$$\mathbf{k}_d = \mathbf{k}_i \pm \mathbf{K} \tag{5.3}$$

where f_i, f_d, and f_a designate, respectively, the frequencies of the incident light, the diffracted light, and the acoustic wave and $\mathbf{k}_i$, $\mathbf{k}_d$, and $\mathbf{K}$ are the corresponding wave vectors. Note that ω_i, ω_d, and Ω are the corresponding radian frequencies. For convenience, Eq. (5.3) is often represented as the wave vector diagram shown in Fig. 5.2. It is to be noted that the Bragg angle θ_B given in Eq. (5.1) is readily derived using Eqs. (5.2) and (5.3) together with the fact that polarization of diffracted light is identical to that of incidental light and that in practice the acoustic frequency f_a is much lower than the optical frequency f_i.

The three desirable characteristics of the basic bulk-wave AO Bragg cell just described, on which numerous applications are based, are now discussed. First, as indicated in Eq. (5.2), the frequency of the diffracted light is either upshifted or downshifted from that of the incident light by the acoustic frequency. The upshift and downshift correspond, respectively, to the cases in which the wave vector of the incident light wave is at an angle greater than and smaller than 90° from the wave vector of the acoustic wave. These frequency shifts can be explained satisfactorily in terms of Doppler effects as well as a

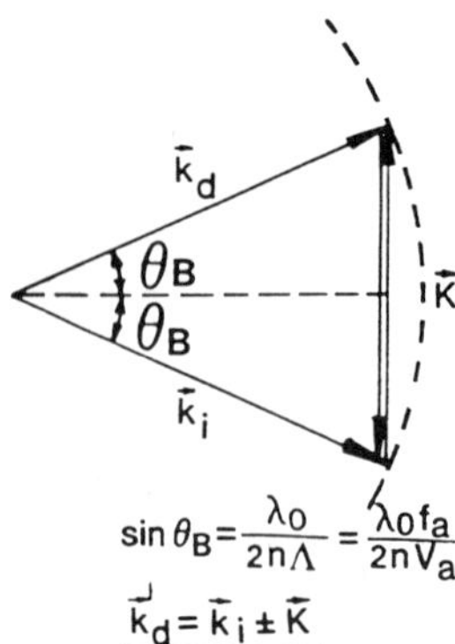

Figure 5.2 Wave vector diagram for isotropic accousto-optical Bragg diffraction.

heuristic picture [16]. This particular characteristic of the AO Bragg cell has been utilized to perform frequency modulation and frequency shifting of lasers.

Second, as indicated in Eqs. (5.1) and Fig. 5.2, the angle of deflection of the diffracted light varies linearly with the acoustic frequency. Thus, by tuning the frequency of the RF driving signal, the AO Bragg cell can be used to perform light beam deflection/scanning and switching. Finally, as mentioned previously, the intensity of the diffracted light is proportional to the power of the acoustic wave or the RF driving signal, provided the AO Bragg cell is not driven into saturation. This particular characteristic makes the AO Bragg cell one of the most practical intensity modulators for lasers. It should be noted that a variety of real-time signal-processing applications such as spectral analysis, correlation, and convolution of wide-band RF signals have also been realized using simultaneously the second and third characteristics.

5.2.2 Anisotropic Acousto-optical Bragg Diffraction

In the preceding discussion we have, for simplicity, considered bulk-wave AO Bragg diffraction in an optically isotropic material. However, since many superior AO materials are optically anisotropic, it is important to summarize the basic characteristics of anisotropic AO Bragg diffraction [6]. Again for simplicity, we shall consider a specific configuration (shown in Fig. 5.3) in which both the incident and the diffracted light as well as the acoustic wave propagate in a plane orthogonal to the *C*-axis of a uniaxial crystal such as $LiNbO_3$. It has been established that conservation of both frequency and the wave vector as expressed in Eqs. (5.2) and (5.3) still holds. However, through optical anisotropy both the refractive index and the polarization of the diffracted light differ from that of the incident or undiffracted light. It can be shown that the wave vector diagram depicted in Fig. 5.3 leads to the following expressions for the incident angle θ_i and the diffraction angle θ_d, which are both measured with respect to the acoustic wavefront:

$$\sin\theta_i = \frac{\lambda_0}{2n_i V_a}\left[f_a + \frac{V_a^2}{f_a\lambda_0^2}(n_i^2 - n_d^2)\right] \tag{5.4}$$

$$\sin\theta_d = \frac{\lambda_0}{2n_i V_a}\left[f_a - \frac{V_a^2}{f_a\lambda_0^2}(n_i^2 - n_d^2)\right] \tag{5.5}$$

where n_i and n_d designate the refractive indices of the incident and the diffracted light, respectively.

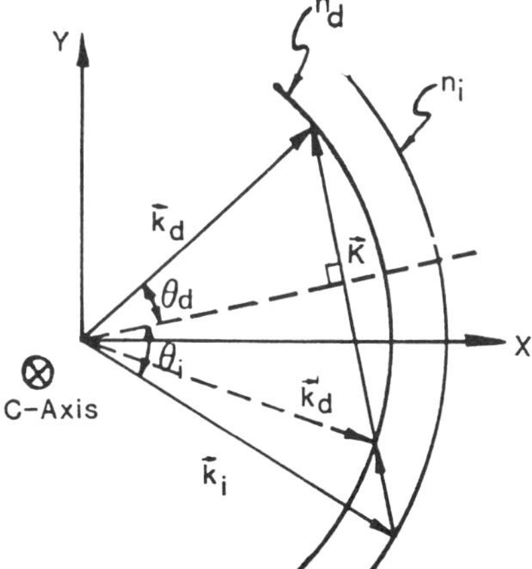

Figure 5.3 Wave vector diagram for anisotropic accousto-optical Bragg diffraction in a plane orthogonal to the C-axis of a uniaxial crystal.

It is readily seen that the general expressions given in Eqs. (5.4) and (5.5) reduce to the simple Bragg condition given by Eq. (5.1) for an isotropic medium in which $n_d = n_i$. Figure 5.4 shows the plots for θ_i and θ_d as a function of the acoustic frequency. As a comparison, θ_i and θ_d $(=\theta_i)$ for isotropic Bragg diffraction are also plotted as a function of the acoustic frequency in Fig. 5.4.

From Fig. 5.4, we identify two special cases of practical importance (shown in Figs. 5.5*a* and *b*. In the first case, which is commonly called optimized anisotropic interaction, wave vector phase matching may be satisfied for a large range of acoustic frequency. As a result, large-bandwidth light beam deflectors and modulators may be constructed utilizing this particular mode of interaction [8b]. For the second case, the collinear geometry (or configuration) involved facilitates construction of electronically tunable AO filters for optical spectra [8b].

The simultaneous availability of a variety of laser sources, piezoelectric transducer technology for efficient generation of very high frequency acoustic waves, and superior

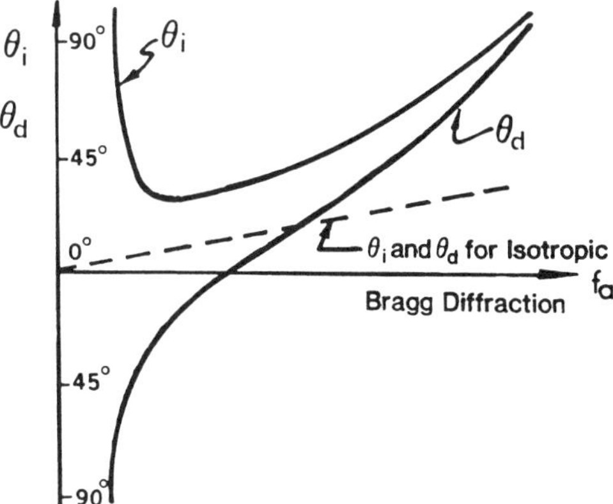

Figure 5.4 Variations of incident and diffraction angles versus acoustic frequency in anisotropic acousto-optical Bragg diffraction.

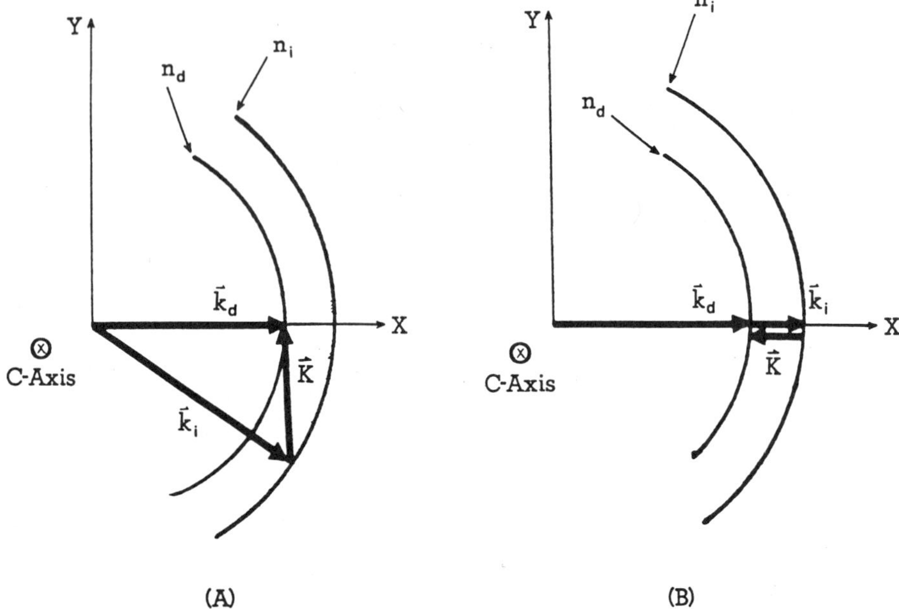

Figure 5.5 Two special cases of anisotropic Bragg diffraction in a uniaxial crystal: (*a*) optimized interaction; (*b*) collinear interaction.

solid materials has enabled realization of various types of bulk AO devices including modulators, scanners, deflectors, Q-switches, mode lockers, tunable filters, spectrum analyzers, and correlators [1–9]. Such bulk AO devices have now been deployed in a variety of commercial and military applications.

5.3 GUIDED-WAVE ACOUSTO-OPTICAL BRAGG DIFFRACTION IN PLANAR WAVEGUIDES

5.3.1 Basic Interaction Configuration and Mechanisms

A basic coplanar guided-wave AO Bragg interaction configuration is shown in Fig. 5.6. Basic elements corresponding to those in Fig. 5.1 for bulk-wave AO Bragg interaction are easily identified. The optical waveguide material should possess desirable acoustic, optical, and AO properties. The SAW is commonly excited by an interdigital electrode transducer (IDT) [26–31] deposited upon the optical waveguide. If the substrate material is sufficiently piezoelectric, such as *Y*-cut $LiNbO_3$, the IDT may be deposited directly on it. Otherwise, e.g., GaAs or a piezoelectric film such as ZnO [29,30,35] must be deposited either beneath or above the interdigital electrode array. The optical waveguide can be either a graded index layer created beneath the substrate or a step index layer deposited on top of the substrate. Examples of graded index guides are $LiNbO_3$ waveguides formed by out- and in-diffusion techniques [36,37] while examples of step index guides are glass [10] or As_2S_3 film [11] waveguides formed by sputtering or deposition. In either case, the waveguide layer is assumed to have an effective thickness or penetration depth *H*. Optical waveguide modes in such structures have been studied in detail [17,20,22,23,38–42].

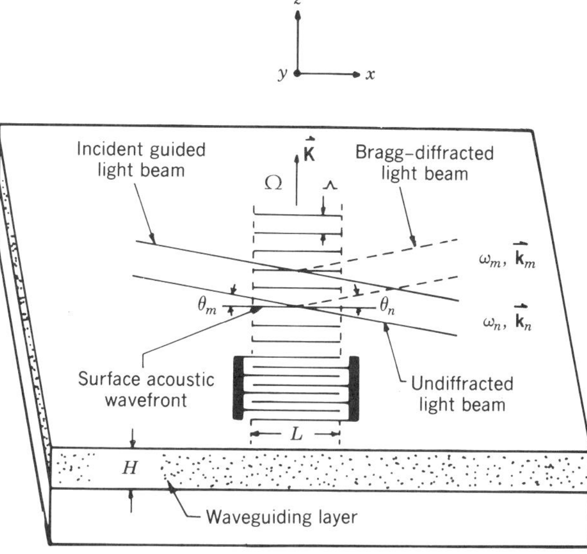

Figure 5.6 Guided-wave acousto-optical Bragg diffraction from a single-surface acoustic wave.

Propagation of the SAW creates both a moving optical grating in the optical waveguide and moving corrugations at both the air–waveguide and waveguide–substrate interfaces. The moving grating and corrugations in turn cause diffraction of an incident guided light wave [10–16]. Thus, except for the corrugations, the underlying mechanisms in coplanar guided-wave AO interactions are analogous to those in bulk-wave AO interactions. While the contribution due to the corrugations can be significant for certain ranges of waveguide thickness and acoustic wavelength, in most practical cases, optical grating is the dominant mechanism in diffraction.

As for bulk-wave AO interaction, diffraction may be either Raman–Nath type or Bragg type. Raman–Nath diffraction consists of a number of side orders when the AO parameter $Q \equiv 2\pi\lambda_0 L/n\Lambda^2$ is less than or equal to 0.3. The symbols λ_0 and Λ designate, respectively, the wavelengths of the guided optical wave (in free space) and the SAW, n is the effective refractive index of the guiding medium, and L designates the aperture of the SAW. When the light wave is incident at the Bragg angle defined in Section 5.2 and Q is larger than 4π, diffraction is of the Bragg type and consists of one side order. We shall limit our discussion to Bragg diffraction because this type of diffraction is capable of higher acoustic center frequency, wider modulation bandwidth, and larger dynamic range and thus greater versatility in application.

It is important, however, to note that while the basic interaction mechanisms for planar guided-wave acousto-optics is analogous to that of bulk-wave acousto-optics, the number of parameters involved in the guided-wave case is greater and the interrelation between them much more complex [15]. For example, the diffraction efficiency is a sensitive function of the spatial distributions of both optical and acoustic waves, which in turn depend, respectively, on the guided optical modes and the acoustic frequency involved. In addition, in the case of a piezoelectric and electro-optical (EO) substrate such as $LiNbO_3$ and ZnO the piezoelectric field accompanying the SAW can be so large that the induced index changes due to the EO effect become very significant [15]. In fact, the contribution from the EO effect that accompanies a Z-propagation SAW in a Y-cut $LiNbO_3$ substrate is larger than that from the AO effect.

Similar to bulk-wave AO Bragg diffraction, the relevant energy (frequency) and momentum (wave vector) conservation relations between the incident light wave, the diffracted light wave, and the SAW are expressed by Eqs. (5.6a) and 5.6b), where m and n designate the indices of the waveguide modes; $\mathbf{k}_n$, $\mathbf{k}_m$, and $\mathbf{K}$ are, respectively, the wave vectors of the diffracted light, undiffracted light, and the SAW; and ω_n, ω_m, and Ω are the corresponding radian frequencies:

$$\omega_n = \omega_m \pm \Omega \tag{5.6a}$$

$$\mathbf{k}_n = \mathbf{k}_m \pm \mathbf{K} \tag{5.6b}$$

Figure 5.7*a* shows the wave vector diagram for the general case in which the diffracted light propagates in a waveguide mode (nth mode) that is different from the incident light (mth mode). The diffracted light may have a polarization parallel or orthogonal to that of the incident light. Again, as in bulk-wave AO Bragg diffraction, these two classes of interaction are called isotropic and anisotropic AO Bragg diffraction. As a special isotropic case, the diffracted light propagates in the same mode and thus has the same polarization as the incident light, as depicted in Fig. 5.7*b* ($n = m$). Many guided-wave AO Bragg diffraction experiments using $LiNbO_3$ and GaAs belong to this particular case [16]. Figure 5.7*c* depicts a particularly interesting case of anisotropic Bragg diffraction in which the wave vector of the diffracted light is perpendicular or nearly perpendicular to that of the acoustic wave, commonly called optimized anisotropic diffraction [15,16]. Note that this is a counterpart to Fig. 5.5*a*. Specific examples for both cases in a Y-cut $LiNbO_3$ substrate as depicted in Figs. 5.7*a* and *c* have been discussed [15,16].

The angles of incidence and diffraction measured with respect to the acoustic wavefront, θ_m and θ_n, are

$$\sin\theta_m = \frac{\lambda_0}{2n_m\Lambda}\left[1 + \frac{\Lambda^2}{\lambda_0^2}(n_m^2 - n_n^2)\right] \tag{5.7a}$$

$$\sin\theta_n = \frac{\lambda_0}{2n_n\Lambda}\left[1 - \frac{\Lambda^2}{\lambda_0^2}(n_m^2 - n_n^2)\right] \tag{5.7b}$$

where n_m and n_n are the effective refractive indices of the undiffracted and diffracted light waves and, as defined earlier, λ_0 and Λ are the optical wavelength in free space and the wavelength of the SAW. Clearly, when the undiffracted and diffracted light propagate in the same waveguide mode (Fig. 5.7*b*), we have $n_n = n_m$. For this particular case Eqs. (5.7a) and (5.7b) both reduce to the well-known Bragg condition in isotropic diffraction [10], namely, $\sin\theta_n = \sin\theta_m = (\lambda_0/2n_m\Lambda) = (\lambda_0\omega/4\pi n_m V_R)$, where $\theta_n = \theta_m$ is the so-called Bragg angle and V_R designates the propagation velocity of the SAW. Thus, the diffraction angle is identical to the incidence angle, and both angles increase linearly with the acoustic frequency because in practice the latter is much lower than the optical frequency.

5.3.2 Coupled-Mode Analysis on Acousto-optical Bragg Diffraction from a Single Surface Acoustic Wave

The most common approach for treatment of AO interactions is the so-called coupled-mode technique [11,13,15,16]. This technique has been employed for the analysis of guided-wave AO Bragg diffraction from a single SAW in a $LiNbO_3$ planar waveguide [13b]. Such an analysis serves to reveal the physical parameters involved and the key device parameters as well as the performance limitations of the resulting devices. Both

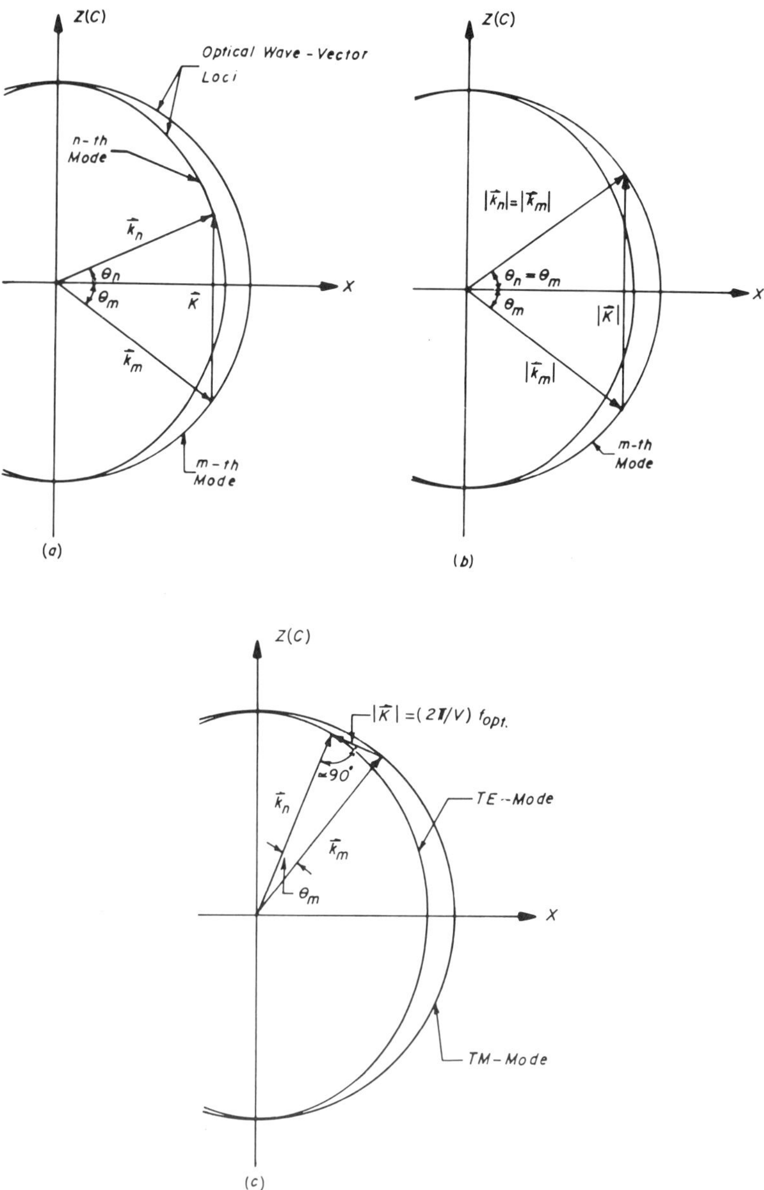

Figure 5.7 Phase-matching diagrams indicating wave vectors of incident light, Bragg-diffracted light, and surface acoustic wave in a Y-cut $LiNbO_3$ waveguide: (a) general case; (b) special isotropic case; (c) optimized anisotropic case.

the analytical procedures and the methodology for numerical computation developed for this simple case can be conveniently extended to the case involving multiple SAWs [13c] as well as other material substrates. The general case involving N SAWs and the special cases involving multiple tilted SAWs, phased SAWs, and their combinations [13c] are briefly mentioned in Section 5.5.

Refer to the coordinate system of Fig. 5.6 and assume that the medium is lossless both optically and acoustically. First, appropriate forms for electric fields $\hat{E}_m(x, y, z, t)$ and $\hat{E}_n(x, y, z, t)$ of undiffracted and diffracted light waves and the strain field $\hat{S}(x, y, z, t)$ of the SAW and its accompanying piezoelectric field $\hat{E}_p(x, y, z, t)$ are taken. Note that the subscripts m and n designate, respectively, the waveguide modes of the undiffracted and diffracted light waves. Next, all the field quantities are substituted into the wave equation for the amplitude of the normalized electric fields of the optical waves $E_m(x)$ and $E_n(x)$ that depend on the x coordinate only. A set of coupled-wave equations for $E_m(x)$ and $E_n(x)$ is then obtained. Finally, these coupled-wave equations are solved subject to the boundary conditions $E_m(0) \equiv 1$ and $E_n(0) \equiv 0$ at the input $x = 0$, resulting in electric fields of undiffracted and diffracted light waves at the output of the interaction region $x = L$ [13b].

5.3.3 Diffraction Efficiency and Frequency Response

From the electric fields of optical waves referred to in Section 5.3.2 the diffraction efficiency $\zeta(f_a)$, defined as the ratio of diffracted light power at the output ($x = L$) and incident light power at the input ($x = 0$) of the interaction region, is found [13b]:

$$\zeta(f_a) = g^2(f_a)\left(\frac{\sin[g^2(f_a) + (K\,\Delta\theta\,L/2)^2]^{1/2}}{[g^2(f_a) + (K\,\Delta\theta\,L/2)^2]^{1/2}}\right)^2 \tag{5.8a}$$

where

$$g^2(f_a) \equiv \frac{\pi^2}{4\lambda_0^2}\,\frac{n_m^3 n_n^3 |\Gamma_{mn}(f_a)|^2 L^2}{\cos\theta_m \cos\theta_n} \tag{5.8b}$$

$$|\Gamma_{mn}(f_a)|^2 \equiv \frac{\left[\left[\int_0^{\infty} U_m(y)U_n(y)\{P\colon SU_a(y) + r\colon E_p U_p(y)\}\,dy\right]\right]^2}{\left[\left[\int_{-\infty}^{\infty} U_m^2(y)\,dy\right]\right]\left[\left[\int_{-\infty}^{\infty} U_n^2(y)\,dy\right]\right]} \tag{5.8c}$$

and

$f_a = \Omega/2\pi$ = frequency of the SAW in cycles per second
$K = 2\pi/\Lambda$ = wave number of the SAW
L = aperture of the SAW
c = velocity of light in frequency space
$\Delta\theta$ = deviation of the incidence angle of the light from the Bragg angle

Also, $U_m(y)$, $U_n(y)$, $U_a(y)$, and $U_p(y)$ designate, respectively, the normalized field distributions (along the waveguide thickness) of the light waves, the acoustic wave, and the piezoelectric field. Finally, the following physical constants with suppressed vector and tensor subscripts are designated: P is the relevant photoelastic constant or constants and r is the relevant EO coefficient or coefficients.

The general expression given by Eq. (5.8a) can be used to calculate the diffraction efficiency and its frequency response as a function of both the polarization and the angle of incidence of the light [15]. Equation (5.8a) reduces to the following simple form when the Bragg condition is satisfied ($\Delta\theta \equiv 0$):

$$\zeta(f_a) = \sin^2 g(f_a) \tag{5.9}$$

It is seen that in contrast to bulk-wave AO Bragg interaction, the diffraction efficiency is a sensitive function of the spatial distributions of the diffracted and the undiffracted light waves and the SAW as determined by the coupling function $|\Gamma_{mn}(f_a)|^2$. Since confinement of the SAW is proportional to the acoustic wavelength, $|\Gamma_{mn}(f_a)|^2$ is also dependent on the thickness of the optical waveguide with respect to the acoustic wavelength. Consequently, $|\Gamma_{mn}(f_a)|^2$ varies strongly with acoustic frequency. An efficient diffraction can occur only in the frequency range for which the confinement of the SAW matches that of the diffracted and undiffracted light waves. The dependence of the diffraction efficiency on acoustic, optical, and AO parameters is further complicated by the accompanying EO effect. Thus, in general, the coupling function $|\Gamma_{mn}(f_a)|^2$ is more complicated than the so-called overlap integral [10,11,15,16]. Only for the special case in which the accompanying EO contribution is either negligible (e.g., in a glass waveguide) or proportional to the elasto-optical contribution will Eq. (5.8c) reduce to one that is proportional to the overlap integral.

For the same reason, it is impossible to define an AO figure of merit as simply as in a bulk-wave AO interaction [7]. However, if we again consider the special case just mentioned and employ a simplified model [14], we may define the total power flow of the SAW, P_a, as

$$P_a \equiv \tfrac{1}{2}\rho V_R^3 L S^2 \int_{-\infty}^{\infty} U_a^2(y)\,dy \tag{5.10}$$

where ρ and V_R designate the density and the acoustic propagation velocity of the interaction medium, S designates the acoustic strain, and $\int_{-\infty}^{\infty} U_a^2(y)\,dy$ carries a dimension in length. Equations (5.8b), (5.8c), and (5.10) are now combined to give the following expression for $g^2(f_a)$:

$$g^2(f_a) = \frac{\pi^2}{2\lambda_0^2}\frac{n_m^3 n_n^3 P^2}{\rho V_R^3} C_{mn}^2(f_a)\frac{L}{\cos\theta_m \cos\theta_n} P_a \tag{5.11a}$$

where the frequency-dependent coupling coefficient $C_{mn}^2(f_a)$ is defined as follows:

$$C_{mn}^2(f_a) \equiv \frac{\left[\int_0^{\infty} U_m(y)U_n(y)U_a(y)\,dy\right]^2}{\left[\int_{-\infty}^{\infty} U_m^2(y)\,dy\right]\left[\int_{-\infty}^{\infty} U_n^2(y)\,dy\right]\left[\int_{-\infty}^{\infty} U_a^2(y)\,dy\right]} \tag{5.11b}$$

Note that $C_{mn}^2(f_a)$ takes a form similar to the overlap integral, with its value depending upon the optical and the acoustic modes of propagation and equal to unity for bulk-wave AO interactions. Also, the factor $n_m^3 n_n^3 P^2/\rho V_R^3$ is similar to the bulk-wave AO figure of merit M_2 [7], henceforth designated by M_{2mn}.

From Eqs. (5.9), (5.10), and (5.11a) it is seen that for the case $g^2(f_a) \ll 1$ and $\Delta\theta \equiv 0$, the diffraction efficiency is approximately proportional to the total acoustic power P_a. The effect of this quasi-linear dependence on the dynamic range of the resulting AO Bragg modulators and deflectors is discussed in Section 5.4. It is also seen that for a fixed total acoustic power and a relatively low diffraction efficiency (e.g., 60% or less) the diffraction efficiency is linearly proportional to the acoustic aperture.

We shall now examine the quantitative dependence of the diffraction efficiency on the acoustic frequency. From Eqs. (5.8a) and (5.9) it is first noted that in contrast to the bulk-wave AO interaction, even at $\Delta\theta \equiv 0$, the diffraction efficiency is a sensitive function of the acoustic center frequency f_0 because of the frequency dependence of the SAW confinement. Thus, the bandwidth of a guided-wave AO Bragg modulator is mainly determined by the frequency dependence of three factors: the transducer conversion efficiency, the Bragg condition (phase matching), and SAW confinement. The individual and combined effects of these three factors on the frequency response can be determined using Eqs. (5.8a) to (5.8c) and a digital computer. For example, to study the effect of SAW confinement alone, we assume that the transducer bandwidth is so large that its frequency dependence can be ignored and the acoustic aperture is so small that the Bragg condition can be satisfied at all frequencies of interest. Similarly, by setting the right side of Eq. (5.8a) equal to one-half of its value at the acoustic center frequency f_0 and $\Delta\theta \equiv 0$, the -3-dB modulator bandwidth $\Delta f_{-3\text{dB,Bragg}}$ as determined by the last two factors can be calculated [13b]. The results for the two types of calculations just described as applied to the Y-cut $LiNbO_3$ waveguides will be summarized in Section 5.3.4.

For the purpose of illustration we now calculate the diffraction efficiency at the acoustic center frequency f_0 for the special case in which the EO contribution is either negligible or proportional to the elasto-optical contribution. This calculation is applicable to the case involving nonpiezoelectric materials such as glass, oxidized silicon, and As_2S_3. We further assume that the power of the SAW is uniformly distributed in a depth of one acoustic wavelength [14]. Thus as a special form of Eq. (5.10), we have

$$P_a = \tfrac{1}{2}\rho V_R^3 S^2 L \Lambda_0 = \tfrac{1}{2}\rho V_R^4 S^2 \frac{L}{f_0} \tag{5.12a}$$

or

$$S = \left(\frac{2P_a}{\rho V_R^4}\right)^{1/2} \left(\frac{f_0}{L}\right)^{1/2} \tag{5.12b}$$

where Λ_0 designates the acoustic wavelength at the center frequency. Substituting Eq. (5.12b) into Eq. (5.8c) and combining Eqs. (5.8b), (5.8c), and (5.9) and restricting to the case of moderate diffraction, we obtain the following expression for the diffraction efficiency:

$$\zeta(f_0) \approx g^2(f_0) = \frac{\pi^2}{2\lambda_0^2} M_{2mn} \frac{f_0 L}{V_R \cos\theta_m \cos\theta_n} P_a \tag{5.13}$$

Equation (5.13) shows that for this special case the diffraction efficiency is proportional to the product of the center frequency and the aperture of the SAW as well as the total acoustic power.

We now turn to the AO Bragg bandwidth. In contrast to the bulk-wave interaction [4], the AO Bragg bandwidth in general cannot be expressed explicitly in terms of the center frequency and the aperture of the acoustic wave. Only for the special case $g^2(f_a) \ll (K\,\Delta\theta\,L/2)^2$ would Eq. (5.8a) lead to an expression similar to that of the bulk-wave case. For this special case the absolute AO Bragg bandwidth for isotropic interaction with the TE_0 mode is given as follows:

$$\Delta f_{-3\text{dB,Bragg}} \approx \frac{1.8 n_0 V_R^2 \cos\theta_0}{\lambda_0 f_0 L} \quad \text{(isotropic)} \tag{5.14a}$$

or

$$\frac{\Delta f_{-3\text{dB,Bragg}}}{f_0} \approx \frac{1.8 n_m V_R \cos\theta_0}{\lambda_0 f_0} \frac{\Lambda_0}{L} \quad \text{(isotropic)} \tag{5.14b}$$

where n_0 and θ_0 designate the effective refractive index and the Bragg angle for the TE_0 mode, and Λ_0 again designates the wavelength of the SAW at the center frequency. Earlier measurements in Y-cut $LiNbO_3$ waveguides have verified Eqs. (5.14a) and (5.14b) [15b]. The absolute AO Bragg bandwidth is inversely proportional to the product of the center frequency and the aperture of the SAW. Equations (5.13) and (5.14) indicate clearly that the diffraction efficiency and the AO Bragg bandwidth impose conflicting requirements on the acoustic frequency and the acoustic aperture. In fact, the diffraction efficiency–Bragg bandwidth product is a constant that is independent of both the center frequency and the aperture of the SAW.

In summary, because of the complicated spatial distributions of the guided-optical waves (GOW) and the SAW as well as the frequency dependence of the latter, numerical calculations using digital computers are required to determine the efficiency and exact frequency response of a guided-wave AO Bragg modulator. The procedure is as follows:

1. Obtain appropriate analytical expressions for the field distributions of the optical waves and the SAW based on their directions and modes of propagation.
2. Include the frequency dependence of the amplitudes, phases, and penetration depth of the SAW.
3. Identify the relevant photoelastic and EO constants.
4. Calculate the diffraction efficiency versus the acoustic frequency using a digital computer, with the acoustic drive power as a parameter, and inserting the information derived in steps 1–3 together with the remaining optical and acoustic parameters into Eq. (5.8).

Note that the frequency response of the SAW transducer can be incorporated in step 2.

5.3.4 Calculated Performances for a *Y*-Cut $LiNbO_3$ Substrate

Detailed numerical calculations and experiments have been carried out for both isotropic (Figs. 5.7*a* and *b*) and anisotropic (Fig. 5.7*c*) Bragg diffraction in Y-cut $LiNbO_3$ waveguides. While the latter is discussed in Ref. 13d, some results of the former are presented in this section. The isotropic Bragg diffraction to be given here involves a He–Ne laser light (0.6328 μm) propagating in the TE_0 mode in Y-cut $LiNbO_3$ in-diffused optical waveguides and a SAW propagating in the $Z(c)$ direction [13]. As indicated earlier, the contribution to Bragg diffraction by the EO effect is very important in a Y-cut $LiNbO_3$ substrate. Spatial distributions of optical fields in such waveguides [38–42] and those of strain fields and accompanying piezoelectric fields are taken from the literature [43–46]. Relevant photoelastic constants and EO coefficients are taken from Ref. 47. Specifically, for the case with a Z-propagation SAW the appropriate photoelastic constants are $P_{31} = P_{32}$, P_{33}. The only relevant EO coefficient is r_{33}.

First, the frequency response as determined by confinements of optical waves and the SAW is calculated. This is equivalent to calculating the coupling coefficient $C_{mn}^2(f_a)$ as a function of the acoustic frequency under the assumptions that the transducer bandwidth is so large that it does not introduce any band-limiting effect and that the Bragg condition ($\Delta\theta \equiv 0$) is satisfied at all frequencies of interest. While the first assumption leads to a constant total acoustic power, the second assumption is equivalent to utilization of a sufficiently small acoustic aperture. Subsequently, from Eqs. (5.8a)–(5.9) the relative Bragg diffraction efficiency as a function of acoustic frequency is calculated with the penetration depth of an optical waveguide mode as a parameter. Figure 5.8 shows the calculated results for the case involving a TE_0 mode for both diffracted and undiffracted light waves.

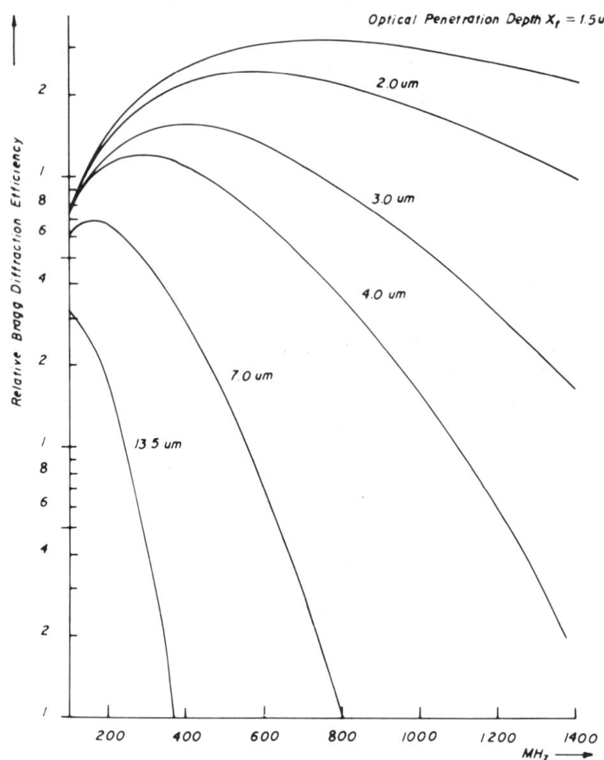

Figure 5.8 Frequency response based on coupling coefficient alone with penetration depth of TE_0 mode in a Y-cut waveguide as a parameter (for Z-propagation SAW).

Note that in these calculated plots the total acoustic power is assumed to be a constant, a consequence of the assumption that the transducer bandwidth is sufficiently large so that it does not introduce any band-limiting effect. These plots clearly show that the smaller the optical penetration depth, the higher will be the optimum acoustic frequency and the diffraction efficiency. It is seen that at the optical wavelength of 0.628 μm the desirable range of optical penetration depth is from 1.0 to 2.0 μm and the optimum range of acoustic frequency should be centered around 700 MHz.

In the preceding discussion, the transducer bandwidth is assumed to be sufficiently large and the acoustic aperture sufficiently small so that no bandwidth limiting could occur. In practice, the finite transducer bandwidth as well as the nonvanishing acoustic aperture must be included in the calculation of the ultimate frequency response of the AO Bragg modulator. For example, a set of frequency responses has been obtained for an optical waveguide of 2.0 μm penetration depth with the fractional transducer bandwidth and the acoustic aperture as parameters [15]. The total acoustic drive power and the peak diffraction efficiency corresponding to each combination of transducer bandwidth and acoustic aperture have also been tabulated [15]. Finally, the corresponding RF drive power is determined by the conversion efficiency of the transducer.

From the aforementioned set of frequency responses a number of conclusions can be drawn. First, for a given acoustic drive power the diffraction efficiency is proportional to the acoustic aperture, and similarly, for a given acoustic aperture the diffraction efficiency is proportional to the acoustic drive power. As indicated earlier, this linear dependence is valid for a diffraction efficiency of up to 60%. Second, the AO Bragg bandwidth decreases as the acoustic aperture increases and may become even smaller than the transducer bandwidth at a sufficiently large acoustic aperture. Also, for a given acoustic aperture the absolute bandwidth of a modulator that utilizes a single SAW of low center frequency is limited mainly by the transducer bandwidth of a periodic interdigital transducer (IDT). On the other hand, for the same acoustic aperture the absolute bandwidth of a modulator that employs a single SAW of high center frequency is limited mainly by the AO Bragg bandwidth. Clearly, such a set of frequency responses provides the basic data and guidelines required for the design of AO Bragg modulators and deflectors at 0.6328 μm optical wavelength using Y-cut $LiNbO_3$ waveguides. A similar set of frequency responses may be generated for AO Bragg modulators and deflectors at the diode laser wavelength.

In summary, the bandwidth of a guided-wave AO Bragg modulator or deflector that employs isotropic Bragg diffraction and a single periodic IDT of relatively large acoustic aperture is rather limited. For example, from the frequency responses referred to in the preceding, a maximum -3-dB modulator bandwidth of only 34 MHz is possible for a device using a Y-cut $LiNbO_3$ in-diffused waveguide of 2 μm penetration depth in which the aperture and the center frequency of the Z-propagation SAW are, respectively, 3 mm and 700 MHz. This small modulator bandwidth (34 MHz) is caused by the small AO Bragg bandwidth associated with the 3 mm acoustic aperture, irrespective of the fractional transducer bandwidth.

5.3.5 Acousto-optical Substrate Materials

Among the many materials that have been explored experimentally for guided-wave AO interactions, Y-cut $LiNbO_3$, nonpiezoelectrics such as oxidized Si, and GaAs have demonstrated the highest potential. Relevant physical parameters of these three AO materials together with a few other potential AO materials are listed in Table 5.1. The excitation and propagation of GOWs and SAWs and the AO interactions involved have been studied most thoroughly for Y-cut $LiNbO_3$ [15] and to a lesser extent for nonpiezoelectric [35] and GaAs [48–50] substrate materials. However, some significant experimental results have only been obtained recently for GaAs [50]. A comparison of these three substrate types now follows.

$LiNbO_3$ Substrate. As shown in Section 5.3.4 and also to be discussed in Section 5.5, efficient and wide-band AO Bragg diffraction can be readily realized in a Y-cut $LiNbO_3$ substrate, as shown in Fig. 5.6. Aside from a relatively high AO figure of merit, $LiNbO_3$ also possesses desirable acoustic and optical properties. As a result of large piezoelectricity, a SAW can be generated efficiently by directly depositing the IDT on the substrate. The typical propagation loss of the SAW is 1–2 dB/cm at 1.0 GHz, which is by far the lowest among all AO materials that have been studied. Optical waveguides can be routinely fabricated using the well-established titanium in-diffusion technique [37]. The measured optical propagation loss is typically 1.0 dB/cm, again the lowest among all existing AO materials. Furthermore, high-quality $LiNbO_3$ crystals of very large size are commercially available. Consequently, $LiNbO_3$ is at present the best substrate material for the realization of wide-band and efficient planar guided-wave AO Bragg cells at gigahertz center frequencies. Unfortunately, despite the many desirable properties referred to, only a partial or hybrid integration of passive and active components has been realized in $LiNbO_3$ thus far. The ultimate advantages of integrated optics cannot be fully realized until a viable technology has been developed that incorporates other suitable materials and thus lasers and detectors into the $LiNbO_3$ substrate to facilitate total or monolithic integration.

TABLE 5.1 Relevant Physical Parameters for AO Materials

Material	Range of Optical Transmission (μm)	n_o	n_e	$n = (n_o + n_e)/2$	ρ (g/cm^3)	Acoustic Wave Polarization and Direction	V_s (10^5 cm/s)	Optical Wave Polarization and Direction	M_2 (10^{-18} s^3/g)	Photoelastic Coefficients
Fused quartz	0.2–4.5	—	—	1.46	2.2	Long.	6.95	$\perp$	1.51	$P_{11} = 0.121$, $P_{12} = 0.270$, $P_{44} = -0.075$
						Trans.	3.76	$\parallel$ or $\perp$	0.467	—
$LiNbO_3$	0.4–4.5	2.29	2.20	2.25	4.7	Long. in [11–20]	6.57	—	6.99	$P_{11} = 0.036$, $P_{12} = 0.072$, $P_{31} = 0.178$, $P_{13} = 0.092$, $P_{33} = 0.088$, $P_{41} = 0.155$
GaAs	1.0–11.0	—	—	3.37	5.34	Long. in [110]	5.15	$\parallel$	104	$P_{11} = -0.165$, $P_{12} = -0.140$, $P_{44} = 0.072$
						Trans. in [100]	3.32	$\parallel$ or $\perp$ in [010]	46.3	—
Gap	0.6–10.0	—	—	3.31	4.13	Long. in [110]	6.32	$\parallel$	44.6	$P_{11} = -0.151$, $P_{12} = -0.082$, $P_{44} = -0.074$
						Trans. in [100]	4.13	$\parallel$ or $\perp$ in [101]	24.1	—
As_2S_3	0.6–11.0	—	—	2.61	3.20	Long.	2.6	$\perp$	433	$P_{11} = 0.277$, $P_{12} = 0.272$
Si	1.5–12.0	—	—	3.435	2.33	—	8.95	—	6.2	—
TeO_2	0.35–5.0	2.26	2.41	2.34	5.99	Long. [001]	4.20	$\perp$	34.5	$P_{11} = 0.0074$, $P_{12} = 0.187$, $P_{13} = 0.340$, $P_{31} = 0.0905$, $P_{33} = 0.240$, $P_{44} = -0.17$, $P_{66} = -0.0463$
						Trans. [110]	0.62	$\parallel$	25.6	
Tl_3AsS_4	0.6–12.0	—	—	—	—	—	—	—	800	—

SiO_2, As_2S_3, or SiO_2–Si Substrates. The second substrate type is comprised of the nonpiezoelectric materials such as fused quartz (SiO_2) [10], arsenic trisulfide (As_2S_3) [11], and oxidized silicon (SiO_2–Si) [35]. Interest in these substrate materials is based on the fact that the first is a common optical waveguide material, the second is an amorphous material with a very high AO figure of merit, and the third may be used to capitalize on the existing silicon technology to further electrical and optical integrations. While it is common to deposit a piezoelectric zinc oxide (ZnO) film on such substrate materials for the purpose of SAW generation, as depicted in Fig. 5.9, the ZnO–SiO_2 composite waveguide has also been utilized to facilitate guided-wave AO interaction because the ZnO film itself also possesses favorable optical waveguiding and AO properties [35,51–53].

Guided-wave AO Bragg diffraction in such substrate materials has also been analyzed in detail using the coupled-mode technique [35c]. The expression for the Bragg diffraction efficiency takes a similar form as that for the $LiNbO_3$ substrate. However, it should be noted that for the case involving the ZnO–SiO_2 composite waveguide the distribution of the SAW fields exhibits a marked variation with the thickness of the ZnO film, while the distribution of the GOW fields is practically unaffected. It was shown that the Bragg diffraction efficiency depends strongly on the thickness of the ZnO film and that for each optical waveguide mode there exists a film thickness for which the diffraction efficiency diminishes. Subsequently, a strain-controlling film of fused quartz was introduced on top of the ZnO film to enhance the diffraction efficiency. In a separate study the ZnO film was used solely to generate the SAW for AO Bragg diffraction experiments in the SiO_2–Si substrate at a SAW frequency of up to 700 MHz [53].

At the present time, measured propagation losses of the SAW in the ZnO–SiO_2–Si composite substrate [35,51] are much higher than those in the $LiNbO_3$ substrate and considerably higher than those in the GaAs substrate. This contrast is accentuated as the acoustic frequency goes beyond 200 MHz. Thus, it may be concluded that until a composite substrate with a greatly reduced SAW propagation loss has been developed, guided-wave AO Bragg devices using the foregoing nonpiezoelectric substrate materials have to be limited to a considerably lower RF center frequency and a smaller bandwidth than those using the $LiNbO_3$ substrate.

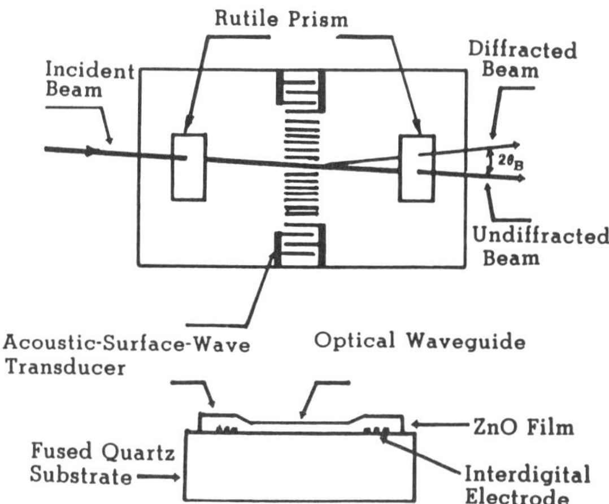

Figure 5.9 Guided-wave acousto-optical Bragg diffraction in a ZnO–SiO_2 composite waveguide.

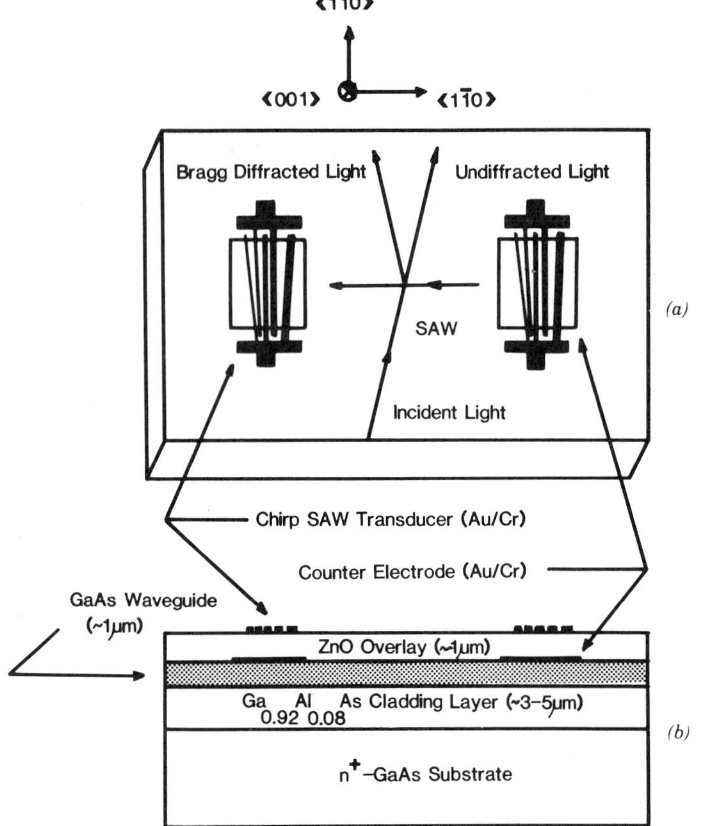

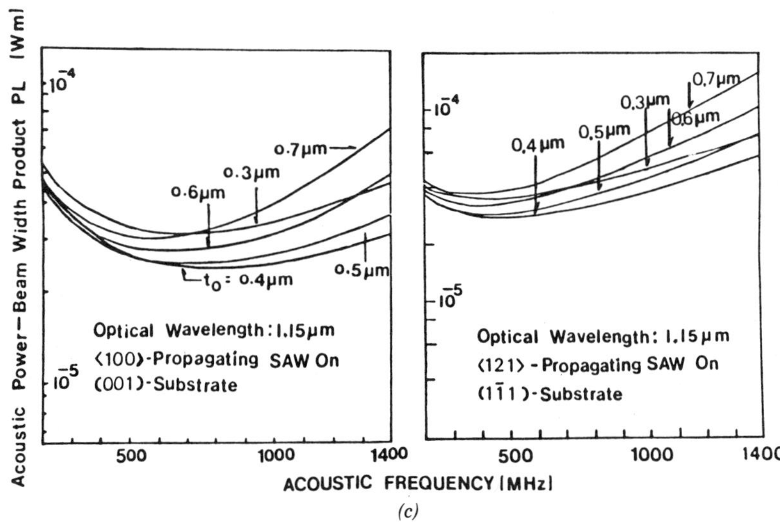

Figure 5.10 (*a*) Guided-wave acousto-optical Bragg diffraction in a *Z*-cut GaAs–GaAlAs planar waveguide (top view). (*b*) Waveguide cross section and ZnO transducer geometry (side view). (*c*) SAW power density required versus acoustic frequency for TE_0–TE_0 100% Bragg diffraction in a single-mode waveguide.

GaAs Substrate. GaAs-based substrate can potentially provide the capability for the total or monolithic integration referred to previously because both the laser sources [54] and the photodetector arrays as well as the associated electronic devices may be integrated in the same GaAs substrate [55]. Clearly, one of the key components in such future GaAs-based monolithic integrated AO modules or circuits is an efficient and wide-band AO Bragg modulator or deflector, as shown in Figs. 5.10*a* and 5.10*b*.

The AO Bragg cells that have been realized most recently utilize the interaction configuration shown in Fig. 5.10*a* in which the SAW propagates in the ⟨110⟩ direction [50a]. A previous theoretical study [49] has predicted an AO Bragg bandwidth as large as 1.6 and 1.4 GHz for the ⟨100⟩ and ⟨110⟩ propagation SAW, respectively, in a *Z*-cut GaAs–GaAlAs waveguide that supports a TE_0 mode at the optical wavelength of 1.15 μm. Figure 5.10*c* shows a set of calculated AO frequency responses with the waveguide thickness as a parameter.

In fabrication, a high-quality GaAs–GaAlAs planar optical waveguide of large size (typically 2.2×2.17 cm^2) was first grown using an in-house liquid phase epitaxy (LPE) system in a 0.35-mm-thick silicon-doped GaAs substrate. The ZnO thin film and the interdigital finger electrode array were deposited subsequently. The cross section of the optical waveguide together with the ZnO thin-film overlay and the transducer geometry as well as their typical dimensions are shown in Fig. 5.10*b*.

The measured performance figures of the Bragg cells at the optical wavelengths of 1.15 and 1.30 μm and at various RF frequency bands are summarized in Table 5.2. Consider the Bragg cell that utilized a tilted-finger chirp transducer centered at 485 MHz. A −3-dB acoustic bandwidth of 250 MHz and a −3-dB AO bandwidth of 245 MHz were obtained. The diffraction efficiency was measured to be 5.0% at 1.0 W RF drive power. Next, consider the Bragg cell that utilizes the parallel-finger synchronous transducer centered at 1100 MHz [50b]. A high diffraction efficiency was also measured in this case. This represents the highest acoustic frequency that has been employed for AO Bragg diffraction in a GaAs planar waveguide. In order to alleviate the resolution limitations with our existing fabrication facility, a parallel double-electrode synchronous transducer [56] was employed at its third harmonic for this purpose. A diffraction efficiency of 19.2% at 1.0 W RF drive power and an AO bandwidth of 78 MHz were

TABLE 5.2 Measured Performances with GaAs Waveguide AO Bragg Cells

Acoustic center frequency, MHz	200	360	400	485	600	800	950	1100
Saw transducer types: parallel-finger synchronous (I), tilted-finger chirp (II)	I	II	I	II	I	I	I	I
Diffraction efficiency (mW acoustic power/mm acoustic aperture), %	1.0	3.2	1.9	—	2.3	—	—	15
−3-dB AO device bandwidth, MHz	10	201	37	240	45	35	40	74
Maximum measured AO diffraction efficiency, %	65	15	28	—	9	9.6	18	30
Measured or projected AO diffraction efficiency at 1.0 W RF power, %	40	6.5	23	5.0	7.3	5.0	6.6	9
RF drive power per megahertz bandwidth at 1.0% diffraction efficiency, mW/MHz	2.5	0.8	1.2	—	3.0	5.7	3.7	1.5
SAW propagation loss, dB/μs	0.32	1.4	1.5	2.0	2.8	—	—	7.72

measured. Based on the aforementioned results, it is reasonable to project realization of GaAs planar waveguide AO Bragg cells that operate at a center frequency of 1.0 GHz and a bandwidth of 500 MHz.

Also important for monolithic (total) integration in GaAs are the waveguide microlens and the linear lens array. For this purpose, fabrication of negative-index-change planar waveguide microlenses in both $LiNbO_3$ [57] and GaAs [58] using ion milling was successfully carried out recently. Thus, similar to titanium in-diffusion proton exchange (TIPE) waveguide lenses in $LiNbO_3$, such ion-milled waveguide lenses should facilitate realization of a variety of monolithic IO device modules and circuits in GaAs and related substrates with applications in communications, signal processing, and computing. Naturally, the resulting monolithic IO signal processor modules will include spectrum analyzers and correlators.

5.4 KEY PERFORMANCE PARAMETERS OF PLANAR GUIDED-WAVE ACOUSTO-OPTICAL BRAGG MODULATOR AND DEFLECTOR

The planar guided-wave AO Bragg diffraction treated in Section 5.3 can be utilized to modulate and/or deflect a light beam. The resulting light beam modulators and deflectors, commonly called AO Bragg cells, can operate at gigahertz center frequencies and over a wide RF band by using a variety of SAW transducer configurations to be presented in Section 5.5 and thus constitute useful devices for integrated and fiber-optical communication, computing, and signal-processing systems. The key device parameters that determine the ultimate performance characteristics of guided-wave AO modulators and deflectors are bandwidth, time–bandwidth product, acoustic and RF drive power, nonlinearity, and dynamic range. A brief discussion of each of these performance parameters now follows.

5.4.1 Bandwidth

As shown in Section 5.3, the diffraction efficiency–bandwidth product of a planar guided-wave AO modulator that employs isotropic Bragg diffraction and a single periodic ID SAW transducer [28] is rather limited. However, if the absolute modulator bandwidth is the sole concern, a large bandwidth can be realized by using either a single periodic ID SAW transducer with small acoustic aperture and small number of finger pairs or a single aperiodic ID SAW transducer with small acoustic aperture and large number of finger pairs (chirp transducers) [59,60]. It should be emphasized, however, that in either case the large bandwidth is obtained at a drastically reduced diffraction efficiency due to the very small acoustic aperture. Consequently, a higher diffraction efficiency will necessarily require large RF or acoustic drive power. Unfortunately, even if the supply of larger RF drive power is not a problem, the risk of transducer failure will increase with the RF drive power. Furthermore, large acoustic drive power will in turn result in an excessive acoustic power density, especially at high acoustic frequency, and thus increase the deleterious effects due to acoustic nonlinearity. A small transducer aperture will also result in a large acoustic radiation impedance [28] and thus in greater complexity in the electrical matching circuit required for optimum electrical-to-acoustic transduction. Therefore, it may be concluded that for applications that require both wide bandwidth and high diffraction efficiency, more sophisticated device configurations must be employed. The unified analysis referred to in Section 5.5 on AO Bragg diffraction from multiple SAWs suggests that this requirement can be met by employing multiple transducers of proper design and placement. Section 5.5.1 lists the five interaction and transducer combinations that have been explored for this purpose. It is now possible to realize high-performance planar guided-wave AO Bragg cells with gigahertz center frequency and gigahertz bandwidth using these wide-band configurations. The design, fabrication, testing, and measured performance figures for a variety of wide-band devices are presented in Sections 5.5.1 and 5.5.2.

5.4.2 Time–Bandwidth Product

The time–bandwidth product of an AO modulator, *TB*, is defined as the product of the acoustic transit time across the incident light beam aperture and the modulator bandwidth [1–16]. It is readily shown that this time–bandwidth product is identical to the number of resolvable spot diameters of an AO deflector, N_R, which is defined as the total angular scan of the diffracted light divided by the angular spread of the incident light. Thus the following well-known identities hold:

$$TB = N_R = \frac{D}{V_R}\,\Delta f = \tau\,\Delta f \tag{5.15a}$$

$$\delta f_R = \frac{V_R}{D} \tag{5.15b}$$

$$\tau = \frac{D}{V_R} \tag{5.15c}$$

where D designates the aperture of the incident light beam, V_R the velocity of the SAW, Δf the device bandwidth, τ the transit time of the SAW across the incident light beam aperture, and δf_R the incremental frequency change required for deflection of one Rayleigh spot diameter. The acoustic transit time τ may be considered as the minimum AO switching time if the switching time of the RF driver is sufficiently smaller than the acoustic transit time. The desirable value for N_R depends upon the individual application. For example, in signal processing it is desirable to have this value as large as possible because this value is identical to the processing gain. Thus, for this particular area of application it is also desirable to have a collimated incident light beam of large aperture. The light beam aperture is limited by the quality of the optical waveguide, the excitation mechanism, and the acoustic attenuation in the waveguide. As indicated in Section 5.3.5, acoustic attenuation is only about 1 dB/cm at 1 GHz in a *Y*-cut $LiNbO_3$ optical waveguide. Using a rutile prism coupler, a guided-light-beam aperture as large as 1.5 cm and good uniformity in this type of waveguide was demonstrated at the author's laboratory. This light beam aperture resulted in an acoustic transit time of about 4.4 μs for a *Z*-propagation SAW ($V_R = 3.488 \times 10^5$ cm/s). It is shown in Section 5.5.2 that a deflector bandwidth of up to 1 GHz can be realized readily using multiple SAW transducers, so that a time–bandwidth product as high as 4400 is achievable. At present the acoustic attenuations measured in all other waveguide materials at 1 GHz are significantly higher and thus considerably limit the maximum time–bandwidth product attainable.

In light modulation and switching applications, on the other hand, it is desirable to have the time–bandwidth product as close to unity as possible so that the highest modulation or switching speed can be achieved. For this purpose, the incident light is focused to a small beam diameter at the interaction region so that the corresponding acoustic transit time is a minimum [61,62]. Since it is possible to focus both the incident light to a spot size of a few micrometers using "titanium in-diffusion proton-exchange (TIPE)" waveguide lens in the *Y*-cut $LiNbO_3$ substrate [33] and the *Z*-propagating SAW in the same substrate using a curved transducer, [63] a switching speed as high as 1 ns can be achieved.

5.4.3 Acoustic and RF Drive Power

Using Eqs. (5.10) and (5.12) and following the common practice of specifying the drive power requirement for 50% diffraction, we simply set $\zeta(f_a)$ of Eq. (5.9) equal to 0.50 to arrive at the following expression for the required acoustic drive power at the center frequency f_0:

$$P_a(50\% \text{ diffraction}) = \frac{\lambda_0^2 \cos\theta_m \cos\theta_n}{8} \frac{1}{M_{2mn,\text{eff}}} \frac{1}{L} \tag{5.16a}$$

where

$$M_{2mn,\text{eff}} \equiv C_{mn}^2(f_0) M_{2mn} \tag{5.16b}$$

Note that $C_{mn}^2(f_0)$ and M_{2mn} have been defined previously. To determine explicitly the total RF drive power P_e, we must first calculate the electrical-to-acoustic conversion efficiency T_c of the transducer used in the modulator or deflector. The frequency response and conversion efficiency of a periodic SAW IDT on $LiNbO_3$ substrates have been studied in detail in terms of one-dimensional equivalent circuits [28]. A detailed treatment is provided in Ref. 28. Accordingly, the total RF drive power at 50% diffraction $P_e(f_0)$ is given by P_a (50% diffraction) divided by T_c. Clearly, both the total acoustic and RF drive powers required are inversely proportional to the acoustic aperture as in the bulk-wave modulator. However, in contrast to the bulk-wave case, because of the coupling factor $C_{mn}^2(f_a)$, the frequency dependence of both drive powers must be considered, even if the conversion efficiency of the transducer remains constant over the frequency band of interest.

Finally, the peak electrical-to-acoustic conversion efficiency of the transducer and thus the total RF drive power can be calculated in terms of the synchronous radiation resistance as well as the impedances of the matching network [64] and the RF generator. As shown in Ref. 28, the synchronous radiation resistance is inversely proportional to the product of the synchronous frequency and the acoustic aperture or inversely proportional to the acoustic aperture measured in terms of the synchronous acoustic wavelength. As will be discussed in Section 5.5, this specific dependence imposes a practical limit on the bandwidth of high-efficiency modulators and deflectors that utilize a single acoustic aperture to considerably smaller than 1 GHz. We shall show that using some of the wide-band transducer configurations to be discussed in Section 5.5.1, the AO Bragg cells requiring only milliwatts of electric drive power per megahertz of bandwidth at 50% diffraction efficiency with a bandwidth approaching 1 GHz can be realized.

5.4.4 Nonlinearity and Dynamic Range

The ultimate dynamic range of an AO modulator is determined by a number of factors including the nonlinearity inherent in the acoustic wave propagation and AO interaction process, nonlinearity in the RF driver circuit, the optical background noises resulting from scattering in the waveguide and in all passive and active optical components, and the dynamic range of the photodetector. In this chapter the discussion is limited to the acoustic and AO interaction nonlinearities as they are independent of the applications involved.

We again focus attention on the acoustic center frequency f_0. It is seen from Eqs. (5.8)–(5.13) that as the diffraction efficiency or the modulation index is increased, the accompanying distortion or nonlinearity is also increased [15]. However, for a very small modulation index it is linearly proportional to the power of the electric drive signal. This linear relationship is the basis of many modulation and signal-processing applications, such as the spectral analysis of RF signals to be discussed in Section 5.6. However, as the RF drive power is sufficiently increased, distortion and nonlinearity become severe and should be included.

In applications such as spectral analysis that involve multiple RF signals the inter- and cross-modulations due to multiple AO diffraction also constitute the sources of nonlinearity. Some measurements on such inter- and cross-modulations in guided-wave diffraction have been made [65]. Note that the accompanying nonlinearity is frequency dependent and thus may be important in applications involving very wide-band RF signals. The AO nonlinearity described sets an upper bound for the strength of the RF signal to be processed and thus limits the dynamic range for large signals. For small signals the dynamic range is limited by the scattered light and the noise current of the photodetector.

Finally, in some cases the effect of acoustic nonlinearity [66] on dynamic range must also be taken into account. Acoustic nonlinearity is particularly important at the higher acoustic frequencies because the absolute penetration depth of the SAW decreases as the frequency increases. As a result, the acoustic power density becomes so large that a significant portion of the fundamental frequency acoustic power is converted to harmonics. In addition, this depletion of fundamental frequency acoustic power increases with the propagation distance. The end result is that an incident light of large aperture will incur significant nonuniformity in the intensity of the diffracted light as well as spurious diffracted light. This in turn degrades the linearity of modulation as well as the spatial resolution in deflection and spectral analysis applications. In fact, this deleterious effect has been observed in an experiment at gigahertz frequencies in a Y-cut $LiNbO_3$ waveguide [15].

5.5 CONSTRUCTION OF WIDE-BAND GUIDED-WAVE ACOUSTO-OPTICAL BRAGG MODULATORS AND DEFLECTORS

As previously indicated, the planar guided-wave AO Bragg modulator and deflector of Fig. 5.6 constitutes a highly useful component in multichannel integrated and fiber-optical systems. Consequently, it is desirable to realize such modulators and deflectors with as large a bandwidth and as high a diffraction efficiency as possible.

The coupled-mode analysis presented in Sections 5.3.2–5.3.4 has shown that the diffraction efficiency–bandwidth product of a planar AO device that utilizes a single SAW is a constant and rather limited. It is intuitively clear, however, that a larger composite bandwidth and thus a larger diffraction efficiency–bandwidth product can be accomplished by employing multiple SAWs that are properly tailored and configured [13,15]. As treated in detail in Ref. 13c, multiple SAWs of staggered center frequency and tilted propagation direction [13a,13b] (Fig. 5.11) as well as phased multiple SAWs of identical center frequency and propagation direction [67] (Fig. 5.12) can be used to achieve this objective. A unified treatment has been developed to analyze the AO Bragg diffraction from N SAWs [13c] (Fig. 5.13). This general approach can be employed to analyze the special cases involving the forementioned multiple tilted and phased SAWs or a combination of both.

In summary, the analysis and methodology for the numerical calculation presented in Sections 5.3.2–5.3.4 can be extended to the case involving N SAWs, and it can be concluded that by using multiple SAWs with proper arrangements in center frequency, propagation direction, and placement, guided-wave AO Bragg devices with a large composite bandwidth and thus a large diffraction efficiency–bandwidth product can be realized. Realization of such wide-band and efficient AO Bragg devices using multiple SAWs is presented in this section.

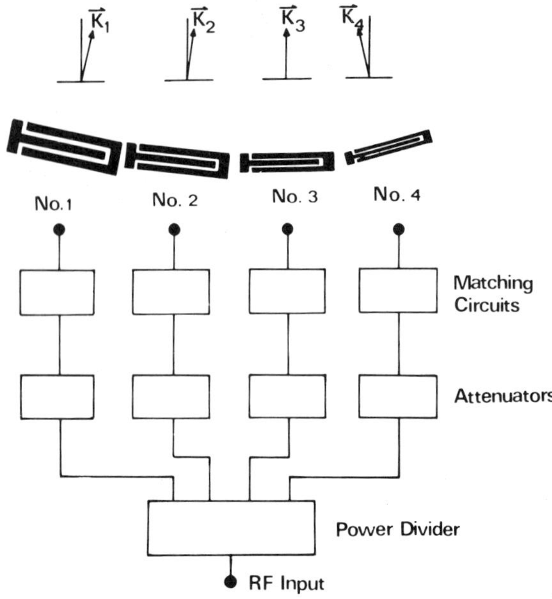

Figure 5.11 Multiple tilted SAW transducers of staggered center frequency and RF driver circuits.

Five wide-band device configurations have been explored for this purpose. These configurations utilize, respectively, the following interaction and SAW transducer combinations: (1) isotropic diffraction with multiple tilted transducers of staggered center (synchronous) frequency [13], (ii) isotropic diffraction with phased-array transducer [67], (iii) isotropic diffraction with a transducer array that combines that of (i) and (ii) [15],

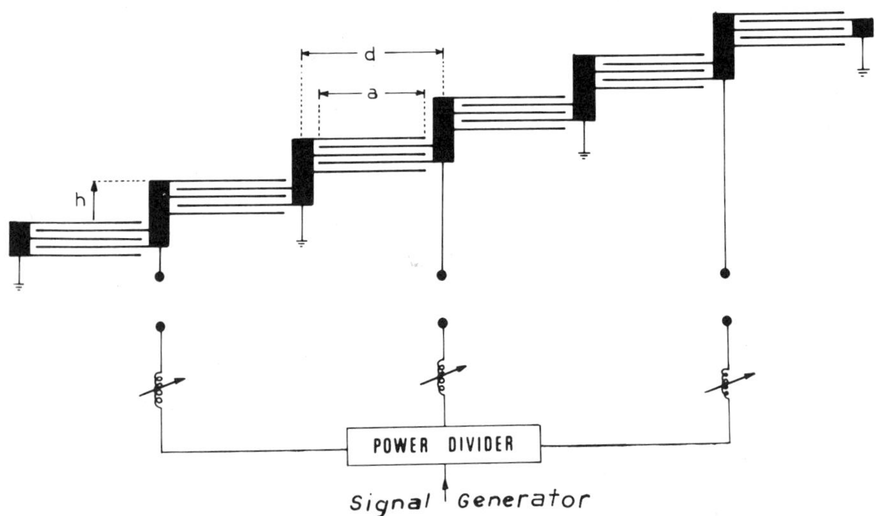

Figure 5.12 Phased-array transducer with matching and driver circuits.

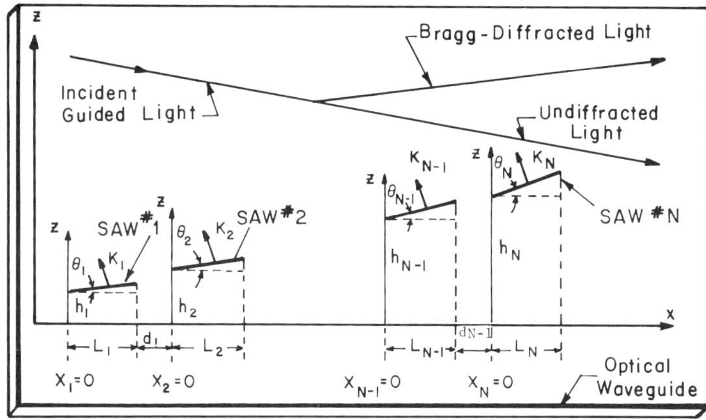

Figure 5.13 General interaction configuration for Bragg diffraction from multiple-surface acoustic waves.

(iv) isotropic diffraction with a single tilted-finger chirp transducer [68] or an array of such transducers, and (v) optimized anisotropic diffraction [69] with multiple transducers of staggered center frequency or a parallel-finger chirp transducer. In earlier work these transducer types and arrangements were fabricated in Y-cut $LiNbO_3$ waveguides using the well-established photolithographic techniques [70]. In more recent work an electron beam lithographic technique [71,72] was used to fabricate transducers of gigahertz center frequency in $LiNbO_3$. In even more recent work double-electrode transducers of gigahertz center frequency in GaAs were fabricated using the photolitho-graphic technique [50b]. The principle of operation and key design parameters and procedure for each wide-band device configuration have been treated in detail in Ref. 15. Because of space limitations only the first and the fourth wide-band device configurations are treated here. Fabrication, testing, and measured results for specific devices in Y-cut $LiNbO_3$ and an update of the performance figures obtained most recently with AO Bragg cells in Z-cut GaAs are then described. Some comments on the relative merits of a single transducer versus multiple transducers for wide-band AO Bragg diffraction are also made.

5.5.1 Design Guideline and Procedure of Wide-Band Device Configurations

Isotropic Diffraction with Multiple Tilted Transducers of Staggered Center Frequency. The transducer arrangement for this wide-band configuration is illustrated in Fig. 5.11. Individual periodic ID transducers are staggered in the center (synchronous) frequency and tilted in the acoustic propagation direction [13]. The tilt angle between each pair of adjacent transducers is set equal to the difference of the two Bragg angles at the two corresponding center frequencies. As indicated in Fig. 5.11, each element transducer is incorporated with a matching network, and the transducers are electrically driven in parallel through a power divider. Individual attenuators may also be incorporated between the outputs of the power divider and the inputs of the matching networks [15] to tailor the peak diffraction efficiency at each center frequency. It is clear that the multiple tilted SAWs generated by such a composite transducer satisfy the Bragg condition in each frequency band and thus enable a broad composite frequency response to be realized.

As indicated earlier, a rigorous analysis for this wide-band modulator–deflector configuration using the unified approach [13c] has been presented. Numerical computation of the frequency response based on this unified approach has also been carried out. The key design parameters and procedure that have been identified are now discussed. The key design parameters for the multiple tilted transducers are the center frequency, bandwidth, and aperture of individual element transducers, the tilt angle between adjacent element transducers, and the relative positions of adjacent element transducers. For each pair of adjacent element transducers the two center frequencies are chosen such that the individual frequency responses intersect at −6 dB down from the peak diffraction efficiency. The aperture of each element transducer must be sufficiently large to ensure efficient diffraction and in the meantime presents a suitable acoustic radiation impedance. Furthermore, the number of finger electrode pairs in each element transducer must be sufficiently small to ensure a fractional transducer bandwidth that is consistent with the required fractional AO Bragg bandwidth.

We now turn our attention to the relative positions of the element transducers. As a result of differences in the phase of SAWs generated by adjacent transducers and in the acoustic propagation path as measured from the front edge of the transducers to the interaction region, individual diffracted lights from adjacent SAWs may differ in the phase for the crossover frequencies. However, this phase difference may be compensated by properly configuring the element transducers through proper choice of both the horizontal separation D_s and the vertical step height h' between adjacent element transducers [15]. For the example involving transducers 1 and 2 shown in Fig. 5.14, horizontal separation D_s and vertical step height h' are

$$D_s = M \frac{2\Lambda_i^2}{\lambda} \tag{5.17a}$$

$$h' = \left(\frac{\lambda}{\Lambda_i} - \frac{\Lambda}{2\Lambda_{01}} \right) D_s \tag{5.17b}$$

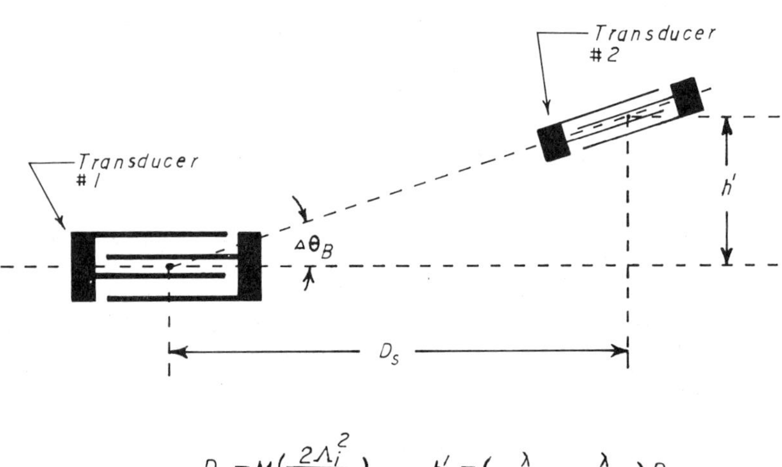

Figure 5.14 Geometrical configuration of element transducers in a frequency-staggered multiple tilted transducer.

where λ = wavelength of the light wave in the waveguide
$\Lambda_{01}, \Lambda_{02}$ = wavelength of the SAW at the center frequency f_{01} for transducer 1 and the center frequency f_{02} for transducer 2, respectively
Λ_i = wavelength of the SAW at the crossover frequency $(f_{01} + f_{02})/2$
M = integer
$\Delta\theta_B$ = difference of the two Bragg angles at f_{01} and f_{02}
$= \sin^{-1}(\lambda/2\Lambda_{02}) - \sin^{-1}(\lambda/2\Lambda_{01})$

Note that a proper choice of M is dictated by the apertures of the adjacent element transducers and that Eqs. (5.17a) and (5.17b) can be successively applied to determine the relative positions of all element transducers in the array.

Based on the aforementioned design guidelines, useful design procedures may be established. For simplicity, it is assumed that the fractional transducer bandwidths and the corresponding fractional AO Bragg bandwidths are identical for all element transducers but with the latter being slightly smaller than the former. In a more refined design the fractional transducer bandwidth and/or the fractional AO Bragg bandwidth may be varied among the element transducers. Thus, given the upper and lower −3-dB points of the modulator's or the deflector's frequency response, the number of tilted transducers and their center frequencies can be determined once the fractional AO Bragg bandwidth is chosen. Subsequently, the aperture of the element transducers can be determined by substituting the center frequency and the fractional AO Bragg bandwidth into Eq. (5.14). By choosing a fractional transducer bandwidth slightly larger than the fractional AO Bragg bandwidth, the number of finger electrode pairs can be determined. Positioning of the element transducers are then determined using Eqs. (5.17a) and (5.17b). Using the center frequency and the aperture of the individual transducers and the number of finger pairs just determined, electrical parameters such as radiation impedance, static capacitance, and conductive resistance for all individual transducers can be calculated. The matching network [64] and the attenuator for each element transducer can be designed accordingly. Finally, based on specifications for the upper and lower −3-dB frequencies the penetration depth (or the thickness) of the optical waveguide can also be determined using the frequency response plots given in Section 5.3.4. It should be noted that in some cases it may be more expedient to follow the design procedures just presented in reverse order. As discussed in the following section, a deflector of 680 MHz bandwidth was realized in a *Y*-cut titanium-diffused $LiNbO_3$ waveguide using the design procedures and the improved transducer geometry just described.

Isotropic Diffraction with a Single Tilted-Finger Chirp Transducer or an Array of Such Transducer. We now return to Fig. 5.11 and consider the situation involving a large number of element transducers of closely spaced synchronous frequencies and each with a single pair of finger electrodes. It is easily shown that this situation results in a composite transducer of varying finger periodicity and tilted angle, as depicted in Fig. 5.15. This composite transducer is called a *tilted-finger chirp transducer* [68,73,74], consistent with the common usage of the term *chirp transducer* for a transducer of parallel finger but linearly varying finger periodicity [59,60]. Like the conventional chirp transducer the acoustic bandwidth of this composite transducer can be very large. Furthermore, with proper design the wavefront of the SAW generated by this composite transducer can be made to track the Bragg condition for the entire frequency band and thus results in a large AO Bragg bandwidth. Thus, to the extent that the bandwidth imposed by the coupling coefficient $C^2_{mn}(f_a)$ is not the limiting factor, this composite transducer should also be capable of providing a large device bandwidth. Verification of the concept described in the preceding was first carried out using a simple tilted-finger chirp transducer

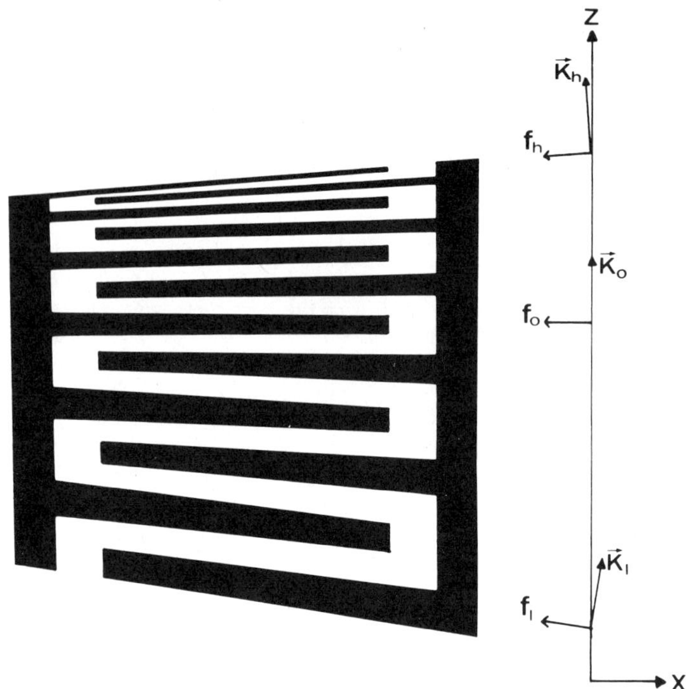

Figure 5.15 Tilted-finger chirp transducer.

fabricated in a titanium-diffused *Y*-cut $LiNbO_3$ waveguide [68]. Also, transducers of more sophisticated finger electrode arrangements was subsequently realized [15].

5.5.2 Design, Fabrication, Testing, and Measured Device Performances

$LiNbO_3$ AO Bragg Modulators and Deflectors. Using the design procedures and guidelines presented in Section 5.5.1, a variety of wide-band AO Bragg modulators and deflectors have been designed and fabricated. For *Y*-cut $LiNbO_3$ substrates, single-mode planar optical waveguides were grown using either out-diffusion [36] or in-diffusion [37] and the interdigital SAW transducers were fabricated using the well-established lift-off method [70].

Testing and performance measurement of the devices were carried out using mostly an He–Ne laser light at 0.6328 μm. Excitation of the optical guided waves at TE_0 and TM_0 modes was accomplished using a rutile prism, and a second rutile prism was used to couple out both the diffracted and the undiffracted light beams for detailed measurement (see Fig. 5.16).

Deflected and undeflected light spots of high beam quality were obtained. For example, Fig. 5.17 shows the undeflected (when no RF power was applied to the device) and deflected light spots, both at the far field, for a device fabricated with the frequency of the driving signal varied from 155 to 410 MHz [13b]. The aperture of the incident light beam employed for this particular experiment was approximately 0.1 cm. The quality of the undeflected light beam (RF power off) was preserved in the deflected light beam, and deflected light beams of satisfactory quality were achieved. Measured values for

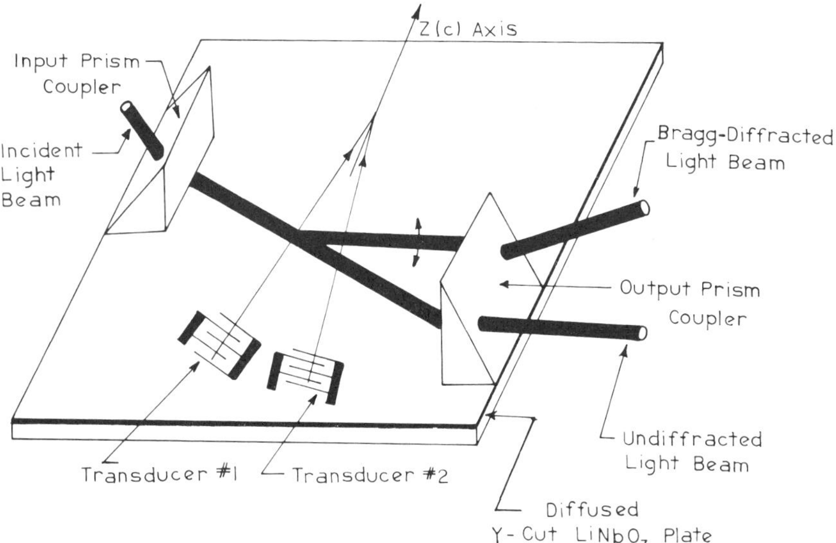

Figure 5.16 Basic experimental setup for measurement of planar guided-wave acousto-optical Bragg diffraction using a pair of prism couplers.

the number of resolvable spot diameters N_R and the incremental frequency change δf_R required for deflection of one Rayleigh spot diameter were found to be in good agreement with those predicted using Eq. (5.15). For example, 400 resolvable frequency channels and a frequency resolution of approximately 0.8 MHz were measured with a deflector of 358 MHz bandwidth using a uniform light beam aperture of 4.5 mm [13c]. A photograph of one of the complete devices is shown in Fig. 5.18. Note that the prism coupler–$LiNbO_3$ plate combination was sitting in the middle of the brass platform. The input and output RF connectors and the matching circuits for the transducers were located at the right and the left of the combination. The complete device was attached to a precision holder (micromanipulator) to facilitate optical alignment and adjustment. The wide-band device configurations described in Section 5.5.1, each demonstrating certain distinct features, were tested, and the device performance was measured. Again, due to space limitations, detailed design specifications and measured performances are given here only for the first and the fourth wide-band device configurations.

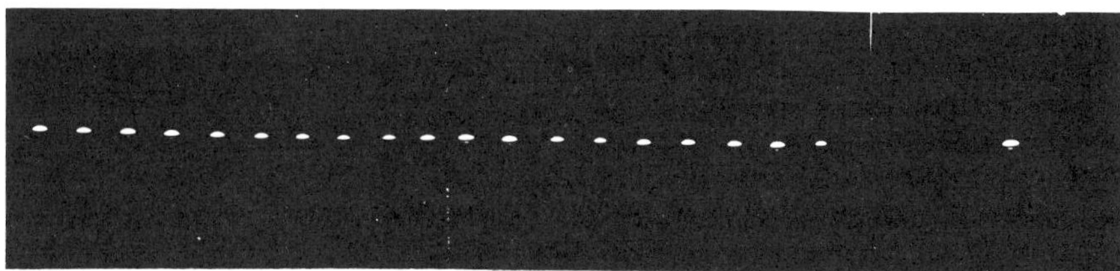

Figure 5.17 Undeflected light spot (left) and deflected light spots (right) obtained using tilted SAW array. SAW frequency is varied from 155 to 410 MHz at 15 MHz per step.

Figure 5.18 Photograph of a guided-wave acousto-optical Bragg deflector using multiple tilted SAW transducers.

Isotropic Device with Multiple Tilted Transducers of Staggered Center Frequency. Using the improved transducer geometry and related design procedures presented in Section 5.5.1, a deflector of 680 MHz composite bandwidth was realized in a *Y*-cut Ti-diffused $LiNbO_3$ waveguide without an electronic phase shifter [68]. The deflector utilized four tilted transducers with center frequencies of 380, 520, 703, and 950 MHz. The corresponding acoustic apertures were 0.64, 0.47, 0.34, and 0.24 mm, respectively, and the tilt angles between adjacent transducers were 5.9, 8.2, and 11.3 mrad, corresponding to the difference in the Bragg angles at the center frequency of the adjacent transducers. In order to obtain as wide an acoustic bandwidth as possible, the number of finger electrode pairs for each transducer was chosen to be as small as two and a half. The measured acoustic bandwidths of approximately 28–35% of the center frequencies were obtained by inserting a single inductance of proper value to each transducer.

The individual transducers were excited in parallel using power dividers. A deflected light beam of high quality was observed. The measured conversion efficiency of the four element transducers were, respectively, −7.5, −7.0, −10, −15 dB. The measured diffraction efficiency was 8% at a total RF drive power of 1 W for the entire −3-dB bandwidth of 680 MHz. Since the measured diffraction efficiency of the Bragg cell with only the first three transducers activated was nearly five times higher, a considerably better diffraction efficiency could be expected if the conversion efficiency of the fourth transducer (the one with the highest center frequency) were improved to that of the first three. Also, once a better transducer conversion efficiency is achieved, a Bragg cell with gigahertz bandwidths can be realized by increasing the center frequencies of the element transducers. Based on this projection, a performance figure of approximately 1 mW electric drive power per megahertz bandwidth with 50% diffraction efficiency and 1 GHz bandwidth should be realizable. In conclusion, the experimental study has verified that efficient Bragg modulators and deflectors of large bandwidth at GHz center frequency can be realized using multiple tilted SAW transducers of staggered center frequency.

Isotropic Device with a Tilted-Finger Chirp Transducer. The first tilted-finger chirp transducer was designed and fabricated in a Y-cut $LiNbO_3$ waveguide and Bragg diffraction experiments carried out using a 0.6328-μm He–Ne laser light propagating at the TE_0 mode [68]. The synchronous frequency of the finger electrodes was designed to vary linearly from 320 MHz (f_l) at one end to 630 MHz (f_h) at the other. The corresponding finger electrode width at the center of the finger aperture varies from 2.7 to 1.4 μm. The transducer contains a total of 51 finger electrodes each with an aperture of 0.55 mm. The transducer was driven directly with an RF signal generator of 50 Ω source impedance. The measured −3-dB transducer bandwidth was 255 MHz. Subsequently, a tilted-finger transducer of improved design using a "dog-leg" configuration was facilitated to realize a deflector bandwidth of 470 MHz [73]. The measured diffraction efficiency was 16% at 200 mW RF drive power. This improved deflector was subjected to 1 W of cw RF drive power without failure. Also, a wide-band transducer of slightly different electrode arrangement with similar performance characteristics was subsequently designed and fabricated [74,75].

GaAs AO Bragg Modulators and Deflectors. Using the same design procedures and guidelines, a variety of AO Bragg modulators and deflectors have also been designed and fabricated in GaAs waveguides. Fabrication steps involved are described in Section 5.3.5. Excitation of the optical guided waves at TE_0 and TM_0 modes was facilitated using edge-coupling via cleaved faces of the GaAs samples. Measured performances have been detailed in Section 5.3.5 and summarized in Table 5.2. It suffices to conclude that it is now possible to realize high-performance guided-wave AO Bragg cells in GaAs waveguides at GHz center frequency and octave bandwidth.

5.5.3 Relative Merits of a Single Transducer versus Multiple Transducers

As demonstrated in Section 5.5.2, both single transducers and multiple transducers (or array transducers) have been utilized successfully to construct wide-band planar AO Bragg cells. It is thus appropriate to comment on the relative merits of a single transducer versus multiple transducers. A Bragg cell that employs a single tilted-finger chirp transducer will require no matching circuits and takes less space than the deflector that employs multiple transducers. However, for the Bragg cell that employs multiple transducers, the availability of multiple and independent RF inputs provides the flexibility for compensations and adjustments after the device has been fabricated. Such compensations and adjustments may be required because of the fabrication errors as well as variations in the physical properties of the AO material and in the quality of the waveguide and the transducers. By means of simple electrical attenuators and filter circuits, such compensations and adjustments can be easily made. Thus, it is reasonable to conclude that each wide-band device configuration has its relative merits, and selection of a particular one among the five would largely depend upon the flexibility and performance required in any specific application.

5.6 APPLICATIONS OF GUIDED-WAVE ACOUSTO-OPTICAL BRAGG CELLS AND MODULES

As shown in Sections 5.5.1 and 5.5.2, it is now possible to realize high-performance guided-wave AO Bragg modulators and deflectors of gigahertz center frequency and large bandwidth in $LiNbO_3$ and GaAs waveguides. Together with the fabrication of miniature laser sources, waveguide lenses, and photodetector arrays, integration of all passive and active components on a single substrate or a small number of substrates has become a reality [16]. The resulting integrated AO device modules or subsystems possess a number of attractive features such as low electrical drive power, small size, light weight, less susceptibility to environmental effects, and potentially low cost. A number of such integrated AO device modules that are being developed will find a number of unique applications [16] in wide-band multichannel optical communications, signal processing, and computing. In this section some of these device modules and some potential applications are described.

5.6.1 Optical Communications

High-speed light beam modulation, deflection and switching, optical frequency shifting, wavelength filtering, and wavelength multiplexing/demultiplexing are among the important functions in future wide-band multichannel integrated and fiber-optical communication systems [15,16]. With regard to modulation, as indicated in Section 5.4, the existing guided-wave AO Bragg cells may provide a maximum modulation bandwidth of around 1 GHz. High-quality waveguide lenses, such as TIPE lenses [33] developed most recently, can readily produce the small beam waist required. In the meantime, a variety of EO modulators of multigigahertz bandwidth requiring very low drive power have been developed [76–83]. Consequently, existing AO Bragg modulators are in general not as competitive as EO modulators for applications that require nanosecond or subnanosecond modulation speed.

However, the situation is quite different with regard to multiport light beam deflection and switching. The AO deflectors discussed herein are capable of deflecting and switching a guided light beam into a large number of ports at moderate speed (microseconds and submicroseconds) and low electrical drive power per port. In contrast, EO deflectors and switches, although capable of a faster switching speed than AO deflectors and switches, can only provide either two ports or a relatively small number of ports per device [15,16]. Consequently, AO deflectors are unique and superior in this particular area of applications. Like bulk-wave AO deflectors, guided-wave AO deflectors can function under either the digital (random-access) or the analog (sequential) mode of operation [15,16]. A brief discussion of projected performance figures and potential applications now follows.

Digital Deflection and Switching. In the digital mode of operation, the frequency of the RF signal applied to the deflector is varied in discrete steps to switch the light beam. The number of resolvable beam spots (N_R), the frequency step required for deflecting one resolvable spot position (δf_R) based on Rayleigh criteria, and the minimum switching time (τ) are determined by Eqs. (5.15).

Equation (5.15c) indicates that the instantaneous scan rate R, that is, the number of resolvable beam positions or ports scanned per second, is given by V_R/D. To appreciate the order of magnitude involved, we consider two specific examples using the $LiNbO_3$ deflector described in Section 5.5.2. We assume that a deflector bandwidth of 500 MHz is used in both examples. In the first example, a light beam aperture of 3.45 mm is used, and we have $N_R = 500$, $\delta f_R = 1$ MHz, and $\tau = 10^{-6}$ s. The corresponding instantaneous scan rate is 10^6 spots/s. In the second example the same deflector bandwidth is used, but a much smaller light beam aperture of 0.134 mm is assumed. We then have $N_R = 20$, $\delta f_R = 25$ MHz, and $\tau = 4 \times 10^{-9}$ s. The corresponding instantaneous scan rate is

25×10^6 spots/s. Thus, for a given deflector bandwidth the number of resolvable ports and the scan rate impose a conflicting requirement on the light beam aperture in a digital AO deflector. Nevertheless, the digital deflectors possess the unique capability for medium-speed random-access deflection and switching. One specific application involves a fiber-optical system with a large number of fan-out ports, as depicted in Fig. 5.19. A single-mode optical fiber that carries optical signals is coupled to an optical waveguide. The light beam is expanded and collimated by the TIPE lens and deflected by the AO Bragg cell. By varying the frequency of the RF driving signal the deflected light beams are then routed and coupled via the TIPE microlens array to any element of the optical fiber array that serves as the output terminal.

Analog Deflection and Switching. A guided light beam may also be scanned and switched using an analog [frequency modulated (FM)] mode of operation in which the frequency of the electrical drive signal varies linearly with time. Using this mode of operation, it is possible to achieve a scan rate much greater than the scan rate that can be achieved using a digital mode of operation [15,16]. This higher scan rate results from the fact that the optical grating created by the linear FM acoustic wave acts as a moving Fresnel zone lens to an incident collimated light beam. Thus, the diffracted light is focused to a small spot size and simultaneously scanned at the acoustic wave velocity to result in a very high scan rate. The relevant parameters include focal length, focused spot size, number of scan spots, and scan rate.

Some potential applications for the analog deflector described in the preceding paragraph include (a) very high data rate multiport switching; (b) very high data rate optical writing and reading in applications such as facsimile; (c) optical pulse compression of radar chirp signal [15]; (d) high-speed parallel-to-serial (spatial-to-temporal) readout for the RF spectrum of integrated optical spectrum analyzers [84]; and (e) time demultiplexing of wide-band multichannel optical pulse trains [15,85]. It is to be noted that the first two applications are self-explanatory.

Frequency Shifting. We now turn to optical frequency shifting. Light sources with electronically tunable frequency offset are needed as local oscillators in heterodyne integrated optical communication and fiber-optical sensor systems [86]. A number of EO

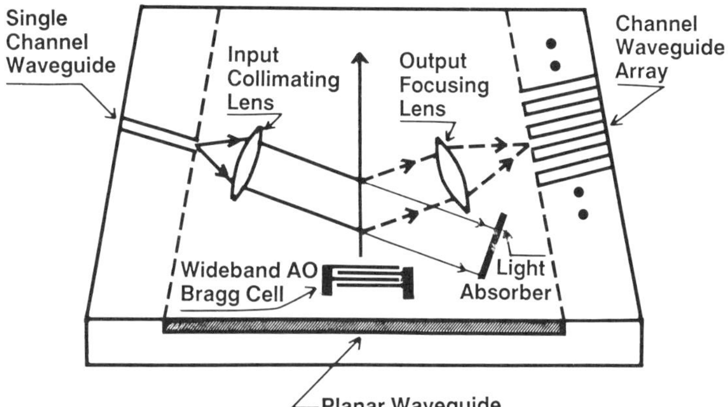

Figure 5.19 Integrated acousto-optical space division demultiplexer using a channel-planar composite waveguide in $LiNbO_3$.

schemes for producing such frequency-shifted light sources in integrated optical format have been reported [16].

An alternate integrated optical scheme would be to utilize AO Bragg diffraction from a traveling SAW as the frequency of the diffracted light is shifted from that of the incident light by the acoustic frequency [15]. Earlier, three distinct device configurations, namely, those involving the planar waveguide [15,16,87], the channel waveguide, and the spherical waveguide [88], were utilized to implement this AO frequency-shifting scheme. Most recently, a pair of tilted and counterpropagating SAWs in Y-cut $LiNbO_3$ planar waveguides were used to facilitate AO Bragg diffractions in cascade, and thus electronically tunable frequency shifting at passband and baseband [89]. The frequency-shifted light propagates in a fixed direction, but spatially resolved from the incident light, irrespective of the magnitude of frequency tuning [89]. These studies have shown that single-sideband frequency shifting using AO Bragg diffraction from the traveling SAW is a viable approach, and the resulting integrated optical modules can be readily and rigidly coupled with single-mode optical fibers to provide simultaneously the unshifted light and the frequency-shifted light sources that are spatially separated.

Optical Wavelength Filtering and Multiplexing/Demultiplexing. With regard to optical wavelength filtering, both passive [90] and active [91–94] guided-wave devices have been explored. Active AO filters that utilize guided-wave anisotropic and noncollinear Bragg diffraction have also been reported [93]. Readers are referred to the literature for detail. Also, guided-wave AO wavelength division multiplexing (WDM) optical switches of high capacity can be constructed [95]. Finally, it is to be noted that guided-wave AO diffraction in $LiNbO_3$ was recently utilized to construct an integrated optical bistable device [96].

5.6.2 Optical Signal Processing

The block diagram of a typical optical signal processor [86] is shown in Fig. 5.20. It consists of an electrical-to-optical transducer (input transducer) that converts the time window of a wide-band electrical signal to a spatial optical display, an optical system that acts on this display to generate the Fourier transform of the signal or the correlation/convolution of the signal with a reference signal, and an optical-to-electrical transducer (output transducer) that reads out this optical Fourier transform or correlation/convolution as an electrical analog signal or digital bit stream. Optical signal processors can process wide-band signals at very high speeds and offer a significant advantage in

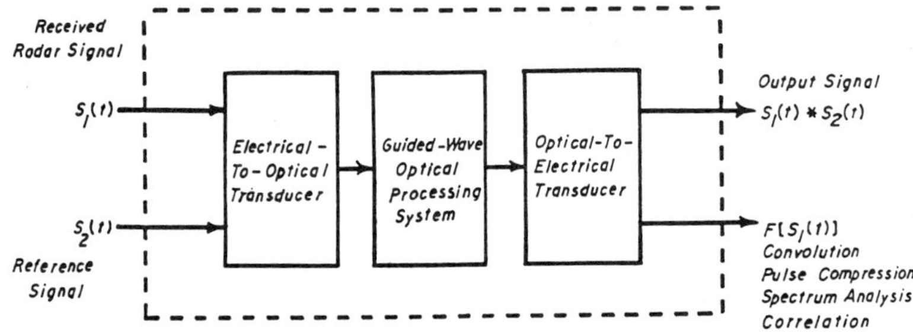

Figure 5.20 Block diagram of a basic guided-wave optical signal processor.

hardware performance (equivalent bits per second per dollar) over digital electronic processors [97].

One of the most important classes of optical signal processors is based on the use of coherent AO interactions. A number of bulk-type AO signal processors have been studied and demonstrated [98]: spatial-integrating correlators, matched filters for chirp radar, convolvers, time-integrating correlators, snap-shot PROMs, and spectrum analyzers. The AO Bragg cell serves as an input transducer and together with the lens pair provides the optical system for all these optical processors. Finally, a photodetector or photodetector array is used as the output transducer. Most bulk-type AO processors for one-dimensional signal processing can be implemented in a guided-wave format [15] using the wide-band AO Bragg cells described in Section 5.5. As indicated earlier, these guided-wave versions possess a number of attractive features. However, it should be noted that an extension to two-dimensional signal processing is more difficult to implement in the guided-wave format than in the bulk-wave format. Again, because of space limitation, only the spectrum analyzers are discussed further.

As mentioned in Section 5.1, development and implementation of RF spectrum analyzers using monolithic or hybrid integrated optical techniques has already become worldwide [99]. A detailed treatment of the $LiNbO_3$-based integrated optical RF spectrum analyzer (IOSA) is given in Refs. 99 and 100. In this section only the principle of operation and some of the major parameters are discussed.

A schematic of a hybrid integrated optical RF spectrum analyzer is depicted in Fig. 5.21. When a spectrum of RF signals is applied to the transducer, each spectral component generates a SAW that deflects the incident light beam in a corresponding direction. As shown in Section 5.3, the deflection angle and the intensity of the Bragg diffracted light are, respectively, proportional to the frequency and the power of the RF or the acoustic signal. Thus, by measuring the linear positions and intensities of the deflected light spots at the focal plane of the transform lens, the power spectral density of the RF signal of

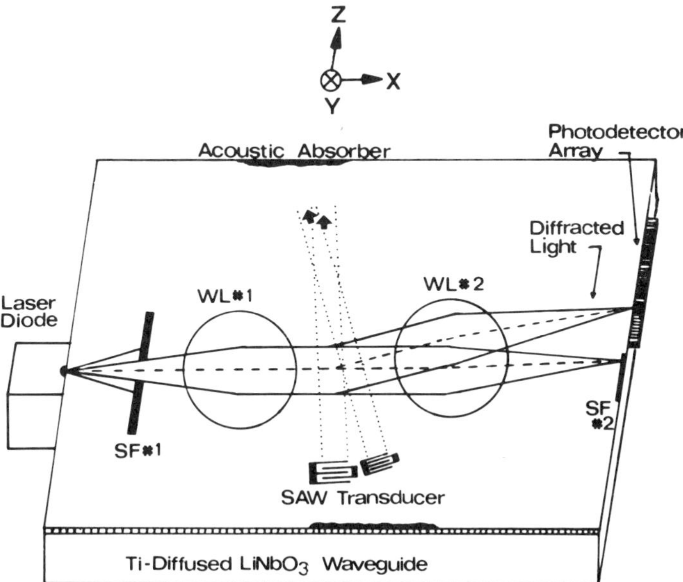

Figure 5.21 Acousto-optical spectrum analyzer using hybrid optical waveguide structure.

interest may be determined. The number of resolvable frequency channels N_R and the corresponding frequency resolution δf_R, based on Rayleigh criteria, are also given by Eqs. (5.15a) and (5.15b).

As described in Section 5.5.2, 400 resolvable frequency channels and a frequency resolution of approximately 0.8 MHz were measured using a deflector of 358 MHz bandwidth and a uniform light beam aperture of 4.5 mm [13b]. For the 680-MHz-bandwidth deflector described in Section 5.5.2, the measured frequency resolution was 0.6 MHz for a truncated Gaussian light beam of 6-mm aperture. This resolution was determined by measuring the half-power width of the deflected light spot [13b]. Thus, based on the measured resolution, this deflector, when used as a spectrum analyzer, would provide 1130 resolvable frequency channels.

In addition to the number of resolvable frequency channels and the frequency resolution, intermodulation and cross-modulation between different frequency components resulting from multiple AO diffraction are among the other important parameters. The extent of such undesirable modulations was measured using two independent RF signals of varying frequency separation [15,16,65]. The measured data show that even for a worst case of as much as 43% diffraction, the intensity of the strongest intermodulation was -38 dB down from those of the two diffracted lights that result from the two fundamental RF frequencies. For the practical cases the diffraction efficiency involved would be much lower than 43%, and the corresponding inter- and cross-modulation would accordingly be much lower. Thus, the dynamic range of guided-wave AO spectrum analyzers is likely to be limited by background noise (due to light scattering) rather than the inter- and cross-modulations due to multiple AO diffraction.

As previously indicated, development and realization of the integrated optical RF spectrum analyzers have become an international effort. Such integrated optical RF spectrum analyzers, when fully developed, are expected to possess two major advantages: (1) increased performance and reduced cost over both currently employed technology and competing technologies and (2) reduced size and increased compactness.

5.6.3 Optical Computing

Realization of optical computing functions in a waveguide substrate has long been considered one of the major objectives of integrated optics. Recently, the prospects for this realization have been significantly advanced through progress in guided-wave microoptic components of active and passive types in both $LiNbO_3$ [33,34] and GaAs [50,58] substrates. In fact, a variety of integrated optic (IO) device modules have been realized in the $LiNbO_3$ substrate, and their counterparts in GaAs substrate are on the way to be developed. In this section, examples of the $LiNbO_3$ device modules together with applications to systolic array computing and programmable correlation of digital sequences are presented followed by a progress report on GaAs device modules.

$LiNbO_3$-based Hybrid Multichannel Integrated Optical Device Modules. Of importance in such device modules are the single-mode microlenses and microlens arrays fabricated in $LiNbO_3$ by the TIPE technique [33]. These microlenses and lens arrays have demonstrated a combination of desirable characteristics. A TIPE microlens array and a large-aperture integrating lens together with a channel waveguide array, a planar waveguide, and an AO Bragg cell were integrated in a $LiNbO_3$ substrate $0.2 \times 1.0 \times 2.0\ \text{cm}^3$ in size [34] (see Fig. 5.22).

The versatile structure that facilitates realization of a variety of IO device modules is shown in Fig. 5.23. First, the channel waveguide array was fabricated in a Y-cut $LiNbO_3$ substrate using the TIPE process. The channel waveguide array is followed by

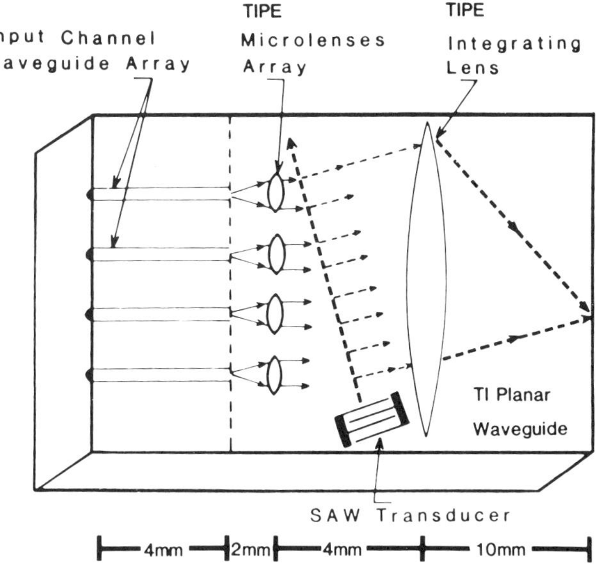

Figure 5.22 Integrated acousto-optical Bragg modulator module using a channel-planar composite waveguide in $LiNbO_3$.

a TIPE linear microlens array fabricated in the planar waveguide. The microlens array is used to capture, expand, and collimate the multiple light beams from the channel waveguide array before incidence upon the AO, the EO, or the AO–EO Bragg diffraction gratings induced in the planar waveguide. Finally, a large-aperture TIPE lens, which is fabricated using the same masking step as the linear microlens array, serves to collect and focus the multiple Bragg-diffracted light beams upon a single photodetector or a photodetector array.

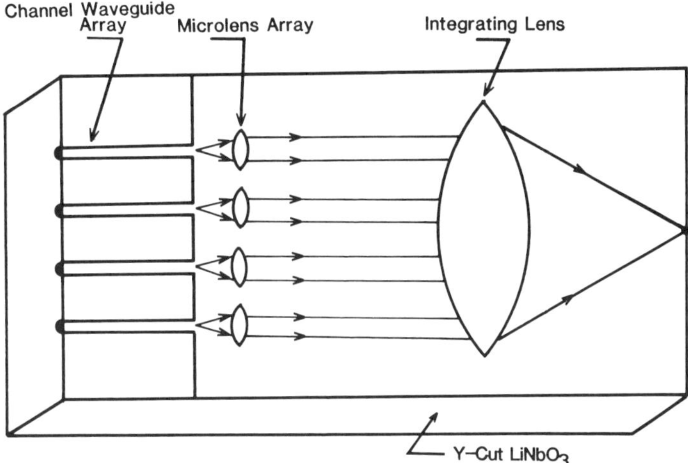

Figure 5.23 Linear microlens array–integrating lens combination in $LiNbO_3$ channel-planar composite waveguide.

The integrated AO device module shown in Fig. 5.22 was used successfully to perform optical computing such as optical systolic array processing [34,101] as it can readily perform the two required basic operations, namely, "multiplication" and "addition." For this particular application multiplication is facilitated by AO Bragg diffraction and addition by the integrating lens. Thus by pulsating the digital data sequences separately into the multiple input light beams (through the channel waveguide array), and the SAW high-speed digital filtering as well as matrix–vector and matrix–matrix multiplication can be performed. A simple experiment on matrix–vector multiplication involving a 2×2 matrix and a two-dimensional vector using two of the channel waveguide arrays and the SAW has been carried out [34]. In the following, two of the most advanced versions of the modules that utilize the basic channel-planar composite waveguides together with the microlenses and the active components are described.

Figures 5.24 and 5.25 show the architectures of these two device modules, each fabricated in a Y-cut $LiNbO_3$ substrate $0.1 \times 1.0 \times 2.0\ cm^3$ in size. The width of each channel waveguide and the separation between adjacent waveguides of the 10-element channel waveguide array (only four elements are shown) fabricated are 6 and 200 μm, respectively. The linear microlens array, which consists of 10 identical lenses each with an aperture of 200 μm, is aligned with the channel waveguide array. Other dimensions along the optical path are shown in Figs. 5.24 and 5.25. A variety of multichannel device modules can be constructed by inserting a single AO Bragg modulator [34a], an array of EO Bragg modulators [34f], or a combination of the two [34g] in the region between the microlens array and the larger-aperture lens. The herringbone Bragg electrode array and the SAW transducer–conventional Bragg electrode array combination used in the two device modules were fabricated using the lift-off technique.

Ideally, the multiple light sources required in the experiments with the two device modules just described should be furnished by a diode laser array. In the absence of such a diode laser array in our laboratory, the multiple light sources were formed by using a large-aperture collimated light beam together with a linear planar microlens array [34e]. The separation between adjacent lenses in this planar microlens array was identical to

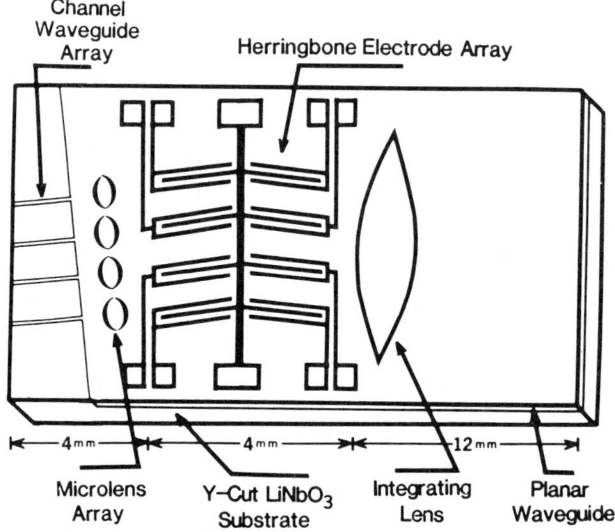

Figure 5.24 Multichannel integrated electro-optical device module.

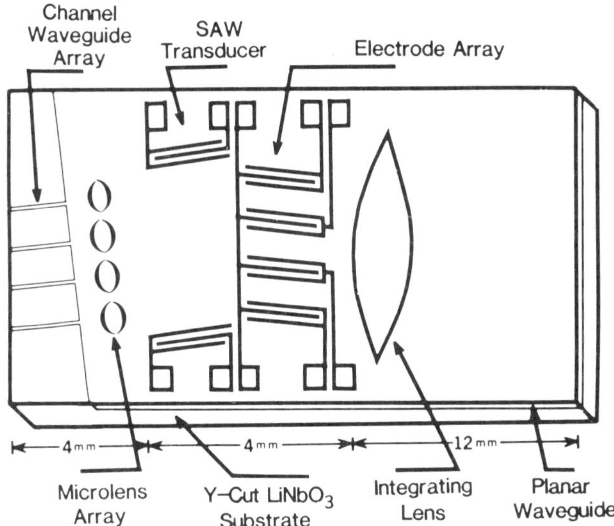

Figure 5.25 Multichannel integrated acousto-optical–electro-optical device module.

that between adjacent channel waveguides, namely, 200 μm. As noted previously, each of the EO Bragg modulator arrays shown in Fig. 5.24 utilizes a herringbone Bragg electrode that consists of two identical conventional electrode arrays (each with a 13-μm periodicity and 1.9-mm aperture) intersecting at twice the Bragg angle. The two separate electrode arrays facilitate engagement of two independent sets of voltages and thus Bragg diffractions in tandem. Efficient and wide-band double Bragg diffractions, namely, 90% diffraction at a drive voltage of 5 V and a base bandwidth of 840 MHz, was measured at the optical wavelength of 0.6328 μm. A dynamic range of 31 dB was also measured. Therefore, high-performance herringbone Bragg electrode arrays have been fabricated successfully to perform multiplication of two independent sets of voltages. Addition of the resulting multiplication products is facilitated by the large-aperture integrating lens. This multichannel EO device module has been used to perform programmable correlation of binary sequences as well as matrix–vector and matrix–matrix multiplication [34,101]. For example, Figs. 5.26*a* and *b* show, respectively, the autocorrelation waveform of the 10-bit binary sequence 1010101010 and the cross-correlation waveform between the sequences 1010101010 and 1000100000 obtained at a data rate of 3 kbits/s. Since the aperture of each EO Bragg modulator can be reduced further from 200 to 100 μm, it should be possible to construct device modules of very large bit capacity (word) (e.g., 100 bits in a $LiNbO_3$ substrate only slightly larger than $0.1 \times 1.0 \times 2.0$ cm^3 in size).

Note that the second device module shown in Fig. 5.25 results from the replacement of the first segment of the herringbone Bragg electrode structure by the AO Bragg diffraction grating [15] induced via the SAW, which is in turn generated by a single-SAW transducer. In order to facilitate AO and EO Bragg diffractions in tandem, the interdigital electrodes of the SAW transducer are tilted from the second segment of the herringbone electrode array by an angle that is equal to twice the AO Bragg angle θ_B. A -3-dB AO bandwidth of 45 MHz centered at 260 MHz and a diffraction efficiency of 90% at an RF drive power of 150 mW were measured at the optical wavelength of 0.6328 μm. Thus, double Bragg diffractions of high efficiency and moderate bandwidth have also been accomplished. Accordingly, this multichannel AO–EO module is being

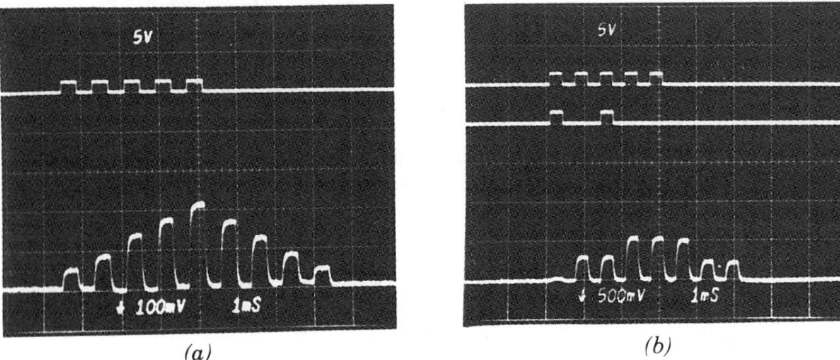

Figure 5.26 (*a*) Auto-correlation waveform of 10-bit binary sequence. (*b*) Cross-correlation waveform of two 10-bit binary sequences.

used to perform systolic array processing and programmable correlation of binary sequences at the data rate of 17.5 Mbits/s via the SAW.

It is important to compare the unique features of the two device modules. The EO module accepts multiple sets of data in parallel format and, therefore, can process the data at a considerably higher rate than is possible with the AO–EO module. However, through the SAW, the AO–EO module accepts data in a pipeline fashion that is desirable in a variety of applications.

GaAs-based Monolithic Multichannel Integrated Optical Device Modules. For the GaAs substrate, as presented in Section 5.3.5, wide-band GaAs waveguide AO Bragg cells that operate in the acoustic frequency range of 300–1200 MHz have been realized. This recent advancement has paved the way for realization of monolithically integrated optical signal processors such as RF spectrum analyzers and correlators in a common GaAs chip. Recently there has also been significant advancement in the realization of multichannel IO modulator modules in GaAs. Specifically, multichannel single-mode EO cutoff modulator arrays [102] and EO Bragg diffraction modulators [103] have been successfully realized in GaAs. Also, waveguide microlenses and linear lens arrays were most recently fabricated in GaAs [58] using the ion milling technique. The waveguide lenses that have been fabricated and tested include single lenses and analog Fresnel lens arrays, chirp grating, and hybrid analog Fresnel/chirp grating types. Near-diffraction-limited spot sizes and good efficiencies have been obtained. Thus, similar to the TIPE waveguide lenses in $LiNbO_3$, such ion-milled waveguide lenses should facilitate realization of a variety of monolithic IO device modules and circuits in GaAs and related substrates, with applications in communications, signal processing, and computing.

Most recently, toward eventual realization of monolithic integrated optic RF spectrum analyzers (IOSA), a wide-band AO Bragg cell was integrated with an ion-milled collimation-Fourier transform lens pair in a ZnO-GaAs-$Al_{0.15}Ga_{0.85}As$ composite waveguide 7×23 mm^2 in size [104].

The detailed structure of the IOSA is shown in Fig. 5.27. A GaAs guiding layer 1.1 μm thick and a ZnO layer 0.3 μm thick were grown to support propagation of the TE_0 and the TM_0 modes at the optical wavelength of 1.3 μm and to faciliate efficient and wide-band AO Bragg diffraction centered at 500 MHz. An incident light beam of 10 mW was coupled into the waveguide, and both the diffracted and the undiffracted light were coupled out of the waveguide via cleaved edges of the waveguide. The f-number of the collimating lens was 5 with an optical aperture of 1.2 mm, while that of the Fourier lens

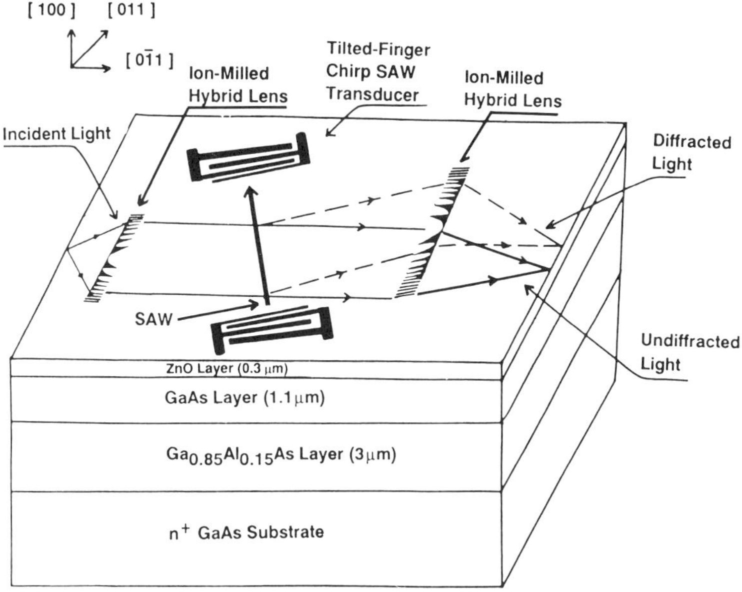

Figure 5.27 Acousto-optical RF spectrum analyzer in a ZnO/GaAs/AlGaAs composite waveguide structure.

was 4.4 with an optical aperture of 1.8 mm. A diffraction efficiency of 7% per watt of RF drive power was measured using the TE_0 mode of propagation. The dynamic range defined as the diffracted light intensity at 50% diffraction efficiency over background noise level was measured to be higher than 16 dB. A large AO bandwidth of 220 MHz was facilitated by the tilted-finger chirp SAW transducer of 1.1 mm aperture.

The focal spot profiles of the undiffracted and diffracted light after porpagating through the waveguide lens pair were measured to be almost identical, namely, a -3 dB focal spot width of 4.5 μm at the optical wavelength of 1.3 μm. The measured scanning of the focused spot of the diffracted light versus the frequency of the input RF signal shows that a scan of 1.03 μm per MHz has been demonstrated. Thus, a frequency increment of 4.5 MHz was needed to produce scan of a focused spot size of 4.5 μm. By simultaneously applying two RF signals of various frequency separation to the SAW transducer, the IOSA was found to be capable of resolving a frequency sparation 5.5 MHz. Thus, a total of 40 channels for the frequencies centered at 500 MHz at a resolution 5.5 MHz have been accomplished with this first GaAs AO RF spectrum analyzer.

REFERENCES

1. C. F. Quate, C. D. W. Wilkinson, and D. K. Winslow, "Interaction of Light and Microwave Sound," *Proc. IEEE*, Vol. 53, pp. 1604–1623, 1965.
2. A. J. DeMaria and G. E. Danielson, Jr., "Internal Laser Modulation by Acoustic Lens-Like Effects," *IEEE J. Quantum Electron.*, Vol. QE-2, pp. 157–164, 1966.
3. A. Korpel, R. Adler, P. Desmares, and W. Watson, "A Television Display Using Acoustic Deflection and Modulation of Coherent Light," *Proc. IEEE*, Vol. 54, pp. 1429–1437, 1966.

4. E. I. Gordon, "A Review of Acoustooptical Deflection and Modulation," *Proc. IEEE*, Vol. 54, pp. 1391–1401, 1966.
5. R. Adler, "Interactions of Light and Sound," *IEEE Spectrum*, Vol. 4, p. 42, 1967.
6. R. W. Dixon, "Acoustic Diffraction of Light in Anisotropic Media," *IEEE J. Quantum Electron.*, Vol. QE-3, p. 85, 1967.
7. (a) R. W. Dixon, "Photoelastic Properties of Selected Materials and Their Relevance for Applications to Acoustic Light Modulators and Scanners," *J. Appl. Phys.*, Vol. 38, pp. 5149–5153, 1967. (b) D. A. Pinnow, "Guide Lines for the Selection of Acoustooptic Materials," *IEEE J. Quantum Electron.*, Vol. QE-6, pp. 223–238, 1970.
8. (a) N. Uchida and N. Niizeki, "Acoustooptic Deflection Materials and Techniques," *Proc. IEEE*, Vol. 61, pp. 1073–1092, 1973. (b) I. C. Chang, "Acoustooptic Devices and Applications," *IEEE Trans Sonics Ultrasonics*, Vol. SU-23, pp. 2–22, 1976.
9. A. Korpel, "Acousto-Optics," in R. Wolfe, Ed., *Applied Solid-State Science*, Vol. 3, Academic, New York, 1972, Chap. 2, pp. 73–179.
10. L. Kuhn, M. D. Dakss, P. F. Heidrich, and B. A. Scott, "Deflection of an Optical Guided Wave by a Surface Acoustic Wave," *Appl. Phys. Lett.*, Vol. 17, p. 265, 1970.
11. Y. Ohmachi, "Acousto-Optical Light Diffraction in Thin Films," *J. Appl. Phys.*, Vol. 44, pp. 3928–3922, 1973.
12. R. V. Schmidt and I. P. Kaminow, "Acoustooptic Bragg Deflection in $LiNbO_3$ Ti-diffused Waveguides," *IEEE J. Quantum Electron.*, Vol. QE-11, pp. 57–59, 1975.
13. (a) C. S. Tsai, L. T. Nguyen, S. K. Yao, and M. A. Alhaider, "High-Performance Guided-Light-Beam Device Using Two-Tilted Surface Acoustic Wave," *Appl. Phys. Lett.*, Vol. 26, pp. 140–142, 1975. (b) C. S. Tsai, M. A. Alhaider, Le T. Nguyen, and B. Kim, "Wideband Guided-Wave Acoustooptic Bragg Diffraction and Devices Using Multiple Tilted Surface Acoustic Waves," *Proc. IEEE*, Vol. 64, pp. 318–328, 1976. (c) B. Kim and C. S. Tsai, "High-Performance Guided-Wave Acoustooptic Scanning Devices Using Multiple Surface Acoustic Waves," *Proc. IEEE*, Vol. 64, pp. 788–796, 1976. (d) C. S. Tsai, I. W. Yao, B. Kim, and Le T. Nguyen, "Wideband Guided-Wave Anisotropic Acoustooptic Bragg Diffraction in $LiNbO_3$ Waveguides," *International Conference on Integrated Optics and Optical Fiber Communications*, July 18–20, 1977, Tokyo, Japan, Digest of Technical Papers, pp. 57–60.
14. R. V. Schmidt, "Acoustooptic Interactions between Guided Optical Waves and Acoustic Surface Waves," *IEEE Trans. Sonics Ultrasonics*, Vol. SU-23, pp. 22–23, 1976.
15. C. S. Tsai, "Guided-Wave Acoustooptic Bragg Modulators for Wideband Integrated Optic Communications and Signal Processing," *IEEE Trans. Circuits Syst.*, Vol. CAS-26, p. 1072, 1979.
16. C. S. Tsai, Ed., *Guided-Wave Acoustooptic Bragg Diffraction, Devices, And Applications*, Springer-Verlag, Berlin, 1990.
17. P. K. Tien, "Light Waves in Thin-Films and Integrated Optics," *Appl. Opt.*, Vol. 10, pp. 2395–2413, 1971.
18. S. E. Miller, "A Survey of Integrated Optics," *IEEE J. Quant. Electron.*, Vol. QE-8, pp. 199–205, 1972.
19. H. F. Taylor and A. Yariv, "Guided-Wave Optics," *Proc. IEEE*, Vol. 62, pp. 1044–1060, 1974.
20. H. Kogelnik, "An Introduction to Integrated Optics," *IEEE Trans. Microwave Theory Tech.*, Vol. 23, pp. 2–16, 1975.
21. I. P. Kaminow, "Optical Waveguide Modulators," *IEEE Trans. Microwave Theory Tech.*, Vol. 23, pp. 57–70, 1975.
22. P. K. Tien, "Integrated Optics and New Wave Phenomena in Optical Waveguides," *Rev. Mod. Phys.*, Vol. 49, pp. 361–423, 1977.
23. T. Tamir, Ed., *Integrated Optics*, Springer-Verlag, Berlin, 1979.
24. A. Yariv, *Introduction to Optical Electronics*, 2nd ed., Holt, Rinehart and Winston, 1976.
25. (a) R. Hunsperger, *Integrated Optics: Theory and Technology*, Springer-Verlag, Berlin, 1982. (b) H. P. Nolting and R. Ulrich, Ed., *Integrated Optics*, Springer-Verlag, Berlin, 1985.
26. R. M. White, "Surface Elastic Waves," *Proc. IEEE*, Vol. 58, pp. 1238–1276, 1970.

27. A. A. Oliner, Ed., *Surface Acoustic Waves*, Springer-Verlag, Berlin, 1979.

28. R. W. Smith, H. M. Gerard, J. H. Collins, T. M. Reeder, and H. J. Shaw, "Design of Surface Wave Delay Lines with Interdigital Transducers," *IEEE Trans. Microwave Theory Tech.*, Vol. MTT-17, pp. 856–873, 1969.

29. G. S. Kino and R. S. Wagers, "Theory of Interdigital Couplers on Nonpiezoelectric Substrates," *J. Appl. Phys.*, Vol. 44, pp. 1480–1488, 1973.

30. F. S. Hickernell and J. W. Brewer, "Surface-Elastic-Wave Properties of DC-Triode-sputtered Zinc Oxide Film," *Appl. Phys. Lett.*, Vol. 21, pp. 389–391, 1972.

31. G. S. Kino, *Acoustic Waves: Devices, Imaging, And Analogy Signal Processing*, Prentice-Hall, Englewood-Cliffs, NJ, 1987.

32. C. V. Verber, R. P. Kenan, and J. R. Blusch, "Correlator Based on an Integrated Optical Spatial Light Modulator," *Appl. Opt.*, Vol. 20, p. 1626, 1981.

33. (a) D. Y. Zang and C. S. Tsai, "Single-Mode Waveguide Microlenses and Microlens Arrays Fabrication Using Titanium Indiffused Proton Exchange Technique in $LiNbO_3$, *Appl. Phys. Lett.*, Vol. 48, pp. 703–705, 1985. (b) D. Y. Zang and C. S. Tsai, "Titanium-indiffused Proton-exchanged Waveguide Lenses in $LiNbO_3$ for Optical Information Processing," *Appl. Opt.*, Vol. 25, pp. 2264–2271, 1986.

34. (a) C. S. Tsai, D. Y. Zang, and P. Le, "Guided-Wave Acoustooptic Bragg Diffraction in a $LiNbO_3$ Channel-Planar Waveguide with Application to Optical Computing," *Appl. Phys. Lett.*, Vol. 47, pp. 549–551, 1985. (b) C. S. Tsai, "$LiNbO_3$-based Integrated-Optic Device Modules for Communication, Computing, and Signal Processing," in *1986 Conference on Lasers and Electro-Optics, Technical Digest*, IEEE Cat. No. 86CH2274-9, 44-46, 1986. (c) C. S. Tsai, "Titanium-indiffused Proton-exchanged Microlens-based Integrated Optic Bragg Modulator Modules for Optical Computing," in H. H. Szu, Ed., Special Issue on Optical and Hybrid Computing, *SPIE*, Vol. 634, pp. 409–421, 1987. (d) C. S. Tsai, D. Y. Zang, and P. Le, "Multichannel Integrated Optic Device modules in $LiNbO_3$ for Digital Data Processing," in *1988 Topical Meeting on Integrated and Guided-Wave Optics*, Santa Fe, NM, Technical Digest, pp. TuA5-1–TuA5-4. (e) P. Le, D. Y. Zang, and C. S. Tsai, "Integrated Electrooptic Bragg Modulator Modules for Matrix–Vector and Matrix–Matrix Multiplications, *Appl. Opt.*, Special issue on optical computing, Vol. 26, pp. 1780–1785, 1988. (f) D. Y. Zang, P. Le, G. D. Xu, and C. S. Tsai, "An Integrated Optic Digital Correlator Module Using Both Acoustooptic and Electrooptic Bragg Diffractions" (in press).

35. N. Chubachi, J. Kushibiki, and Y. Kikuchi, "Monolithically Integrated Bragg Deflector for an Optical Guided Wave Made of Zinc-Oxide Film," *Electron. Lett.*, Vol. 9, pp. 193–194, 1973. (b) N. Chubachi, "ZnO Films for Surface Acoustooptic Devices on Nonpiezoelectric Substrates," *Proc. IEEE*, Vol. 64, pp. 772–774, 1976. (c) N. Chubachi and H. Sasaki, "Surface Acousto-Optic Interaction in ZnO Thin Films," *Wave Electron.*, Vol. 2, p. 379, 1976.

36. I. P. Kaminow and J. R. Carruthers, "Optical Waveguiding Layers in $LiNbO_3$ and $LiTaO_3$," *Appl. Phys. Lett.*, Vol. 22, pp. 326–329, 1973.

37. R. V. Schmidt and I. P. Kaminow, "Metal-diffused Optical Wave Guides in $LiNbO_3$," *Appl. Phys. Lett.*, Vol. 25, pp. 459–460, 1974.

38. D. Marcuse, "TE Modes of Graded-Index Slab Waveguides," *IEEE J. Quantum Electron.*, Vol. QE-9, pp. 1000–1006, 1973.

39. E. M. Conwell, "Modes in Optical Waveguides Formed by Diffusion," *Appl. Phys. Lett.*, Vol. 26, pp. 328–329, 1973.

40. P. K. Tien et al., "Optical Waveguide Modes in Single Crystalline $LiNbO_3$–$LiTaO_3$ Solid-Solution Films," *Appl. Phys. Lett.*, Vol. 24, pp. 503–506, 1974.

41. H. Kogelnik, "Theory of Dielectric Waveguides," in T. Tamir, Ed., *Integrated Optics*, Springer-Verlag, Berlin, 1975, Chap. 2.

42. G. B. Hocker and W. K. Burns, "Modes in Diffused Optical Waveguides of Arbitrary Index Profile," *IEEE J. Quantum Electron.*, Vol. QE-11, pp. 270–276, June 1975.

43. G. W. Farnell, "Properties of Elastic Surface Waves," in W. P. Mason and R. N. Thurston, Eds., *Physical Acoustics*, Vol. 6, Academic, New York, 1970, pp. 109–166.

44. R. N. Spaight and G. G. Koerber, "Piezoelectric Surface Waves on $LiNbO_3$," *IEEE Trans. Sonics Ultrasonics*, Vol. SU-18, pp. 237–238, 1971.

45. A. J. Slobodnik, E. D. Conway, and R. T. Delmonico, Eds., *Microwave Acoustics Handbook*, Vol. 1A, *Surface Wave Velocities*, under AFCRL-TR-73-0593, October 1, 1973, Air Force Cambridge Reseach Laboratories, L.G. Hanscom Field, Bedford, MA.

46. B. A. Auld, *Acoustic Fields and Waves in Solids*, Vol. 2, Wiley, New York, 1973.

47. R. J. Pressley, Ed., *Handbook of Lasers with Selected Data of Optical Technology*, Chemical Rubber Co., Cleveland, OH, 1971.

48. K. W. Loh, W. S. C. Chang, W. R. Smith, and T. Grudkowski, "Bragg Coupling Efficiency for Guided Acoustic Interaction in GaAs," *Appl. Opt.*, Vol. 15, pp. 156–166, 1976.

49. O. Yamazaki, C. S. Tsai, M. Umeda, L. S. Yap, C. J. Lii, K. Wasa, and J. Merz, "Guided-Wave Acoustooptic Interactions in GaAs–ZnO Composite Structure," *1982 IEEE Ultrasonics Symp. Proc.*, IEEE Cat. No. 82CH1823-4, pp. 418–421.

50. (a) C. J. Lii, C. S. Tsai, and C. C. Lee, "Wideband Guided-Wave Acoustooptic Bragg Cells in GaAs–GaAlAs Waveguide," *IEEE J. Quantum Electron.*, Vol. QE-22, pp. 868–872, 1986. (b) Y. Abdelrazek and C. S. Tsai, "High-Performance Acoustooptic Bragg Cells in ZnO-GaAs Waveguide at GHz Frequencies," *Optoelectronices–Devices and Technologies*, Vol. 4, pp. 33–37, 1989.

51. D. Mergerian, E. Malarkey, B. Newman, J. Lane, R. Weinert, B. R. McAvoy, and C. S. Tsai, "Zinc Oxide Transducer Array for Integrated Optics," *1978 Ultrasonics Symp., Proc.*, IEEE Cat. No. 78-CH1344-ISU, pp. 64–69.

52. S. K. Yao, R. R. August, and D. B. Anderson," Guided-Wave Acoustooptic Interaction on Nonpiezoelectric Substrates," *J. Appl. Phys.*, Vol. 49, pp. 5728–5730, 1978.

53. N. Mikoshiba, "Guided-Wave Acoustooptic Interactions in ZnO Film on Nonpiezoelectric Substrates," in C. S. Tsai, Ed., *Guided-Wave Acoustooptic Bragg Diffraction, Devices, and Applications*, Springer-Verlag, Berlin, 1988, Chap. 6.

54. (a) A. Yariv and P. Yeh, *Optical Waves in Crystals*, Wiley, New York, 1984, Chap. 11. (b) Y. Suematsu, "Advances in Semiconductor Lasers," *Physics Today*, Vol. 32, pp. 32–39, 1985.

55. (a) J. L. Merz, R. A. Logan, and A. M. Sergent, "GaAs Integrated Optical Circuits by Wet Chemical Etching," *IEEE J. Quantum Electron.*, Vol. QE-5, pp. 72–82, 1979. (b) Y. R. Yuan, L. Perillo, and J. L. Merz, "Monolithic Integration of Curved Waveguides and Channeled-Substrate DH Lasers by Wet Chemical Etching," *J. Lightwave Tech.*, Vol. LT-1, pp. 630–636, 1983. (c) N. Bar-Chaim, S. Margalit, A. Yariv, and I. Ury, "GaAs Integrated Optoelectronics," *IEEE Trans. Electron. Devices*, Vol. ED-29, p. 1372, 1982. (d) A. Yariv, "The Beginning of Integrated Optoelectronic Circuits," *IEEE Trans. Electron. Devices*, Vol. ED-3, p. 1956, 1984. (e) O. Wada, T. Sakurai, and T. Nakagami, "Recent Progress in Optoelectronic Integrated Circuits," *IEEE J. Quantum Electron.*, Vol. QE-22, p. 805, 1986.

56. T. W. Bristol, W. R. Jones, P. B. Snow, and W. R. Smith, "Applications of Double Electrodes in Acoustic Surface Wave Device Design," *1972 IEEE Ultrasonics Symp. Proc.*, IEEE Cat. No. 72 CH0708-8SU, p. 343.

57. T. Q. Vu, J. A. Norris, and C. S. Tsai, "Formation of Negative Index Changes Waveguide Lenses in $LiNbO_3$ Using Ion Milling," *Opt. Lett.* Vol. 13, pp. 1141–1143, 1988.

58. T. Q. Vu, J. A. Norris, and C. S. Tsai, "Planar Waveguide Lenses in GaAs Using Ion Milling," *Appl. Phys. Lett.* Vol. 54, pp. 1098–1100, 1989.

59. R. H. Tancrell and M. G. Holland, "Acoustic Surface Wave Filters," *Proc. IEEE*, Vol. 59, pp. 393–409, 1971.

60. W. R. Smith, H. M. Gerard, and W. R. Jones, "Analysis and Design of Dispersive Interdigital Surface-Wave Transducers," *IEEE Trans. Microwave Theory Tech.*, Vol. MTT-20, pp. 458–471, 1972.

61. H. V. Hance and J. K. Parks, "Wideband Modulation of a Laser Beam Using Bragg-Angle Diffraction by Amplitude Modulated Ultrasonics Waves," *J. Acoust. Soc. Amer.*, Vol. 38, pp. 14–23, 1965.

62. D. Maydan, "Acoustooptical Pulse Modulators," *IEEE J. Quantum Electron.*, Vol. QE-6, pp. 15–24, 1970.

63. T. Van Duzer, "Lenses and Graded Films for Focusing and Guiding Acoustic Surface Waves," *Proc. IEEE*, Vol. 58, pp. 1230–1237, 1970.

64. G. L. Matthaei, L. Young et al., *Microwave Filters, Impedance Matching Networks and Coupling Structures*, McGraw-Hill, New York, 1964.

65. C. S. Tsai, Le T. Nguyen, B. Kim, and I. W. Yao, "Guided-Wave Acoustooptic Signal Processors for Wideband Radar Systems," Special Issue on Effective Utilization of Optics in Radar Systems, *Proc. SPIE*, Vol. 128, pp. 68–74, 1978.

66. E. G. Lean and C. C. Tseng, "Nonlinear Effects in Surface Acoustic Waves," *J. Appl. Phys.*, Vol. 41, pp. 3912–3917, 1970.

67. Le T. Nguyen and C. S. Tsai, "Efficient Wideband Guided-Wave Acoustooptic Bragg Diffraction Using Phased-Surface Acoustic Wave Array in $LiNbO_3$ Waveguides," *Appl. Opt.*, Vol. 16, pp. 1297–1304, 1977.

68. C. C. Lee, K. Y. Liao, C. L. Chang, and C. S. Tsai, "Wideband Guided-Wave Acoustooptic Bragg Deflector Using a Tilted-Finger Chirp Transducer," *IEEE J. Quantum Electron.*, Vol. QE-15, pp. 1166–1170, 1979.

69. C. S. Tsai, I. W. Yao, B. Kim, and Le T. Nguyen, "Wideband Guided-Wave Anisotropic Acoustooptic Bragg Diffraction in $LiNbO_3$ Waveguides," *International Conference on Integrated Optics and Optical Fiber Communications*, July 18–20, 1977, Tokyo, Japan, *Digest of Technical Papers*, pp. 57–60.

70. H. I. Smith, F. J. Bachner, and N. Efremow, "A High-Yield Photolithographic Technique for Surface Wave Devices," *J. Electrochem. Soc.*, Vol. 118, pp. 822–825, May 1971.

71. T. Suhara, T. Shiono, H. Nishihara, and J. Koyama, "An Integrated-Optic Fourier Processor Using an Acoustooptic Deflector and Fresnel Lenses in As_2S_3 Waveguide," *J. Lightwave Technol.*, Vol. LT-1, pp. 624–630, 1983.

72. C. Stewart, G. Serivener, and W. J. Stewart, "Guided-Wave Acoustooptic Spectrum Analysis at Frequencies Above 1 GHz," *1981 International Conference On Integrated Optics And Optical Fiber Communication, Technical Digest*, April 1981, p. 122.

73. K. Y. Liao, C. L. Chang, C. C. Lee, and C. S. Tsai, "Progress on Wideband Guided-Wave Acoustooptic Bragg Deflector Using a Tilted-Finger Chirp Transducer," *1979 Ultrasonics Symp. Proc.*, IEEE Cat. No. 79CH1482-9SU, pp. 24–27.

74. T. R. Joseph, "Broadband Chirp Transducers for Integrated Optics Spectrum Analyzers," *1979 Ultrasonics Symp. Proc.*, IEEE Cat. No. 79CH1482-9SU, pp. 28–32.

75. D. Gregoris and V. M. Ristic, "Wideband Transducer for Acousto-optic Bragg Deflector," *Can. J. Phys.*, Vol. 63, pp. 195–197, 1985.

76. H. Kogelnik and R. V. Schmidt, "Switched Directional Couplers with Alternating $\Delta\beta$," *IEEE J. Quantum Electron.*, Vol. QE-12, p. 396, 1976.

77. R. F. Alferness, "Guided-Wave Devices for Optical Communication," *IEEE J. Quantum Electron.*, Vol. QE-17, p. 946, 1981.

78. (a) C. S. Tsai, B. Kim, and F. R. El-Akkari, "Optical Channel Wave-Guide Switch and Coupler Using Total Internal Reflection," *IEEE J. Quantum Electron.*, Vol. QE-14, p. 513, 1978. (b) C. L. Chang, F. R. El-Akkari, and C. S. Tsai, "Fabrication and Testing of Optical Channel Waveguide TIR Switching Networks," *Proc. SPIE*, Vol. 239, p. 147, 1981. (c) C. S. Tsai, C. C. Lee, and P. Le, "A 8.5 GHz Bandwidth Single-Mode Crossed Channel Waveguide TIR Modulator and Switch in $LiNbO_3$," *Technical Digest of Postdeadline Papers, 1984 Topical Meeting on Integrated and Guided-Wave Optics*, IEEE Cat. No. 84CH1997-6, pp. PD 5-1–PD 5-4.

79. K. Wasa, H. Adachi, T. Kawaguchi, K. Ohji, and K. Setsune, "Optical TIR Switches Using PLZT Thin Film Waveguides on Sapphire," *International Conference on Integrated Optics And Optical Fiber Communications*, June 27–30, 1983, Tokyo, Japan, *Technical Digest*, pp. 356–357.

80. A. Neyer, "Electro-optic X-switch using single-mode Ti:$LiNbO_3$ channel waveguides," *Electron. Lett.*, Vol. 19, p. 553, 1983.

81. H. Nakajima, I. Sawaki, M. Seino, and K. Asama, "Bipolar Voltage Controlled Optical Switch Using Ti:$LiNbO_3$ Intersecting Waveguides," in *Proc. 4th Int. Conf. IO IOFC*, Tokyo, Japan, 1983, paper 29C4-5, pp. 363–366.

82. M. Papuchon, Y. Comemale, X. Mathieu, D. B. Ostrowsky, L. Reiber, A. M. Roy, B. Sejourne, and M. Werner, "Electrically Switched Optical Directional Coupler: COBRA," *Appl. Phys. Lett.*, Vol. 27, pp. 289–291, 1975.

83. (a) O. Mikami and H. Nakagome, "Waveguided Optical Switch in InGaAs/InP Using Free-Carrier Plasma Dispersion," *Electron. Lett.*, Vol. 20, No. 6, pp. 228–229, 1984. (b) K. Tada and Y. Okada, "Bipolar Transistor Carrier-injected Optical Modulator/Switch: Proposal and Analysis," *IEEE Electron. Device Lett.*, Vol. EDL-7, No. 11, pp. 605–606, 1986. (c) K. Ishida, H. Nakamura, and H. Matsumura, "In GaAsP/InP Optical Switching Using Carrier Induced Refractive Index Change," *Appl. Phys. Lett.*, Vol. 50, pp. 141–142, 1987. (d) J. P. Lorenzo and R. A. Soref," 1.3 μm Electro-optic Silicon Switch," *Appl. Phys. Lett.*, Vol. 51, pp. 6–8, 1987.

84. C. C. Lee and C. S. Tsai, "An Acoustooptic Readout Scheme for Integrated Optic RF Spectrum Analyzer," *1978 Ultrasonics Symp. Proc.*, IEEE Cat. No. 78 CH1344-ISU, pp. 79–81.

85. (a) C. S. Tsai, "The Increase of Bragg Diffraction Intensity Due to Acoustic Resonance and Its Application for Demultiplexing and Multiplexing in Laser Communication," *Appl. Opt.*, Vol. 10, pp. 215–218, 1971. (b) S. K. Yao and C. S. Tsai, "Acoustooptic Bragg-Diffraction with Application to Ultrahigh Data Rate Laser Communication Systems, Part I—Theoretical Considerations of the Standing Wave Ultrasonic Bragg Cell," *Appl. Opt.*, Vol. 16, pp. 3032–3043, 1977. (c) C. S. Tsai and S. K. Yao, Part II—Experimental Results of the SUBC and the Acoustooptic Multiplexer/Demultiplexer Terminals," *Appl. Opt.*, Vol. 16, pp. 3044–3060, 1977. (d) C. S. Tsai, S. K. Yao, M. A. Alhaider, and P. Saunier, "High-Speed Guided-Wave Acoustooptic and Electrooptic Switches," *1974 International Electronic Devices Meeting, Technical Digest*, pp. 85–87.

86. (a) J. L. Davis and S. Ezekiel, "Closed-Loop, Low-Noise Fiber-Optic Rotation Sensor," *Opt. Lett.*, Vol. 6, p. 505, 1981. (b) R. F. Cahill and E. Udd, "Phase-Nulling Fiber-Optic Laser Gyro," *Opt. Lett.*, Vol. 4, p. 93, 1979.

87. C. S. Tsai and Z. Y. Cheng, "A Novel Optical Frequency Shifting Scheme using Guided-Wave Acoustooptic Bragg Diffraction in Cascade," *Proc. 1988 IEEE Ultrasonics Symp.*, Oct. 3–5, Chicago, II (in press).

88. (a) C. S. Tsai, C. L. Chang, C. C. Lee, and K. Y. Liao, "Acoustooptic Bragg Deflection in Channel Optical Waveguides," *1980 Topical Meeting on Integrated and Guided-Wave Optics, Technical Digest of Post-Deadline Papers*, IEEE Cat. No. 80CH1489-4QEA, pp. PD7-1–PD7-4, C. S. Tsai, C. T. Lee, and C. C. Lee, "Efficient Acoustooptic Diffraction in Crossed Channel Waveguides and Resultant Integrated Optic Module," *1982 IEEE Ultrasonics Symp. Proc.*, IEEE Cat. No. 82CH1823-4, pp. 422–425. (b) C. S. Tsai and Q. Li, "Wideband Optical Frequency Shifting Using Acoustooptic Bragg Diffraction in a $LiNbO_3$ Spherical Waveguide," *Proc. 5th Int. Conf. Int. Opt. Optic. Fiber Communi.*, Venezia, Italy, October 1–4, *1985*, pp. 129–132.

89. (a) C. S. Tsai, and Z. Y. Cheng,"Novel Guided-Wave Acousto-optic Frequency Shifting Scheme Using Bragg Diffractions in Cascade," *Appl. Phys. Lett.*, Vol. 54, pp. 1616–1618, 1989. (b) Z. Y. Cheng and C. S. Tsai, "Baseband Integrated Acoustooptic Frequency Shifter," Presented at *1990 Conference on Lasers and Electrooptics*, Anaheim, CA, May 21–25; *Technical Digest Series*, Vol. 7, pp. 152–153 (Optical Society of America, Washington, D. C.), IEEE Cat. No. 90CH2850-6.

90. (a) D. C. Flanders, H. Kogelnik, R. V. Schmidt, and C. V. Shank, "Grating Filters for Thin Film Optical Waveguides," *Appl. Phys. Lett.*, Vol. 24, pp. 194–196, 1974. (b) A. C. Livanors, A. Katzir, A. Yariv, and C. S. Hong, "Chirped-Grating Demultiplexers in Dielectric Waveguides," *Appl. Phys. Lett.*, Vol. 30, pp. 519–521, 1977. (c) W. J. Tomlinson, "Wavelength Multiplexing in Multimode Optical Fibers," *Appl. Opt.*, Vol. 16, pp. 2180–2194, 1977. (d) H. W. Yen, H. R. Friedrich, R. J. Morrison, and G. L. Tangonan, "Planar Rowland Spectrometer for Fiber-Optic Wavelength Demultiplexing," *Opt. Lett.*, Vol. 6, pp. 639–641, 1981. (e) T. Suhara, Y. Handa, H. Nishihara, and J. Koyama, "Monolithic Integrated Microgratings and

Photodiodes for Wavelength Demultiplexing," *Appl. Phys. Lett.*, Vol. 40, pp. 120–122, 1982. (f) E. Voges, "Multimode Planar Devices for Wavelength Division Multiplexing and Demultiplexing," presented at the 1983 International Conference on Integrated Optics and Optical Fiber Communications, Tokyo, Japan, June 27–30, Paper No. 29A1-5. (g) G. Winzer, "Wavelength Multiplexing Components—A Review of Single-Mode Devices and Their Applications," *J. Lightwave Technol.*, Vol. LT-2, pp. 369–378, 1984. (h) J. Lipson, W. J. Minford, E. J. Murphy, T. C. Rice, R. A. Linke, and G. T. Harvey, "A Six-Channel Wavelength Multiplexer and Demultiplexer For Single Mode Systems," *J. Lightwave Tech.*, Vol. LT-3, pp. 1159–1163, 1985.

91. (a) L. Kuhn, P. F. Heiderich, and E. G. Lean, "Optical Guided Wave Mode Conversion by an Acoustic Surface Wave," *Appl. Phys. Lett.*, Vol. 19, pp. 428–430, 1971. (b) R. C. Alferness and R. V. Schmidt, "Tunable Optical Waveguide Directional Coupler Filter," *1978 Topical Meeting on Integrated- and Guided-Wave Optics, Technical Digest*, Paper TuA3; R. C. Alferness, "Efficient Electro-optic TE–TM Mode Converter/Wavelength Filter," *Appl. Phys. Lett.*, Vol. 36, pp. 513–515, 1980.

92. Y. Ohmachi and J. Noda, "$LiNbO_3$ Te–TM Mode Converter Colinear Acoustooptic Interaction," *IEEE J. Quantum Electron.*, Vol. QE-13, pp. 43–46, 1977.

93. B. Kim and C. S. Tsai, "Thin-Film Tunable Optical Filtering Using Anisotropic and Noncollinear Acoustooptic Interaction in $LiNbO_3$ Waveguides, *IEEE J. Quantum Electron.*, Vol. QE-15, pp. 642–647, 1979.

94. V. P. Hinkov, R. Opitz, and W. Sohler, "Collinear Acoustooptical TM–TE Mode Conversion in Proton Exchanged Ti:$LiNbO_3$ Waveguide Structures," *J. Lightwave Tech.*, Vol. 6, pp. 903–908, 1988.

95. C. S. Tsai and Z. Y. Cheng, "Wavelength-Division-Multiplexing Optical Switch Using Guided-Wave Acoustooptic Bragg Cells" (unpublished).

96. H. Jerominek, J. Y. D. Pomerleau, P. Tremblay, and C. Delisle, "An Integrated Acousto-Optic Bistable Device," *Opt. Commun.*, Vol. 51, pp. 6–10, 1984.

97. K. Preston, Jr., "A Comparison of Analog and Digital Techniques for Pattern Recognition," *Proc. IEEE*, Vol. 60, pp. 1216–1231, 1972.

98. (a) R. W. Damon, W. T. Maloney, and D. H. McMahon, in W. D. Mason and R. N. Thurston, Eds., *Physical Acoustics*, Vol. 7, Academic, New York, 1970, p. 273. (b) R. A. Sprague, "A Review of Acoustooptic Signal Correlators," Special Issue on Acoustooptics, *Opt. Eng.*, Vol. 16, pp. 467–478, 1977. (c) A. Vander Lugt, Ed., Special Section on Acousto-Optic Signal Processing, *Proc. IEEE.*, Vol. 69, pp. 48–92, 1981.

99. (a) M. C. Hamiltion, D. A. Wille, and W. J. Miceli, "An Integrated Optical RF Spectrum Analyzer," *Proc. 1976 Ultrasonics Symp.*, IEEE Cat. No. 74CH11120-5SU, pp. 218–222. (b) D. B. Anderson, J. T. Boyd, M. C. Hamilton, and R. R. August, "An Integrated-Optical Approach to the Fourier Transform" *IEEE J. Quantum Electron.*, Vol. QE-13, p. 268, 1977. (c) C. S. Tsai, "Guided-Wave Acoustooptic Bragg Modulators for Wideband Integrated Optic Communications and Signal Processing," *IEEE Trans. Circuits Syst.*, Vol. CAS-26, pp. 1072–1098, 1979. (d) M. K. Barnoski, B. Che, T. R. Joseph, J. Y. M. Lee, and O. G. Ramer, "Integrated-Optic Spectrum Analyzer," *IEEE Trans. Circuits Syst.*, Vol. CAS-26, pp. 1113–1124, 1979. (e) D. Mergerian, E. C. Malarkey, R. P. Pautienus, J. C. Bradley, G. E. Marx, L. D. Hutcheson, and A. L. Keller, "Operational Integrated Optic RF Spectrum Analyzer," *Appl. Opt.*, Vol. 19, pp. 3033–3034, 1980. (f) E. T. Aksenov, N. A. Esepkina, A. A. Lipovskii, and A. V. Pavenko, "Prototype Integrated Acoustooptic Spectrum Analyzer," *Sov. Tech. Lett. (USA)*, Vol. 6, pp. 519–520, 1980. (g) V. Neuman, C. W. Pitt, and L. M. Walpita, "An Integrated Acoustooptic Spectrum Analyzer Using Grating Components," *Proc. 1st Eur. Conf. Int. Opt.*, London, England, September 14–15, 1981, pp. 89–92. (h) T. Suhara, H. Nishihara, and J. Koyama, "A Folded-Type Integrated Optic Spectrum Analyzer Using Butt-coupled Chirped Grating Lenses," *IEEE J. Quantum Electron.*, Vol. QE-18, pp. 1057–1059, 1982. (i) V. M. Ristic and S. A. Jones, "Theoretical Considerations Related to Development of an Integrated Acousto-Optic Receiver," *Can. Electr. Eng. J.*, Vol. 7, pp. 7–18, 1982. (j) R. L. Davis and F. S. Hickernell, "Application of Wideband Bragg Cells for Integrated Optic Spectrum Analyzer," *Proc. SPIE, Int. Soc. Opt. Eng.*, Vol. 321, pp. 141–148, 1982. (k) V. M. Ritic, S. A. Jones, and

G. R. Dubois, "Evaluation of an Integrated Acousto-Optic Receiver," *Can. Electr. Eng. J.*, Vol. 8, pp. 59–64, 1983. (l) S. Valette, J. Lizet, P. Mottier, J. P. Jadot, S. Renard, A. Fournier, A. M. Grouillet, P. Gidons, and H. Denis, "Integrated Optical Spectrum Analyzer Using Planar Technology on Oxidized Silicon Substrate," *Electron. Lett.*, Vol. 19, pp. 883–885, 1983. Also *IEEE Proc. H.*, Vol. 131, pp. 325–331, 1984. (m) T. Suhara, T. Shiono, H. Nishihara, and J. Koyama, "An Integrated-Optic Fourier Processor Using an Acoustooptic Deflector and Fresnel Lenses in As_2S_3 Waveguide," *J. Lightwave Technol.*, Vol. LT-1, pp. 624–630, 1983. (n) C. Stewart, G. Serivener, and W. J. Stewart, "Guided-Wave Acoustooptic Spectrum Analysis at Frequencies Above 1 GHz," *IOOC 1981, Third International Conference on Integrated Optics and Optical Fiber Communication, Technical Digest*, April 1981, p. 122.

100. M. Hamilton and A. Spezio, "Spectrum Analysis Using Integrated Acoustooptics," in C. Tsai, Ed., *Guided-Wave Acoustooptic Interactions, Devices, And Applications*, Springer-Verlag, Berlin, 1990, Chap. 7.
101. H. J. Caulfield, W. J. Rhodes, M. J. Foster, and S. Horvitz, "Optical Implementation of Systolic Array Processing," *Opt. Commun.*, Vol. 40, p. 86, 1981.
102. R. T. Chen and C. S. Tsai, "GaAs–GaAlAs Heterostructure Single-Mode Channel-Waveguide Cutoff Modulator and Modulator Array," *IEEE J. Quantum Electron.*, Vol. QE-23, pp. 2205–2209, 1987.
103. X. Cheng and C. S. Tsai, "Electrooptic Bragg-Diffraction Modulators in GaAs/AlGaAs Heterostructure Waveguides," *J. Lightwave Tech.*, Vol. 6, pp. 809–817, 1988.
104. Y. Abdelrazek, C. S. Tsai, and T. Q. Vu, "An Integrated Optic RF Sectrum Analyzer in GaAs Waveguide," *1990 Conference on Lasers and Electro-Optics*, May 21–25, Anaheim, CA, *Technical Digest Series*, Vol. 7, pp. 446–447 (Optical Society of America, Washington, D. C.). IEEE Cat. No. 90CH2850–8.

6

OPTICAL MODULATION: MAGNETO-OPTICAL DEVICES

Alan E. Craig[†]

U.S. Naval Research Laboratory
Washington, D.C.

6.1 INTRODUCTION

Modern uses of magnetic phenomena for optical modulation depend almost exclusively on the Faraday effect (linear with internal magnetic field strength), the phenomenological description of the polarization rotation of linearly polarized light as it propagates through a Faraday-active medium. (The Cotton–Mouton effect also causes optical polarization rotation, with strength depending on the square of the internal magnetic field.) The interaction of spinning, and orbiting, electrons and their response to externally imposed or impinging fields are the bases of all effects that occur in magnetic materials. Only a small subset of magnetic materials exhibits optical properties conducive to propagation of light in the visible or near-infrared spectrum, however. A review of the description of magnetic materials followed by an examination of those that exhibit desirable optical properties will precede discussion of the various magneto-optical interactions applied in devices.

Most devices that incorporate the Faraday effect change their modulation status very slowly (spatial light modulators) or not at all (optical isolators), but some applications invoke magnetic fields that oscillate at microwave frequencies. All three of these types are being developed currently. Intermittently modulated devices envisioned include two-dimensional spatial light modulators for optical linear algebra operations; crossbar switches for communication networks; and perhaps most importantly, erasable disk memories. Planar waveguide optical isolators will undoubtedly find extensive service in optical fiber telecommunication links where laser diodes need protection from back reflection to ensure stable operation. Magneto-optical modulation at microwave frequencies, where certain magnetic materials support propagation of magnetostatic waves, may supplant acousto-optical devices (which are severely limited by acoustic attenuation at high frequencies) for spectrum analysis and other similar functions.

[†] Present address: Air Force Office of Scientific Research, Bolling Air Force Base, Washington, D.C. 20332-6448

Some magnetic phenomena are purely quantum mechanical in origin, in particular those linked to the orbital and spin momenta attributed to the electron. However, taking these phenomena and their associated magnetic dipole moments as empirical fact, much of magnetism can be understood in the classical domain; the discussion here will pursue the subject mainly from this point of view. Similarly, optical interactions are in reality quantum mechanical. Considerable understanding can be extracted from classical and macroscopic descriptions of the response of light waves to modification of material properties via models such as the Faraday effect and the optical permeability and permittivity tensors.

6.2 UNITS

Although the International System of Units (SI) has attained widespread usage in many branches of electromagnetics, the centimeter-gram-second (cgs) system remains dominant in magnetics literature and will be used here, with exceptions noted.

The definition of units in magnetics is somewhat unintelligible without some rudimentary accompanying physical description. Although the phenomena discussed later derive mostly from quantum-mechanical observations of the behavior of materials, the entities used to describe them originated from observation of a macroscopic phenomenon, namely the attractive force between two parallel wires carrying current in the same direction. This force (in dynes), for two elementary currents (namely, moving charges) is given by the expression

$$\mathbf{F} = \frac{k(q_1 q_2)[\mathbf{v}_1 \times (\mathbf{v}_2 \times \mathbf{n})]}{|\mathbf{r}_1 - \mathbf{r}_2|^2}$$

where q designates the charge of a particle, $\mathbf{v}$ its vector velocity, $\mathbf{r}$ the vector spatial coordinate of the particle, and $\mathbf{n}$ the unit normal vector from particle 1 to particle 2 ($\mathbf{n} = (\mathbf{r}_1 - \mathbf{r}_2)/|\mathbf{r}_1 - \mathbf{r}_2|$). The unit of the elementary charge becomes an issue here: if $k = 1$ is chosen, q is in electromagnetic units (e.m.u.); if $k = 1/c^2$ is chosen ($c = 3 \times 10^{10}$ cm/s), q is in electrostatic units (e.s.u.), the standard in the cgs system; if $k = \frac{1}{100}$ is chosen, q is in coulombs, the mks (or SI) standard (1 e.m.u. = 10 C = 3×10^{10} e.s.u.). Using e.s.u., the differential force on a moving charged particle resulting from the presence of a filamentary current element $i\,d\mathbf{l}$ is

$$d\mathbf{F} = \frac{q_1 \mathbf{v}_1}{c} \frac{i\,d\mathbf{l} \times \mathbf{n}}{c|\mathbf{r}_1 - \mathbf{r}_2|^2}$$

From this expression the magnetic induction $\mathbf{B}$ is defined as

$$\mathbf{B} = \frac{i\,d\mathbf{l} \times \mathbf{n}}{c|\mathbf{r}_1 - \mathbf{r}_2|^2}$$

where i is in e.s.u. per second and $\mathbf{B}$ is in gauss. For the special case of the field a distance r from a straight wire carrying current I,

$$|\mathbf{B}| = \frac{2I}{c \times 10r} \quad (\text{e.s.u./cm}^2)$$

where I is in e.s.u. per second, or

$$|\mathbf{B}| = \frac{2I}{10r} \quad (\text{A/cm})$$

where I is in amperes; in both cases $\mathbf{B}$ is in gauss. Figure 6.1 illustrates these fields. Given a current I in a circular wire loop of radius R, the field at the center is

$$|\mathbf{B}| = \frac{2\pi I}{cR} \quad (\text{G})$$

where I is in e.s.u. per second, or

$$|\mathbf{B}| = \frac{2\pi I}{10R} \quad (\text{G})$$

where I is in amperes. A solenoid, an infinite stack or spiral of these loops, having radius R and current I per loop, produces a field on its axis $|\mathbf{B}| = 4\pi n(I/c)$ gauss (cgs units), or $|\mathbf{B}| = 4\pi n(I/10)$ gauss (I in amperes), where n is the number of loops or turns per unit length.

Various conventions are adopted by authors for defining the remaining magnetic quantities and their interrelations, in part depending on the system of units adopted. These quantities account for the empirical behavior of materials in magnetic fields and for the response of charged particles moving in materials that have internal magnetization. In the cgs system used here, the magnetic field $\mathbf{H}$ is substituted for $\mathbf{B}$ external to magnetic materials. (While $\mathbf{B}$ is in gauss, the precisely equivalent unit for $\mathbf{H}$ is the oersted.) The preceding equations could justifiably have been written using $\mathbf{H}$. In fact, common usage employs $\mathbf{H}_i$, with $\mathbf{B} = \bar{\bar{\mu}}\mathbf{H}_i$, to discuss the magnetic field internal to a magnetic material. Further definition of $\bar{\bar{\mu}}$ is introduced below; the use of $\mathbf{H}_i$ will be seen in the section on spin waves. Inside magnetic materials the magnetization (or magnetic moment per unit

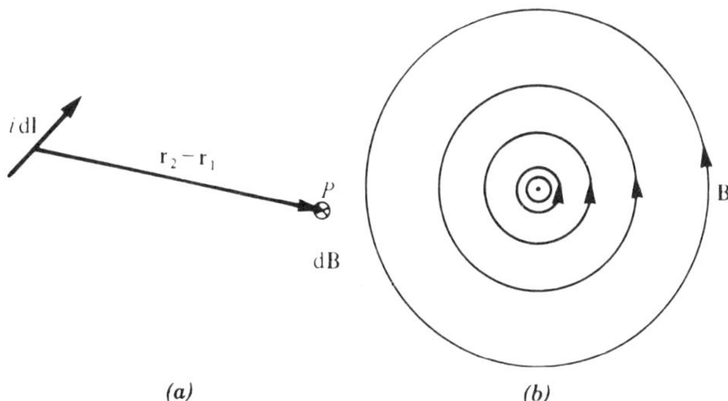

Figure 6.1 (*a*) Current element $i\,d\mathbf{l}$ is situated at the point $\mathbf{r}_1$, and the contribution of this to the induction at the point $P(\mathbf{r}_2)$ is directed normal to the diagram, away from the observer. (*b*) The induction represented by concentric circles around a current directed toward the observer. (From D. J. Craik, "Magnetostatic Principles," *Structure and Properties of Magnetic Materials*, Pion Limited, London, 1971, p. 1. Copyright 1971, used by permission.)

volume) **M** contributes to the magnetic induction according to $\mathbf{B} = \mathbf{H} + 4\pi\mathbf{M}$. The factor 4π arises by definition, but via the concept of equivalent currents.

A current loop placed in a uniform magnetic field **B** experiences a torque $\mathbf{T} = -(I/c)|\mathbf{a}|\hat{\mathbf{a}} \times \mathbf{B}$, in which I/c is the numerical value of the current in the loop (I in e.s.u. per second), $|\mathbf{a}|$ is the scalar area of the loop, and $\hat{\mathbf{a}}$ is the unit vector normal to the plane of the loop. The quantity $(I/c)|\mathbf{a}|\hat{\mathbf{a}}$ is called the magnetic dipole moment of the loop and designated **m**, with units of ergs per gauss (the energy of a magnetic dipole in magnetic induction **B** is $E = -\mathbf{m} \cdot \mathbf{B}$). Magnetic materials also can be described in terms of their magnetic dipole moments because they experience the same torques in the presence of magnetic induction ($\mathbf{T} = -\mathbf{m} \times \mathbf{B}$). Thus any material magnetic dipole moment can be represented by an equivalent current loop. Figure 6.2 shows the field of a magnetic dipole.

The magnetization **M** is just the magnetic dipole moment per unit volume, $\mathbf{m}/V$. A section of magnetic material in which magnetic dipoles, each having scalar area $|\mathbf{a}|$, are aligned in a string back-to-back looks like a solenoid. Then **M** is given by $(\mathbf{m}/|\mathbf{a}|)n$, where n is the number of dipoles per centimeter along the string, or equivalently the number of solenoid turns per centimeter. The current–dipole equivalent is a solenoid with $|\mathbf{M}| = |(\mathbf{m}/|\mathbf{a}|)n| = n(I/c)$, and comparison to the expression for **B** due to a solenoid (see the preceding) necessitates the inclusion of the 4π factor for the contribution of material magnetization to the magnetic induction.

A few other names are sometimes invoked as magnetic units. The maxwell is a measure of *lines of force*, or magnetic flux; one oersted is equivalent to one maxwell per square centimeter. The weber is also a measure of magnetic flux; one weber equals 10^8 maxwells. Last, the tesla is equivalent to 10^4 gauss.

Two parameters, both tensors, describe the response of materials to applied magnetic fields: the susceptibility, $\bar{\bar{\chi}}$, and the permeability, $\bar{\bar{\mu}}$. The susceptibility $\bar{\bar{\chi}}$, with $\mathbf{M} = \bar{\bar{\chi}}\mathbf{H}$, characterizes the proclivity of the magnetic dipoles of a material to align with an applied magnetic field. It generally is dependent on both the magnitude and direction of $\mathbf{H}$: $\bar{\bar{\chi}} = \bar{\bar{\chi}}(\mathbf{H})$. For high values of **H**, $\bar{\bar{\chi}}$ tends asymptotically to zero, and the material is said to saturate. At this point, $\mathbf{M} = \mathbf{M}_s$, the saturation magnetization. ($M_s = |\mathbf{M}_s|$ will denote its magnitude.) The permeability $\bar{\bar{\mu}}$, with $\mathbf{B} = \bar{\bar{\mu}}\mathbf{H}$, characterizes the total flux density produced by a given applied magnetic field. As can be seen by manipulation of $\mathbf{B} = \mathbf{H} + 4\pi\mathbf{M}$, $\bar{\bar{\mu}} = 1 + 4\pi\bar{\bar{\chi}}$; thus it also depends on $\mathbf{H}$: $\bar{\bar{\mu}} = \bar{\bar{\mu}}(\mathbf{H})$. Often $\bar{\bar{\chi}}$ and $\bar{\bar{\mu}}$ are diagonal and isotropic and so can be written as scalars.

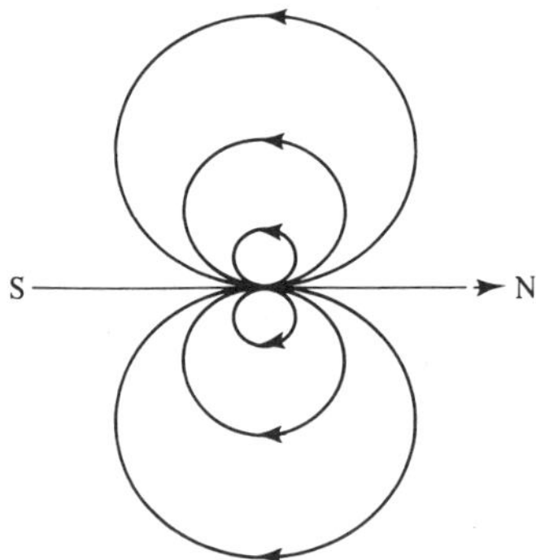

Figure 6.2 Field of a magnetic dipole. (From B. D. Cullity, "Definitions and Units," *Introduction to Magnetic Materials*, Addison-Wesley, Reading, MA, 1972, p. 8. Copyright 1972, used by permission.)

6.3 MATERIAL PHYSICS OF MAGNETICS

The magnetic moment of a material derives from the consolidated effects of the charge and the angular moments of its electrons. An electron has spin and orbital angular momentum, both of which contribute to its magnetic moment. These properties are fundamentally quantum mechanical; however, a phenomenological description provides a reasonable basis for understanding their magnetic contributions.

The contribution of the orbital angular momentum of an electron to the magnetic moment **m** can be thought of classically in terms of a current loop surrounding an area, as in the preceding. For a single electron, the current is $(e/c)v/2\pi r$, where v is the electron velocity around a loop of radius r, and the area is πr^2; so

$$|\mathbf{m}| = \text{current} \times \text{area} = \frac{(e/c)vr}{2}$$

The quantum mechanically allowed values of the angular momentum are $m_e vr = n(h/2)\pi$, in which m_e is the electron mass and $h = 6.626 \times 10^{-27}$ erg-second is Planck's constant. Substituting the allowed values of velocity from this expression, the Bohr magneton is defined as $|\boldsymbol{\mu}_B| = [\mathbf{m}] = (e/c)h/(4\pi m)$ in the orbit for which $n = 1$. (Its value is $|\boldsymbol{\mu}_B| = 0.927 \times 10^{-20}$ ergs/Oe. This value turns out to be correct for the orbital magnetic moment even though the model presented here to aid visualization is naive.) There is no reasonable analog to aid envisioning the spin angular momentum contribution of the electron; a spinning top having charge distributed around its equator is cumbersome. However, its value is found experimentally to be exactly the same as that of the orbital magnetic moment.

The magnetic moment contribution of a single electron to an atom is the sum of these orbital and spin moments. For various reasons, either of these moments may be quenched to some degree by the presence of other electrons and by the influence of the atomic environment. The total magnetic moment of an atom results from summing the (modified) contributions of all of its electrons. As already alluded to, the magnetic moments depend on the orbital and spin angular moments, whose allowed values fall under the jurisdiction of quantum mechanics. For free atoms, electrons that inhabit filled orbitals contribute very little since each electron having spin up is matched by one having spin down. The Pauli exclusion principle, which allows only one electron of each spin quantization in any energy level, governs the spin contribution in unfilled levels. In solid materials, electron orbitals are distorted by bonding of neighboring atoms (orbit–lattice coupling) so that the orbital contribution to the magnetic moment often becomes small. That is, electrons that in a free atom might experience distorted orbitals in response to an applied magnetic field will be restrained in a solid by the strong preferential direction of the bonds. Spin–spin and spin–orbit coupling is generally much weaker; consequently spin-originating magnetic dipole moments provide most of an atom's response to a magnetic field in a solid.

A parameter called the g-factor has been devised to indicate the degree to which the magnetic moment derives from the electron's spin or orbital angular momentum. The g-factor represents the ratio of the number of Bohr magnetons of magnetic moment in an atom to the number of units of angular momentum, measured in $h/2\pi$ (h is Planck's constant, $h = 6.626 \times 10^{-27}$ erg-s.) Its value varies from $g = 1$ for orbital dependence alone to $g = 2$ for the solely spin-dependent case. Further details can be obtained in Refs. 1–4.

An external applied magnetic field exerts a torque on the total magnetic dipole. The torque must equal the rate of change of the angular momentum and consequently a change in the magnetic moment of the dipole. (The torque causes no change in momentum in the direction of the field, however.) The equation that describes this process, called the gyromagnetic equation, is $d\mathbf{m}/dt = \gamma(\mathbf{m} \times \mathbf{H})$. Solution of this equation gives sinusoidal precession in the plane perpendicular to the field at the gyromagnetic frequency: $\omega_0 = \gamma|\mathbf{H}|$. (Inside a material, $\mathbf{H}$ would be replaced by $\mathbf{H}_i$.)

With this rudimentary description of electron magnetic dipole moments in mind, the magnetic behavior of solid materials can be addressed.

Only a small subset of magnetic materials is important for optical interactions. A general description of the broad classifications of magnetic materials delineates this subset. The physical mechanisms that produce the various responses will be described; thorough analysis of these mechanisms is deferred to the references [1–3].

Several terms are invoked to describe the macroscopic behavior of materials when subjected to external magnetic fields. These are diamagnetic, paramagnetic, ferromagnetic, antiferromagnetic, and ferrimagnetic. The dipole characteristics that correspond to many of these classifications are illustrated in Fig. 6.3. Although a particular material will respond predominantly according to the features of only one of these classes, more than one mechanism is almost always operating; in addition, the class of response changes with temperature for many materials.

The physical mechanism that produces diamagnetism is the analog at an atomic scale to Lenz's law, which describes the electromotive force (emf) produced in a macroscopic wire loop when a change occurs in the magnetic flux encircled (emf $= -d\phi/dt$; $\phi = |\mathbf{H}|A$ volts; ϕ in maxwells, $|\mathbf{H}|$ in oersteds, A in square centimeters). The emf establishes a current loop magnetic dipole field that opposes the change in the intercepted flux. An electronic magnetic dipole responds similarly, except that whereas the macroscopic loop current persists only as long as the flux is changing, the electronic dipole state corresponds to the applied magnetic field strength and does not decay, as if its path were without resistance, or superconducting. (This feature is attributable to the quantized nature of the electron orbitals.)

For diamagnetic behavior to dominate in a material a balance must exist between the orientations of dipoles, so that if they are randomly aligned the net magnetic moment is zero; if their alignment is governed (by the orientation of atomic orbitals or by that of interatomic bonds), every spin-up dipole is compensated by one oriented spin down. The Pauli exclusion principle ensures this condition in the inert gas atoms as well as in a great many molecular gases in which atomic orbitals are filled by sharing electrons.

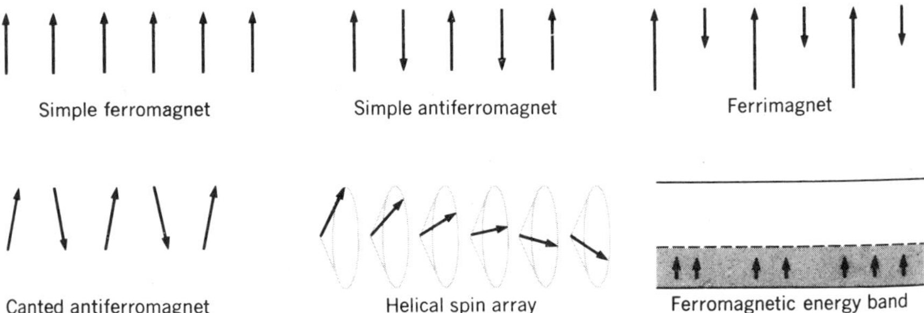

Figure 6.3 Possible ordered arrangements of electron spins, in which one particular type of cooperative behavior dominates. (From Charles Kittel, "Ferromagnetism and Antiferromagnetism," *Introduction to Solid State Physics*, 4th ed., Wiley, New York, 1971, p. 528. Copyright 1971, used by permission.)

Many solids, in which bonding satisfies the full-orbital condition, also exhibit diamagnetic behavior. Regardless of whether diamagnetic behavior is dominant, all materials exhibit diamagnetic response; the other magnetic behaviors are stronger effects, however, and occlude diamagnetism when they are present.

Paramagnetism occurs in materials in which all magnetic moment contributions are not fully compensated, and the atoms, molecules, or crystalline unit cells have net magnetic dipole moments. Many ions are paramagnetic since their valence electrons are unpaired. Metals that have incomplete outer shells also tend to paramagnetism. Most transition elements and rare earths, either in isolation or incorporated in crystal structures, have large intrinsic magnetic dipole moments because of their incomplete inner shells.

Paramagnetism results from the competition between the tendency of the uncompensated magnetic dipoles to coalign with an applied magnetic field and the disorienting effects of thermal phonons or interatomic collisions. The Maxwell–Boltzmann distribution governs the thermal randomization, with the energy term given by the dipole energy in the magnetic field, $E = -\mathbf{m} \cdot \mathbf{H}$. Classically, alignment of a dipole with the field is allowed to occur to any degree whatsoever and to assume any angular orientation; quantum mechanics, permitting only discrete values of angular momentum, constrains the relative orientation angles and prohibits complete alignment. The resulting expression for the dependence of bulk magnetization on the ratio of magnetic field to temperature is called the Langevin function ($\mathbf{M}/M_s = \mathbf{L}[(\mathbf{m} \cdot \mathbf{H})/kT]$, where $\mathbf{L} = \mathbf{L}(x)$ is the Langevin function and M_s is the saturation magnetization) in the case of a classical derivation and the Brillouin function when quantum-mechanical arguments have been used. Curves plotted from the two functions are similar; their slopes, $\frac{1}{3}$ at the origin, decrease gradually with increasing field (or decreasing temperature) to zero, corresponding to an asymptotic approach to magnetic saturation. See Fig. 6.4. The mechanism of paramagnetism, being temperature dependent, obeys the Curie–Weiss law for the susceptibility: $\chi = C/(T - \theta)$, where C is the Curie constant, T is the temperature, and χ is the scalar magnitude of the susceptibility, assumed isotropic. The parameter θ has temperature units, and varies from one material to another depending on the interaction between electrons that share orbitals or otherwise pass regularly within close proximity. This form of the law is a low-field, low-temperature linear approximation of the expression derived by either a classical or a quantum-mechanical analysis.

Pierre Weiss discovered that the linear susceptibility relationships of paramagnetics in small fields do not intercept the zero magnetization line at the origin in many cases. He postulated that the responsive dipoles do not act totally independently but interact through creation of an interdipole magnetic field he called the molecular field. (In truth, the effect is more nearly atomic in origin.) The Weiss temperature parameter, θ in the preceding, results from the mathematical introduction and inclusion in the phenomenological theory of this magnetic "molecular field," denoted $\mathbf{H}_m$, which is proportional to the material's magnetization ($\mathbf{H}_m = \lambda\mathbf{M}$); that is, $\mathbf{H}_m$ is proportional to the amount of alignment among the dipoles caused by the external field. The effect is highly cooperative and self-augmenting. When this conceptual molecular field is added to the externally applied field, the actual susceptibility is accurately modeled. The parameter θ can be positive or negative and is usually small in magnitude for paramagnetics, less than about 10 K.

Weiss's molecular field is explainable in terms of exchange energy. Electrons that share orbitals, more precisely bonding orbitals, are constantly interchanging responsibilities and atom allegiances. An energy is connected to this interchange, called the exchange energy. Generally the exchange energy is negative when the electron spins are antiparallel. Then, as expected from the Pauli exclusion principle applied to the bonding orbitals of molecules, the energy is lower when the electrons involved have opposite

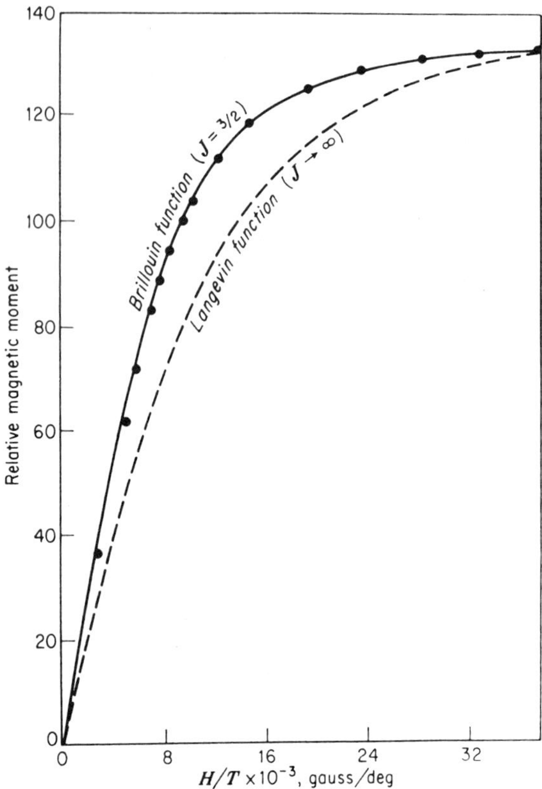

Figure 6.4 Plot of relative magnetic moment versus H/T for potassium chromium alum at $T = 1.29$ K. The solid line is a Brillouin curve for $g = 2$ and $J = S = \frac{3}{2}$ (spin only; no orbital contribution). The broken line is a Langevin curve. (From Benjamin Lax and Kenneth J. Button, "Paramagnetism and Ferromagnetism," *Microwave Ferrites and Ferrimagnetics*, McGraw-Hill, New York, 1962, p. 61. Copyright 1962, used by permission.)

spins, and bonding is promoted. Van der Waal forces are the result of exchange, for example. In some materials the exchange energy is negative when the spins of the exchange electrons are parallel, although this situation applies to electrons inhabiting inner, partially filled orbitals. The important class of materials for which this is true is the ferromagnetics, to be discussed.

The expression for the exchange energy is

$$E_{ex} = -2J_e(\mathbf{S}_i \cdot \mathbf{S}_j)$$

where $\mathbf{S}_i$ and $\mathbf{S}_j$ are the spin vectors of two electrons. The exchange integral J_e indicates the overlap of the orbitals of the two electrons (or equivalently, of their charge distributions). The relation between the exchange energy and the molecular field is then $E_{ex} = -\mathbf{m} \cdot \bar{\bar{\boldsymbol{\mu}}} \cdot \mathbf{H}_m = -\mathbf{m} \cdot \bar{\bar{\boldsymbol{\mu}}} \cdot \lambda\mathbf{M}$. Crystals of different elements and compounds have various separations between those constituent atoms responsible for their magnetic character. When two atoms whose electron orbitals exhibit exchange effects are widely separated, the integral is positive but negligibly small. As the atomic separation decreases from specie to specie, the integral grows larger, passes a maximum, and then becomes progressively smaller, turning negative for very small atomic separations. Antiferromagnetic materials are found in this region of negative exchange integral, while ferromagnetics are located in the region of the maximum. Large separation corresponds to paramagnetic response.

Examining once more the dipoles that constitute the paramagnetic material leads to a few other points. First, quantization of the total angular momentum prohibits complete alignment of a magnetic dipole with an applied magnetic field, even at absolute zero temperature. (In fact, because of the angular momentum restriction on dipole orientation, realignment with the field can only occur by means of collisions or phonon interactions, in which energy is exchanged. Otherwise only a precession around the field direction occurs, at the gyromagnetic frequency; it is superimposed on the dipole's orientation, which is determined by its total angular momentum.) A dipole is always free to precess around the field direction, which is driven by the torque of the dipole–field coupling, a useful picture for envisioning the spin waves described in a later section. Second, the contribution of exchange energy—responsible for the molecular field (and the Weiss temperature θ)—to paramagnetic susceptibility, which appears only during the application of an external magnetic field, is not necessarily large compared to the exchange interaction that occurs in an orbital in the absence of the applied field. Exchange is normally acting continuously to affect the strength of interatomic bonding according to the spin orientations of the contributing electrons (or conversely, to determine the spin orientations of electrons in common, bonding orbitals). The exchange that produces the Weiss temperature is generally acting between inner-shell electrons on differing atoms that are barely close enough to exert common influence and do not participate in bonding. It is certainly a secondary effect compared to the paramagnetic realignment of magnetic dipoles in an external magnetic field. Unlike the basic paramagnetic alignment process, however, the dipole may be influenced by exchange without resort to collision or phonon jostling since here the field influence is internal to the atom, changing the energy levels of its orbitals. (Incorporation of the electron magnetic dipole—magnetic field interaction energy into the Hamiltonian perturbs the allowed energy levels of electrons in an atom.) Revisiting the macroscopic picture, paramagnetism is the state assumed by ferromagnetic, antiferromagnetic, and ferrimagnetic materials at elevated temperatures.

Understanding ferromagnetism is a difficult undertaking; experts on magnetism promote various theories that explain, each in part, the mechanisms that cause a material to be ferromagnetic. In fact, a general theory that predicts the ferromagnetic response of all materials may be strictly unattainable. The reality is that ferromagnetism results partly from the interaction of electrons having unpaired spins, and situated on particular atomic sites, via the exchange energy; and partly from the interaction of electrons (sometimes called itinerant) that are not fixed in the material but are to some extent free to move through bands of energy levels, a theory not yet visited here. The aspects of the theories that must be invoked to rationalize the ferromagnetism of any particular material are in some ways unique; the selection depends on the way in which the atomic orbitals of the elements combine into bands of energy states in a solid and to what extent the electrons responsible for the magnetic response occupy the bands.

A qualitative, somewhat cursory description follows. Ferromagnetism is not particularly important for magneto-optics except insofar as ferromagnets normally are used to provide the required biasing fields. For more detailed discussion, read the references [1–3].

Ferromagnetics comprise the three commonly known ferromagnetic transition element metals—iron, cobalt, and nickel—along with a selection of rare earth elements at low temperature (gadolinium below 16°C) and various compounds and alloys, many of which contain transition elements or rare earths that may not be ferromagnetic in their pure forms.

The classical argument for ferromagnetism provides for a molecular field in the absence of any applied field. This molecular field is very strong, strong enough to cause saturation by coaligning neighboring dipoles. A sample is proscribed from exhibiting bulk magnetization by the subdivision of the coaligned dipoles into small volumes called domains, which orient randomly when no external field is applied. A small applied external field, by aligning the domains, can cause the material to exhibit a very large bulk magnetization. (An applied magnetic field of as low as 50 Oe suffices to achieve technical saturation, all domains aligned, for pure crystalline iron. Fields on the order of a few hundred to at most a few thousand oersteds are required for other ferromagnetics.)

The combination of molecular field theory with the Langevin theory, which describes the thermal statistics of magnetization caused by coaligning dipoles, produces a mathematical argument for the spontaneous magnetization of domains in ferromagnetic materials. Recall that the Langevin function, $\mathbf{M}/M_s = \mathbf{L}[(\mathbf{m} \cdot \mathbf{H})/kT]$, has a slope that equals $\frac{1}{3}$ at the origin and decreases to zero for greater values of the function argument. Meanwhile, the molecular field expression can be inverted to read $\mathbf{M} = \mathbf{H}_m/\lambda$. Whenever λ is greater than $3kT/(|\mathbf{m}|M_s)$ the curves representing the two functions intersect. This infers that the molecular field can to a considerable extent align the dipoles of a domain in the absence of an external applied field. For high enough temperature, λ is insufficient to ensure intersection of the two curves, and the domains revert to paramagnetic behavior. The temperature at which the transition occurs is called the Curie temperature for a material and corresponds very closely to the Weiss temperature θ introduced to parameterize the molecular field effect in paramagnetics. For ferromagnetic materials, θ may be many hundreds of degrees; it is always positive, aiding dipole alignment with the applied field.

The way in which the exchange energy produces the molecular field has been shown in a preceding section. That the exchange energy must be responsible rests on two observations: first, the magnetic dipole moment in ferromagnetics is almost entirely caused by electron spin, not orbital, momentum; second, the internal fields caused by the environment of coaligned dipoles surrounding any particular dipole are simply not strong enough by several orders of magnitude to spontaneously magnetize the domains. As mentioned earlier, for ferromagnetics a negative exchange energy corresponds to parallel dipoles in the domains.

The band theory of solids can help to clarify the details of ferromagnetism. As previously stated, ferromagnetic materials are often formed of elements in which incompletely filled inner shells exist. These are the transition metals and rare earth elements. When these elements form solids, their outer (s) shells and inner (d or f) shells form bands of allowed states in which the energy levels lose their degeneracy. The energies of the s bands overlap those of the d or f bands. The way that electrons fill the overlapping bands is staggered, and electrons can with relative ease transfer between the states of the two bands. The density of states in the bands is also a factor in determining their electron occupancy; there are five times as many d states as s and seven times as many f states as s. In these bands electrons normally would be in the lowest energy configuration if they were paired, one spin up, one spin down, in each level. Exchange forces in the three ferromagnetic metals—iron, cobalt, and nickel—distort the electron occupancy; the s band loses some fraction of its average population and the balance between spins in the d band is destroyed. The exchange energy is least if the d band is half filled with electrons of one spin, the remaining electrons adopting the parity spin and filling states as high as their numbers permit. The end result is ferromagnetism. The theory does not hold for neighboring lighter transition metals in the periodic table (they are not ferromagnetic). Apparently exchange forces can persuade electrons to take on only so much extra potential energy in bands holding progressively fewer of them.

One final clarification: the exchange energy depends strongly on physical separation of the participating electron orbitals. For large separation, the energy is small but is in favor of parallel dipoles. As separation decreases, the energy first increases and then decreases to zero. Even smaller separations see the exchange energy reemerge, but now in favor of antiparallel spins. This arrangement is antiferromagnetism.

Antiferromagnetism depends on forces working in the same way as those that produce ferromagnetism but on a different type of material. Most antiferromagnetics are ionic compounds in which atoms that have intrinsic magnetic moments are not neighbors in the lattice. This separation influences the exchange energy to be negative when nearby spins are antiparallel. For crystalline materials the result is that two interlaced sublattices form, in each of which all dipoles are parallel but which orient in opposite directions as groups.

Above the critical temperature, called the Néel temperature (T_N) for antiferromagnetics, each sublattice acts separately as a paramagnetic in the presence of a molecular field, except that the exchange energy influencing an atom in the sublattice originates from the surrounding atoms of the interpenetrating sublattice. (The assumption of negligible interaction with atoms on sites in the same sublattice is not quite accurate. In some crystal structures, some nearest-neighbor atoms are on the same sublattice. However, exchange forces between atoms on differing sublattices dominate those between atoms in the same sublattice for antiferromagnetics.) Since the sublattices are composed of identical atoms, the resulting total magnetization is the sum of that of the sublattices individually, that is, twice that for one alone. In this paramagnetic temperature region, the material appears to have a negative value of the Weiss temperature parameter. Figure 6.5 is illustrative.

Below the Néel temperature the exchange energy dominates the thermal randomizing effects. Magnetic dipole moments on the two sublattices align antiparallel, and the

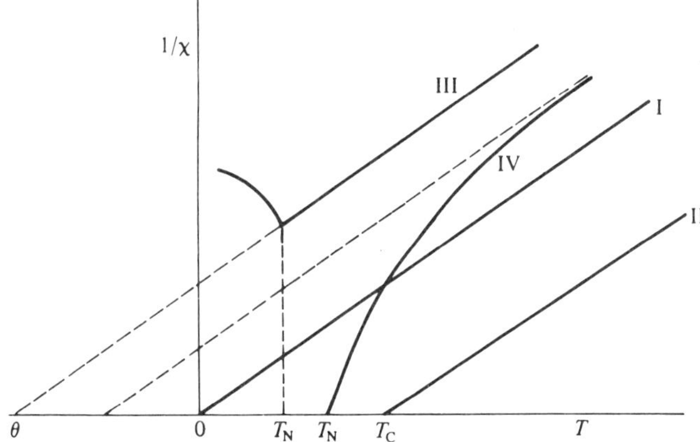

Figure 6.5 Reciprocal of susceptibility versus temperature: (I) Simple paramagnetic with no interactions; Curie law. (II) Curie–Weiss paramagnetic with positive interaction term T_C, indicating ferromagnetic ordering below T_C (the Curie temperature). (III) Antiferromagnetic with negative interaction constant, θ. The Néel temperature below which the interaction is effective in causing antiferromagnetic ordering is T_N; χ falls below T_N because the moments are effectively locked into a particular orientation in the lattice and an applied field must overcome this anisotropy as well as the ordering. (IV) Ferrimagnetic. (From D. J. Craik, "Magnetic Dipoles in Applied Fields," *Structure and Properties of Magnetic Materials*, Pion Limited, London, 1971, p. 39. Copyright 1971, used by permission.)

spontaneous magnetization becomes zero. Small amounts of magnetization may be induced by an external field; the amount depends on the relative orientations of the field and the crystal axis that defines the sublattice orientations. The field causes a small magnetization in each sublattice, which in turn creates a counteracting molecular field in the other sublattice; for each sublattice, a complicated reevaluation of the Brillouin function, which governs the balance between the exchange energy and the thermal randomizing statistics, establishes the resultant total magnetization. For a field perpendicular to the crystal axis the dipoles experience a slight rotation; for a parallel field, a small displacement occurs. In the perpendicular case, the susceptibility becomes temperature independent; in the parallel case, it goes to zero with vanishing temperature.

The ferrimagnetic materials remain for consideration. These materials also have characteristics similar to those of ferromagnetics but derive from a different type of structure. Like the antiferromagnetics, they comprise various complex oxides and are electrically insulating, meaning that the magnetic elements are separated so that bands of electron energies do not form. Local spin interactions dominate, and the molecular field theory provides an accurate description. Unlike the antiferromagnetics, the magnetic ions are of more than one element and are located on more than one type of crystalline site in the material. The overall magnetic character depends on the imbalance of magnetic moments between the elements and on the influence exerted by the site-dependent molecular fields on the intercoupling of the moments.

Several crystal types produce ferrimagnetics. Most are spinel-like structures incorporating iron oxides, known as cubic ferrites. Two types of sites exist for the metallic ions in spinels: one is tetrahedrally coordinated by oxygen atoms (*A* site); the other is octahedrally coordinated (*B* site). Materials in which the two different metallic ions are segregated according to site type—divalent on *A*, trivalent on *B*—are called normal spinels. (Ferrites of this structure are generally paramagnetic.) When the divalent ions sit on *B* sites, forcing the trivalent ions onto either *A* sites or some mixture of the two sites, the spinel is called inverted; these are often ferrimagnetic.

Hexagonal ferrites are also represented among the ferrimagnetics. This crystal structure differs from the cubic spinel mainly in the stacking of the crystalline planes; face-centered-cubic and hexagonally close-packed structures are otherwise similar. The iron atoms that provide the magnetic moments occupy sites of three different coordinations: tetrahedral, octahedral, and hexahedral.

Another structure that produces ferrimagnetics is garnet. Three different metallic ion sites exist in these materials. They are tetrahedrally, octahedrally, and dodecahedrally coordinated. (See Fig. 6.6. In addition, Ref. 5 exhibits an excellent illustration of the garnet unit cell and the arrangement of eight formula units of atoms that in various orientations comprise it.) Usually, iron ions are found on the first two site types; they interact strongly via exchange to cause antiparallel alignment of the moments on these two sites. The third site type can be occupied by an ion having no magnetic moment. An imbalance of the number of iron ions on its two site types establishes the net moment. The most important of these generally weakly ferrimagnetic materials is yttrium iron garnet (YIG), which propagates several types of magnetic dipole spin waves at microwave frequencies; in particular, magnetostatic waves propagate with relatively low absorption.

Molecular field theory describes ferrimagnetism. Because at least two metallic ions are present, often occupying more than one site type, particular use of the theory is formidable. Marked, yet still useful, simplification results from assuming the presence of only a single ion type. The magnetization of ions on each site type affects the molecular field acting on ions at the same site type, as well as on ions at alternate site types. The algebra is arduous, the result being that above the critical (Curie) temperature, the dependence of the susceptibility on temperature is hyperbolic, appearing paramagnetic

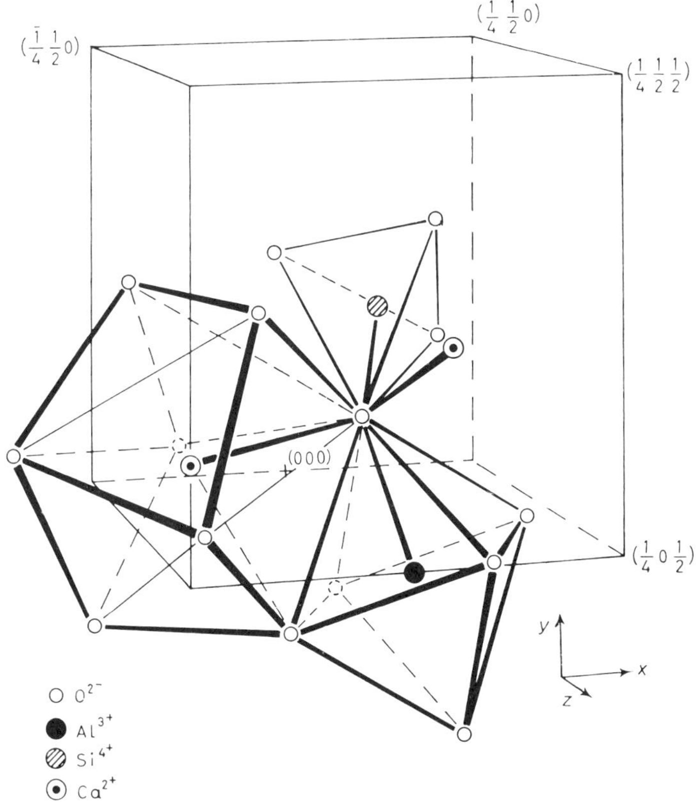

Figure 6.6 Coordination about an oxygen ion in a garnet. (From S. Geller, "Crystal and Static Magnetic Properties of Garnets," *Physics of Magnetic Garnets*, North-Holland, Amsterdam, 1978, p. 4. Copyright 1978, used by permission.)

at temperatures far above critical but exhibiting increased susceptibility near the Curie temperature (where it is infinite). Below the Curie temperature the ferrimagnetic behavior is the sum of the behavior of the several contributing sublattices. Each of these is determined, as for ferromagnetics and antiferromagnetics, by the simultaneous satisfaction of the thermal distribution of dipole alignment described by the Brillouin function and the spontaneous magnetization determined by the molecular field expression. The sublattice contributions to the magnetization are antiparallel, and various behaviors can be found; in some materials there exists a temperature at which the component contributions exactly cancel, called the compensation point (see Fig. 6.7).

Certain other properties are common to many magnetic materials. Under the influence of an external field, the consequent alignment of magnetic dipoles in a material affects the internal field in a way that depends on the physical shape of the specimen; the diamagnetic response is constrained by the geometry of the specimen's boundary conditions. (An alternative view considers the contribution of fields produced by magnetic poles induced at the specimen boundaries to the total internal field.) This effect reduces the internal field below the applied field value and produces nonuniform (fringing) fields near surfaces and edges. The phenomenon is denoted shape-factor demagnetization. The problem of determining a functional form of the internal field, assuming arbitrary boundary conditions, is realistically intractable, but for ellipsoids of revolution the contribution

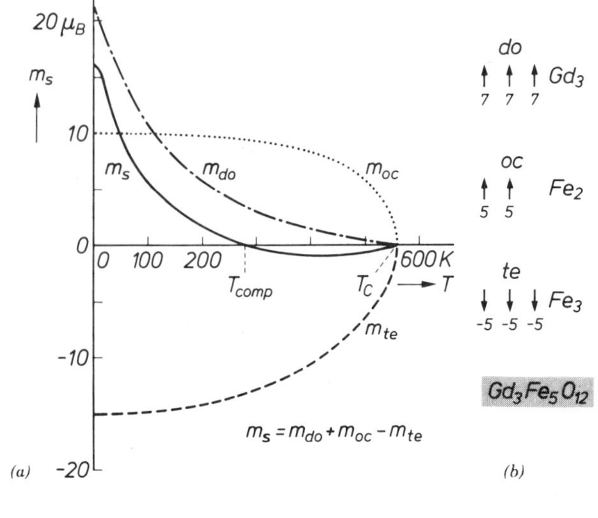

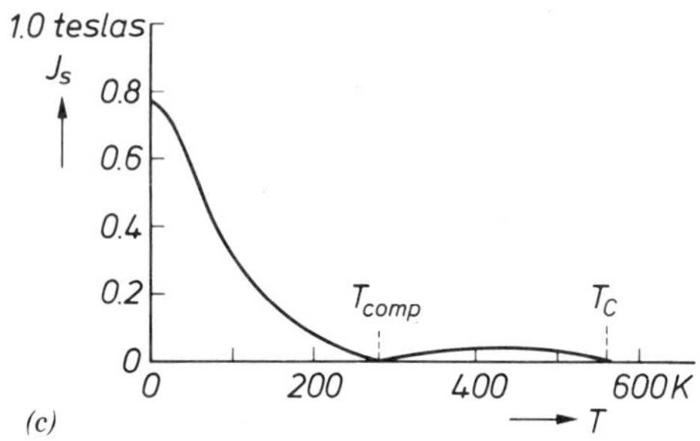

Figure 6.7 (*a*) Magnetic saturation moment m_s per unit of $Gd_3Fe_5O_{12}$ (an eighth part of a unit cell) in Bohr magnetons μ_B as a function of absolute temperature T in a gadolinium–iron garnet film. The saturation moment m_s is the sum of the magnetic moments m_{oc}, m_{te}, and m_{do} originating from the spins of the Fe^{3+} ions at the octahedral and the tetrahedral lattice sites and of the GD^{3+} ions at the dodecahedral lattice sites. The compensation temperature is T_{comp}; when this temperature is exceeded and an external magnetic field is applied, the direction of m_s rotates through 180°. The Curie point is T_C. (*b*) Origin of the total saturation magnetization m_s as the sum of the magnetic moments of the sublattices for a unit $Gd_3Fe_5O_{12}$ at absolute zero. (*c*) Magnetic saturation polarization J_s in teslas (the magnetic saturation moment $j_s = \mu_0 m_s$ weber meters per unit volume) derived from (*a*) for the material as a function of the absolute temperature T. The saturation polarization is $J_s = \mu_0 |m_s| \mu_B 8/a^3$ teslas, where $\mu_B = 0.927 \times 10^{-23}$ Am^2 is the magnetic moment associated with a Bohr magneton, and a is the lattice constant. (From P. Hansen et al., "Optical Switching with Bismuth-Substituted Iron Garnets," *Philips Tech. Rev*., Vol. 41, No. 2, 1983/84, p. 37. Published by N. V. Philips. Copyright 1985, used by permission.)

of surface poles to the internal field is uniform throughout the specimen when the externally applied field is large enough to ensure saturation. The contribution is described by $\mathbf{H}_i = \mathbf{H}_o - (\mathbf{N} \cdot 4\pi\mathbf{M})(\mathbf{M}/M)$. The demagnetizing factor $\mathbf{N}$ has components N_x, N_y, and N_z that sum to 1 in all cases; for a sphere, $N_x = N_y = N_z = \frac{1}{3}$; for a thin slab with surface perpendicular to the z direction, $N_x = N_y = 0$, $N_z = 1$.

Shape-factor demagnetization contributes to the magnetic anisotropy field when a sample is not saturated. The demagnetizing energy affects the spontaneous domain structure in small (nonsaturating) fields but is not a factor under saturation conditions. In unsaturated planar samples, anisotropy caused by shape-factor demagnetization is in-plane as a consequence of the lower magnetostatic energy for this alignment. This alignment is counter to the contribution of shape-factor demagnetization in saturated samples; in-plane saturation encounters no internal demagnetizing effect, while perpendicular saturation must supersede a demagnetizing internal field of $4\pi M_s$.

The symmetries of the crystallographic structure of a magnetic material generally confer crystalline magnetic anisotropy. It can be uniaxial, as it is in crystals having hexagonal symmetry, but often has much more complicated structure depending on the number of distinguishable directions in the unit cell of the crystal. Anisotropy results from the influence of directional bonding on the orientation of the orbital angular momentum and on the derivative spin dipole orientation caused by spin–orbit coupling. When the internal magnetic field lies parallel to the crystal axis along which the dipoles naturally align (the easy axis), it is enhanced by the anisotropy field. The magnetic anisotropy energy is the difference between the energy required to orient all dipoles along the hard direction and that required to orient them along the easy direction. For a uniaxial crystal the energy follows $E_a = K_1 \sin^2 \theta$ to first order, reflecting the energy acquired by a dipole reorienting in the presence of an aligning torque. In a cubic crystal, with the cosine of the angle between the domain orientation and each crystal axis i given by α_i,

$$E_a = K_1(\alpha_1^2\alpha_2^2 + \alpha_2^2\alpha_3^2 + \alpha_3^2\alpha_1^2)$$

In either case, the total anisotropy energy is proportional to the volume of material considered. Susceptibility curves (B vs. H) for anisotropic materials vary with specimen alignment in the external field. As discussed further in what follows, the anisotropy energy balances the exchange energy in determining the thickness of the Bloch walls that separate magnetic domains. The anisotropy contribution also changes the (field-dependent) frequency structure of magnetic resonances in a material.

Other sources of magnetic anisotropy can also contribute to material behavior. Strain can be introduced to a sample in several ways. For epitaxial films of magnetic media, crystal mismatch between film and substrate often introduces strain. Alternatively, applying a magnetic field to a sample causes its dimensions to change slightly as a result of the realigning of the dipoles, an effect called magnetostriction. The magnetization of the material is affected by this stress, introducing magnetoelastic energy. For isotropic media the magnetostriction energy contribution to material anisotropy is given by $E_{\mathrm{me}} = \frac{3}{2}\lambda T \sin^2 \theta$, in which λ is the magnetostriction constant, T the temperature, and θ the angle between the strain and magnetization directions.

Crystal growers often acknowledge another source of magnetic anisotropy, which they call growth anisotropy. This is apparently caused by imperfect crystallization of a sample during the growth process; disordered or irregular location of large atoms in a crystal, for example, induces local strain fields. Often, postgrowth annealing can relax the irregularities and virtually eliminate this source of anisotropy.

Surface states can inhibit magnetic dipoles from assuming arbitrary orientation. This phenomenon is called spin pinning and is responsible for boundary conditions different from those that would be invoked considering formal truncation of an infinite medium. Spin pinning can contribute strongly to coupling between various modes of the RF magnetic field in finite samples.

Magnetic materials that contain unpaired magnetic dipoles, that is, ferromagnetics and ferrimagnetics, generally organize them in coaligned clumps called domains. Several energy factors contribute to determining the size of the domains, their relative orientation, and the characteristics of the boundaries, called Bloch walls, which separate them. The three most obvious of these are the magnetostatic energy, the exchange energy, and the anisotropy energy. Consider a crystal in which all the magnetic dipoles are aligned in parallel. As previously discussed, the material exhibits an internal magnetic field contribution called the demagnetizing field. The dipoles of the material that align in this field take on an energy, the magnetostatic energy, equal to one-half the dot product of the field and the magnetization:

$$E_m = \tfrac{1}{2}\mathbf{H} \cdot \mathbf{M}$$

The magnitude of E_m when $\mathbf{H}$ is parallel to $\mathbf{M}$ is $E_m = \frac{1}{2}NM^2$ (N is the demagnetizing factor; it equals 4π, e.g., when the field is perpendicular to the surface of a thin slab.) The factor of $\frac{1}{2}$ discounts the influence of the magnetic dipoles on themselves. This energy can be reduced substantially (by nearly a factor of 2) by dividing the crystal conceptually into halves and reorienting one-half in the antiparallel direction; the total field at any point will be the sum of contributions from the two segments. Calculating the domain magnetostatic energy is difficult, but it is apparently approximately equal to $E_m = LM^2$ per unit area, where L is the sample thickness. When many antiparallel domains are present, the energy per unit area decreases by a factor of about the inverse of the number of domains n_d, giving the proportionality dependence $E_m = M^2/n_d$ (see Fig. 6.8). Repeating this conceptual subdivision of the dipoles into counteroriented domains to diminish the internal energy could continue indefinitely except for the influence of the other two energy contributors to domain formation. These (the exchange energy and the anisotropy energy) result from the skew in alignment between neighboring dipoles, and between dipoles and crystallographically preferred directions, and therefore contribute not via the domains themselves but via the interdomain walls. Minimizing the sum of the energy of the domains

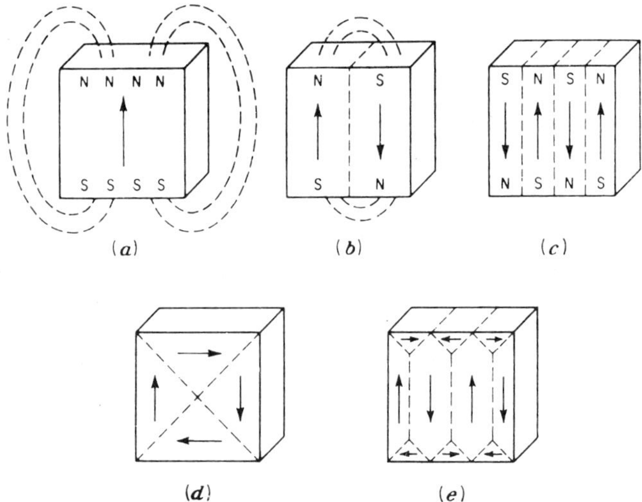

Figure 6.8 Several arrangements of domains that may occur in single crystals. The magnetostatic energy associated with the single domain shown in (*a*) is greater than that of (*b*) and (*c*). Domains of closure are shown in (*d*) and (*e*), and the energy of demagnetization is zero. (From B. Lax and K. J. Button, "Paramagnetism and Ferromagnetism," *Microwave Ferrites and Ferrimagnetics*, McGraw-Hill, New York, 1962, p. 76. Copyright 1962, used by permission.)

(according to their size and orientation) and the energy of the domain walls (according to their thickness and separation) governs the magnetic partitioning. The total energy, $E_T = M^2/n_d + n_d E_w$, is minimized for $n_d = M/\sqrt{E_w}$, where E_w, the wall energy, is not yet specified.

Consider temporarily a material without anisotropy, so that the effects of exchange energy are isolated. The exchange energy in ferromagnetic and ferrimagnetic materials acts to influence the majority of dipoles to coalign. Reducing exchange energy and reducing magnetostatic energy are thus countervailing criteria, and the size of domains is determined by the configuration of dipoles that minimizes the total energy in a specimen. Recall that the exchange energy for electron-populated orbitals in any two neighboring atoms is given by the product of the exchange integral and the dot product of the spin angular moment of the two electrons. The exchange energy finds the source of its contribution for the domain-organizing problem in the character of the regions that separate antiparallel domains, namely, in the domain walls (Bloch walls). From one domain to its neighbor the dipoles change orientation by 180° (see Fig. 6.9). (Actually, domain wall rotations of 90°, as well as other angles according to the specifics of crystal anisotropy, also occur regularly.) The energy required to make this transition can be decreased if many layers of atoms are invoked, with small rotation between consecutive layers. The exchange energy is then given by

$$E_{ex} = -\tfrac{1}{2} J_e S^2 \cos^2 \theta$$

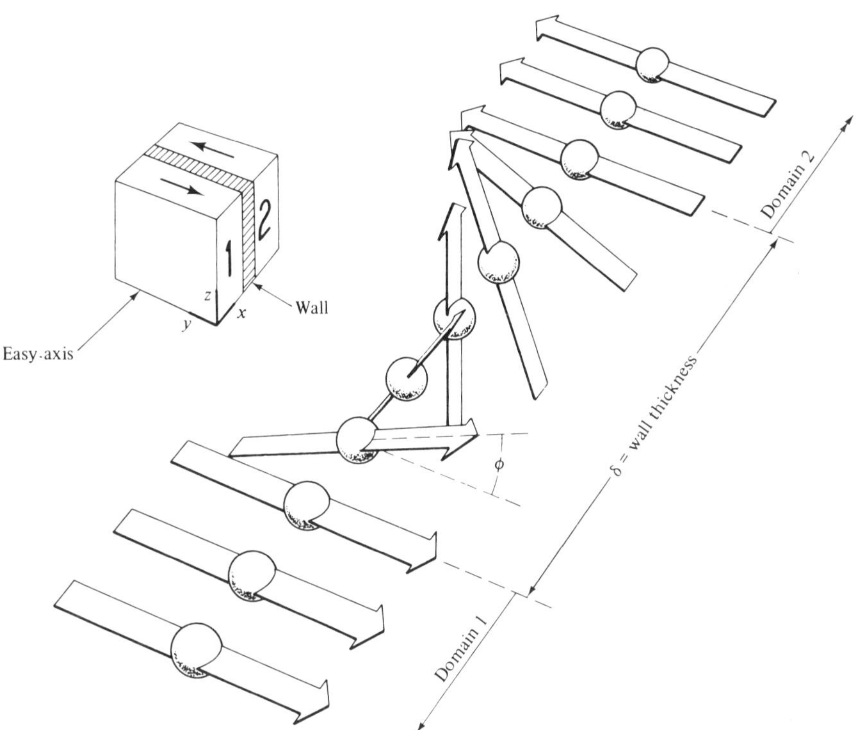

Figure 6.9 Structure of a 180° wall. The axis of easy magnetization lies along the *y*-axis. (From B. D. Cullity, "Domains and the Magnetization Process," *Introduction to Magnetic Materials*, Addison-Wesley, Reading, MA, 1972, p. 289. Copyright 1972, used by permission.)

for an angle θ between neighboring spins. For a small angle θ the difference in E_{ex} from zero rotation is equal to

$$E_{ex} = J_e S^2 \theta^2$$

For a 180° wall, $\theta = \pi/n$, where n atomic layers are invoked to constitute the wall. If the atomic separation is d, then the total exchange energy per unit area of a wall is $J_e S^2 \pi^2 / nd^2$, where d^2 represents the atoms per unit area and n is the number of atomic layers through the wall.

The exchange energy could go to zero for very thick walls. This proclivity is tempered by the anisotropy energy, which, per unit area, increases proportionally with the wall thickness. The total wall energy per unit area is

$$E_w = E_{ex} + E_a = \frac{J_e S^2 \pi^2}{nd^2} + K_1 nd$$

It is minimized when the wall thickness is

$$nd = \sqrt{\frac{J_e S^2 \pi^2}{K_1 d}}$$

The exchange energy equals kT_c at the Curie temperature. (Above the Curie temperature the exchange energy is dominated by thermal fluctuations and the material becomes paramagnetic. Domain structure and domain (Bloch) walls cease to exist.) The wall thickness nd then varies as $\sqrt{T_c/K_1}$.

6.4 SPIN WAVES

Within a domain or in a material in which the domains have been saturated by a uniform biasing magnetic field (whether an external bias field or an intrinsic field, e.g., an anisotropic field), the magnetic dipoles will precess around the direction of the internal magnetic field. This is because the force of the field on the dipole is in the form of a torque. (Dipoles come near to alignment with the field by means of energy transfer to lattice vibrations. As discussed previously in the section delineating paramagnetic materials, the aligning influence of the torque is balanced by thermal agitation and is ultimately restricted by the quantum-mechanical constraints imposed by the uncertainty principle. The angle between a dipole and the field is never precisely zero.) The coupling between neighboring dipoles—interdipole fields, the source of the magnetostatic energy already discussed, are responsible—tends to cause them to precess with a common phase. This is called the uniform precession mode. The application of an RF electromagnetic field to the material will affect the precesssion in a way that depends on the directions of propagation and polarization of the electromagnetic disturbance with respect to the magnetization direction in the material. The dipole response to this perturbation is added linearly (for small electromagnetic wave amplitudes) to the intrinsic precessional motion.

The torque-type action of the RF (and bias) magnetic fields on the dipoles is represented in a macroscopic mathematical model by introducing imaginary-valued, off-diagonal elements into the RF permeability (or susceptibility) tensor $\overline{\overline{\mu}}$ (or $\overline{\overline{\chi}}$). Both the diagonal and off-diagonal elements become complex valued in the presence of damping, or loss, in a material.

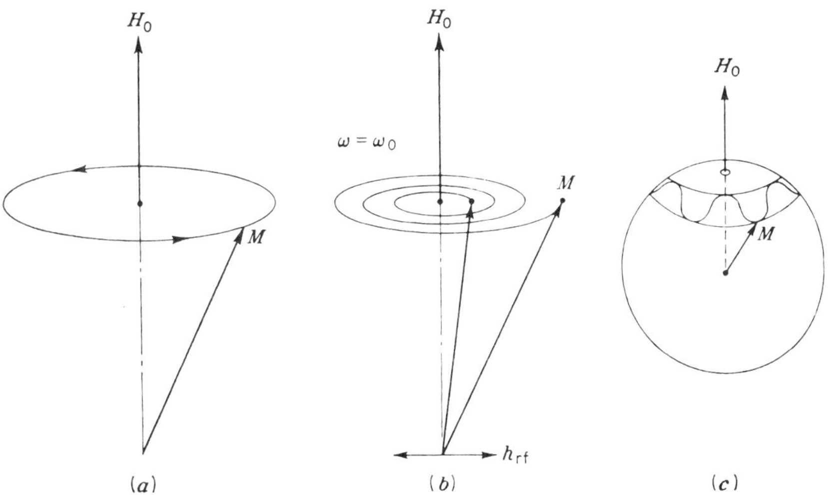

Figure 6.10 Precession of the magnetization vector (*a*) in a constant magnetic field grows in amplitude (*b*) when the frequency of the alternating magnetic field h_{rf} is near the natural precession frequency. A perspective sketch of the oscillation of the magnetization vector when $\omega \neq \omega_0$ is shown in (*c*). The magnitude of the magnetization vector is a constant, so that the oscillations shown in this figure actually take place on the surface of a sphere of radius $|M|$. The exaggerated oscillatory component ω is superimposed on the precessional motion. (From B. Lax and K. J. Button, "Ferromagnetic Resonance," *Microwave Ferrites and Ferrimagnetics*, McGraw-Hill, New York, 1962, p. 148. Copyright 1962, used by permission.)

To examine the reaction of magnetically saturated materials to magnetic field perturbation, assume a linearly polarized electromagnetic wave impinging, of which the magnetic field component is important because of its singular ability to interact with the spin contribution to the dipole moment. A dipole will in general experience a torque coercing it to align with the instantaneous RF magnetic field. Disregarding damping and inertia, the dipole will track the component of the RF magnetic field that is perpendicular to the saturating internal field. (A precessing dipole points nearly in the direction of the saturating field, so it has little moment to respond to a perturbing magnetic field polarized* in the longitudinal direction.) An RF transverse magnetic wave propagating parallel to the constant (or static) internal field drives the angle between the dipole and the bias field to successively greater and smaller values. An RF magnetic wave propagating perpendicular to the bias field can similarly interact with the dipoles if its magnetic polarization is transverse to the bias field. If the wave is polarized so that the magnetic component is parallel to the static internal field, the material response exhibits no magnetic features, and the wave propagates in it as if in a nonmagnetic dielectric. Applying a torque to a vector quantity cannot change its magnitude, only its direction. The locus of the dipole moment therefore oscillates on the surface of a sphere, the same one that delimits the intrinsic precessional motion (see Fig. 6.10).

* Standard usage connects the term *polarization*, or *polarized*, only with the electric field component of an electromagnetic wave. In this section it will often be used to indicate the direction in which the magnetic field component is pointing, but always with the distinguishing magnetic designation in *magnetic polarization* (or *magnetically polarized*). The text reverts to standard usage in the following section.

When the frequency of the magnetic field approximates the natural precession frequency of the dipole (which depends on the bias field strength by way of the gyromagnetic equation), energy can be transferred to the dipole out of the RF field if the required phase relationship exists. The bandwidth of this resonant absorption is governed by a Lorentzian lineshape, which results from the decay of excited spins to lattice phonons, directly or via coupling to the reservoir of other dipoles. Experiments that expose this line width are critical to modeling the attributes of magnetic materials.

The mathematical description of the interaction of magnetic dipoles with RF magnetic fields invokes both Maxwell's equations (viz. $\nabla \times \mathbf{h} = d\mathbf{E}/dt$ and $\nabla \cdot \mathbf{b} = 0$; $\mathbf{b} = \bar{\bar{\boldsymbol{\mu}}} \cdot \mathbf{h}$) and the equation of motion for magnetic dipoles in a magnetic field $[d\mathbf{m}/dt = \gamma(\mathbf{m} \times \mathbf{H}_i)]$.* For the appropriate RF magnetic polarization perpendicular to the saturating static magnetic field, the material response depends on the wavelength of the internal RF magnetic field relative to the separation of the contributing magnetic dipoles. (The internal wavelength depends on several parameters: the strength of the static saturating magnetic field, the frequency of the perturbing RF electromagnetic wave, and the material constants, often including the sample's dimensions and orientation in the static field. The wavelength of the vacuum electromagnetic wave is almost always much greater than the size of the sample; at 3 GHz the vacuum wavelength is 100 cm.) For a long internal RF magnetic wavelength the material response is minimal, and the propagation of the RF wave occurs as if in a dielectric. Alternatively, for a very short RF magnetic wavelength the driving field varies in strength (via its phase excursion) over the distance separating neighboring dipoles, and the field attempts to modulate neighboring dipoles out of phase. In this case, along with the magnetostatic force that acts to coalign the dipoles, the exchange force becomes effective. The restoring force acting to align the dipoles is much stronger in this regime. The resonant frequency shifts upward, but more importantly a new type of wave becomes viable in the material, the exchange-dominated spin wave. An additional term is appended to the right side of the equation of motion, so that now

$$\frac{d\mathbf{m}}{dt} = \gamma(\mathbf{m} \times \mathbf{H}_i) + \gamma H_m d^2 \frac{\mathbf{M} \times \nabla^2 \mathbf{M}}{M},$$

in which $H_m = |\mathbf{H}_m| = \lambda|\mathbf{M}|$ is the magnitude of the exchange field, λ is the molecular field coefficient, and d is the separation between neighboring dipoles. The new solution for the resonant frequency is greater by a term $\omega_{ex} d^2 k^2$, in which ω_{ex} is the frequency corresponding to the exchange energy ($\omega_{ex} = \gamma H_m$), and k is the wave number in the magnetic medium.

If the RF magnetic field is applied to dipoles on one side of a specimen, neighboring dipoles follow the perturbation of their orbits, and the disturbance propagates across the material. (See Fig. 6.11.) In samples of finite dimension, coupling of vacuum RF waves to exchange-dominated spin waves occurs either in the presence of a nonuniform static internal magnetic field, such as can be caused by shape factor demagnetization (recall that only an ellipsoid of rotation has a uniform internal magnetic field), or when surface states cause dipoles near an interface to be held, or "pinned," in a particular orientation so that they are unable to rotate in response to the RF perturbation. In these cases skew between neighboring dipoles is inevitable and exchange-dominated spin waves result.

An intermediate wavelength regime also exists. In this range the exchange field is weak enough compared to the internal field that its influence can be ignored (recall that

* Lowercase representation of the magnetic field vectors **b** and **h** indicate rapidly time-varying components.

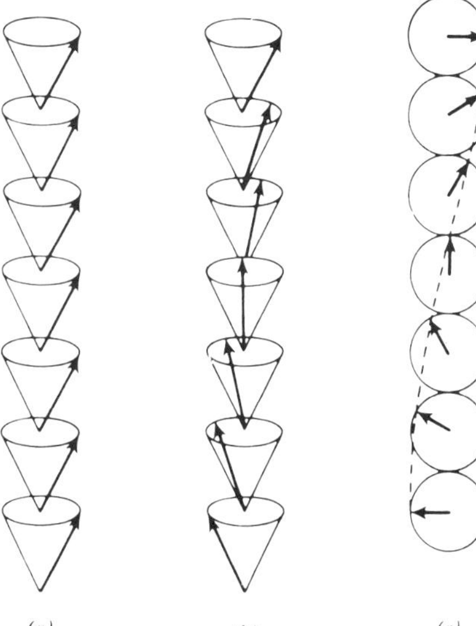

Figure 6.11 Schematic illustration of spins precessing with negative phase differences to give spin waves: (*a*) coherent mode; (*b*) one half wavelength; (*c*) spins in plan. (From D. J. Craik, "Effects of Crystal Size and Shape Rotational Processes," *Structure and Properties of Magnetic Materials*, Pion Limited, London, 1971, p. 187. Copyright 1971, used by permission.)

the exchange energy contribution increases as k^2; here k is small). Moreover, in this range the simultaneous solution of Maxwell's equations and the magnetic equation of motion lead to expressions in which the RF electric field components in the material may be very small. Depending on the propagation direction with respect to the static internal magnetic field, the transverse RF magnetic field components may also be negligible, and the wave is essentially longitudinal–magnetic polarized. This behavior will be explained somewhat more thoroughly in what follows.

This type of wave is called, somewhat imprecisely, a magnetostatic wave. The wave field still oscillates with time at RF. But since the magnetic field polarization is quasi-longitudinal, the oscillating component is described approximately by the equation $\nabla \times \mathbf{h} = \mathbf{0}$, which in light of Maxwell's equation, $\nabla \times \mathbf{H} = d\mathbf{E}/dt$, confers the nomenclature magnetostatic wave. Incorporating the magnetostatic approximation makes the simultaneous solution of the gyromagnetic equation of motion and Maxwell's equations more tractable for intermediate values of k ($10 < k < 10^5$ cm^{-1}) without severely altering its form compared to the exact solution.

The partitioning of RF magnetic perturbation phenomena outlined in the preceding into uniform precession, magnetostatic, and exchange-dominated spin waves according to wave number magnitude, and in particular the nature of magnetostatic waves, can be difficult to intuit. Propagation of magnetic-field-dominated waves is very sensitive to propagation direction and magnetic polarization. A delineation of the characteristics of RF magnetic plane waves in infinite magnetic media may aid understanding.

The polarization of the magnetic field for which spin waves are stimulated (uniform precession, magnetostatic, or exchange dominated) can be described in the language of optics as the extraordinary wave. The orthogonal magnetic polarization, for which the interaction of the RF magnetic field and the dipole spins are negligible (magnetic polarization parallel to static internal field), is analogous to an ordinary wave. The propagation and dispersion relation for the ordinary wave magnetic polarization are just those of an RF field in a dielectric. (Remember: magnetic fields act on dipoles via torque; no torque exists between coaligned forces and moments.) Ordinary RF magnetic waves will be mentioned only once again, when their straight-line dispersion serves as a reference for the complicated dispersion curves of the extraordinary waves. For extraordinary waves, solution of the preceding equations give rise, in an infinite medium, to two different dispersion relations. The applicable form depends on whether the spin wave propagates parallel or perpendicular to the static internal field. (Intermediate directions exhibit dispersion relations bounded by these orthogonal forms.)

The dispersion relation for the perpendicular propagation direction has two branches separated by a zone that is forbidden in the absence of exchange energy considerations (see Fig. 6.12). A graph of the low-frequency branch is coincident with the dielectric dispersion at the origin but curves gently away to become asymptotic to a line at constant frequency at high wave numbers (the bottom edge of the zone). This asymptote occurs at a frequency that is related to the gyromagnetic frequency, $\omega_0 = \gamma|\mathbf{H}_i|$, where $\mathbf{H}_i$ is the internal field due to an applied external field, and to the equivalent magnetic saturation frequency, $\omega_m = \gamma \times 4\pi|\mathbf{M}_s|$, where $\mathbf{M}_s$ is the material's saturation magnetization, by $\omega = \sqrt{\omega_0(\omega_0 + \omega_m)}$. The graph of the high-frequency branch is coincident with and parallel to the upper edge of the forbidden zone at very low wave numbers and curves gently to become asymptotic to the straight-line dielectric dispersion (the RF magnetic ordinary wave dispersion) at high frequency. The upper edge of the zone occurs at $\omega = \omega_0 + \omega_m$.

The dispersion relation for propagation parallel to the static internal field also has two branches (see Fig. 6.12). Their graphical appearance is very similar to that of the

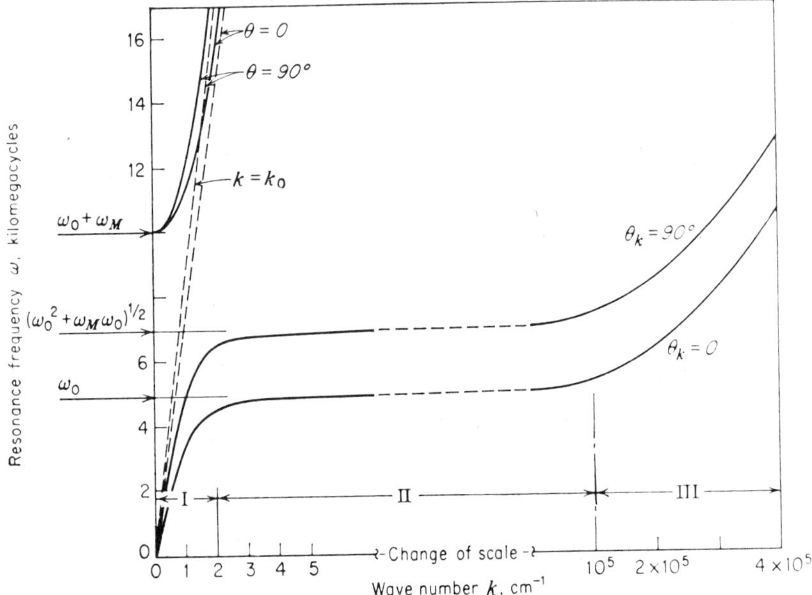

Figure 6.12 Resonance frequency in the infinite medium as a function of wave number, where $k = 0$ corresponds to the frequency of uniform precession. The regions indicated are (I) electromagnetic propagation, (II) magnetostatic modes, and (III) spin wave modes. The two dashed lines are the ordinary wave dispersion relations. Curves in upper left are upper branches of extraordinary mode. Curves across center are lower branches of extraordinary mode. Here $4\pi M_s = 2000$ Oe. (From B. Lax and K. J. Button, "Plane-Wave Propagation," *Microwave Ferrites and Ferrimagnetics*, McGraw-Hill, New York, 1962, p. 319. Copyright 1962, used by permission.)

perpendicularly propagating waves, but mathematically their derivation is different, and the physical interpretation of the waves that obey them differs appreciably. One branch has two segments that mimic the appearance of the graph of the perpendicularly propagating wave just described, except that the lower edge of the forbidden zone occurs at a lower frequency, $\omega = \omega_0$. The upper segment appears, at zero wave number, at the same point coincident with the upper edge of the forbidden zone at $\omega = \omega_0 + \omega_m$ as the high-frequency part of the perpendicular dispersion relation. It is asymptotic to the alternate branch of the parallel dispersion relation rather than to the dielectric dispersion. This alternate branch is nearly linear and differs appreciably in appearance from the dielectric dispersion only in that the mean velocity, derivative from its slope, is slightly less.

These two branches of the parallel dispersion relation describe propagation of circularly polarized magnetic waves; the nearly straight branch corresponds to left circularly polarized (LCP) propagation, and the two segments of the other branch correspond to right circularly polarized propagation (RCP). (For RCP propagation, a stationary observer looking in the direction of propagation would see on a screen in front of him the magnetic field vector rotate in a clockwise sense with increasing time, that is, in the direction that the fingers of a right hand would curl when the right thumb is pointed along the propagation direction.) The LCP straight branch exhibits nearly no interaction with the magnetic dipoles, while the RCP branch is strongly dispersive. This behavior can be anticipated since dipole precession is also clockwise around the static magnetic field. At a certain frequency the RCP wave will be resonant with the dipoles, a situation that can never occur for the LCP wave. These two branches of the parallel propagation dispersion relation have different slopes, demonstrative of different wave velocities for RCP and LCP waves. This phenomenon is commonly known as Faraday rotation. (As will be shown in what follows, its basic mechanism differs from that for optical Faraday rotation, even though the phenomenological effects are similar.)

The mathematics of the derivations that produce these dispersion relations (not reproduced here) show that for perpendicular propagation, at intermediate values of wave number, both transverse components of the RF magnetic field are negligibly small (the component along the static field direction is identically zero), as are all components of the RF electric field. This wave has become essentially longitudinal–magnetic polarized, and the magnetostatic approximation $\mathbf{\nabla} \times \mathbf{H} = 0$ is valid. Conversely, the parallel propagating waves have transverse magnetic omponents that are equal in magnitude and electric field components that are not vanishingly small. These waves do not satisfy the magnetostatic approximation.

Consider the perpendicularly propagating waves. Throughout the magnetostatic region of intermediate wave number magnitudes their dispersion relation is nearly flat, with phase velocity near zero. According to the discussion so far, perpendicular RF magnetic waves exist only along the single dispersion curve below the forbidden zone. Experimentally, perpendicular waves propagate on a manifold of dispersion curves in this region and in the forbidden zone as well. The parallel propagating waves maintain a single dispersion curve, although they too transgress the forbidden zone. The analysis of plane waves in unbounded media requires adjustment to represent physical reality.

The first adjustment reincorporates exchange energy considerations at high wave numbers. Recall that they introduce an additional term in the gyromagnetic equation of motion and an additive term in the resonant frequency expression that depends on k^2. This k^2 term causes the dispersion curve to turn parabolically across the forbidden zone toward higher frequency at high wave number values ($k > 10^5\ \mathrm{cm}^{-1}$).

The addition of boundary conditions to the problem additionally complicates the mathematics by introducing shape factor demagnetization terms to the internal magnetic

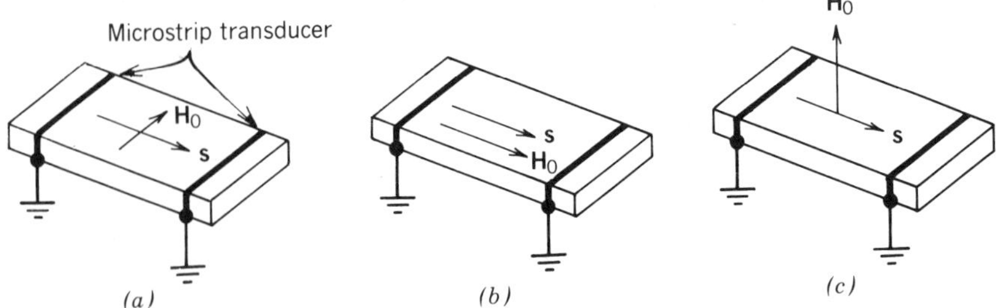

Figure 6.13 The required orientation of the static bias magnetic field $\mathbf{H}_0$ for each of the three types of magnetostatic wave. The *s* vector indicates the MSW propagation direction. Note that surface waves (*a*) are nonreciprocal; they propagate only on the particular surface of the film where $\mathbf{H}_0 \times \hat{\mathbf{n}}$ is parallel to $\mathbf{s}$ ($\hat{\mathbf{n}}$ normal to the surface). Also shown are backward (*b*) and forward (*c*) volume wave modes.

field. The shape factor was discussed in Section 6.3 on magnetic materials. For thin planar samples, the demagnetization factor is perpendicular to the surface. When the applied magnetic field (which saturates the material) is also perpendicular to the surface, the effective internal magnetization is reduced by $4\pi\mathbf{M}_s$, and the gyromagnetic resonant frequency is reduced by $\gamma \times 4\pi\mathbf{M}_s$ (i.e., by ω_m). This shifts the dispersion relation and its attendant forbidden zone to lower frequencies. To see how much, replace ω_0 by $\omega_0 - \omega_m$ everywhere it occurs in the preceding discussion. This allows the forbidden zone to be bisected by the frequency of uniform precession at ω_0, and coupling between uniform precession modes and exchange-dominated spin wave modes, for which the dispersion curves bend across the zone, can occur. Of course, if the planar sample is tilted with respect to the applied field, the demagnetization contribution is reduced to N_i times ω_m, where N_i is the demagnetization component in the direction of the field. If the field is in the plane of the specimen, the contribution is zero.

To analyze the propagation of spin wave modes in bounded media, it is convenient to invoke the magnetostatic approximation ($\nabla \times \mathbf{H} = \mathbf{0}$). Then $\mathbf{H} = \nabla\psi$, with ψ the magnetic potential. The gyromagnetic and Maxwell equations become straightforward, if still complicated, differential equations in ψ, and propagation in a bounded media becomes a boundary value problem. The mathematical derivations of the various solutions will not be presented here (see Ref. 3 or 4), but a description follows of the types of waves that will propagate for various orientations of a thin, planar sample (or film) in the static field. The eigensolutions are found when the sample is oriented either perpendicular or parallel to the static field (see Fig. 6.13), and only these conditions will be discussed.

Both surface and bulk waves exist. (A preview of the dispersion manifolds for the various modes of each possible wave type appears in Fig. 6.14.) When the static field is in the plane of the sample but perpendicular to the propagation direction, solution of the boundary problem produces surface waves whose magnetic field amplitude decays exponentially with distance away from the surface in both directions. A single dispersion curve applies to these waves for any specified static magnetic field strength. Propagation of the surface waves is nonisotropic; they travel along one surface of the sample in one direction and the opposite surface in the reverse direction, specified according to $\mathbf{H}_i \times \hat{\mathbf{n}}$, where $\hat{\mathbf{n}}$ is the normal vector to the surface under consideration.

The dispersion relation for surface waves is

$$\exp(2\beta_{\text{MSW}}d) = \frac{(2\pi M_s)^2}{(|\mathbf{H}_i| + 2\pi M_s)^2 - \omega/\omega^2}$$

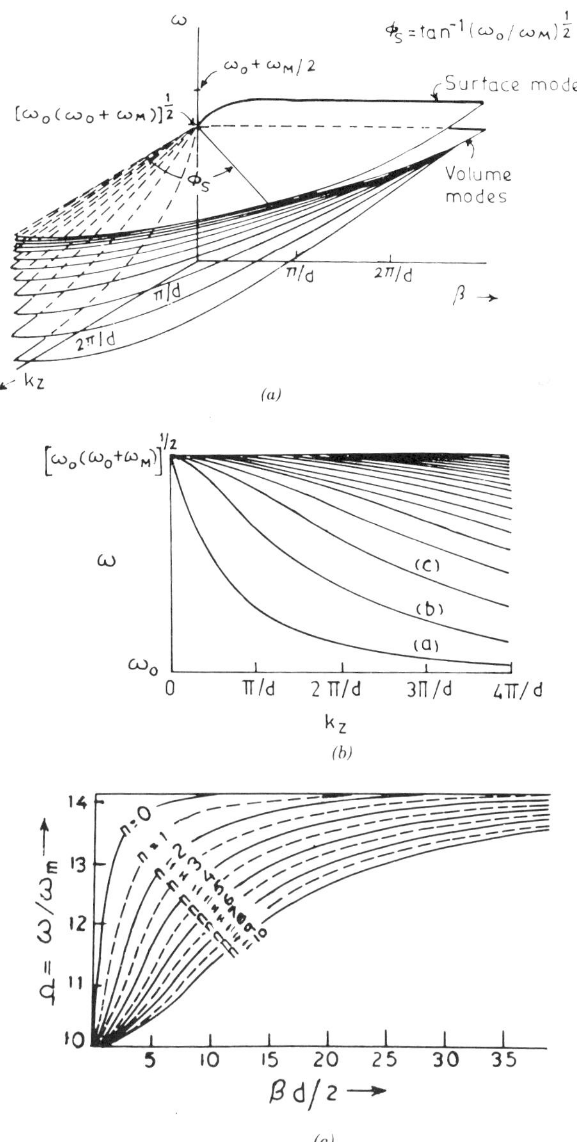

Figure 6.14 (*a*) General magnetostatic mode spectrum for a ferrimagnetic slab magnetized parallel to its faces. The surface and bulk modes are simultaneously allowed only when the angle between the directions of propagation and biasing field exceeds $\phi_s = \cot^{-1}(\omega_m/\omega_0)^{1/2} = \tan^{-1}(\omega_0/\omega_m)^{1/2}$. (From M. S. Sodha and N. C. Srivastava, "Magnetostatic Waves in Layered Planar Structures," *Microwave Propagation in Ferrimagnetics*, Plenum, New York, 1981, p. 173. Copyright 1981, used by permission.) (*b*) Dispersion curves for the case of magnetostatic wave propagation along the biasing field. The curves designated by *a*, *b*, *c*, etc., correspond to $n = 0, 1, 2, \ldots$. The slope of ω-versus-k_z is negative throughout the frequency range for all mode orders. Thus the modes are backward waves, for which the directions of energy flow and phase propagation are mutually opposite. (From M. S. Sodha and N. C. Srivastava, "Magnetostatic Waves in Layered Planar Structures," *Microwave Propagation in Ferrimagnetics*, Plenum, New York, 1981, p. 172. Copyright 1981, used by permission.) (*c*) Magnetostatic mode spectrum for a normally magnetized slab. The phase velocity is smaller for the higher-order modes. (From M. S. Sodha and N. C. Srivastava, "Magnetostatic Waves in Layered Planar Structures," *Microwave Propagation in Ferrimagnetics*, Plenum, New York, 1981, p. 183. Copyright 1981, used by permission.)

In this equation, β_{MSW} is the component of the MSW wave vector parallel to the film surface, d is the film thickness, ω is the MSW frequency, and γ is the gyromagnetic frequency.

When the static field is in the sample plane but parallel to the propagation, backward-volume waves propagate. These are waveguide-confined, bulklike waves for which the group velocity is negative. A manifold of dispersion curves exists for backward-volume waves that appears approximately as a mirrored replica of the manifold for forward-volume waves, reflected across the center of the frequency bandwidth.

The dispersion relation for backward-volume waves is

$$2\cot(\alpha\beta_{\text{MSW}}d) = \alpha - \frac{1}{\alpha}$$

Here $\alpha = \sqrt{-1/\mu}$, where μ is the scalar magnitude of the permeability, assumed isotropic; other variables are as previously defined.

Forward-volume waves exist when the static field is perpendicular to the planar sample. Their propagation is isotropic with the direction in the film. The magnetic field amplitude for these waveguide-confined, bulklike, forward-volume waves is a standing-wave pattern across the thin planar sample; the component of the wave vector transverse to the sample is real inside the film. Although the forward-volume waves submit readily to an RF magnetic wave boundary condition analysis, they can also be analyzed by a procedure analogous to that used for optical guided-wave modes. The approach is complicated by the fact that the spin waves in the reflecting-ray or zig-zag ray model propagate neither perpendicular nor parallel to the sample boundary. The applicable dispersion relation depends on the angle between the propagating direction and the surface of the sample, as does the value of the phase change experienced by the wave on reflection. Either approach results in the existence of a different dispersion curve for each forward-volume wave mode, where a mode is distinguished by the number of standing-wave nulls found between the film surfaces.

The dispersion relation for forward-volume waves is

$$\tan\frac{\beta_{\text{MSW}}}{2\alpha} = \alpha$$

The manifold of dispersion curves resulting from the two bulk-wave solutions (forward- and backward-volume waves) shifts to higher frequency when the static field strength is increased. The frequency bounds for the manifold are

$$\omega_0 < \omega < \sqrt{\omega_0(\omega_0 + \omega_m)}$$

for the upper and

$$\omega_0 = \gamma|\mathbf{H}_i|$$

for the lower. This lower bound is at the gyromagnetic frequency, which is also the lower edge of the forbidden zone for plane RF magnetic waves propagating parallel to the static magnetic field in an unbounded medium. The upper bound is the same as the lower edge of the forbidden zone for propagation perpendicular to the static field. The geometric constraints have induced propagation in frequency ranges that do not support waves in the infinite medium.

The single dispersion curve for the surface wave also shifts to a higher frequency for increased static fields. It is bounded in frequency by

$$\sqrt{\omega_0(\omega_0 + \omega_m)} < \omega < \omega_0 + \tfrac{1}{2}\omega_m$$

This frequency band is adjacent to and above that for the bulk waves, with its upper limit still within the forbidden zones of the analysis of plane waves in an infinite medium.

In recapitulation, three types of spin waves—uniform precession, magnetostatic, and exchange dominated—can be stimulated in saturated magnetic materials by externally impressed, oscillating magnetic (electromagnetic) fields. The uniform precession modes occupy the very low wave number region of frequency–wave number space. Magnetostatic waves are intermediate in wave number. Exchange-dominated spin waves exist only at high wave numbers where the exchange force is appreciable. There exists a forbidden zone for plane waves in an infinite isotropic magnetic medium, but it is breached on inclusion of exchange interactions at high wave numbers and by the effect of shape factor demagnetization on internal fields. Complicated rules emerge from the gyromagnetic and Maxwell equations to govern the propagation of three subtypes of magnetostatic waves in planar film samples.

A few other details merit mentioning. Anisotropy fields in magnetic materials also change the location of the allowed frequency bands for the various spin wave disturbances. Their exact effect depends on the orientation of the anisotropy in the sample and the orientation of the sample in the static field (see Ref. 4).

These spin waves can be stimulated in several ways. Usual approaches comprise two alternatives. A sample can be placed in a microwave waveguide (transmission line) or in a cavity terminating such a line. The degree to which the symmetry of the RF magnetic field in the waveguide matches that required for a particular spin wave mode determines the efficiency of excitation of the mode. Alternatively, for thin-film samples, a simple metallic microstripline antenna carrying RF current and placed near the surface of the sample will stimulate any of the three magnetostatic wave mode types very efficiently.

A technique for determining the attenuation of propagating magnetostatic waves (and exchange-dominated spin waves, perforce, since their resonant frequency regions can overlap) employs the microwave waveguide approach to stimulating the modes in a thin-film sample. Usually the sample is placed near an antinode of the RF magnetic field in a waveguide or cavity, with its surface perpendicular to both the RF magnetic field and an externally imposed static magnetic field (from an electromagnet). A small coil carries a small-amplitude, low-frequency alternating current that modulates the static field slightly. The large field supplied by the electromagnet is swept slowly on a linear ramp, which shifts the magnetostatic and exchange-dominated spin wave spectra commensurately, so that the spin wave modes pass successively through resonance with the RF magnetic field. The tickler magnetic field supplied by the small coil allows lock-in amplifier detection of changes in the RF power in the waveguide after insertion of a small probe. The differential line width of the spin wave mode is detected, from which the propagating absorption can be calculated. (Figure 6.15 illustrates this apparatus schematically.) First the real line width is calculated from the detected derivative line by assuming a Lorentzian lineshape. The absorption is then $\alpha = 76\Delta|\mathbf{H}|$, where $\Delta|\mathbf{H}|$ is the sweep range of the large field over which the resonance occurs, and α is in decibels per microsecond. Exchange-dominated spin wave modes do not always appear if the sample has been grown so that the surface spins remain unpinned and if the sample is situated in the waveguide so that the RF field is uniform over its dimensions.

When the wave numbers of magnetostatic waves overlap those for which uniform precession or, particularly, exchange-dominated waves exist, coupling to them occurs, similar to the coupling between uniform precession and exchange-dominated modes previously mentioned. This is the natural evidence of phase-matched, coupled oscillators. The frequency passband of magnetostatic waves stimulated in a planar sample exhibits notches at the corresponding frequencies.

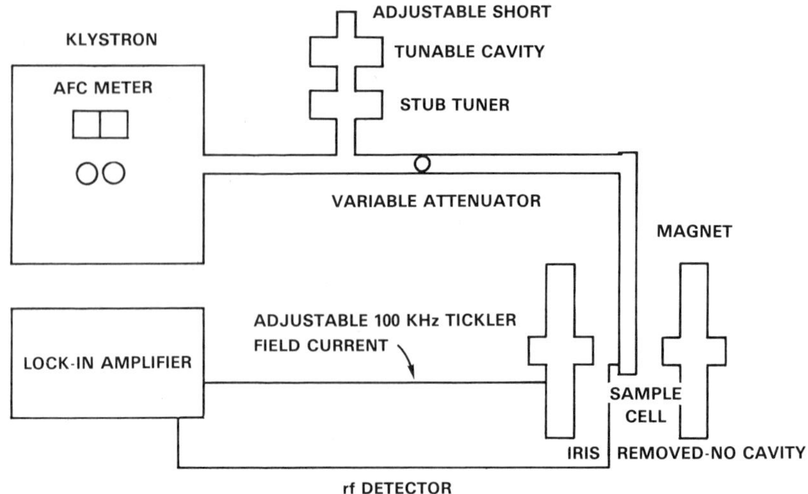

Figure 6.15 Ferrimagnet resonance line width measurement.

6.5 MATERIAL PHYSICS OF OPTICS

Optical waves are generally recognized as those transverse electromagnetic disturbances governed by Maxwell's equations that have wavelengths in vacuum between approximately 300 and 30,000 nm. Most light-modulating devices are concerned with wavelengths in a considerably narrower range, from 400 to 1600 nm, partly because of the paucity of coherent sources outside this range but more critically because of the reduced efficiency or response speed of detectors and the difficulties encountered in controlling thermal noise for wavelengths beyond these limits on one end or the other. This section presents those characteristics and descriptive models of the propagation of optical waves in materials that pertain particularly to magneto-optical interaction. Many details of optical propagation in materials that are generically useful for description of device operation will be assumed familiar or can be acquired elsewhere. These include, for example, light ray refraction according to Snell's law, as well as the effect on propagation velocity, caused by a changing refractive index; a thorough description, via Bessel function expansion of a sinusoidally varying phasor, of the diffraction of light from thin and thick phase gratings; and a detailed derivation of the equations that delineate optical propagation in planar waveguides. Several books are listed in the bibliography that provide this background material.

Magnetic interactions with light induce changes in the optical polarization. (This is the polarization of the electric field.) The polarization of a transverse electromagnetic wave at some point along its propagation path can be described by any number of pairs of orthogonal states: linear polarizations at any angular orientation; right and left circular polarizations; and elliptical polarizations of opposite handedness, orientation, and ellipticity (elliptical states are complicated, and declaring two elliptical polarizations to be orthogonal requires careful analysis). Generally there is one pair that renders any particular analysis most tractable. Polarization changes can be described (i) as differences in phase change on propagation (i.e., differences in velocity), (ii) by differences in absorption, or (iii) by coupling between the orthogonal states, depending on the mechanism. Some comments on light polarization states follow.

Elliptical polarization is the most general description of polarized light. Linear and circular polarizations are specific cases of elliptical polarization. Elliptical polarization can be constructed from two orthogonal, linearly polarized waves of equal magnitude, one delayed by a fixed phase with respect to the other. This is the easiest construction to visualize. The ellipse drawn on a plane perpendicular to the propagation direction by the tip of the electric field vector will always have its major axis oriented at 45° to the two component linear polarization directions. The ratio of minor to major axis lengths will depend on the phase delay. Zero phase delay makes the ratio zero (ellipticity 0), corresponding to linear polarization; and 90° phase delay produces ellipticity 1, for circular polarization.

Many other pairs of orthogonally polarized waves can replicate any particular ellipse, however, which complicates the mapping of the resultant polarization to a unique description. As an example, consider the following. For any particular ellipticity, the same ellipse can be produced on the perpendicular plane by any pair of linearly polarized waves oriented along the same axes as in the preceding but having different magnitudes in the two component polarizations, provided that the sum of the squares of their magnitudes equals that for the preceding description and their relative phase delay is properly adjusted. Thus a polarization-dependent absorption can appear to produce the same elliptical polarization as a polarization-dependent phase delay. In reality, the electric field vector in the second situation does not rotate at a uniform angular rate, and the appropriate description is not clear. An experimental measurement could easily produce confusing results.

Some authors invoke a geometrical artifice to catalogue and analyze changes in polarization states called the Poincaré sphere. Readers who are avidly interested in polarization states should consult the literature.

The polarization eigenstates of photons emitted by electric dipole transitions are circularly polarized. This physical reality corresponds to the quantum-mechanical rules governing the change of angular momentum that must accompany an electron transition between orbitals in an atom.

Changes in electromagnetic field polarization can be described, rigorously, in terms of either the magnetic permeability tensor or the dielectric permittivity tensor. Mathematically, use of either tensor requires identical manipulation; they occur as multiplicands in the same term of the wave equation. Considering the physical processes that occur in materials, however, it seems reasonable to identify the magnetic permeability with lower-energy magnetic dipole transitions and the dielectric permittivity with higher-energy electric dipole transitions [6]. At optical frequencies there are still residual effects of magnetic exchange interactions that take place at the lower frequencies (in the gigahertz range), discussed in preceding sections, but most features in the optical frequency range are artifacts of resonances that involve electronic orbital transitions. Magnetically induced effects then alter these electronic transition energies. To reflect the nature of the predominant physical processes that produce features in the optical frequency range, the dielectric permittivity tensor will be used in the macroscopic description of magneto-optical effects. In so doing, to avoid confusion, the unitary tensor is assigned to the magnetic permeability tensor. The off-diagonal terms of this tensor govern magneto-optical behavior in materials.

The dielectric permittivity can be separated into a nonmagnetic part and a much smaller part that is dependent on the magnetization vector $\mathbf{M}$ so that

$$\bar{\bar{\varepsilon}} = \bar{\bar{\varepsilon}}_{nm} + \bar{\bar{\varepsilon}}_{m}$$

Wettling [7] makes a careful argument, introducing Hermitian and anti-Hermitian parts to $\bar{\bar{\varepsilon}}_m$ to discover the symmetries of its real and imaginary parts. Then he invokes the Onsager relation: terms of the permittivity tensor that are symmetrical with respect to the diagonal have dependences on the magnetization **M** that differ by a minus sign. Next, the permittivity tensor elements are expanded in a power series in the Cartesian components of **M**, retaining only the linear and quadratic terms. (The linear terms have anti-Hermitian real and Hermitian imaginary parts; the quadratic terms are real Hermitian and imaginary anti-Hermitian.) Combining the restrictions discovered through the analysis outlined in the preceding, the linear and quadratic magnetic dependence of the nonzero elements of the permittivity tensor can be derived in view of the symmetries of any chosen crystal class. Only results for cubic crystals seem to be extant in the literature; presumably other crystal classes are more difficult to analyze, and the exercise contributes little to understanding.

The resulting tensor for **M** directed along the z-axis (the 3 direction) and light also propagating in the z direction is

$$\bar{\bar{\varepsilon}} = \begin{bmatrix} \varepsilon_{11} + G_{12}M_z^2 & KM_z & 0 \\ -KM_z & \varepsilon_{11} + G_{12}M_z^2 & 0 \\ 0 & 0 & \varepsilon_{11} + G_{11}M_z^2 \end{bmatrix}$$

where ε_{11} is the value of the diagonal terms of $\bar{\bar{\varepsilon}}_{nm}$, KM_z is the magnitude of the off-diagonal term that is linearly dependent on **M**, and $G_{12}M_z^2$ and $G_{11}M_z^2$ are the terms quadratic in **M** that can be seen to be diagonal contributors. (Some authors use an alternative convention in which the off-diagonal elements are written with a preceding multiplier $j = \sqrt{-1}$. This difference is resolved by interchanging the assignment of the real and imaginary parts of the tensor elements to phase and absorption effects.)

(Wettling [7] points out some details about optical propagation in other directions that most other sources do not acknowledge: the pertinent G coefficients change with the direction. The parameter $\bar{\bar{\mathbf{G}}}$ is really a four-tensor, of which many elements are degenerate, and many zeros. See Ref. 7 for details.)

This tensor is more tractable in a simpler form:

$$\bar{\bar{\varepsilon}} = \begin{bmatrix} \varepsilon_1 & \varepsilon_2 & 0 \\ -\varepsilon_2 & \varepsilon_1 & 0 \\ 0 & 0 & \varepsilon_3 \end{bmatrix}$$

The tensor elements generally are complex, reflecting the refractive index and absorption effects of the material. The preceding tensor expression does not include terms for uniaxial or biaxial birefringence or for optical activity, all phenomena that occur in some crystals in the absence of a perturbing imposed field. Birefringence would be evidenced by unequal values of the field-independent terms along the tensor diagonal. Optical activity, also called magnetic optical rotation, contributes to field-independent off-diagonal terms. The symmetry of these field-independent terms is even about the tensor diagonal, in contrast to the behavior of the field-dependent contributions. Here, all materials will be assumed free of these effects unless explicitly stated.

Elements of the tensor that are zero above may be nonzero when static magnetic fields are present along directions that do not correspond to cubic crystal axes or when RF magnetic fields are superimposed. The selection of tensor elements that is effected by superimposed RF magnetic fields depends on the magnetic polarization of those fields. In general, the RF magnetic fields of spin waves have components both transverse and longitudinal to the propagation direction (although none are parallel to the static saturating field). To first order in the RF magnetic fields, the elements of the tensor that are populated above do not change. Each remaining term becomes nonzero: the (1, 3) element (first row, third column) becomes $-Km_y + G_{44}M_zm_x$; the (2, 3) element becomes $+Km_x + G_{44}M_zm_y$; the (3, 1) element becomes $+Km_y + G_{44}M_zm_x$; and the (3, 2) element becomes $-Km_x + G_{44}M_zm_y$ (see Ref. 8).

The microscopic model of the source of the permittivity tensor's off-diagonal elements comprises several different processes. Each of them relies on a difference in the interaction of a particular electronic transition with the RCP and LCP polarization components of the light. The two transition types, discussed next, are illustrated in Fig. 6.16.

One process, which gives paramagnetic response, derives from an electronic transition in which a single excited state (or energy degenerate pair of excited states) is accessed from a ground state split by the Zeeman effect. This occurs in materials for which the spin–orbit coupling is strong. (The Zeeman effect is phenomenological; application of a magnetic field splits the energy of a transition into two levels, one for RCP and one for LCP light. Classically, this dependence can be calculated by incorporating the Lorentz force into the equation of motion for the electron, with the result that the natural resonance frequency in the expression for the dielectric constant is either augmented or diminished by the Larmor precession frequency, $\omega_L = e\mu_0|\mathbf{H}|/2m_e$, wherein e is the electronic charge, μ_0 is the free-space permeability, and m_e is the electron mass.) The energy difference between states split by the Zeeman effect is actually very small, and its influence on off-diagonal tensor elements, or equivalently on the difference in the refractive index for RCP and LCP light via shifts in their respective resonant frequencies, is dominated by the following quantum-mechanical effect.

Transitions to the excited state can occur for RCP light only when the total magnetic quantum number (spin and orbital quantum numbers summed) is incremented by 1 (the

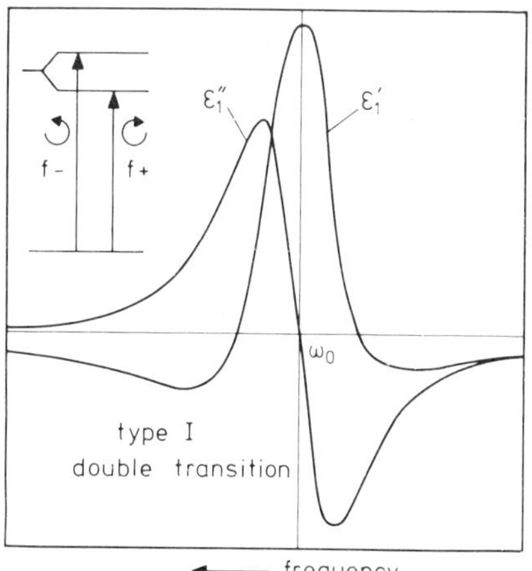

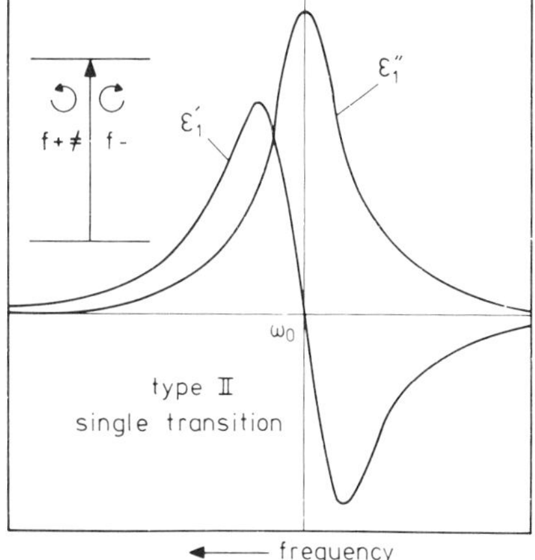

Figure 6.16 The calculated spectra of ε_1' and ε_1'' for a so-called double transition with a "diamagnetic" line shape (type I) and for a so-called single transition with a "paramagnetic" line shape (type II). (From S. Wittekoek et al., "Magneto-Optic Spectra and the Dielectric Tensor Elements of Bismuth-Substituted Iron Garnets at Photon Energies between 2.2–5.2 eV," *Phys. Rev. B*, Vol. 12, No. 7, 1975, p. 2783. Published by the American Physical Society. Copyright 1971, used by permission.)

lower energy transition) and for LCP light only when it is decremented by 1 (higher energy transition); a magnetic dipole transition accompanies that of the electric dipole. (These restrictions on allowable transitions are called selection rules.) The average transition probability for each polarization depends on the number of dipoles in each of the two pertinent electronic ground states, that is, the number having the appropriate magnetic dipole state to access the excited state. That number depends on two factors, both of which influence the magnetization of the dipole. The difference in energy between the two Zeeman-split states causes their respective populations to differ, as required by the Boltzmann thermal distribution function at temperatures above absolute zero. The populations also depend on the magnitude of the static internal magnetic field evidenced by the magnetization. These effects on magnetic dipole alignment were discussed earlier and apply to paramagnetic materials—hence the designation of this effect. Since it applies to the balance between two transitions of nearly equivalent energy, the paramagnetic effect exhibits a dispersive lineshape near the transition resonance. Its amplitude is determined by the imbalance and is proportional to the Verdet constant. When the temperature is lowered or the static internal field is raised enough to magnetically saturate the material, the paramagnetic effect also saturates.

The mechanism described in the previous paragraph prevails when the material is characterized by strong spin–orbit coupling. In the case of a saturated material (no further effect of applied field) or for a material in which the spin–orbit coupling is weaker, a paramagnetic response may be caused through a different mechanism. In these materials the orbital angular momentum contributes very weakly to the magnetization, and the magnetic field interacts almost exclusively with the spin. Light interacts only with the electric dipole moment, however. The spin–orbit coupling may still be strong enough to transfer the energy change effected by the magnetic field on the spin levels to the orbital energy levels where it can influence the RCP and LCP electric dipole transitions. In the particular case of ferromagnetics and ferrimagnetics, the exchange energy causes level splitting between the spin states that may exceed the magnetic field effect on the dipole transition energy by a factor of 1000. Spin–orbit coupling plays a vital role in transferring the effect of a magnetic field to the values of the off-axis elements of the permittivity tensor.

A different type of response is elicited when the material has a single ground state and the excited state is split. The magnetic field influences the electric dipole transition via strong spin–orbit coupling, weaker spin–orbit transfer of exchange spin-level splitting, or Zeeman splitting of the upper state of the transition pair. In this case the average probability of a transition is not affected by the field since electrons reside predominantly in the ground state, which is unaffected. Instead, the RCP and LCP resonant frequencies are fractionally offset. The difference in the individually dispersive responses is approximately derivative and has an absorption-type lineshape, which characterizes the off-axis permittivity tensor elements for frequencies in the vicinity of the resonance. This type of response is called diamagnetic; it applies to materials that have no intrinsic magnetic dipole moment (diamagnetic materials) and contributes to the response of saturated magnetic materials.

6.6 MAGNETO-OPTICS

Section 6.5 addressed the description of polarized light, outlined the basic physical processes that produce the phenomenological manifestations of magneto-optical interactions, and introduced the tensor formalism that connects the physical processes to empirical observations. This section describes the macroscopic phenomena according to their several nomenclatures and surveys the types of materials in which each effect is important.

Magneto-optical effects in solids can be divided into transmissive and reflective classes. Both depend on the magnitudes of the elements of the dielectric permittivity tensor. Transmissive effects also depend on the thickness of the material transited by the light.

Consider propagation of light through a medium in a direction parallel to the internal magnetization. The effects on its polarization, both in terms of phase retardation and absorption, will be determined by the dielectric permittivity tensor. The simpler of the two forms for the permittivity tensor shown in the preceding will suffice to explain the interaction. Remember that each tensor element is in general a complex value.

Inserting this tensor into the electromagnetic wave equation (obtained from Maxwell's equations) and setting the determinant of coefficients of the Cartesian components of the electric field equal to zero reveals an expression for the refractive index in terms of the tensor elements. It is quartic in n, the refractive index. For propagation parallel to the z-axis and collinear with the static magnetic field (or magnetization $\mathbf{M}$), two solutions result:

$$n_+^2 = \varepsilon_1 + j\varepsilon_2$$
$$n_-^2 = \varepsilon_1 - j\varepsilon_2$$

Substituting these values of the refractive index into the wave equation produces the respective proper, or eigen-, modes of the electric field:

$$\mathbf{E}_+ = (E_x, E_y, E_z) = E_0(1, j, 0)$$
$$\mathbf{E}_- = (E_x, E_y, E_z) = E_0(1, -j, 0)$$

where E_0 is an arbitrary constant. The convention for assigning the nomenclature right and left circularly polarized to a propagating wave in optics is converse to that in radio or microwaves. In optics, RCP is assigned to light whose electric field vector rotates clockwise on a stationary plane placed perpendicular to the propagation direction, viewed into the beam, against the propagation direction. Thus the field labeled with the plus sign is called RCP, and that labeled with the minus sign is the LCP.

The real part of the difference in refractive indices for RCP and LCP propagation through a sample is called magnetic circular birefringence (MCB). The MCB multiplied by the vacuum wave number,

$$k = |\mathbf{k}| = \frac{2\pi}{\lambda_0}$$

(where λ_0 is the vacuum wavelength), and by the sample thickness d is known as the Faraday rotation ϕ_F:

$$\phi_F = \frac{2\pi}{\lambda_0}(n_+ - n_-)d$$

where $(n_+ - n_-)(n_+ + n_-) = n_+^2 - n_-^2 = j2\varepsilon_2$, which implies $n_+ - n_- = j\varepsilon_2/n_0$, where n_0 is the average refractive index, $n_0 = (n_+ + n_-)/2$. (Here ε_2 is complex. Its imaginary part applies to MCB and the Faraday rotation.) Strictly, the Faraday designation should be reserved for materials in which the MCB is proportional to the applied magnetic field, in which case $\varepsilon_2 = KM_z = K|\chi|\mathbf{H}_z$ (here $|\chi|$ is the magnitude of the diagonal tensor component in an isotropic medium for a static magnetic field, and K is an empirical constant); in this case the equation for the Faraday rotation ϕ_F is often written $\phi_F = VH_zd$, where V is the Verdet constant. In practice, the Faraday designation is also used to describe the rotation experienced by light that transits saturated magnetic materials as well as those that have natural rotatory power (optical activity), in which the phenomenon should be distinguished as magnetic optical rotation (MOR).

Linearly polarized light incident on a material sample that imparts Faraday rotation is decomposed into equal-amplitude components of RCP and LCP light. At any point along its propagation path in the sample the two circular polarizations may be recombined, conceptually, to give a linearly polarized resultant. If the resultant polarization is rotated along a right-handed helical path from its point of entry to the sample, the material is designated dextrorotatory. For left-handed helical propagation the material is designated lavorotatory.

With the sign convention used here, the imaginary part of the off-diagonal tensor element ε_2 gives birefringence (MCB), as just shown. The real part of ε_2 imbues differential absorption to RCP and LCP propagating waves. This effect is called magnetic circular dichroism (MCD) and is responsible for Faraday ellipticity (linearly polarized light is made elliptically polarized, the rotated linear polarization axis becoming the ellipse's major axis).

A similar analysis can be performed for light propagating perpendicular to the magnetic field. There are two eigenmodes. For one the electric field is linearly polarized nearly perpendicular to the magnetic field (its associated electric displacement is exactly perpendicular). If the magnetic field is directed in the z direction and optical propagation is in the y direction, the electric field is given by

$$\mathbf{E}_{\text{perp}} = (E_x, E_y, E_z) = E_0(\varepsilon_1, \varepsilon_2, 0)$$

with the corresponding refractive index given by

$$n_{\text{perp}}^2 = \frac{\varepsilon_1^2 + \varepsilon_2^2}{\varepsilon_1}$$

The second eigenmode is linearly polarized parallel to the magnetic field:

$$\mathbf{E}_{\text{para}} = (E_x, E_y, E_z) = E_0(0, 0, 1)$$

with its corresponding refractive index given by

$$n_{\text{para}}^2 = \varepsilon_3$$

The birefringence implied by the difference in n_{perp} and n_{para} applies to linearly polarized eigenmodes and so is called magnetic linear birefringence (MLB):

$$\begin{aligned}(n_{\text{para}} - n_{\text{perp}})(n_{\text{para}} + n_{\text{perp}}) &= n_{\text{para}}^2 - n_{\text{perp}}^2 \\ &= \varepsilon_3 - \frac{\varepsilon_1^2 + \varepsilon_2^2}{\varepsilon_1}\end{aligned}$$

which implies

$$n_{\text{para}} - n_{\text{perp}} = \frac{\varepsilon_3 - \varepsilon_1 - \varepsilon_2^2\varepsilon_1}{2n_0}$$

where n_0 is the average refractive index, as before. Since ε_3 and ε_1 vary as the square of the magnetic field, whereas ε_2 varies linearly with $|\mathbf{H}|$, the birefringence has quadratic dependence on the magnetic field.

The MLB depends on the real parts of the permittivity tensor elements (recall that the MCB depends on the imaginary parts). The related quadratic effect to MLB, dependent on the imaginary parts of the tensor elements, is a second form of differential absorption called magnetic linear dichroism (MLD).

Magnetic linear dichroism is the basis in solids (crystals) for the Voigt effect, a response that relies on electronic transitions. The Voigt effect will be revisited later in the discussion of magneto-optical interactions in gases. This quadratic interaction is also often referred to, imprecisely, as the Cotton–Mouton effect. Strictly, the Cotton–Mouton effect applies to realignment of magnetic molecules in liquids; it too is quadratic in dependence on applied magnetic field.

In summary, phenomena that are linear in magnetic field strength occur when light propagation is parallel to the modulating magnetic field component, that is, MCB and MCD for the rotatory and absorptive effects, respectively. The optical normal modes for this interaction are right and left circularly polarized. Phenomena that are quadratic in magnetic field occur when light propagation is perpendicular to the magnetic field component responsible for modulation. These are the MLB and MLD. The normal modes for optical propagation in this case are linearly polarized.

The same magnetic-field-induced differences in refractive index that impart Faraday rotation via MCB and MCD are also responsible for a group of reflection phenomena (see Fig. 6.17). These fall under the designation of Kerr effects, of which there are three types. An analysis by Freiser [9] is instructive. Two significant concepts are proffered. The first concerns the eigenmodes for reflection. As are those for transmission, these eigenmodes are polarized neither circularly nor linearly completely, except at normal incidence angle and when propagating in Cartesian directions with respect to the magnetization. The second concerns the amplitude reflectivity from an interface. The same Fresnel reflection laws apply as for a common dielectric interface, except that the expressions for the refractive indices are modified according to the magneto-optical permittivity.

For external reflection (incident on the magneto-optical material surface from the vacuum or air side), the continuity of tangential electric field across an interface ensures that the angle of reflection equals the angle of incidence regardless of the relative directions of light propagation, sample orientation, and magnetic field. (The component of the propagation **k** vector tangential to the interface is always conserved.) Internal reflection presents a more difficult case. The complicated boundary conditions that accompany reflection from a magnetic material interface cause the light's polarization state in the sample to change. The refractive index, which governs the propagation direction and velocity of the reflected light according to its polarization, does not in general match

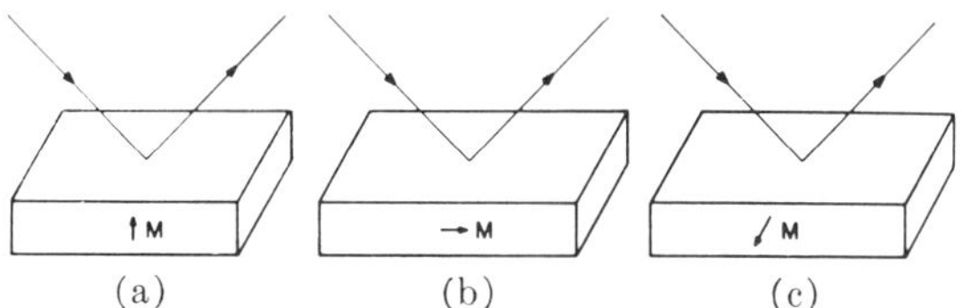

Figure 6.17 Kerr effect configurations: (*a*) polar; (*b*) longitudinal; (*c*) equatorial. (From M. J. Freiser, "A Survey of Magnetooptic Effects," *IEEE Trans. Magn.*, Vol. MAG-4, No. 2, 1968, p. 156. Published by the Institute of Electronics Engineers. Copyright 1968, used by permission.)

that which governs the incident wave. After reflection the eigenmodes are generally of mixed polarization and propagate independently away from the interface, just as they would at non-Cartesian directions in transmission.

The analysis of internal reflection is purportedly horrendous (Dillon [10] references a thesis by C. C. Robinson at MIT). Internal incidence reflection will not be described here.

The polar Kerr effect describes reflection from a sample magnetized perpendicular to its surface. For normal incidence RCP light and LCP light are transformed into each other on reflection. Transmission in this orientation occurs with circularly polarized eigenmodes; since these are possible modes of the vacuum, normal reflection can be described in terms of the n_+ and n_- previously shown. The reflection coefficient is given in terms of the circular eigenmodes by

$$r_+ = -\frac{n_+ - 1}{n_+ + 1} \quad \text{and} \quad r_- = -\frac{n_- - 1}{n_- + 1}$$

These are complex values; the difference in their phases gives the polar Kerr rotation. Linearly polarized incident light can be analyzed as the sum of equal-amplitude RCP and LCP waves. Then the Kerr rotation is one-half the differential phase retardation accumulated on reflection. Because of the simplicity of these expressions, the polar Kerr effect is most readily related to the values of the permittivity tensor elements.

The longitudinal Kerr effect describes reflection when the magnetization is both in the plane of the surface and in the plane of incidence of the light. At normal incidence the eigenmodes are linearly polarized. At nonnormal incidence elliptical polarization is introduced to linearly polarized incident light on reflection. This effect is important for detection in magnetic disk memories since domains in these materials align in the plane of the surface.

The transverse or equatorial Kerr effect is the third reflection regime. The magnetization is in the plane of the surface but perpendicular to the plane of incidence of the light. For normal incidence the transverse and longitudinal effects are degenerate. For nonnormal incidence, the transverse Kerr effect has minimal effect on light polarized perpendicular to the plane of polarization but produces a small change in reflected amplitude that depends on the direction of the magnetization (to the right or left with respect to the light's incident direction). Magnetization states for strongly magnetized materials can be analyzed without using a polarization analyzer when the equatorial Kerr effect is employed.

Freiser [9] lists the expressions for reflectivity in each of these three cases for nonnormal incidence. Deriving them is a tedious task that originates in assuring the continuity of tangential electric field components and normal magnetic field components across the interface.

Other familiar magneto-optical effects were originally observed in radiating or absorbing gases in applied magnetic fields. The names given them have in some cases also become associated with solids; in either case, the magneto-optical response is determined by magnetic field influences on electronic dipole transitions. The most important of these is the Zeeman effect. The observed traits of the effect change with viewing direction. Viewed parallel to the magnetic field, a natural resonant emission line splits into a doublet, one line up-shifted and LCP, the other down-shifted and RCP. A classical physics description such as that provided by Jenkins and White [11] suffices to explain this splitting. Viewed perpendicular to the magnetic field, a triplet of lines appears. The central line is unshifted in frequency and is linearly polarized along the direction of the magnetic field. The up-shifted and down-shifted lines are linearly polarized perpendicular to the magnetic field. These two lines imply a birefringence for linear polarization; the birefringence is proportional to the square of the magnetic field strength and is called the Voigt effect.

A description of this behavior is straightforward. As already mentioned, in a classical view the Lorentz force of a magnetic field on a moving charge is responsible. In an assemblage of oscillating electric dipoles, in the absence of a magnetic field, linear vibration will occur in evenly distributed random directions. The vibration of any particular dipole may be analyzed into a component parallel to the direction of a magnetic field (to be imposed presently) and two mutually orthogonal components both perpendicular to the field direction. The projection of the vibration along the three directions will vary from dipole to dipole. A sum over the statistical distribution is required to discover the strength of radiation of the assemblage of dipoles, with a particular polarization, in a particular direction. However, the frequency of radiation of any dipole depends only on the polarization and propagation direction of the radiation with respect to the magnetic field, not on its magnitude. Consider the component of a dipole oscillation parallel to a magnetic field (now turned on). The Lorentz force has no effect on this oscillation. Viewed against the direction of the magnetic field, this oscillation has no transverse component; no radiation is observed. Viewed perpendicular to the direction of the magnetic field, linearly polarized light will be seen, and it will have no frequency shift from the natural line center (no magnetic field). (This is the center line of the transverse triplet.) Now consider the component of a dipole oscillation that originally lies along one of the axes perpendicular to the magnetic field. The Lorentz force causes the path of the vibration to curve in the plane perpendicular to the field, counterclockwise, looking against the direction of the field for a negatively charged oscillating particle (an electron). The resulting path is rosette-like, and the resultant emitted ratiation will mimic this behavior. It is plane polarized but continually rotating. This is the behavior of MCB, in which the frequency of the LCP light exceeds that of the RCP light by twice the Larmor frequency (the frequency shift imposed by the Lorentz force; recall $\omega_L = e\mu_0|\mathbf{H}|/2m_e$). Viewed against the direction of the magnetic field, both LCP and RCP light will be seen, the LCP up-shifted by the Larmor frequency, the RCP down-shifted by the same amount. Viewed perpendicular to the field, the rosette-like oscillation will still occur at the difference frequency of twice the Larmor frequency. However, the oscillation is in the plane of observation, so the two fields will appear linearly polarized perpendicular to the field. These two lines form the satellites of the transverse triplet, evenly spaced in frequency on either side, and are responsible for the Voigt effect.

Similar effects appear in absorption; these are called the inverse Zeeman effect.

6.7 REAL MATERIALS

Materials of importance for magneto-optical devices fall into two basic categories, with some variations and exceptions. The first of these contains materials that are reasonably transparent in at least some portion of the optical spectrum and that have large Verdet constants (when not magnetically saturated) or large values of Faraday rotation (when saturated) or, occasionally, large values of Voigt rotation (which is usually called the Cotton–Mouton effect) in the same wavelength region. The second category contains materials that have high optical reflectivity in at least some segment of the optical spectrum, where they also introduce significant Kerr rotation to reflected incident light.

6.7.1 Transmissive Magneto-optical Materials

In transmissive magneto-optical materials, both optical absorption and the magneto-optical effects are generally strongest for optical frequencies (wavelengths) in the vicinity of resonances; longer incident wavelengths are generally accompanied by weaker magneto-optical effects and by weaker absorption (greater transparency). One useful figure of merit is the ratio of Faraday rotation to optical absorption, $Q = \phi_F/\alpha$; this ratio is strongly wavelength dependent near absorption bands in crystalline materials. (Unless otherwise indicated, attention will be confined henceforth to Faraday rotation, the most commonly used transmissive effect.)

Magneto-optically important transmitting materials are of two types: glass doped with paramagnetic or diamagnetic atomic species and crystalline (or polycrystalline) materials, usually garnets. (Some crystals of the ferrite family were investigated previously, but the properties of these materials render them noncompetitive with garnets.) A brief look at glass materials will be indulged first; lower figures of merit, for which a low Verdet constant or Faraday rotation, rather than high absorptivity, is responsible limit their applications somewhat.

A representative paramagnetic glass is terbium-doped borosilicate [12]. Its Verdet constant (0.0058 deg/cm-Oe at 531 nm) is about 200 times that of quartz. (Koshizuka and Okuda indicate that either terbium-doped glass or terbium–aluminum garnet may have a Verdet constant as high as 0.015 deg/cm-Oe at 800 nm. See Ref. 13.) The terbium exists in its Tb^{3+} ionization state in the glass. Its paramagnetic response is evinced by a susceptibility curve that follows $\chi = C/T$ for temperatures of 5–300 K. Judging from data included in the Hoya Corporation magneto-optical products catalogue, which lists Faraday effect isolators and rotators that use paramagnetic and diamagnetic glass, the optical absorption of these glass materials is low, and their refractive indices are high. (Absorption is less than 0.25 dB/cm, and the refractive index is near 3 at a wavelength of 633 nm.) Paramagnetic glass has higher Faraday rotation by a factor of almost 3 but is temperature sensitive, whereas diamagnetic glass is not.

A discussion of garnets, which exhibit Faraday rotation, is incomplete without inclusion of the effects that the parameters and constraints of the growing process impress on their properties. First, the magnetic, optical, and magneto-optical properties that may be important to control are listed. The magnetic properties comprise the magnetic uniaxial and crystalline anisotropy constants K_1 and K_μ and the orientation of the anisotropy; the saturation magnetization $\mathbf{M}_s$; the Curie temperature, the compensation temperature if the material has one, and the rate of variation of the susceptibility in the vicinity of these temperatures; the Bloch wall energy; and the field required to saturate the material in consideration of its saturation magnetization and the magnitude and direction of its anisotropy. Optical properties are optical absorption and its wavelength dependence; optical birefringence; scattering from surface pits; and inclusion defects. Magneto-optical parameters include the Verdet constant and its saturation value, which corresponds to the maximum Faraday rotation, and its wavelength dependence.

Bulk samples of magneto-optical materials may be pulled by the Czochralski method or grown by the float-zone technique. Relatively pure bulk crystals can be expected, with only minute contamination by the crucible (platinum). Devices that can make use of these crystals are relatively large, however, and to accommodate the motivations of laser diode and optical fiber compactness and integrated optical configurations, growing garnets as thin films has been investigated. Viable growth techniques include liquid phase epitaxy (LPE) from fluxes (currently most productive), RF sputtering, and vapor deposition. These are sometimes attended by annealing in a hydrogen, oxygen, or inert atmosphere. The problems encountered in growing garnet films will be discussed briefly here.

To create a garnet that has the desired properties generally entails beginning with a simple, well-known structure and then modifying it by substituting elements that have similar size and chemical proclivities to those of the basic constituents in the crystal. Because of the widely varying sizes of ions that share chemical valences, the degree to which one element substitutes for another may be limited by the integrity of the crystal structure. Examples will bc given.

To grow substituted garnets as thin films by LPE, a flux must be found that will hold the appropriate fractions of film constituent atoms in solution and render them to the film in the desired ratio; hopefully the flux material proves reticent to enter the film itself. Among variations in the growth parameters that influence the composition of an LPE film are the fractional quantities of the various elements dissolved in the flux, the temperature of the flux (which determines the degree of supersaturation of the solvents) during film growth, and the nature of agitation, or mixing, of the flux during the growth process.

The films must be grown on a substrate. The substrate must not melt in the flux at the growth temperature. It must have the same crystal structure and lattice constant as the expected film material to prevent stress effects from arising. It should have nearly the same coefficient of expansion so that lattice constants that match at room temperature will also match at growth temperature. For optical waveguide applications, the refractive index of the substrate should be less than that of the film. Generally, the substrate material should be nonmagnetic. It should take a smooth, flat polish while retaining good crystal structure on the surface.

The crystal type that serves as the base structure for almost all significant magnetic and magneto-optical garnets is a rare earth iron garnet (recall Fig. 6.6). (Trivalent iron, a transition metal ion, has a large intrinsic magnetic dipole moment, about 5 Bohr magnetons per ion for temperatures from absolute zero to near 300 K.) Yttrium iron garnet (often denoted YIG), whose chemical formula is $Y_3Fe_5O_{12}$, is representative of this general class of garnets having the composition $R_3A_2B_3O_{12}$. The crystal is cubic. It has three different types of cation sites. The R sites are generally occupied by rare earth ions such as yttrium, lutetium, or gadolinium (large ions). Each of the three trivalent ions located on R sites is surrounded by an irregular dodecahedron (triangular faces) of eight oxygen ions. Each of the two A ions is surrounded by an octahedron (triangular faces) of six oxygen ions, and each of the three B ions by a tetrahedron (also triangular faces) of four oxygens. These sites, like the R sites, are generally and desirably occupied by trivalent ions. (The Fe^{2+} has a somewhat smaller dipole moment than Fe^{3+}; more important here, it contributes strongly to optical absorption.) In YIG both the A and B sites are occupied by iron ions. The block of material defined by the 20 ions of the YIG formula has an axis of rotational symmetry through a cube diagonal. Eight of these formula cubes, with their diagonals at various orientations, are stacked together in a larger cube to form the crystallographic unit cell.

In the garnet structure, magnetic ions on the A and B sites are antiferromagnetically coupled. If identical ionic species occupy these sites, as in YIG, their unequal number renders the crystal ferrimagnetic. In YIG the R site is occupied by a nonmagnetic ion, yttrium (no intrinsic magnetic dipole moment), and the entire magnetic behavior is determined by the iron ions. Since the tetrahedrally coordinated B sites outnumber the octahedrally coordinated A sites 3 to 2, the net magnetic moment aligns with the B sites in this crystal. The R site may be occupied by magnetic ions in other similar crystals, however. In this case their dipole moments are influenced to lie canted with respect to the octahedral (A) sites but so that their net contribution aligns with that of the octahedrally coordinated A site ions. The total magnetic dipole moment in the crystal is then determined by summing the contributions of the ions on the three site types. This value generally varies with temperature; the dipole moments of the ions occupying each site type decrease with increasing temperature (all three contributions go to zero at the Curie temperature) but with differing behaviors, depending on the ionic species and the influence its site location in the crystal asserts on the magnetic contribution of its orbital magnetic moment. Many garnets that have magnetic ions on all three sites exhibit a total magnetic moment that passes through zero at a temperature well below the Curie temperature. This is called the compensation point, an important feature for most magneto-optical memory media, as will be discussed later (recall Fig. 6.7).

Yttrium iron garnet is a representative garnet having a simple structure that exhibits good, if not optimum, values of many parameters. Continued discussion of this crystal will illuminate the capabilities, variations, and limitations of the garnet class.

As a thin film YIG is most often grown by LPE methods, although RF-sputtered [14–16], hydrothermal process [17], and chemical vapor deposition (CVD) [18,19] films have been grown. The most common substrate is gadolinium gallium garnet (GGG), $Gd_3Ga_5O_{12}$, which presents a good lattice match along with good thermal, surface, and optical refractive index characteristics. When grown by LPE, a flux combining lead oxide with boron oxide (PbO and B_2O_3) has historically been used, although preventing the incorporation of lead in films grown from this flux is problematic for optical absorption.

Magnetically, YIG has a nominal saturation magnetization value of $4\pi M_s = 1750$ Oe. The crystalline anisotropy field is small, on the order of $2K_1/M_s = 80$ Oe. The crystalline anisotropy constant, K_1, in ergs/cm^3-G, can be calculated through understanding that $M_s = 1750/4\pi$. (In YIG the saturation field is 1750 Oe; $K_1 = 5570$ ergs/cm^3). The anisotropy constant is positive, which imparts in-plane anisotropy to films grown on (111)-oriented substrates. Its Curie temperature is near 600 K, and it has no compensation point in its pure state. Optically, it is very absorptive throughout the visible wavelength region, with peaks broadly localized around 600 and 910 nm. The effects of octahedral and tetrahedral iron crystal field transitions responsible for the absorption dwindle substantially for wavelengths beyond 1100 nm. Faraday rotation is appreciable in various wavelength regions, exceeding 25,000 deg/cm at 430 nm and at 530 nm and nearing 10,000 deg/cm at 680 nm. Unfortunately, in these wavelength regions optical absorption is mostly oppressively high and consequently produces a poor figure of merit. Windows of higher Q occur at 560, 780, and 1100 nm for visible and near-visible wavelengths of interest, but near-IR wavelengths between 1100 and 1500 nm must be considered for best performance.

Another material parameter that is of paramount importance for a particular device application and so should be mentioned here is the ferrimagnetic resonance line width discussed earlier. Pure YIG leads all competitors in exhibiting a narrow line width (indicative of low microwave absorption for spin waves, of which magnetostatic waves are a special case), with values as low as 0.15 Oe occasionally demonstrated at an RF field frequency near 10 GHz and values below 0.8 Oe routinely available. Retaining a low line width value while changing the constituents of the garnet films to enhance other parameters has proved challenging to materials specialists.

A garnet system that vies with YIG in some applications is gadolinium iron garnet, GdIG. (See Ref. 5.) Gadolinium occupies the dodecahedrally coordinated R site in the garnet lattice. Unlike yttrium, gadolinium is paramagnetic, and in the presence of the strongly magnetizing iron garnet crystal structure it displays a large intrinsic magnetic dipole moment, 7 Bohr magnetons per ion at 0 K. This value diminishes rapidly with increasing temperature to about 1.5 Bohr magnetons at room temperature. Combined with the slowly diminishing iron ion moments on the other two lattice sites, a compensation point appears at about room temperature (thus the saturation magnetization is near zero). The Faraday rotation at 546 nm at room temperature for GdIG is 5000 deg/cm.

The problem of substituents will now be superficially addressed. Good reviews of their effects can be found in Refs. 20 and 21. By including various rare earth and transition metal elements in the flux solution, garnets can be grown that incorporate them and whose properties vary from those of YIG or GdIG appreciably. (Substituting the bismuth ion into the garnet structure acts to greatest benefit in many applications.) By intelligent combination of several substituents, many properties of garnet materials can be adjusted simultaneously to fit an application. The saturation magnetization can be lowered. The Faraday rotation can be enhanced. Compensation and Curie temperatures can be adjusted. The anisotropy constant can sometimes be controlled. Of course, efforts to incorporate substituent ions in garnet films require adaptations in flux composition and growth condition parameters and sometimes produce unexpected or undesired changes in the other structurally related parameters as well.

The list of substituents is rather long; it begins with the rare earth elements and seems to effectively end with bismuth, as will be discussed. The radii of the trivalent ions of the rare earths contract fairly smoothly with increasing atomic number, from 1.11 Å for cerium to 0.94 Å for ytterbium (Ref. 22). (Lutetium is also quite small.) Generally one of these is substituted into a garnet structure either to provide a strong magnetic dipole moment (as, e.g., praseodymium or neodymium will contribute) or to compensate by its small size for the large size of another substituent ion (lutetium is commonly employed in this capacity). These ions generally occupy the dodecahedral site in the garnet structure. Cobalt, a transition metal neighboring iron in the periodic table, has been investigated recently for its efficacy in enhancing Faraday rotation in iron garnets. It substitutes for iron on tetrahedrally coordinated sites. Various other elements Ca, Mg, Va, Ga, Ge, Si, and Al have been incorporated into garnets to fine tune the magnetic and magneto-optical properties, provide charge compensation, or stabilize the crystal structure by means of their ionic radii in specific materials development approaches. Bismuth, and to some extent for the same purposes lead, can be substituted into the dodecahedrally coordinated R site, displacing yttrium in YIG or gadolinium in GdIG. Through a complicated interaction with the garnet lattice, these produce the greatest enhancement of the Faraday rotation.

Saturation magnetization at room temperature can be changed by two substitutional approaches. A diamagnetic ion can be included that displaces iron on the tetrahedral site. Gallium is often employed, with aluminum an alternative trivalent choice; germanium or silicon is occasionally used sparingly to give charge compensation. As the number of formula units of this ion is increased from zero to 1, the antiferromagnetically coupled iron ions on the remaining tetrahedrally coordinated B sites and the octahedrally coordinated A sites become more nearly balanced in number. Consequently, the net magnetic moment of the crystal (the difference of the iron contributions when the dodecahedrally coordinated ion is nonmagnetic) approaches zero as the substituent ion concentration nears one formula unit. (Complete compensation is still temperature dependent since the temperature dependence of each iron dipole moment contribution depends on its lattice site.) Conversely, an ion having a permanent intrinsic magnetic dipole moment that will occupy the dodecahedrally coordinated lattice site can be used, provided it aligns parallel to the octahedrally coordinated iron sites. A prominent example of this type of compensation was discussed in the preceding in describing the traits of GdIG. Praseodymium, neodymium, and dysprosium are among other paramagnetic ions used as R site substituents, but not for magnetization compensation. The alignment of their dipoles, unlike that of gadolinium, is parallel to the tetrahedrally coordinated B site iron ions, which outnumber the A site iron ions. These ions therefore increase the room temperature saturation magnetization somewhat.

Considerable effort has been devoted over the years to increasing the magneto-optical effects in garnets: Faraday and Kerr rotation (transmissive and reflective effects, respectively). Substituting paramagnetic rare earth ions in iron garnets is a reasonably useful approach. (The larger of these rare earth ions—those of La, Pr, and Nd, which have lower atomic numbers—cannot form pure iron garnets with the formula $R_3Fe_5O_{12}$

because their large ions distort the would-be crystal structure excessively.) (See Ref. 21.) Praseodymium, Nd, Sm, Dy, and Eu all have been employed with some degree of success. They enter garnets on the dodecahedrally coordinated sites, displacing yttrium in YIG or gadolinium in GdIG. Each of these ions has its own intrinsic magnetic dipole moment. Optical transitions in materials containing these ions therefore exhibit magnetic field dependence, which appears phenomenologically as Faraday, or Kerr, rotation at longer wavelengths. As a starting point for investigating the influence of praseodymium on magneto-optical effects in garnets, the interested reader should consult Ref. 23. Some additionally useful discussion that pertains to the class of rare earth substituted garnets and examines the mechanisms by which their properties are produced is contained in Ref. 21.

Although substituted rare earth ions are effective, the discovery of the effects on magneto-optical parameters of substitution of one of two particular nonmagnetic ions relegated them to a subordinate role. The persuasive ions are those of bismuth and lead. They substitute on the dodecahedrally coordinated garnet sites. The bismuth ion is trivalent, while the lead ion is divalent. (This observation may in part explain why lead incorporation exacerbates optical absorption to the extent that while it enhances Faraday rotation strongly, the quality factor Q for lead-substituted garnets is so poor that it is regarded as an obstinate impurity for most Faraday rotation device applications; Pb is the predominant solvent in the growth flux and thus is difficult to eliminate.) Substituted in YIG, either of these ions changes the Faraday rotation dramatically. At room temperature and at 633 nm wavelength, Ref. 20 discloses that the Faraday rotation changes at a rate of -2.06×10^4 deg/cm per formula unit for bismuth or -1.84×10^4 deg/cm per formula unit for lead. It should be noted that the Faraday rotation of unsubstituted YIG is positive, on the order of 200 deg/cm for wavelengths in the near-IR. Substitution of bismuth (or lead) drives the Faraday rotation negative with increasing concentrations; it passes through zero at a small level of substitution and then grows to high negative values as the substitution level nears or exceeds one formula unit. Bismuth makes a similar contribution to Faraday rotation when substituted into GdIG, but lead is not nearly as effective in this host. The details of bismuth-substituted YIG (Bi:YIG) and its variants will be outlined next.

Trivalent bismuth is a relatively large ion, appreciably larger than the yttrium ion it displaces. Bismuth may be substituted to about 1.88 formula units (about 60% of the three R sites) in YIG when grown as a bulk crystal, but to substitute bismuth to greater than about one formula unit in an epitaxial film requires size compensation to prevent stress cracking as the worst case and stress anisotropy and substrate lattice mismatch in any event. To avoid stress anisotropy in an epitaxial film, mismatch of film and substrate lattice constants must not exceed about 1 pm (0.001 nm). Commonly, lutetium is incorporated to relieve the stress; it too occupies the dodecahedrally coordinated R site. In fact, $Bi_xLu_{3-x}IG$ has been grown, excluding the yttrium component altogether. Alternatively, lattice match with the substrate can be achieved by altering the substrate composition. Gadolinium gallium garnet is the time-tested substrate material. Its lattice constant matches that of YIG almost perfectly at 1.237 nm. Replacing some or all of the gadolinium on the R site in GGG with samarium or neodymium increases its lattice constant: NdGG has a lattice constant of 1.249 nm. Another successful approach is to replace some of the gallium ions with zirconium and magnesium ions. Zirconium is tetravalent while magnesium is divalent, so charge balance is retained. Although this technique produces a larger lattice constant, its value seems to drift as the crystal is grown, perhaps through unequal incorporation of the two ions. Incorporation of an appropriate amount of divalent calcium on the dodecahedrally coordinated site stabilizes the lattice constant; to retain the charge balance, the magnesium content must be reduced commensurately. Varying the amount of zirconium and the relative amounts of calcium and magnesium allows substrates to be grown that have lattice constants throughout the range 1.245–1.25241 nm (Ref. 5).

A brief digression will elucidate the magnetic properties of rare earth gallium garnets vis-à-vis rare earth iron garnets. The rare earth ions contribute paramagnetically to the magnetic properties of materials at room temperature. In the iron garnet crystal, their environment exhibits a strong magnetization because of the ferrimagnetic alignment of the A site and B site iron ions. They align in the magnetization (except for gadolinium, their magnetic moments lie parallel to the magnetization of the tetrahedrally coordinated trio of iron ions) and contribute to the magnetic character of the crystal, as was discussed for GdIG. Now consider the effects of replacing iron ions with trivalent gallium ions. Gallium ions enter the crystal, in small concentrations, on tetrahedrally coordinated sites. (This lowers the Curie temperature by tens of degrees kelvin and raises the compensation temperature dramatically; it disappears in $Gd_3Fe_{5-y}Ga_yO_{12}$, having surpassed the Curie temperature, when y exceeds about 0.6.) The saturation magnetization of the crystalline environment decreases as the tetrahedrally coordinated iron is replaced. Beyond some gallium concentration, octahedrally coordinated sites are also commandeered, and eventually GGG results from the complete absence of iron ions in the growth melt and the resulting crystal. Paramagnetic ions that are substituted for gadolinium on the R site or replace gadolinium entirely no longer adopt a ferrimagnetic posture or contribute to material magnetization.

Besides the stress-induced anisotropy caused predominantly by film-to-substrate lattice mismatch via magnetostriction, Bi:YIG exhibits magnetic anisotropy induced by other mechanisms. As in YIG itself and most magnetic materials, the bonds that establish the crystal structure influence the orbital angular momentum and through it the magnetization direction according to the degree of spin–orbit coupling in the magnetic ions. The resulting magnetic alignment is called crystalline anisotropy. In garnets, it shows cubic symmetry. Bismuth and yttrium have no intrinsic magnetic moment, so the crystalline anisotropy of Bi:YIG depends only on the spin–orbit coupling of the iron ions, which is weak.

Since YIG is very nearly a perfectly cubic crystal, its uniaxial anisotropy is also very small. The Bi:YIG has a rather large positive uniaxial anisotropy, however, induced in the iron ions by the noncubic (irregular) distribution of the bismuth ions. This is also commonly referred to as growth-induced anisotropy, since the nonuniform distribution of bismuth ions on the R sites, which results from the growth process, is responsible for the distorted bonding orbitals of the iron ions. The possibility of moderating this anisotropy by the inclusion of praseodymium ions, the sole rare earth ion that induces negative anisotropy in garnets, has been explored in Ref. 23. It is apparently very sensitive to constituent concentrations in the growth flux, and to the growth conditions (temperature and sample rotation rate). The effects of substituting particular diamagnetic ions on the tetrahedrally coordinated and octahedrally coordinated sites generally occupied by iron have been studied, showing prospects for controlling anisotropy by this means. (See Ref. 24.) Of course, these diamagnetic substitutions replace iron, the ion that is responsible for the desirably strong magnetic and magneto-optical properties of Bi–YIG.

This uniaxial anisotropy of Bi–YIG may prove troublesome to the reliable manufacture of devices, not so much because of its magnitude as because of the difficulty in accurately controlling, or reproducing, the anisotropy value in the growth process.

Saturation magnetization is not appreciably affected by bismuth substitution in YIG. It increases slightly as bismuth formula fraction increases.

Optically, Bi:YIG has an absorption edge in the near IR, like YIG, but somewhat beyond 1100 nm. Doormann et al. present a conscientious measurement of the refractive index and absorption coefficient of LPE films of YIG substituted with formula fractions of bismuth ranging up to 1.42 [25]. Although their films have lead concentrations up to 0.08 formula fraction, which may contribute significantly to absorption, the indicated trend is that the absorption edge for Bi:YIG films extends slightly farther into the near IR as bismuth concentration is increased. A measurement on one sample in which the

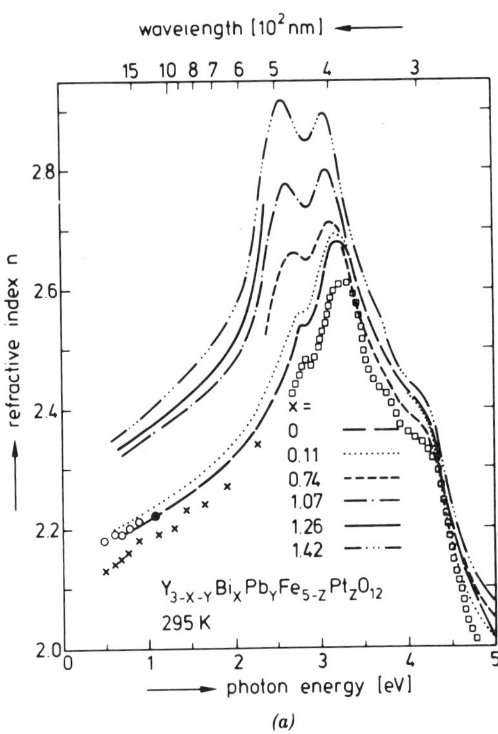

(a)

Figure 6.18 (*a*) Refractive index spectra of Bi:YIG films as measured by thin-film interference for $\lambda \gtrsim 0.52\ \mu m$ and by ellipsometry for $\lambda \lesssim 0.52\ \mu m$. Crosses refer to interference spectrometry on single-crystal wafers, open circles to the refraction of a single-crystal YIG prism, filled circles to *m*-line spectroscopy, and open squares to a Syton polished thick water. For references to original experiments, consult V. Doormann et al., *Appl. Phys.*, Vol. A34, p. 223, 1984. (From V. Doormann et al., "Measurement of the Refractive Index and Optical Absorption Spectra of Epitaxial Bismuth Substituted Yttrium Iron Garnet Films at uv to Near-ir Wavelengths," *Appl. Phys. A Solids Surf.*, Vol. A34, No. 4, 1984, p. 227. Published by Springer-Verlag. Copyright 1984, used by permission.) (*b*) Optical absorption spectra of Bi:YIG materials as measured by thin-film interference for $\lambda \gtrsim 0.52\ \mu m$ and by ellipsometry for $\lambda \lesssim 0.52\ \mu m$. The border of applicability of the two spectroscopic techniques is indicated by the horizontal broken lines. Good overlap has been confirmed by ellipsometry on the bulk YIG sample. (From V. Doormann, "Measurement of the Refractive Index and Optical Absorption Spectra of Epitaxial Bismuth Substituted Yttrium Iron Garnet Films at uv to Near-ir Wavelengths," *Appl. Phys. A Solids Surf.*, Vol. A34, No. 4, 1984, p. 228. Published by Springer-Verlag. Copyright 1984, used by permission.) (*c*) Faraday rotation versus wavelength for (1) $Y_3Fe_5O_{12}$; (2) $Bi_{0.6}Tm_{2.4}Fe_{3.9}Ga_{1.1}O_{12}$ (LPE); (3) $Bi_{0.8}Yb_{2.2}Fe_{3.9}Ga_{1.1}O_{12}$ (LPE); (4) $Bi_{1.0}Tm_{2.0}Fe_{3.9}Ga_{1.1}O_{12}$ (LPE); (5) $Bi_{0.9}Sm_{2.1}Fe_{3.9}Ga_{1.1}O_{12}$; and (6) $Bi_{1.1}Eu_{1.9}Fe_{4.3}Ga_{0.7}O_{12}$. (From G. B. Scott and D. E. Lacklison, "Magnetooptic Properties and Applications of Bismuth Substituted Iron Garnets," *IEEE Trans. Magn.*, Vol. MAG-12, No. 4, 1976, p. 292. Published by the Institute of Electrical and Electronics Engineers, Copyright 1976, used by permission.)

bismuth content is quite high, 1.42 formula fraction, shows remanent absorption of about 50 cm^{-1} at a wavelength of 1900 nm. For wavelengths shorter than 1100 nm, bismuth increases the absorption above that of YIG in a complicated way, introducing or enhancing several absorption resonances, most of them intrinsic to iron ion transitions. These transitions are also the root of the Faraday rotation enhancement and appear as the result of a magnetically coupled, second-order effect called super exchange. Measurements of refractive index, optical absorption, and Faraday rotation for various bismuth-substituted garnets are reproduced in Fig. 6.18.

Superexchange is the mechanism by which the magnetically dependent energy levels in one ion influence those of a second ion that is its next-nearest neighbor. The superexchange process, like the simple exchange process, ascribes to an electronic state a magnetic dipole whose strength depends on the spin orientation of the occupying electron. This is equivalent to magnetically induced energy-level splitting of an electronic state. Electronic states involved in superexchange may interact through partial overlap of the bonding orbitals of the affected magnetic ions in the vicinity of the intermediary ion; or

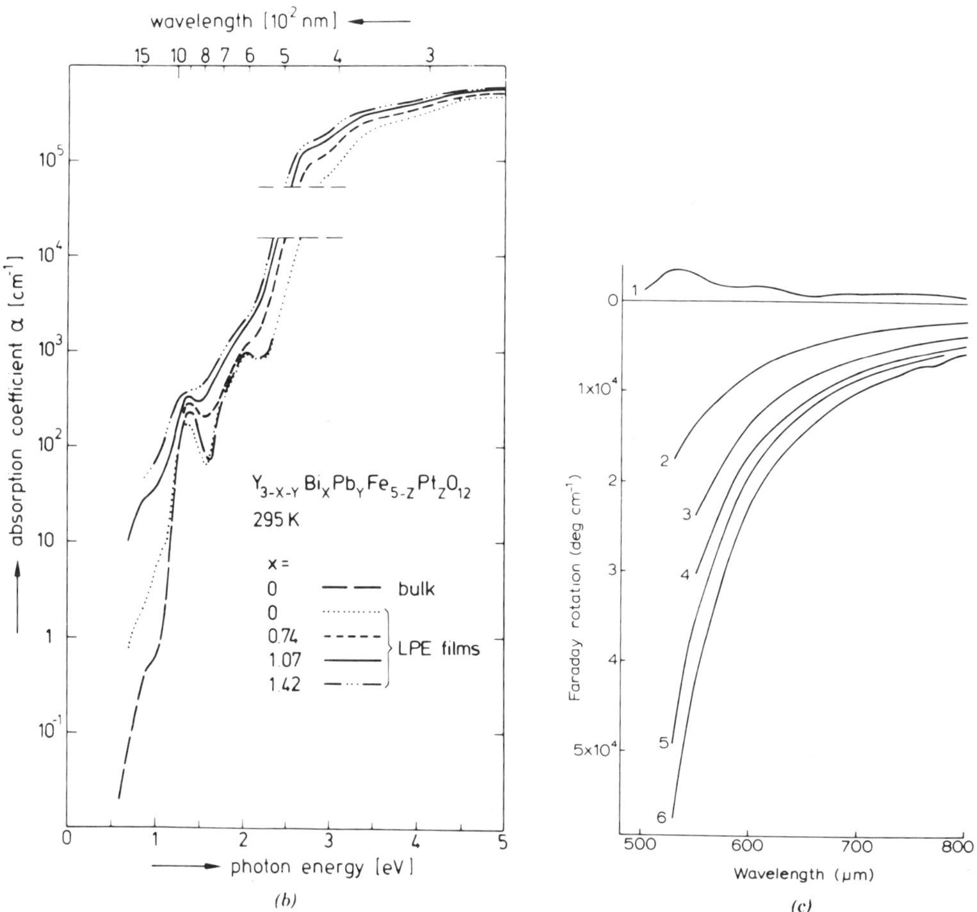

Figure 6.18 (Continued)

they may interact more obliquely by their individual influence on the electron spins in the bonding orbitals of the intermediary ion.

In the garnet lattice, superexchange applies to interaction between the energy levels of two iron ions located on differing site types (one on an A site and one on a B site) that are separated by and bonded to an oxygen ion. The presence of bismuth in the lattice structure apparently enhances their magnetic coupling by distorting the angles and ion center-to-center separations along the iron ion–oxygen ion bonds and by introducing a perturbation in the electronic wave function around the oxygen ion with its own bonds to the oxygen, the participating bismuth orbitals being strongly spin–orbit coupled. (That is, bismuth increases the superexchange coupling.) In bismuth-substituted YIG, the superexchange coupling is suspected of promoting two types of transitions. (The paper by Scott and Lacklison contains possibly the most current analysis of the electronic transitions responsible for increased absorption and enhanced Faraday rotation in bismuth-substituted YIG. See Ref. 26.) These are, first, an excitation of a bi-exciton (i.e., two simultaneously created excitons in such close proximity that they interact to lower their total energy, much as molecular hydrogen forms to decrease the total energy of its atomic electrons) in a superexchange coupled electronic transition in next-nearest-neighbor iron ions and, second, a charge transfer transition between states associated with the two differing site iron ions. Each of these mechanisms is responsible for one of the two transitions whose strength increases most dramatically when bismuth is substituted in YIG, with the bi-exciton transition dominating. These transitions occur at wavelengths of 433 nm (the charge transfer transition) and 391 nm (the stronger, bi-exciton transition). Both of these transitions have a diamagnetic lineshape, indicating a split excited state. A curve that is often published (see Fig. 6.19), from work by Wittekoek et al. [27], shows the imaginary part of ε_1^2 between 250 and 550 nm and its dramatic increase near 390 nm for bismuth-substituted films compared to that for pure YIG films.

Controlling the magnitude of the optical absorption has begun to be addressed. As already mentioned, lead incorporated in YIG films contributes strongly to absorption. To eliminate lead from YIG films, fluxes free of PbO and based on Bi_2O_3 instead have been used with preliminary indications of success. In addition, divalent and to some extent tetravalent iron ions contribute strongly to optical absorption. Borrowing a technique used to suppress the optical absorption and accompanying optical damage caused by divalent iron ions in lithium niobate, researchers have incorporated divalent magnesium ions into bismuth-substituted LPE films [namely, into LPE $(BiLu)_3Fe_5O_{12}$]. See Ref. 28.

Several paragraphs earlier, the enhancement of Faraday rotation was named as the motivation for considering bismuth substitution in LPE garnet films. Although Bi:YIG demonstrates very high values of Faraday rotation, the values of Faraday rotation for equal substitutional composition in Bi:GdIG are even higher. Consequently, most of the literature that addresses bismuth substitution for enhanced Faraday rotation centers its discussion around GdIG. Bismuth substitution in YIG (or, alternatively, the combined bismuth–lutetium iron garnet system) remains important because these formulations may also prove capable of transmitting microwaves, in the guise of magnetostatic waves, with low absorption, a task for which the GdIG host is apparently less suitable. This aspect will be dealt with later in the discussion of ferrimagnetic resonance line width.

The properties of bismuth-substituted iron garnets are comprehensively presented by Hansen et al. [5], but with emphasis devoted particularly to Bi:GdIG. Two figures from their paper are reproduced here (Fig. 6.20). Figure 6.20*a* shows the variation with wavelength of the Faraday rotation for various substitutional levels of bismuth. Note that the unsubstituted GdIG has positive Faraday rotation at all wavelengths and that bismuth substitution makes it negative, passing through zero for a bismuth content of about 0.1

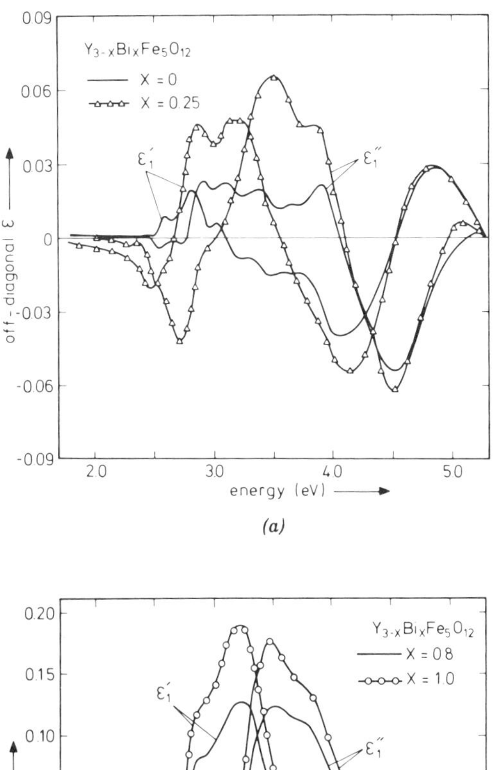

Figure 6.19 (*a*) Off-diagonal tensor elements ε_1'' and ε_1'' for $Y_{3-x}Bi_xFe_5O_{12}$ for $x = 0, 0.25$. (From S. Wittekoek et al., "Magneto-Optic Spectra and the Dielectric Tensor Elements of Bismuth-Substituted Iron Garnets at Photon Energies Between 2.2–5.2 eV," *Phys. Rev. B*, Vol. 12, No. 7, 1975, p. 2780. Published by the American Physical Society. Copyright 1975, used by permission.) (*b*) Off-diagonal tensor elements ε_1' and ε_1'' for $Y_{3-x}Bi_xFe_5O_{12}$ for $x = 0.8, 1.0$. (From S. Wittekoek et al., "Magneto-Optic Spectra and the Dielectric Tensor Elements of Bismuth-Substituted Iron Garnets at Photon Energies Between 2.2–5.2 eV," *Phys. Rev. B*, Vol. 12, No. 7, 1975, p. 2780. Published by the American Physical Society. Copyright 1975, used by permission.)

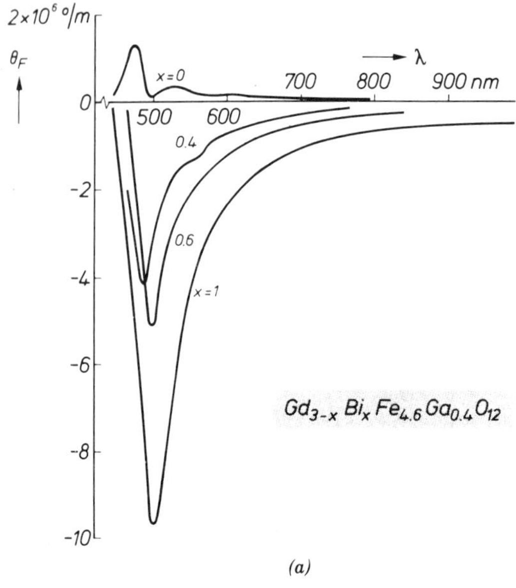

(a)

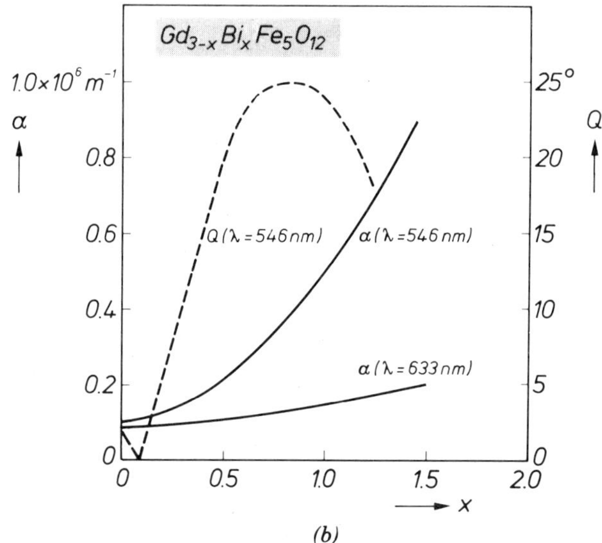

(b)

Figure 6.20 (*a*) Specific Faraday rotation θ_F in degrees per meter of $Gd_{3-x}Bi_xFe_{4.6}Ga_{0.4}O_{12}$ as a function of wavelength λ in nanometers of incident light, with the bismuth content x as parameter. (From P. Hansen et al., "Optical Switching with Bismuth-Substituted Iron Garnets," *Philips Tech. Rev.*, Vol. 41, No. 2, 1983/84, p. 39. Published by N. V. Philips. Copyright 1985, used by permission.) (*b*) Results of measurements of the figure of merit $Q = 2|\theta_F|/\alpha$ in degrees (at wavelength of 546 nm) and the absorption coefficient α in reciprocal meters (at wavelengths of 546 and 633 nm) as a function of the bismuth content x in $Gd_{3-x}Bi_xFe_5O_{12}$. The curve for Q is shown dashed. (From P. Hansen et al., "Optical Switching with Bismuth-Substituted Iron Garnets," *Philips Tech. Rev.*, Vol. 41, No. 2, 1983/84, p. 40. Published by N. V. Philips. Copyright 1985, used by permission.)

formula fraction. Unsubstituted YIG also has positive Faraday rotation that becomes negative with bismuth substitution. The peak Faraday rotation is near 530 nm in the Bi:GdIG system; for Bi:YIG it is nearer the superexchange transitions responsible for the large effect, around 400 nm. Although not directly extractable from the figure, the Faraday rotation contribution due to the bismuth incorporation depends linearly on the bismuth fraction, almost uniformly without regard to wavelength (differing slopes apply). Figure 6.20*b* shows the figure of merit Q for Bi:GdIG when the wavelength is 546 nm, as a function of bismuth content. The optimum bismuth content depends on wavelength, but it is broadly peaked, which allows for some adjustment of composition to satisfy other requirements.

Values of Faraday rotation and absorption for wavelengths in the near IR are difficult to ascertain from these curves, and only recently have measurements appeared specifically at wavelengths of lasers that emit in that range: around 1150, 1300, and 1500 nm. At 1150 nm, Ohno and his coworkers [29] discuss a theoretical model for the rate of change of Faraday rotation with bismuth content that is based on observed behavior in the range of 500–800 nm. They estimate $d\phi_F/dX = -1977.2X$ in YIG and $d\phi_F/dX = -15{,}900X$ in GdIG, where X is the formula fraction of bismuth substitution. Since the transitions that produce the large effects are almost exclusively at wavelengths on the short end of this range, the estimate seems reasonable. Experimental corroboration would be useful.

Hibiya et al. provide some data in Ref. 30. Values of Faraday rotation at 1300 nm as high as -800 deg/cm for Bi:YIG with bismuth formula fraction $X = 0.49$ and as high as -1500 deg/cm for Bi–GdIG with bismuth formula fraction $X = 0.58$ are shown in one set of figures. Elsewhere in the paper their Figure 7 shows Faraday rotation at 1300 nm of -2880 deg/cm for Bi:Gd(FeAlGa)G grown from a different melt composition that produced the bismuth content $X = 1.56$. (This film had Faraday rotation of $-12{,}800$ deg/cm at 780 nm and $-23{,}600$ deg/cm at 633 nm.) In addition, Faraday rotation is shown to increase approximately linearly with bismuth content; at 1300 nm the slope $d\phi_F/dX = -2000X$ deg/cm in Bi:GdIG and Bi:Gd(FeAlGa)G. Finally, the authors report the minimum optical absorption value measured on their films at 1300 nm to be 1.6 cm^{-1}; this film was of the Bi:Gd(FeAlGa)G variety and had Faraday rotation of -1880 deg/cm, so its quality factor $Q = 267$ deg/dB, apparently the highest value for LPE garnets measured to date.

Stress-induced (photoelastic) optical birefringence, caused by lattice constant mismatch of film and substrate, can be untenably large in bismuth-substituted garnets. Presumably, differential thermal expansion rates produce stress on cooling from the oven even with substrates that have adjusted lattice constants. This birefringence has been shown to be linearly proportional both to lattice constant mismatch [31] and to bismuth substitution concentration [32]. Birefringence greater than $\Delta n = 1 \times 10^{-3}$ was reported in each report; in addition, the analysis of Ref. 31 indicates that stress alone cannot account for the measured birefringence in some films.

The ferrimagnetic resonance line width is not often discussed in characterizations of magneto-optical garnet materials; it is a property relevant to the propagation of microwave frequency (1–30 GHz) magnetic spin waves. The interaction of magnetic spin disturbances with light is somewhat removed from the common magneto-optical applications for display, isolation, and low-frequency modulation but holds promise for future high-frequency light modulation applications.

To implement the microwave magnetic modulation, low absorption of spin waves must be assured. As discussed in Section 6.4, microwave magnetic spin waves, in particular the various types of magnetostatic waves (MSWs), require static magnetic bias fields to coalign the internal dipoles. In addition, the dispersion relation for any particular MSW type shifts in frequency as the bias field strength is varied. At any predetermined frequency, the MSW wavelength depends on the strength of the static bias magnetic field. The technique commonly used to ascertain spin wave absorption places a microwave cavity supplied with microwaves from a klystron source (usually at a frequency of about 9.3 GHz) between the poles of a strong electromagnet. (This technique was described in a preceding section, but further details may be of interest. Recall the apparatus of Fig. 6.15.) A carefully prepared rectangular or circular sample of garnet film (or a sphere of bulk material when bulk properties are of interest) is mounted on a nonmagnetic rod in the microwave cavity and oriented so that the static magnetic field direction is appropriate for the type of MSW to be generated (volume waves are generally invoked). The magnitude of the bias field is then slowly swept so that the wavelength of the MSW in the sample varies until it fills the transverse dimension of the sample with a near integral number of half-wavelengths. Near this resonant condition, the energy of the MSW in the garnet sample increases, as does the energy extracted from the microwave cavity via absorption in the sample. This depletion of the cavity RF field can be measured by a microwave probe. The range of MSW wavelengths over which the sample absorbs appreciable energy maps linearly onto the swept bias magnetic field. This range is narrow, as for any Lorentzian resonance, when the absorption is low. Since the range of the bias magnetic field over which the absorption occurs is directly observable, it has become the custom to report the line width in terms of the swept field range ΔH, in oersteds, at 9.3 GHz. Translation of the ferrimagnetic resonance (FMR) line width ΔH into more traditional terms can be accomplished using $\alpha_{\text{MSW}} = 76\ \Delta H$ decibels per microsecond. The dispersion relation must be consulted for the MSW velocity to convert to decibels per centimeter. Sample thickness and ground plane location play roles in determining the shape of the dispersion curve and thus the dependence of the MSW velocity on frequency; values near 50 mm/μs are typical in the range where MSWs are useful for optical interaction, which gives a rough approximation of absorption as $\alpha_{\text{MSW}} = 1.5\,\Delta H$ decibels per millimeter. It can be appreciated that garnet films that have values of $\Delta H > 2$ Oe are unimportant in this application.

At room temperature, the FMR line width in epitaxial YIG films has been measured as low as about 0.15 Oe, and values around 0.5 Oe are commonplace. This variation depends on a number of factors deriving from the film growth process: the supersaturation temperature of the flux; the base solvent composition and balance of constituents in the flux; the agitation rate of the substrate; the lattice match between film and substrate; and the film surface quality, which depends on the care with which remnant flux is spun or etched off following film growth. No other material can be grown as a thin film with such a low line width, including other garnets as well as ferrites. Yttrium iron garnet films have been grown by methods other than LPE in attempts to find processes more amenable to production, namely sputtering and vapor deposition. Sputtering tends to produce films composed of crystallites in which the line widths are considerably larger, although Faraday rotation is unaffected.

Gadolinium iron garnet has been discussed previously since it assumes the highest values of Faraday rotation, particularly when substituted with bismuth. It has a considerably broader line width than YIG, as do all the garnets that incorporate ions that have intrinsic magnetic dipole moments (Pr, Nd, etc.). The magnetic dipole moment of this type of ion is strongly coupled both to the lattice, via spin–orbit coupling, and to the magnetic dipoles of the neighboring iron ions, by exchange interaction. In addition, the rare earth ions have unusually short relaxation times. In combination, the consequence of these two attributes is that even minute amounts of rare earth ions in garnets increase the spin wave absorption immensely.

Introducing substituents (e.g., Bi) into garnet films, into YIG particularly, can act to broaden the ferrimagnetic resonance line width even when nonmagnetic ions are introduced. Consider an ion that preferentially substitutes for yttrium ions on the R site. Assume, as a best case, that exactly one substituent ion per formula unit is introduced. At least three conditions exist that can produce local distortion in the garnet crystal structure. Substituent ions will not all reside on the sites that they prefer due to the statistics of the crystallization process. Likewise, on a statistical basis, not every unit cell will have its exact apportionment of substituent ions (recall that the unit cell in YIG consists of eight formula units); some will have none, others more than one. Last, the distribution of substituent ions on R sites within unit cells or within formula units need not be exactly lock-step. When substituent ions do not constitute an integral number of formula units, at least some of these events are bound to occur. This atomic-scale local distortion of the lattice broadens the ferrimagnetic resonance line width as it subtly distorts the bond angles and perturbs the bond strengths of the iron ions, consequently changing the exchange and superexchange coupling of their magnetic dipole moments.

In spite of these considerations, substituted garnet films having bismuth concentration exceeding 0.8 formula units have recently been grown with quite narrow FMR line widths [33]. A film having composition $Bi_{0.85}Lu_{2.15}Fe_5O_{12}$ had a line width of 1.5 Oe and that of a film having composition $Bi_{1.4}Y_{1.6}Fe_5O_{12}$ was measured to be 0.73 Oe. The critical procedure for producing films having such low line widths was the use of a lead-free flux consisting of Bi_2O_3 alone.

6.7.2 Reflective Magneto-optical Materials

Reflective magneto-optical materials are used essentially in a single application, namely, magneto-optical recording in thin films, and their important parameters are defined in terms of the needs of this technology. These parameters are high polar Kerr rotation of the polarization of reflected light; high uniaxial anisotropy perpendicular to the film plane; a moderate value of saturation magnetization far from the compensation temperature (too high and it exceeds the coercivity, too low and the Kerr effect is small, even though the Kerr effect magnitude must depend predominantly on one of the sublattices alone, since magnetization is near zero around T_{comp}) and relatively high magnetic remanence and coercivity (also called coercive field, required to demagnetize a domain) near the operating temperature; a magnetic compensation temperature (or a Curie temperature, a rarer occurrence) near the ambient temperature; and the capability of forming small domains to enable high-density recording. (Many of these magnetic properties are indicated in Fig. 6.21.) Thermal properties contribute to the dynamics of domain formation in such a way that low heat capacity and high thermal conductivity are desired to keep data domains small. Substantial optical absorption at the operating wavelength (e.g., 50%) is also important for recording data on magneto-optical media since all writing mechanisms depend on raising the local temperature in the material.

Near-normal light incidence is required for most reflective magneto-optical applications. The polar Kerr effect, rather than either the longitudinal or transverse effect, must be invoked since of the three it alone is nonzero for normal-incidence light. The polarization rotation available from polar Kerr rotation is not large, always less than 1° in known materials when used in the absence of enhancing surface coatings (e.g., films in which the incident light is resonant).

Materials that induce high values of polarization rotation in reflected light consist almost exclusively of rare earth–transition metal (RE–TM) alloys. Amorphous films are desired over polycrystalline films since the signal-to-noise ratio of light reflected from oriented domains in this type of film is greater. For example, MnBi was initially considered a promising material since its Kerr rotation is high (0.7°) but its polycrystalline films give high background scattering.

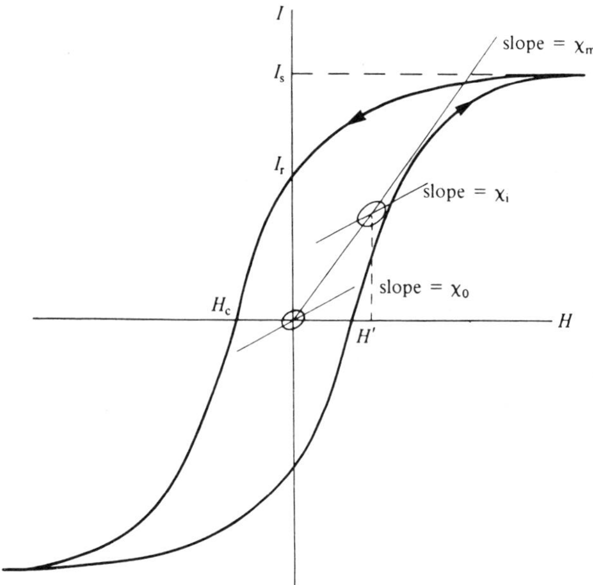

Figure 6.21 Schematic of I versus H or hysteresis loop for a ferromagnetic (or ferrimagnetic) material, defining the following: I_s, saturation magnetization; I_r, remanence or remanent magnetization; H_c, coercivity or coercive field. The slopes of the lines indicated define: χ_m, maximum susceptibility; χ_i, incremental susceptibility in the bias field H'; χ_0, initial susceptibility (for a minor loop of vanishingly small amplitude with no bias field, as found by extrapolation). The differential susceptibility χ_d is the slope, at any point, of the loop itself. (From D. J. Craik, "Magnetic Dipoles in Applied Fields," *Structure and Properties of Magnetic Materials*, Pion Limited, London, 1971, p. 35. Copyright 1971, used by permission.)

The RE–TM alloys are generally Néel-type ferrimagnetic; the RE and TM atoms are coupled antiferromagnetically. The RE magnetization is larger at low temperature while the TM magnetization dominates at higher temperatures. Thus a compensation point exists; its exact value can be adjusted by varying the material composition, particularly in ternary and quaternary alloys.

Near the compensation temperature the magnetization components of the RE and TM sublattices nearly cancel, and the total magnetization is about zero (see Fig. 6.22). In this condition a saturated domain has very little net magnetic moment, so the torque exerted on the domain by an applied magnetic field is small, and reducing the magnetization of the domain to zero requires a high coercive field. Magnetized domains thus are stable with respect to small variations in local applied magnetic field and temperature when the ambient temperature is in the vicinity of the compensation temperature.

The magnetization of any particular domain can be inverted by changing the local temperature appreciably from the compensation temperature in the presence of a magnetic field. (The field may be externally applied or it may result from a geometric factor, viz., shape factor demagnetization, or another mechanism that induces uniaxial anisotropy, e.g., growth-induced stress.) The coercive field H_c required to reduce domain magnetization to zero decreases with temperature excursion from the compensation temperature [$H_c = H_c(T)$ has a pole at $T = T_{\text{comp}}$]. When the coercive field decreases sufficiently that the local magnetic field exceeds it, the domain orientation may be flipped. Similarly,

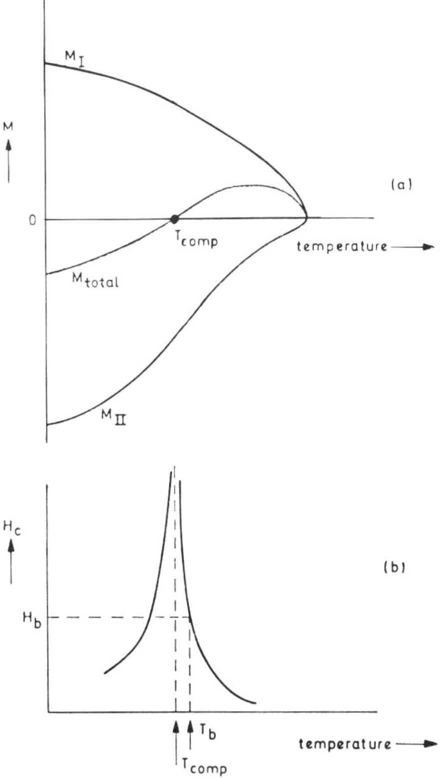

Figure 6.22 (*a*) Temperature dependence of the magnetization of a ferrimagnetic garnet and its sublattices and (*b*) temperature dependence of the coercive force. (From S. Middlehoek, P. K. George, and P. Dekker, "Beam Accessible memories," *Physics of Computer Memory Devices*," Academic, New York, 1976, p. 354. Copyright 1976, used by permission.)

domain magnetization may be imposed or inverted when the local temperature is raised above the Curie temperature since the domain then loses all intrinsic magnetization. This approach to domain reversal is practical in materials that have been devised to exhibit very low Curie temperatures, within a few hundred degrees of room temperature.

A high value of uniaxial anisotropy perpendicular to the film surface is important for maintaining high coercivity and rectangular hysteresis loops (magnetization vs. magnetic field) and, when the uniaxial anisotropy field exceeds the saturation magnetization, for inducing domain reorientation when the local temperature is elevated. Perpendicular uniaxial anisotropy may be induced by two types of mechanisms. The first are spin–orbit mechanisms, represented by atomic-level effects like single-ion anisotropy, and anisotropic exchange and magnetostriction. These occur mostly in alloys containing rare earth ions whose outer electrons are not *S* state (e.g., Tb, which is not a spherically symmetric ion). The second are magnetic dipolar interactions, of which pair ordering (in GdCo) is exemplar and to which growth process stress (differential contraction on cooling or substrate lattice mismatch) is a contributor. These are listed in Ref. 34 and discussed more thoroughly in Ref. 35 and the references therein.

A few other material properties are important in manufacturing magneto-optical films for memory applications. Domain formation and size are determined primarily by the thermal properties of the film material and substrate. Specific heat and thermal conductivity determine the dynamics of raising the local temperature sufficiently to impose domain magnetization and record data, as well as the rate of cooling in the local region, which determines domain size. Equally important is the saturation magnetization, since local demagnetizing fields increase dramatically as the size of the magnetized domain shrinks. The film material must be designed so that this local demagnetizing field does not exceed the coercive field. The saturation demagnetization must not be too large. Finally, for commercial viability the films should be producible using a continuous processing technology and be comprised of affordable elements.

Investigation of several representative reflective magneto-optical materials is reported in the current literature. Most of these are variations of quaternary RE–TM alloys or of their ternary subsets. Achievable values of Kerr rotation near room temperature are in the range 0.20°–0.35°. Uniaxial anisotropy constants can be obtained greater than 5×10^5 ergs/cm^3, equivalent to greater than 7000 Oe anisotropy field in materials that have high values of saturation magnetization ($H_a = 2K_\mu/M_s$). The coercive field can be maintained greater than 300 Oe more than 40°C from the compensation temperature without unreasonable difficulty.

For extensive details on particular RE–TM materials systems, the following references may be useful. For the GdTbFeCo system, see Refs. 36–43. Several papers discuss the PtMnSb system; see Refs. 44–46. For the (Nd, Pr)–(Fe, Co) system, see Ref. 47. A related system is Tb(Nd, Pr)Co; see Ref. 48. And rejuvenating the MnBi system is Ref. 49.

In addition to these discussions of the properties of specific materials, a good review of the requirements of magneto-optical recording systems is Ref. 50. Also, see Ref. 51.

6.8 MODERN MAGNETO-OPTICAL DEVICES

The list of magneto-optical devices currently commercially available or under development in laboratories is not long. By far the greatest resources are apportioned to development of magneto-optical data storage media and system components. Optical addressing of these media allows writing and reading at densities as high as or higher than competing technology (all-magnetic addressing). Prospects for erasure and rewrite capabilities of magneto-optical media are being explored extensively. Second in commitment of resources is the development of magneto-optical polarization-rotation isolators. Their use is envisioned in particular for isolation of laser diodes from reflective feedback of their emitted light, incident on optical fiber end-faces or on lenses positioned to couple light from laser to fiber. Otherwise, commercially proffered devices comprise Faraday rotator modulators and optical isolators, which perform light-ray-controlling functions, and a two-dimensional spatial light modulator for light wavefront addressing and modulating functions. A few other devices are under development in various laboratories. A magneto-optical crossbar switch directs optical signals between multiple communication ports in one system. Several laboratories are investigating the diffraction of wave-guided optical beams from magnetostatic waves in thin garnet films. This new Bragg diffraction technology may provide spectral analysis, convolution functions, and optical modulation at microwave frequencies between 1 and 20 GHz.

Magneto-optical data-recording technology is advancing along two different paths, although both envision erasable media as their competitive advantage over current optical technology and increased data density as the advantage over current nonoptical, magnetic technology. The first of these employs amorphous, or fine-grain polycrystalline, RE–TM films as the magneto-optical medium. The films are generally ternary or quaternary metal alloys; the multiplicity of constituents allows control of various device parameters simultaneously. Most approaches design the films to operate around the ferrimagnetic compensation point, the temperature at which the magnetization components of the antiferromagnetically aligned RE and TM sublattices balance and cause the net internal magnetization to vanish. This temperature is generally near or somewhat above ambient (room) temperature. The coercivity of the films becomes very large in the vicinity of the compensation temperature. (The net magnetic dipole moment disappears; the torque applied to a magnetic domain depends on the dot product of the local magnetic field and the net moment.) The magnetic orientation of a data-storing domain is thus stable with respect to small variations in the environmental magnetic field and temperature. To write data in a magnetic domain, a pulse of light is directed at the domain location, where the energy accrued from absorption of the pulse raises the local temperature. The increased temperature restores the net magnetic dipole moment in the domain and decreases the coercivity. A magnetic field at the domain site that exceeds the coercive field value can then reverse the domain's orientation. Cooling at the domain location freezes in the new data. Figure 6.23 illustrates a domain switching process as well as sketches of several types of domains that can exist in magnetic media under various conditions of material parameters, magnetic field strength, and temperature.

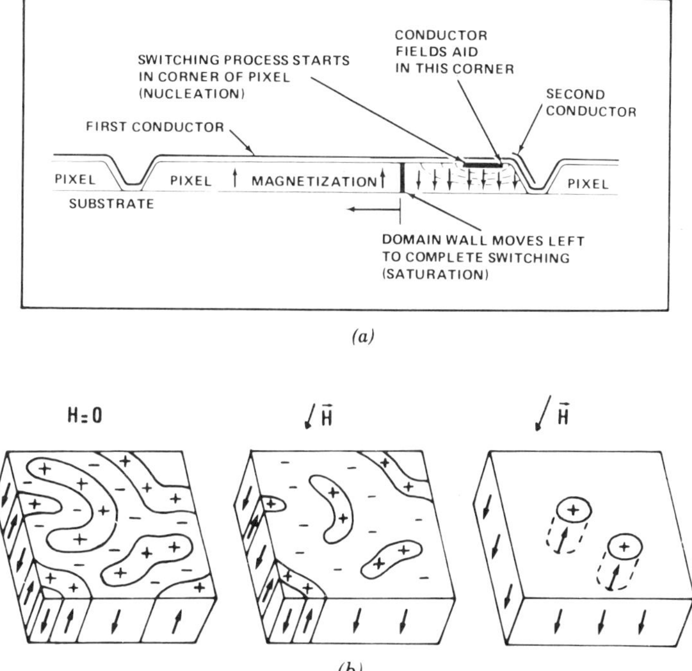

Figure 6.23 (*a*) Process of nucleation and switching a domain's orientation. (From William E. Ross et al., "Two-Dimensional Magneto-Optic Spatial Light Modulator for Signal Processing," *Opt. Eng.*, Vol. 22, No. 4, 1983, p. 486. Published by the Society of Photo-Optical Instrumentation Engineers. Copyright 1983, used by permission.) (*b*) Equilibrium domain alignment at various applied magnetic field strengths. (From P. Paroli, "Magneto-Optical Devices Based on Garnet Films," *Thin Solid Films*, Vol. 114, No. 1/2, 1984, p. 190. Published by Elsevier Sequoia S.A. Copyright 1984, used by permission.)

The magnetic field at the domain site can be derived from one of a number of sources. In the most general case the magneto-optical film that supports the data storage need not have uniaxial anisotropy. Geometric influences (shape factor anisotropy) cause the magnetic dipoles of this type of film to lie in the film plane. The local magnetic field that drives the domain orientation perpendicular to the film plane is then supplied solely by an external magnet. Usually this is an electromagnet, so that the magnetic field can be modulated on and off to correspond to particular write, read, erase, and overwrite requirements. The direction and magnitude of the magnetic field applied during the optical pulse, when the local temperature is elevated, saturates the dipoles of the region in a perpendicular orientation; the magnetic remanence of the material sustains this orientation when the temperature falls, the coercivity becomes large again, and the external magnetic field is removed. The written domain corresponding to a data bit then appears as a region magnetically oriented perpendicular to the film plane surrounded by material with magnetization in the film plane. In fact, since the ambient temperature is near the compensation temperature, neither the domain itself nor the surrounding in-plane-magnetized dipoles exhibits appreciable magnetization. Employing optical readout requires that the magneto-optical Kerr rotation depend substantially on only one of the ferrimagnetic sublattices in the film. Thus, although the net magnetization of the oriented domain is negligible, the domain may nevertheless be read optically via Kerr rotation. (See Ref. 52, pp. 350–360.)

Alternatively, the film may exhibit strong perpendicular uniaxial anisotropy. The entire film of dipoles then exhibits magnetization perpendicular to the film plane, with various regions aligned antiparallel. If such a film were cooled adiabatically from above the Curie temperature, or annealed, in the absence of an applied magnetic field, many stripe domains would result. The size of the domains in annealed films depends on the balance of energy minimization effects: shape factor anisotropy counteracts the aligning forces of dipole–dipole influences, tending to produce high-density contradirectional domains; uniaxial anisotropy and exchange effects account for the energy and consequent width of domain walls; and the interaction of these energy contributions determine, in the absence of external fields, the density and patterns of domains. These mechanisms have been discussed in Section 6.3. Externally applied perpendicular fields encourage the orientation of domains in one direction over that in the other, causing stripe domains of one magnetization direction to shrink to cylinders for moderate fields and to collapse altogether in higher fields.

Uniaxially anisotropic films for magnetic recording are cooled in the presence of an external magnetic field or saturated by application of a sufficiently high field so that in the unaddressed state all dipoles are aligned in one direction. (In these films the uniaxial anisotropy generally far exceeds the saturation magnetization due to shape factor demagnetization.) With the application of a reversing magnetic field, a small cylindrical domain flips magnetization when heated by an optical pulse. To understand the permanence of these small cylindrical domains, it is important to realize that a true equilibrium condition does not exist over the entire film area. Otherwise, cylindrical domains would not be stable in the absence of an external perpendicular magnetic field; rather, stripes would mediate the domains, as described in the preceding paragraph. Only because the brief optical pulse and the accompanying thermal transient temporarily change the local magnetization and coercivity can these domains be created and persist.

Shieh has devised a procedure to accomplish write, read, and overwrite functions in the absence of any externally applied magnetic field [53]. The shape factor demagnetizing field is invoked to nucleate the cylindrical domain that records a data bit; along with careful control of the optical power incident on the domain site, the various mechanisms of domain dynamics can implement write and erase functions. Once again, domain formation and stability rely on the balance between the domain energy and the domain wall energy. Domain wall energy varies directly as the square root of the product of the exchange constant and the anisotropy constant and inversely as the domain diameter for cylindrical domains. Wall energy thus contributes forces tending to reduce domain size. Both factors in the wall energy product decrease as the temperature is raised; this weakens the retaining forces of the wall. Conversely, demagnetization acts to increase the cylindrical domain size, whereby regions oriented antiparallel may more nearly cancel the total self-induced field. (Recall that the film outside the domain is uniformly magnetized perpendicular to the surface.) This force increases with temperature above the compensation point, where the internal magnetization increases in response to the growing dominance of one of the ferrimagnetic sublattices. So both wall and domain energy contributions act to increase the domain diameter when the local temperature is elevated. When their sum added to that of any applied field exceeds the stabilizing contribution of the coercive field, the domain expands. Expansion is limited first by equilibrium and then by the short duration of the heating optical pulse. When the domain again cools to ambient, the expansive forces diminish, the coercive field reassumes its overriding influence, and the domain stabilizes. Further detail may be discovered in Ref. 53.

Writing a cylindrical domain in the absence of an external magnetic field proceeds in several steps. First, the temperature of the irradiated region of the film increases from below to above the compensation temperature of the material. This causes the local magnetization to change sign with respect to the surrounding saturated medium and the sum of the intrinsic magnetic fields (demagnetizing and anisotropic) to exceed the coercivity. This small (typically 1–2-μm-diameter) region establishes its own demagnetizing field, which is now oriented coparallel with the surrounding medium (because of the flip in magnetization of the small region caused by the local temperature's crossing through the compensation temperature). The magnetization in the heated region is greater in the center since more optical power is absorbed there. The demagnetizing field can thus nucleate a domain in the center of the small region, with accompanying cylindrical domain walls. The domain walls expand until they reach the edge of the small, heated region. The region cools, whereupon the temperature falls through the compensation temperature, the local magnetization reverses sign again, and the data-storing domain remains, with magnetization opposite to the surrounding film. Erasure follows an almost identical progression, with the added feature that the second expanding domain wall collides with the first and they coannihilate. The power, or the duration, of the erase pulse must be carefully limited so that a third nucleation is not initiated. For clarifying details, see Ref. 54.

Several other papers extant discuss the desirable material parameters for RE–TM films, fabrication techniques for obtaining them, and the dynamics of domain formation. For example, see Refs. 55 and 56.

The Curie point writing approach to magneto-optical data storage previously mentioned has fallen into disregard. The primary material that can support this approach is MnBi, which presents several problematic characteristics that are difficult to surmount. First, films of MnBi tend to be polycrystalline rather than amorphous. Polycrystalline films scatter reflected light more strongly, giving high background light levels and poorer signal-to-noise ratio. Second, MnBi has two structural phases. The Curie temperature (T_C) for the low-temperature phase is above the phase transition temperature (360°C). The Curie point for the high-temperature phase, however, is below the phase transition temperature (at 190°C). To write a domain by the Curie point technique, the local temperature must be raised above 360°C. The MnBi then becomes paramagnetic, since its temperature is above the Curie temperature of the high-temperature phase. If it is cooled quickly (quenched) to below 190°C, the domain orientation induced in the paramagnetic state will persist. The high-temperature structural phase will be partially frozen in, however, so that a mixed-phase structure results. This structure slowly relaxes to the low-temperature phase. The high-temperature phase demonstrates saturation magnetization 30% lower than the low-temperature phase, so the magneto-optical read signal is initially reduced. Last, MnBi decomposes at above 446°C, only 20% above T_C. The MnBi does have a high Kerr rotation, near 0.4°. It can also be used in transmission as a Faraday rotation material, although its high absorptivity requires thin films whose growth must be carefully monitored since pit inclusion is common. (See Ref. 52.) Research on this material continues sporadically. (See Ref. 57.)

The second magneto-optical data-recording technology employs ferrimagnetic garnets, invoking Faraday rotation in transmission. The large possible Faraday rotation in these materials has been discussed in preceding sections. Growing the films by an expeditious technique is problematic; LPE is a slow process for mass production and requires expensive crystalline substrates; sputtering or vapor deposition techniques using glass substrates are preferable provided attractive film properties can be retained. The transmissive geometry is not as compliant as the reflective one to system layout, but it can be accommodated.

Domain writing in ferrimagnetic garnets takes place in the same manner in the vicinity of the compensation temperature as in RE–TM alloys. Iron garnets incorporating bismuth grow with perpendicular uniaxial anisotropy, and the substitution of gallium for some iron allows tailoring of the compensation temperature throughout the range 50–500°C via substitution on the predominating coordination site, as described in Section 6.7.1 The anisotropy is primarily growth induced and is due to the nonuniform distribution of bismuth ions on the dodecahedrally coordinated sites; a smaller contribution to the anisotropy is stress induced, caused by lattice mismatch and differential thermal contraction with respect to the substrate.

Write and read procedures are identical to those employed for reflective media. For analytical discussion of the various transmissive techniques, see Ref. 58. These transmissive media can sometimes be designed to operate in a double-pass configuration incorporating reflection from a back-surface coating and generally with optical access through the substrate. This approach increases the total polarization rotation (Kerr rotation often accompanies the reflection) for a given film thickness and recasts the system layout into a reflective format. For both the transmissive and reflective materials, front-surface quarter-wave film layers have been envisioned to increase the polarization rotation incurred in the magneto-optical interaction. Balasubramanian and Macleod have done computer analysis to demonstrate that Kerr rotation can be enhanced to about 2° using surface films; see Ref. 59.

Representative discussions of garnet films prepared by sputtering techniques are contained in Refs. 60–62. Vapor deposition is a newer approach; see Ref. 63.

The theoretical aspects of domain formation and stability for magneto-optical data storage are formulated expertly in several papers by Mansipur; see Refs. 64–66. These papers invoke a complicated computer model of domain dynamics over an extended area, incorporating demagnetization, anisotropy, exchange, film thickness, coercivity, and temperature dependencies of all parameters, complete with a graphic visualization.

Rare earth iron garnets are being investigated for other applications as well. One of these is as Faraday rotators in optical isolators. Both bulk and thin-film waveguide embodiments exist. Bulk isolators are generally constructed of rods of YIG, although paramagnetic Faraday glass is sometimes used at wavelengths shorter than 1.0 μm because of the high absorption of YIG there, saturated along the optical propagation direction in a static magnetic field imposed by a permanent magnet (see Fig. 6.24). The YIG rod in an isolator must be of appropriate thickness to rotate the polarization of incident light by 45°. Since the Faraday rotation of YIG varies with wavelength, these isolators cannot be broadband. Some commercially available isolators can be tuned over

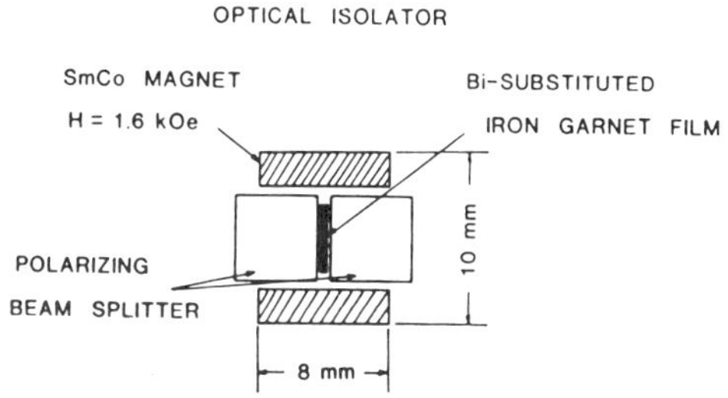

Figure 6.24 Structure of an optical isolator. (From N. Koshizuka et al., "Application of Bi-Substituted Iron Garnet Films to 0.8 μm Range Optical Isolator," *IEEE Translat. J. Magn. Jpn.*, Vol. TJMJ-I, No. 1, 1985, p. 103. Published by the Institute of Electrical and Electronics Engineers. Copyright 1985, used by permission.)

a wavelength range of 250 nm by moving the YIG specimen longitudinally into a higher or lower magnetic field. In these devices, the field is not strong enough to saturate the magnetization, and the polarization rotation induced by the crystal is governed by the expression $\theta_F = \int_0^L (VH)\, dx$, in which θ_F is the total cumulative Faraday rotation, V is the Verdet constant at the wavelength of interest, $H = H(x)$ is the local applied magnetic field strength, and dx and L are the differential and total lengths of the crystal, respectively. At constant temperature, isolation can exceed 20 dB over approximately a 60-nm wavelength excursion, peaking at a little more than 30 dB. The value of the Faraday rotation is temperature sensitive; at the design (center) wavelength, however, isolation can exceed 25 dB over a temperature range of 30°C (from 10 to 40°C).

A variation in operation of the bulk isolator characteristic produces a magneto-optical modulator (Fig. 6.25). The Faraday rotation of the transmitted optical beam polarization is modulated, via the Verdet dependence on the applied longitudinal magnetic field, by a signal current driving an electromagnet. One version of this modulator operates at rates of up to 10 kHz; it uses paramagnetic or diamagnetic glass, the former for higher rotation (by a factor of almost 3), the latter when stability against temperature drift is required.

Optical polarization isolators are less frequently found in a waveguide geometry, particularly with the advent of bismuth-substituted iron garnets (e.g., Bi:YIG and Bi:GdIG). These materials have such large Faraday rotation (near or exceeding 2000 deg/cm at 1300 nm wavelength) and such good figure of merit, particularly at optical wavelengths greater than 1.0 μm where absorption is low (θ_F/α greater than 250 deg/dB at 1300 nm), that films only a few hundred micrometers thick can achieve 45° rotation when used at perpendicular incidence. Doping with divalent ions such as calcium and magnesium reduces, optical absorption at 0.8 μm wavelength by preventing formation of the highly absorptive divalent iron ion; see Ref. 67. Saturation magnetization can be reduced to a few hundred oersteds. Further, techniques have been developed that circumvent the strong temperature dependence of Faraday rotation in bismuth garnets. Among these are the sandwiching of pairs of materials whose change of Faraday rotation with temperature have opposite sign (Ref. 68 suggests several alternatives) and the intermixing of these two kinds of crystal in a kind of solid-solution composite.

Other limitations of optical, waveguide geometry isolators are that they carry only a single spatial mode, restricting polarization isolation to temporally modulated optical beams, and they are somewhat tedious to couple to laser diode sources since use of butt coupling, end firing, or prism or grating surface couplers is required to insert the optical beam into the isolator.

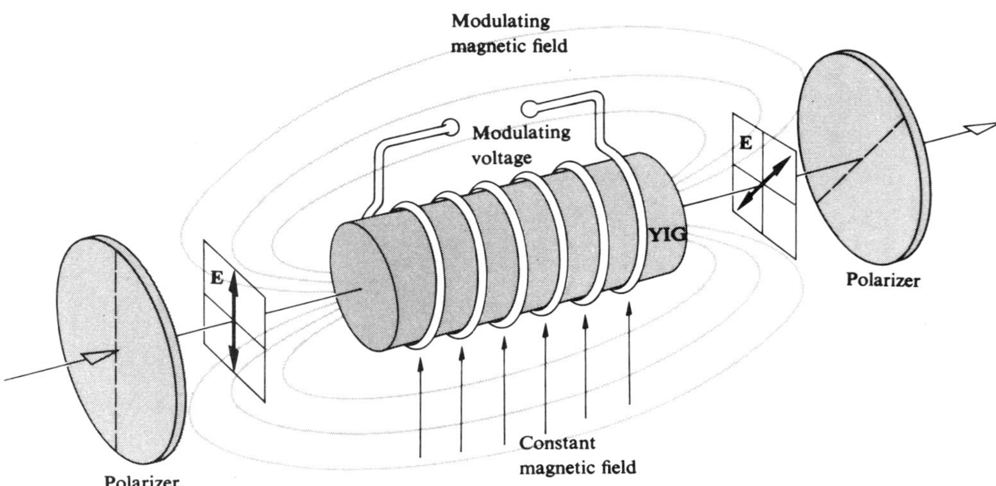

Figure 6.25 Faraday effect modulator. (From E. Hecht and A. Zajac, "Polarization," *Optics*, Addison-Wesley, Reading, MA, 1974, p. 263. Copyright 1974, used by permission.)

Nevertheless, waveguide isolators continue to be investigated in light of the prospects of integrated optical devices and their anticipated manufacturability. These waveguide isolators must avoid films in which the propagation constants for the equal-mode-number TE and TM optical polarizations are excessively disparate. (Even in the absence of material birefringence, optical waveguide solutions indicate differing velocities along the propagation direction for the TE and TM polarizations. Velocity—or equivalently, wave vector—mismatch impedes the efficiency of polarization conversion because of imperfect phase matching in the coupled-mode interaction.) Intramodal TE–TM phase matching can be approached by making the waveguide weakly guiding, that is, fabricating the waveguide with a small refractive index difference between guide and cladding. Otherwise the wave vector mismatch can often be compensated by material birefringence, which occurs routinely as a growth or structural feature of most unannealed bismuth-substituted garnets.

Several novel concepts that compensate geometrical and material birefringence have been explored for achieving polarization rotation in waveguides (see Refs. 68–70). One conception for a waveguide isolator that combines these effects envisages placing equivalent-rotation sections of Faraday, nonreciprocal rotation material back to back with birefringent, reciprocal-rotation material; the rotation contributions add when the light passes in the direction of the magnetic field applied to saturate the Faraday rotator segment and cancel for reverse propagation. (See Ref. 68.) A second device compensates for the combined birefringence of geometry and materials simultaneously by reversing the in-plane magnetization periodically so that it is alternately parallel, then antiparallel, to the optical propagation direction. The periodicity of the magnetization reversal is established permanently in the magneto-optical waveguide material by local heating (laser light absorption is used) in the presence of a saturating external magnetic field. The field is reversed in direction when magnetizing neighboring regions to produce alternately magnetized strips, in much the same way as domains are written in data-recording applications. The magnetization periodicity equals that of the birefringence phase delay between the TE and TM light; the magnetization direction is switched whenever the phase difference between light in the two polarizations reaches a half wave. Reference 70 describes this device.

Another signal-switching component, a circulator, has been constructed as an optical waveguide device (see Fig. 6.26). This device receives signal data on one (or several) input port(s) and selectively transmits it on a single (or multiple) designated output port(s). Selection of an output port is achieved when signal data arrives there along two different device paths in phase. Combinations of reciprocal and nonreciprocal phase-shifting elements, of which Faraday rotators represent the former, along with mode-selecting waveguide channel junctions can be devised to perform this function. In addition, output port selection can be made flexible by overlaying the Faraday rotating sections of the waveguide structure with metalization to carry current and induce modulating magnetic fields. See Refs. 71 and 72 for details.

Wafers of ferrimagnetic garnet have been incorporated into spatial light modulators capable of binary modulation on a 128×128 array of pixels. Two different versions have been offered commercially. Both use a bismuth-substituted iron garnet film on which a regular two-dimensional array of mesas has been etched. The iron garnet film exhibits perpendicular magnetic anisotropy, and each mesa can support a single, large, magnetically saturated domain oriented either parallel or antiparallel to the direction of propagation of an incident optical beam whose wavefront illuminates the entire array. According to the magnetization of a mesa pixel, the polarization of transmitted light is rotated clockwise or counterclockwise a few tens of degrees by the Faraday effect. (The exact rotation depends on the optical wavelength since the Faraday rotation is spectrally

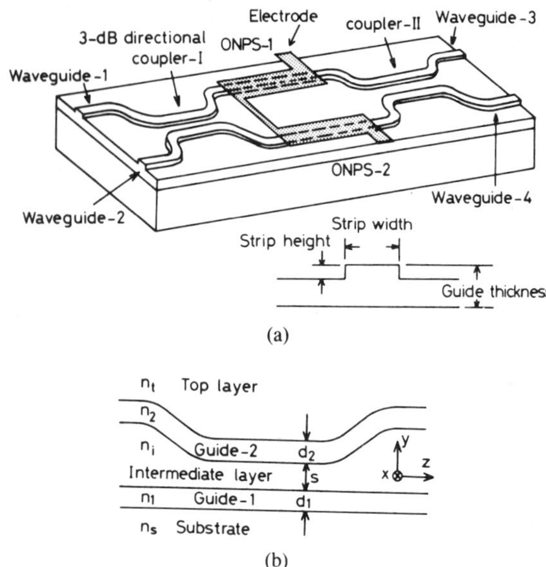

Figure 6.26 Schematic drawings of waveguide-type optical circulators. (*a*) Optical waveguide consists of a rib guide, and dc electric current flowing along an electrode yields a magnetic field. (*b*) Light propagates in the *z*-axis direction, and an external magnetic field is applied along the *x*-axis. (From T. Mizumoto et al., "Measurement of Optical Nonreciprocal Phase Shift in a Bi-Substituted $Gd_3Fe_5O_{12}$ Film and Application to Waveguide-Type Optical Circulator," *J. Lightwave Tech.*, Vol. LT-4, No. 3, 1986, p. 350. Published by IEEE. Copyright 1986, used by permission.)

dependent.) A design compromise is required since thicker films producing higher rotation also inflict higher absorption. The array is sandwiched between polarizers, with the analyzer generally oriented perpendicular to the output polarization produced by one of the pixel magnetization states (see Fig. 6.27); this orientation provides the highest contrast ratio.

The two devices differ in the mechanism employed to switch the pixels; to support each particular switching approach, the characteristics of their garnet films also differ somewhat. One device switches the mesa domains by the same mechanism discussed for magneto-optical recording in materials for which the compensation temperature is near ambient. The temperature of an addressed pixel is elevated so that the combined magnetic field of the saturation magnetization, the anisotropy, and the external field exceeds the coercivity. Heating results from current conduction through a transparent resistive element that overlies the pixel of interest. The pixel is addressed by a pair of electrodes that, lying in the grooves between rows and columns of mesas, are connected by the resistive element. An externally applied magnetic field is required to reverse the mesa domain orientation for this device, and its direction must be switchable to accommodate the desired orientation of an addressed domain. Operation of this device is described in Ref. 73.

The other device switches the domains using small nucleating magnetic fields produced by circulating currents near the pixel of interest. As in the preceding device, electrodes lie in the channels between rows and columns of mesas. In the activation state the current that passes along one of the wires is insufficient to nucleate a domain alone, but when current passes in the appropriate direction along each electrode of the pair that crosses at a pixel, an inverted domain is nucleated. This nucleation is aided by ion implantation,

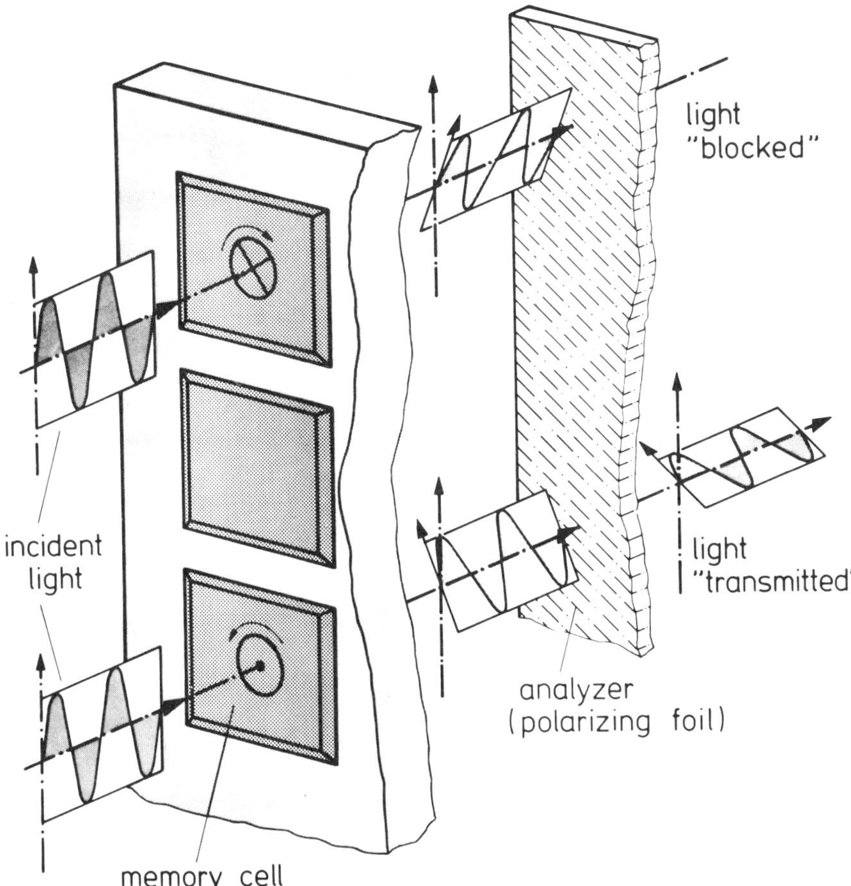

Figure 6.27 Viewing of magnetic domains using the Faraday effect and polarization optics. (From B. Hill and K. P. Schmidt, "Fast Switchable Magneto-Optic Memory—Display Components," *Philips J. Res.*, Vol. 33, Nos. 5/6, 1978, p. 214. Published by N. V. Philips. Copyright 1978, used by permission.)

in the corner of the pixel nearest the crossing electrodes, of impurities that lowers the magnetic anisotropy. Once nucleated, the new domain is driven across the mesa by the nucleating field. Further description of this device may be found in Ref. 74.

The stability of domains in these spatial light modulators is discussed in Ref. 75. As shown therein, the perpendicular anisotropy reduced by the saturation magnetization must exceed the field required to saturate the film in order that large domains will be stable. Equivalently, the perpendicular anisotropy must exceed twice the saturation magnetization.

A large-aperture, diffractive, magneto-optical deflector is being developed by a team of scientists at Unisys. The device implementation has seen several revisions, but the physical operation and its foreseen variations are recurrent. A thin film of Bi–YIG is grown with perpendicular anisotropy, so that the dipoles organize naturally into an array of stripe domains, alternatingly oriented up and down with respect to the wafer surface. These stripe domains can be aligned parallel to an externally applied, in-plane, magnetic field. The strength of this field determines the width of the stripes: the stronger the field, the narrower the stripes. Figure 6.28 provides a drawing of the deflector.

The basic device operation relies on the counterrotation of the polarization of light passing through adjacent stripe domains. Light transmitted by a stripe domain acquires a polarization component that lies in the plane orthogonal to that of the incident light polarization. Light transmitted by a neighboring stripe whose polarization is rotated in the opposite sense also acquires a polarization component in that plane but pointing in the opposite direction. The relative phase between light transmitted by neighboring stripes is thus 180° for the polarization component orthogonal to the incident polarization. Ideally the rotation is 90° in each stripe for optimal diffraction efficiency, although considerably smaller rotations can be effective (with lower efficiency). Diffraction as from a phase grating occurs and multiple diffraction orders are produced. Enhanced performance can be obtained by employing nonnormal incidence Bragg diffraction to eliminate all but a single diffracted order. Techniques employing a resonant cavity for either the input light alone or for both the input and diffracted light have been explored that produce increased efficiency; 11% at 800 nm and 14% at 1060 nm have been realized. Rotating the alignment, and consequently the diffraction direction, of the stripe domains and varying the width of the stripes through some small range by adjusting the applied magnetic field strength and orientation provides control for steering of the diffracted beam between multiple ports. For further details see Ref. 76. In this device, as in others described earlier, increasing the ratio of Faraday rotation to absorption is important for

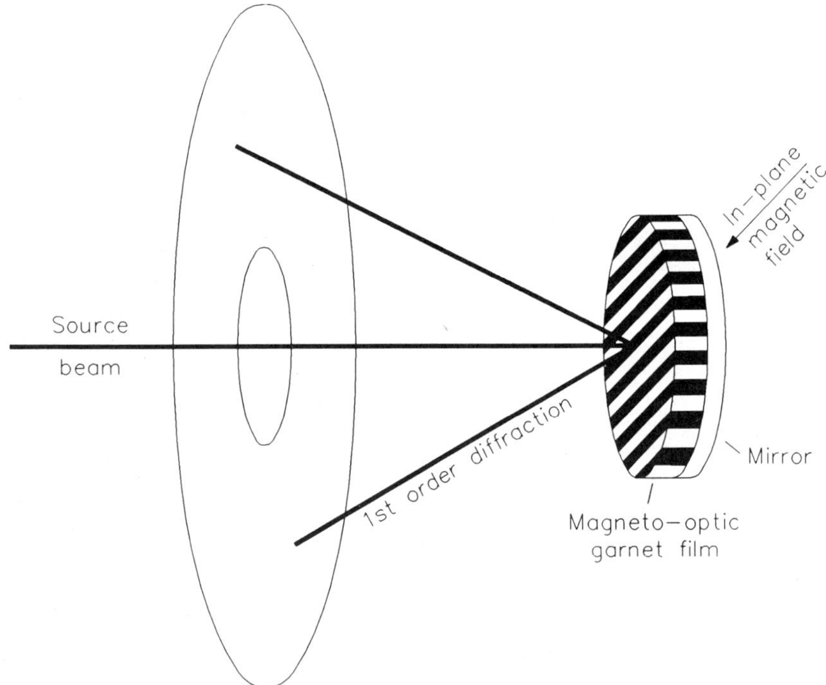

Figure 6.28 Diffractive magneto-optical deflector. An applied in-plane magnetic field controls the stripe domain grating orientation and periodicity to provide a two-dimensional field of view. (From G. L. Nelson et al., "Stripe Domain Light Deflection for Intersatellite Communications," *Communications Networking in Dense Electromagnetic Environments, Proc. SPIE*, Vol. 876, 1988, p. 122. Published by the Society of Photo-Optical Instrumentation Engineers. Copyright 1988, used by permission.)

high efficiency, and material properties delimit its success. An interconnect switch for an optical communications network is envisioned.

A ferrimagnetic garnet film can also support optical diffraction from a moving magnetic disturbance, such as an MSW. An MSW propagating in a thin film of ferrimagnetic garnet induces a periodic perturbation in the off-diagonal elements of the optical permittivity tensor. The perturbation is thus a traveling-wave, microwave frequency modulation of the Faraday rotation. This type of traveling, periodic, polarization-rotating grating can be produced by each of the three types of magnetostatic waves and can cause Bragg regime diffraction of an intersecting optical wave provided that the process is phase matched. Details of the interaction are somewhat complicated.

The magnetostatic wave modes of a thin film (a waveguide) have been discussed in Section 6.4. Three types of MSW propagate. Volume waves, MSWs whose propagation is isotropic in the plane of the film, are excited when the field that saturates the magnetic dipoles is perpendicular to the film surface. Surface waves are excited when the saturating field is perpendicular to the MSW propagation direction but in the plane of the film; they are localized on the top film surface when propagating in one direction and on the bottom surface when propagating in the opposite direction. Backward-volume waves are excited when the saturating magnetic field is parallel to the propagation direction; the group velocity for these MSWs is antiparallel to the phase velocity. The microwave component of the magnetization for each of these types of waves can be visualized lying perpendicular to the dipole-saturating magnetic field and circulating as the dipole precesses around that direction, approximately circumscribing a circle. Recall that light polarization can be rotated by the Faraday effect only when a component of the magnetic field is parallel to the optical propagation direction. Thus, to first order, forward-volume waves can cause optical polarization rotation at microwave frequencies for light propagating either collinear or perpendicular with respect to the MSW propagation direction, whereas surface and backward waves induce microwave frequency Faraday rotation in light only when its propagation is collinear with the MSW. (Several mechanisms can cause higher-order effects. Magnetostatic wave precession can be saturated or elliptical because of anisotropy fields; high microwave power can introduce the Cotton–Mouton effect or allow multiple optical interactions with various permittivity perturbations to occur.)

Magnetostatic waves propagate as guided waves in thin ferrimagnetic garnet films. Thus for an optical wave to intersect them and diffract from them, the light must also propagate as a mode of the waveguide formed by the garnet film. (An illustration of the layout of magnetic fields, MSW excitation and propagation, and optical insertion, propagation, and diffraction in a YIG film waveguide is provided in Fig. 6.29.) Note that the waveguide criteria for MSWs is not identical to that for waveguiding light. The former requires a low-loss magnetic film medium bounded by nonmagnetic media or, alternatively, media having differing saturation magnetization values from the guiding layer, so that with the predetermined MSW wavevector and external bias field, the excitation lies outside the MSW frequency band in the bounding media. The latter requires that the guiding film have an optical refractive index value greater than that of either bounding medium so that light can be completely trapped inside the film by total internal reflection; i.e., it can be incident on the film surfaces at angles exceeding the critical angle. Fortuitously, a YIG or Bi:YIG film grown on GGG (or one of its variants) satisfies both of these criteria.

In isotropic thin films (no birefringence), light can be guided in one of two polarizations: an electric field parallel to the surface, called transverse electric (TE), or a magnetic field parallel to the surface, called transverse magnetic (TM). Guiding is allowed when two conditions are met.

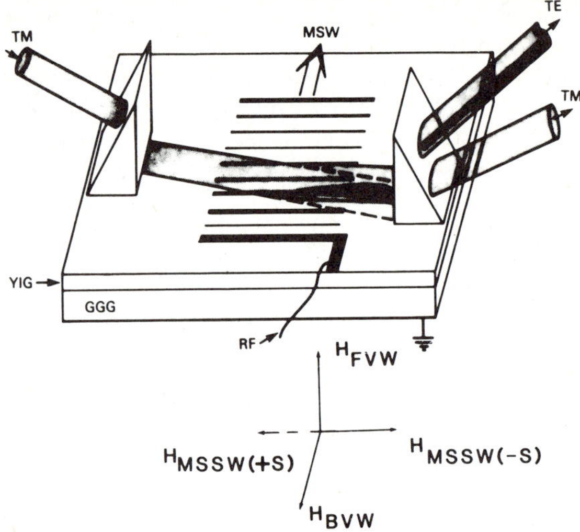

Figure 6.29 Bragg diffraction of a guided optical wave by an MSW induces conversion between orthogonally polarized optical modes. Lower section shows the magnetic field bias direction required to produce each of the four types of MSWs. (From A. D. Fisher, "Optical Signal Processing with Magnetostatic Waves," *Circuits, Syst. Signal Process.*, Vol. 4, No. 1–2, 1985, p. 267. Published by Birkhauser Boston. Copyright 1985, used by permission.)

First, light within the film must be incident at each film boundary (or film–cladding interface) at an angle greater than the critical angle, that is, the maximum angle at which light propagating in a higher refractive index medium can be partially transmitted across an interface with a medium of lower refractive index. This causes the light to be totally internally reflected. The value of the critical angle can be derived from Snell's law of refraction:

$$n_1 \sin\theta_1 = n_2 \sin\theta_2$$

where n_1 and n_2 are the refractive indices of the transmitted light medium and the incident light medium, respectively, with $n_2 > n_1$, and θ_1 and θ_2 are the respective angles of propagation of the light at the interface in the two media, measured with respect to the surface normal. Setting $\theta_1 = 90^\circ$ gives θ_2, the value of the critical angle.

Second, the light must be resonant in the film structure, that is, in propagating from a point in the center of the waveguiding film through a reflection at each of the two interfaces and returning to the center of the film, the total accumulation of phase must be an integral number of wavelengths of the light in the film. Accurately, the phase of interest is the sum of that accrued by the propagation component directed perpendicular to the waveguide surfaces and the phase changes incurred on reflection at each interface. (Constructive interference of overlying optical wavefronts is required.) To better understand this concept, it is instructive to write the optical wave vector as the sum of two orthogonal components: $\mathbf{k} = \boldsymbol{\kappa} + \boldsymbol{\beta}$. In this decomposition $\boldsymbol{\kappa}$ is the component of $\mathbf{k}$ that is perpendicular to the waveguide surface and $\boldsymbol{\beta}$ is the component that is parallel to it: $|\boldsymbol{\kappa}| = |\mathbf{k}|\cos\theta$ and $|\boldsymbol{\beta}| = |\mathbf{k}|\sin\theta$. The characteristic equation that governs the resonant propagation of light in a waveguide film as it reflects alternately from the top and bottom surfaces is

$$2nd\cos\theta + 2\phi_{21} + 2\phi_{23} = m\lambda$$

where n is the refractive index of the film and d is its thickness; θ is the angle of incidence of the light on the film–boundary interface ($\theta = 0$ is perpendicular to the surface); λ is the vacuum optical wavelength; and m is the number of the waveguide mode occupied by the light or, equivalently, the number of nodes of the standing wave formed by the light as it bounces back and forth across the film, propagating down the waveguide. The ϕ parameters are somewhat more complicated. When light reflects from an interface at which its incidence angle exceeds the critical angle, it garners a phase increment, as if it had traveled an additional distance. This phase increment depends on the optical polarization, the angle of incidence, and the refractive indices of the interface media; the quantities ϕ_{ij} in the preceding expression represent these increments for the two interfaces of the waveguide, the subscript i indicating the waveguide side of the interface and j the cladding side. For a TE mode,

$$\phi_{ij} = \tan^{-1} \frac{\sqrt{n_i^2 \sin^2 \theta_i - n_j^2}}{n_i \cos \theta_i}$$

For a TM mode,

$$\phi_{ij} = \tan^{-1} \frac{(n_i^2/n_j^2)\sqrt{n_i^2 \sin^2 \theta_i - n_j^2}}{n_i \cos \theta_i}$$

Because these expressions are not identical, the characteristic equation produces differing propagation angles θ for TE and TM modes having the same mode number m. Consequently, their longitudinal propagation velocities (group velocities) in the waveguide differ. Equivalently, the components of their wave vectors parallel to the waveguide surface, conventionally called β, differ; $\beta_{\mathrm{TM}} < \beta_{\mathrm{TE}}$ for the same mode number. For waveguide diffraction calculations, these βs are equivalent to propagation wavevectors for diffraction calculations in bulk devices.

The number of modes supported by a waveguide structure is determined by the values and degree of symmetry of the refractive indices of the guiding film and the cladding media and by the thickness of the guiding film. Waveguide structures that are symmetric in refractive index always support at least a single TE mode, although not always a TM mode. Lower-order modes, those with fewer nodes in the standing-wave pattern across the guiding film, propagate at higher angles, that is, more nearly down the axis of the waveguide. Light in higher-order modes is incident on the boundaries at angles nearer the critical angle. As the mode number is increased, the characteristic equation eventually fails to have a bound solution; this is called the cutoff condition for the waveguide. For YIG films, which form asymmetric refractive index structures on their GGG substrates in air, the lowest-order ($m = 0$) TM mode appears when the thickness exceeds about 0.75 μm (the refractive index values are 2.212 for YIG and 1.94 for GGG at an optical wavelength of 1150 nm). For a review of waveguide optics, see Ref. 77.

Bragg diffraction of waveguide-confined light from an MSW requires that several criteria be met. Because of the magneto-optical nature of the interaction, polarization rotation is induced. However, light propagating in a waveguide is constrained to occupy either a TE or a TM mode; its polarization cannot be arbitrary, as would be the case for bulk implementation. Diffraction therefore requires the waveguide light to transition from a TE to a TM mode polarization, or vice versa. Further, the diffraction mechanism requires that phase-matching criteria be met, or at least approximated, as in any optical interaction, particularly those high-efficiency interactions that depend on long interaction distances. Specifically, the MSW wave vector should be approximately of magnitude and direction such that the vector sum of the TM light wave vector and the MSW wave vector equal the TE wave vector:

$$\boldsymbol{\beta}_{\text{TM}} + \mathbf{K}_{\text{MSW}} = \boldsymbol{\beta}_{\text{TE}}$$

Physically, solutions to this equation are governed by the observation that for the same mode number $|\boldsymbol{\beta}_{\text{TM}}| < |\boldsymbol{\beta}_{\text{TE}}|$. That is, in diffracting, the wave vector of the light changes magnitude. Thus the usual requirement of bulk diffraction (in the Bragg, i.e., phase-matched, regime) that the angle of incidence on the diffraction grating equal the angle of diffraction does not pertain.

From Ref. 78, the intensity diffraction efficiency for low MSW power is given by

$$\frac{I_d}{I_0} = \frac{4\kappa^2}{4\kappa^2 + (\Delta\beta)^2} \sin^2\left[\tfrac{1}{2}y\sqrt{4\kappa^2 + (\Delta\beta)^2}\right]$$

where I_0 and I_d are the initial and diffracted intensities, respectively; $\Delta\beta$ is the phase mismatch; κ is the coupling constant; and y is the interaction length. In detail,

$$\Delta\beta = |\boldsymbol{\beta}_{\text{TE}} - (\boldsymbol{\beta}_{\text{TM}} + \mathbf{K}_{\text{MSW}})|$$

is the mismatch magnitude of the sum of wave vectors, which equals zero when the diffraction is perfectly phase matched. Additionally

$$\kappa = \frac{\varepsilon_{31} k_0 \sqrt{\varepsilon}}{2\varepsilon_0 \varepsilon},$$

where k_0 is the free-space optical wave vector; ε is the permittivity so that $\sqrt{\varepsilon} = n$, the refractive index; ε_0 is the permittivity of free space; and ε_{31} is the value of the permittivity tensor at the (3, 1) position, which couples the TM and TE modes. These expressions may be recognized as belonging to solutions to a coupled-mode formalism. They show that for perfectly phase-matched interaction, the intensity diffraction efficiency is proportional to $\sin^2(\kappa y)$. For short interaction lengths (or less desirably, for small κ), $\sin^2(\kappa y)$ can be approximated by $(\kappa y)^2$. So I_d/I_0 is proportional to κ^2, which is in turn proportional to ε_{31}. Rewriting $\varepsilon_{31} = \varepsilon_0 fm$ and substituting for the linear coefficient f,

$$f = \frac{2\sqrt{\varepsilon}\phi_F}{k_0 M}$$

where m is the applicable component of the microwave frequency magnetization associated with the MSW, ϕ_F is the Faraday rotation coefficient, and M is the strength of the bias magnetic field, a new expression for κ can be realized:

$$\kappa = \phi_F \frac{m}{M}$$

The intensity diffraction efficiency is proportional to the square of the Faraday rotation coefficient and to the square of the local microwave frequency magnetization. This realization drives the exploration for materials to support this technology.

First, the material must have a high value of ϕ_F. Second it must have low MSW attenuation (measured as low ferrimagnetic resonance line width) so that m, the RF magnetization, is not negligible in the MSW–optical interaction region. Ancillary concerns are low optical attenuation at the wavelength of interest and repeatable values of saturation magnetization, magnetic anisotropy, and optical birefringence from a particular growth sequence. A technique for measuring MSW attenuation was described in the preceding section treating spin waves. The measurement of ϕ_F is straightforward, incorporating an electromagnet to adjust the magnetic field strength, a laser source and detector, and a polarizer–analyzer set of polarizing prisms. Collimated light is transmitted through the polarizer, perpendicularly through the garnet sample, and then through the analyzer and onto the detector. The sample is placed between the poles of the electromagnet so that the optical propagation is parallel to the magnetic field. This apparatus is described in Ref. 79 and illustrated in Fig. 6.30. A technique for measuring optical waveguide attenuation is also described in Ref. 79 (see also Fig. 6.30) in which light scattered from the waveguide surface is detected as a function of propagation distance. Assuming isotropically distributed absorption and scattering centers in the garnet film, an estimate of total waveguide optical attenuation can be derived. Garnet films tend to scatter very little waveguide light since they are nearly perfect crystals, and scattered optical power is often unsatisfactory for accurate measurements. Greater accuracy can be obtained, with care, using the three-prism technique described in Ref. 80 (see Fig. 6.30 again).

The magnitude of the MSW wave vector, $|\mathbf{K}_{\mathrm{MSW}}|$, is generally not large. Although some range of selectable values is available at any particular frequency via adjustment of the biasing static magnetic field to shift the MSW dispersion curve, it is common to obtain wave vectors in the vicinity of 500 cm^{-1}. (Magnetostatic wave velocity is about 5×10^6 cm/s; this wave vector value is approximate around 2 GHz frequency.) Since this value is very small compared to the wave vectors for light in either a TE or TM waveguide mode (about $6.7 \times 10^4\ \mathrm{cm}^{-1}$) and is even somewhat smaller than the usual difference between wave vector amplitudes for TE and TM modes having the same mode number, the angle of deflection of the diffracted light is only a few degrees, almost always less than 10°. Also, diffraction in the low MSW power, linear regime is almost without exception between TE and TM modes of the same value of m and the least difference in β. Low-order modes (between the zeroth and fifth) of waveguides that support 10 or more optical modes are best used. (Such films are about 10 μm thick). Films of thickness on the order of 10 μm constitute a good compromise between the desired attributes of few optical modes, influencing design toward films nearer to 1 μm thick, and that of efficient MSW insertion. An MSW is launched from a microstrip antenna placed near or on the garnet surface; thicker garnet films present a greater capture volume for the microwave field radiating from the antenna.

The efficiency of diffraction in YIG films on GGG substrates has generally been observed to be small when the MSW power has been restrained below the nonlinear, saturating region, where only MSW-induced (dynamic) Faraday effects can be invoked to describe the interaction. Diffraction efficiency for transverse interaction of light with low-power, forward-volume MSWs has been in the range of 0.01% (somewhat less than 10% per microwave watt at about 2 mW of inserted MSW power). Diffraction efficiency for collinear interaction of light and surface MSWs having power near the saturation level has reached a few percent (again at about 5% per microwave watt). See Refs. 78, 81, and 82.

The instantaneous bandwidth of the low-power interaction in a uniform bias magnetic field is about 30 MHz. Phase matching is lost over wider bandwidths when the interaction region is long enough to maintain reasonable efficiency; in YIG, the minimum length is about 7 mm. A geometric view of phase matching assists in understanding this limitation. In optically isotropic media, the wave vector loci for TM and TE polarizations are both circles in the plane of the waveguide of radii $|\boldsymbol{\beta}_{\mathrm{TM}}|$ and $|\boldsymbol{\beta}_{\mathrm{TE}}|$, respectively. Ideally, the tail

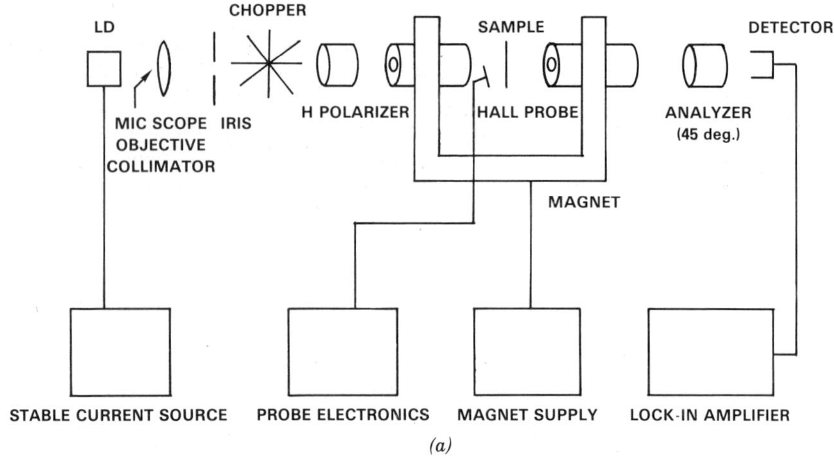

(a)

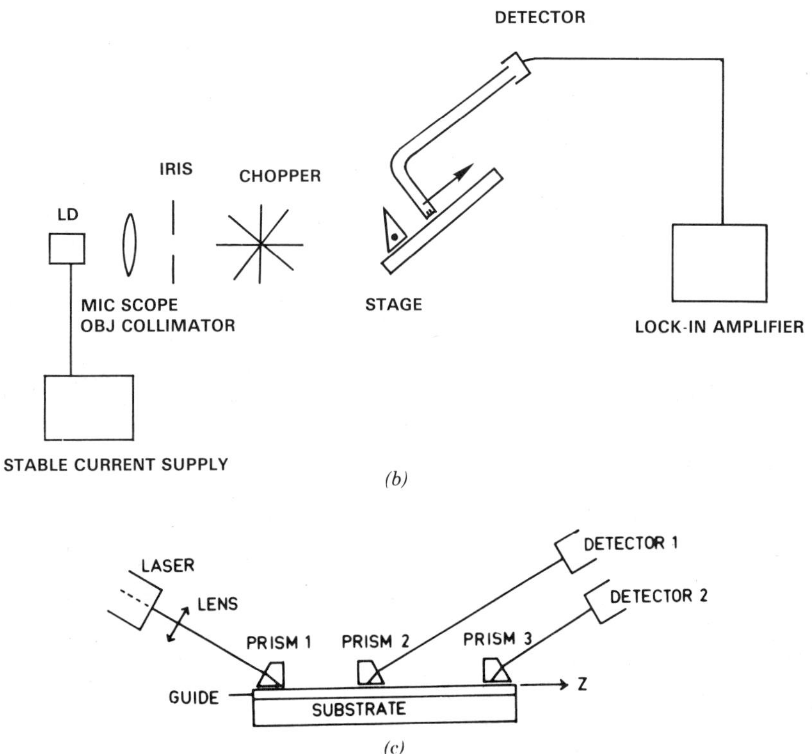

(b)

(c)

Figure 6.30 (*a*) Faraday rotation measurement. (*b*) Optical waveguide attenuation measurement. (*c*) Schematic of the three-prism method. [(*c*) From Y. H. Won et al., "Three-Prism Loss Measurements of Optical Waveguides," *Appl. Phys. Lett.*, Vol. 37, No. 3, 1980, p. 269. Published by the American Institute of Physics. Copyright 1980, used by permission.)

of the MSW wave vector lies on one circle (e.g., the outer one, the TE circle), touching the head of one optical wave vector, $\boldsymbol{\beta}_{TE}$, while the MSW wave vector head lies on the other circle (the inner one, TM), touching the head of the other optical wave vector, $\boldsymbol{\beta}_{TM}$. The bandwidth of the diffraction interaction reflects the allowable mismatch in magnitude and/or direction of $\mathbf{K}_{MSW}$. As can be seen from the preceding diffraction efficiency equation, for a small coupling constant the efficiency per unit length of interaction depends on $\Delta\beta$ as $\operatorname{sinc}^2(\Delta\beta y/2)$ using the definition $\operatorname{sinc} x = (\sin x)/x$. (This expression for diffraction efficiency indicates the roll-off penalty for nonzero mismatch.) The MSW wave vector is generally so short in practice that it connects the two optical wave vectors in a nearly radial direction (hence the small observed diffraction angles). Substantial bandwidth gains could be realized by finding a way for the MSW wave vector to connect the two optical wave vectors while lying perpendicular to the shorter of them ($\boldsymbol{\beta}_{TM}$), that is, tangent to the inner circle; this condition is called tangential phase matching. With this alignment $|\mathbf{K}_{MSW}|$ can vary over a larger range without abandoning the phase-matched region. Tangential phase matching is illustrated in the center diagram of Fig. 6.31. (A means of accomplishing tangential phase matching has been investigated and will be discussed briefly in what follows.) A substantial investigation of the use of gradient bias magnetic fields to invoke different portions of the (nonlinear) MSW dispersion curve across the optical–MSW interaction region has shown that the interaction bandwidth may be increased to about 300 MHz using this technique (again, in YIG). (See Refs. 83 and 84.) Varying relative delays across the band, complicated beam-steering effects on the MSWs, and nonspecific diffraction directionality discourage use of this technique in practical devices, however.

The desirable attributes for garnet materials in which MSW–optical diffraction can occur efficiently were already outlined: high Faraday rotation, low MSW attenuation, and low optical absorption. Recently, scientists have been able to grow bismuth-substituted YIG having these characteristics along with reasonably low optical absorption. Using bismuth-substituted iron garnet films, a research group has recently shown dramatic results in high-efficiency MSW–optical diffraction (Ref. 28). The LPE-grown garnets were grown on GGG and consisted of (BiLu) IG with 0.89 formula units bismuth and 2.11 formula units lutetium. The flux from which the iron garnet films were grown was solely Bi_2O_3, relieving the concern of incorporating lead in the films. This proportion of bismuth produced Faraday rotation of -140.0 deg/nm at an optical wavelength of 1300 nm. Guided-wave optical attenuation at this wavelength was low in the film for which results are reported (only 0.6 dB/cm), as was MSW attenuation; the FMR line width was 0.7 Oe. Best diffraction results were obtained at a microwave frequency of 4.0 GHz, at which 47% diffraction was achieved with a microwave power of 0.8 W in an interaction region of length 7 mm. Backward-volume magnetostatic waves were employed since forward-volume and surface MSWs were found to couple strongly to the exchange-dominated spin wave manifold in the films used, whereas the backward-wave dispersion relation does not coincide with that for the exchange-dominated spin waves and does not couple.

Development of the highly bismuth-substituted garnet films grown from a lead-free flux to provide low optical attenuation is the primary capability enabling demonstration of this landmark result. The initial report on the material characteristics of the films appears in Ref. 85. The absence of lead apparently promotes low FMR line width as well, possibly because lead may induce 2+ and 4+ ionization states of the iron ions in the lattice. Reference 86 instructs that a few percent MgO was incorporated in films grown later, presumably to maintain the appropriate 3+ ionization of the iron ions. Scientists working on reducing optical damage in lithium niobate, which was believed to be caused by the presence of inappropriately ionized iron, showed several years ago

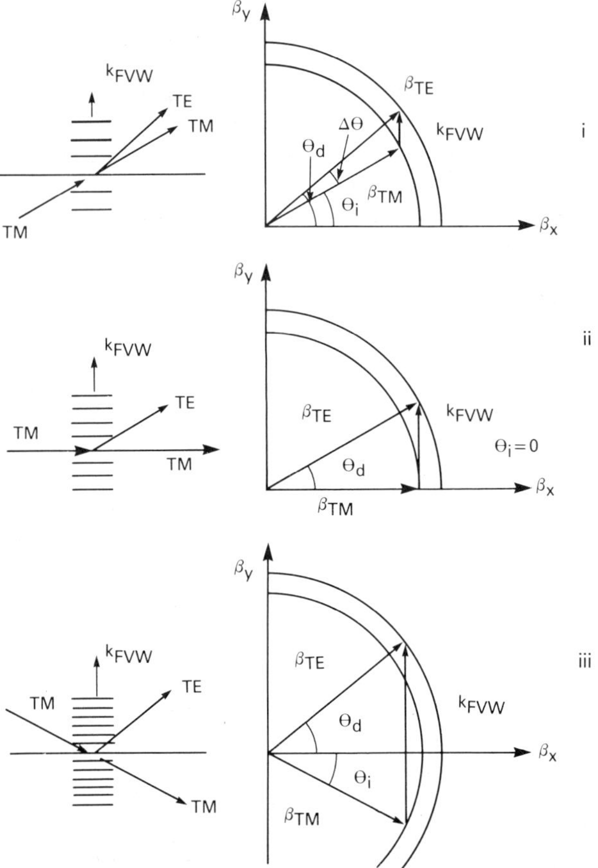

Figure 6.31 Geometries for guided-wave optical diffraction from magnetostatic forward volume waves. Three different wave vector magnitudes and their corresponding wave vector surface configurations for perfect phase matching are shown. Center sketch shows tangentially phase-matched case.

that incorporation of MgO was effective in controlling iron charge states. Perhaps the same type of charge compensation mechanism is responsible for success here.

Higher MSW power (greater than tens of milliwatts) causes multiple interactions to occur, which depend on static Faraday rotation and static and dynamic Cotton–Mouton effects. (In brief, the dynamic Faraday effect is that caused by the microwave frequency magnetic field. The static Faraday effect is caused by the static or biasing magnetic field. Likewise, the dynamic and static Cotton–Mouton effects depend on microwave and static fields.) The static effects occur even in the absence of MSWs. Even though the process is not well phase matched, some conversion of polarization between TM and TE modes of the same mode number occurs in the presence of a static magnetic field. It is not deflected and does not carry the frequency shift (a Doppler shift caused by diffraction from a moving grating) of MSW-diffracted light. The multiple interactions that occur at higher powers seem capable of significantly increasing the instantaneous bandwidth and also possibly of increasing the diffraction efficiency. In addition, interactions that are forbidden or negligible in the linear MSW power regime, such as transverse diffraction from surface MSWs, become possible. This work is described in Refs. 86–89.

At the instigation of this author, Stancil has recently investigated theoretically several concepts that may enhance viability of MSW–optical interactions in the linear power regime [90]. The first relies on two facts. First, an MSW has an evanescent tail that extends several wavelengths into the media bounding its waveguiding host film; this tail can modify the magneto-optical properties of the neighboring medium. Second, the optical phase change accumulated on reflection by an optical beam depends on the refractive index of the reflecting medium as well as that of the medium in which the light propagates. As is evident from the characteristic equation, a change in one of the ϕ_{ij} terms may be as effective in perturbing the optical wave as is a change in the propagating phase term caused by a perturbed refractive index. The ϕ_{ij} perturbation is particularly sensitive to changes in the interfacial refractive index near the critical angle of incidence (near cutoff). Modulating the refractive index of the reflecting medium can therefore cause diffraction in a similar way as does modulating the refractive index of the propagation medium. The combined impact of these observations is that the optical guided waves and MSWs need not propagate in the same garnet film to effect diffraction but can propagate in neighboring films whose optical and magnetic properties need not be the same. Material requirements for each of the neighboring films can be relaxed considerably. The second concept incorporates periodic variations in the properties of the film that supports the MSW oriented parallel to the MSW propagation direction. The periodic perturbation assists in satisfying the phase-matching criteria for efficient optical diffraction by enhancing the effective wave number magnitude of the MSW; the tangential phase-matching condition may be approached using this technique.

A few other efforts pertain to MSW–optical interactions. Reference 91 discusses the theory of TM to TE mode coupling caused by diffraction from MSW forward-volume waves. In particular, Huahui et al. predict from theory that coupling between differing mode numbers should dominate (over polarization-switching diffraction within the same mode number) at MSW frequencies that invoke large wave vector values, nearer to the high end of the frequency band for a given bias magnetic field value. (These wave vectors are more difficult to excite experimentally.) Further, they write that coupling is restricted to pairs of TM–TE modes that have the same symmetry (odd or even) in the waveguide. This constraint results from the nature of the expression for the coupling constant, which is an overlap integral between the electric field profiles across the waveguide of the polarization-rotated input mode and the diffracted mode. These field profiles are nearly orthogonal for modes of opposite symmetry; the overlap integral and the coupling constant are then negligible.

Probing of MSWs by Brillouin scattering of light incident on garnet films perpendicular to their surfaces has been used as a diagnostic tool for observing MSW propagation and dispersion. See Ref. 92.

Theoretical analysis of the correlation between static and dynamic magneto-optical effects on one hand and elastic and inelastic spin–photon scattering by electric and magnetic dipoles on the other has been presented by Le Gall and associates. For readers engaged in quantum interpretations, these papers provide particular insight to the physics of spin–particle interactions with photons. See Ref. 93 and the references therein.

Finally, Solomko and associates published a succession of papers of both theoretical and more recently experimental focus, detailing their work on MSW–optical interactions. Many of these papers are listed in Refs. 94 and 95.

REFERENCES

1. B. D. Cullity, *Introduction to Magnetic Materials*, Addison-Wesley, Reading, MA, 1972.
2. C. Kittel, in *Introduction to Solid State Physics*, 4th ed., Wiley, New York, 1971, Chaps. 14–17.
3. B. Lax and K. J. Button, *Microwave Ferrites and Ferrimagnetics*, McGraw-Hill, New York, 1962.
4. M. S. Sodha and N. C. Srivastava, *Microwave Propagation in Ferrimagnetics*, Plenum, New York, 1981.
5. P. Hansen et al., "Optical Switching with Bismuth-Substituted Iron Garnets," *Philips Tech. Rev.*, Vol. 41, p. 33, 1983/84.
6. J. F. Dillon, Jr., "Magneto-Optical Properties of Magnetic Garnets," in A. Paoletti, Ed., *Physics of Magnetic Garnets*, North-Holland, New York, 1978, p. 379.
7. W. Wettling, "Magneto-Optics of Ferrites," *J.* Magnet. Magn. Mat., Vol. 3, p. 147, 1976.
8. A. D. Fisher, "Optical Signal Processing with Magnetostatic Waves," *Circ., Syst., Sig. Process.*, Vol. 4, No. 1–2, p. 265, 1985.
9. M. J. Freiser, "A Survey of Magnetooptic Effects," *IEEE Trans. Magn.*, Vol. MAG-4, No. 2, p. 152, 1968.
10. J. F. Dillon, "Magneto-Optical Properties of Magnetic Crystals," in J. Smit, Ed., *Magnetic Properties of Materials*, McGraw-Hill, New York, 1971, Chap. 5.
11. F. A. Jenkins and H. E. White, *Fundamentals of Optics*, 4th ed., McGraw-Hill, New York, 1976.
12. M. A. Butler and E. L. Venturini, "High Frequency Faraday Rotation in FR-5 Glass," *Appl. Opt.*, Vol. 26, p. 1581, 1987.
13. N. Koshizuka and T. Okuda, "The Problems in the Development of Magneto-Optical Devices," *IEEE Trans. J. Magn. Jpn.*, Vol. TJMJ-1, p. 1044, 1985.
14. J.-P. Krumme, et al., "Bismuth-Substituted Iron Garnet Films Prepared by RF Diode Sputtering," *IEEE Trans. Magn.*, Vol. MAG-20, p. 983, 1984.
15. J.-P. Krumme et al., "Bismuth Iron Garnet Films Prepared by RF Magnetron Sputtering," *J. Appl. Phys.*, Vol. 57, p. 3885, 1985.
16. T. Okuda et al., "Faraday Rotation in Highly Substituted Yttrium Iron Garnet Films Prepared by Ion Beam Sputtering," *IEEE Trans. Magn.*, Vol. MAG-23, p. 3491, 1987.
17. Y. Toudic and M. Passaret, "Croissance por Voie Hydrothermale de Films Ferrimagnetiques; Epitaxies sur des Substrats de GdGaG," *J. Cryst. Growth*, Vol. 24/25, p. 621, 1974.
18. G. R. Pulliam, "Chemical Vapor Growth of Single Crystal Magnetic Oxide Films," *J. Appl. Phys.*, Vol. 38, p. 1120, 1967.
19. K. Maeto et al., "Bi-substituted DyIG Thin Films Prepared by Chemical Deposition for Magneto-Optic Recording Medium," *IEEE Trans. J. Magn. Jpn.*, Vol. TJMJ-2, p. 839, 1987.
20. P. Hansen and J.-P. Krumme, "Magnetic and Magneto-Optical Properties of Garnet Films," *Thin Solid Films*, Vol. 114, p. 69, 1984.
21. S. Geller, "Crystal and Static Properties of Garnets," in A. Paoletti, Ed., *Physics of Magnetic Garnets*, North-Holland, New York, 1978, p. 1.
22. C. Kittel, *Introduction to Solid State Physics*, 4th ed., Wiley, New York, 1971, p. 507.
23. P. Hansen et al., "Magnetic and Magneto-Optic Properties of Praseodymium- and Bismuth-Substituted Yttrium Garnet Films," *J. Appl. Phys.*, Vol. 60, p. 721, 1986.
24. V. J. Fratello et al., "Effect of Diamagnetic Substitution on Growth-Induced Anisotropy in $(YBi)_3Fe_5O_{12}$," *J. Appl. Phys.*, Vol. 60, p. 718, 1986.
25. V. Doormann et al., "Measurement of the Refractive Index and Optical Spectra of Epitaxial Bismuth Substituted Yttrium Iron Garnet Films at UV to Near-IR Wavelengths," *Appl. Phys. A*, Vol. 34, p. 223, 1984.
26. G. B. Scott and D. E. Lacklison, "Magnetooptic Properties and Applications of Bismuth Substituted Iron Garnets," *IEEE Trans. Magn.*, Vol. MAG-12, p. 292, 1976.

27. S. Wittekoek et al., "Magneto-Optic Spectra and the Dielectric Tensor Elements of Bismuth-Substituted Iron Garnets at Photon Energies between 2.2–5.2 eV," *Phys. Rev. B*, Vol. 12, p. 2777, 1975.

28. H. Tamada et al., "TM–TE Optical-Mode Conversion Induced by a Transversely Propagating Magnetostatic Wave in a $(BiLu)_3Fe_5O_{12}$ Film," *J. Appl. Phys.*, Vol. 64, p. 554, 1988.

29. H. Ohno et al., "Faraday Rotation of LPE Bi-Substituted Magnetic Garnet Films," *IEEE Trans. J. Maan. Jpn.*, Vol. TJMJ-1, p. 93, 1985.

30. T. Hibiya et al., "Growth and Characterization of Liquid-Phase Epitaxial Bi-Substituted Iron Garnet Films for Magneto-Optic Application," *Japan. J. Appl. Phys.*, Vol. 24, p. 1316, 1985.

31. A. Murata et al., "Optical Birefringence in BiNdLuAlIG Thin Film Waveguides," *IEEE Trans. J. Magn. Jpn.*, Vol. TJMJ-1, p. 91, 1985.

32. M. Imamura et al., "Bi-Content Dependence of Birefringence in LPE–YIG and GdIG Films," *IEEE Trans. J. Magn. Jpn.*, Vol. TJMJ-1, p. 95, 1985.

33. H. Tamada et al., "Bi-Substituted LPE Garnet Films with FMR Linewidth as Small as YIG," *IEEE Trans. J. Magn. Jpn.*, Vol. 3, p. 98, 1988.

34. G. Bate, "Materials Challenges in Metallic, Reversible, Optical Recording Media: A Review," *IEEE Trans. Magn.*, Vol. MAG-23, p. 156, 1987.

35. P. Hansen and M. Hartmann, "Magneto-Optical Properties of Iron Garnets and Amorphous Alloys and Their Use in Device Applications," in J. Rauluszkiewicz et al., Eds., *Proceedings of the Second International Conference on Physics of Magnetic Materials*, World Scientific Publishing, Singapore and Philadelphia, 1985, p. 158.

36. C. D. Wright et al., "Stability Phenomena in Amorphous Rare Earth–Transition Metal Films," *IEEE Trans. Magn.*, Vol. MAG-23, p. 162, 1987.

37. H.-P. D. Shieh and M. H. Kryder, "Magneto-Optic Recording Materials with Direct Overwrite Capability," *Appl. Phys. Lett.*, Vol. 49, p. 473, 1986.

38. M. H. Kryder et al., "Control of Parameters in Rare Earth–Transition Metal Alloys for Magneto-Optical Recording Media," *IEEE Trans. Magn.*, Vol. MAG-23, p. 165, 1987.

39. H.-P. D. Shieh and M. H. Kryder, "Operating Margins for Magneto-Optic Recording Materials with Direct Overwrite Capability," *IEEE Trans. Magn.*, Vol. MAG-23, p. 171, 1987.

40. F. Tanaka et al., "Magneto-Optical Recording Characteristics of TbFeCo Media by Magnetic Field Modulation Method," *Japan. J. Appl. Phys.*, Vol. 26, p. 231, 1987.

41. S. Tsunashima et al., "Magneto-Optic Kerr Effect of Amorphous Gd–Fe Films," *IEEE Trans. Magn.*, Vol. MAG-23, p. 3205, 1987.

42. F. Kirino et al., "Gd-Base Amorphous Magneto-Optical Films with Large Kerr Rotations," *IEEE Trans. J. Magn. Jpn.*, Vol. TJMJ-2, p. 1110, 1987.

43. M. Tanaka et al., "Magnetic Properties of Compositionally Modulated Tb/FeCo Films," *IEEE Trans. Magn.*, Vol. MAG-23, p. 2955, 1987.

44. R. Ohyama et al., "Preparation of PtMnSb Thin Films and Magneto-Optical Properties," *IEEE Trans. J. Magn. Jpn.*, Vol. TJMJ-1, p. 122, 1985.

45. T. Inukai et al., "Magneto-Optical Properties of Substituted Pt–Mn–Sb Thin Films," *IEEE Trans. J. Magn. Jpn.*, Vol. TJMJ-2, p. 1102, 1987.

46. A. Ito et al., "Magnetic and Magneto-Optical Properties of PtMnSb Thin Films," *IEEE Trans. J. Magn. Jpn.*, Vol. TJMJ-2, p. 1100, 1987.

47. T. Suzuki et al., "Magnetic and Magneto-Optical Properties of (Nd,Pr)–(Fe,Co) Alloy Amorphous Films with Huge Perpendicular Magnetic Anisotropy," *IEEE Trans. Magn.*, Vol. MAG-23, p. 2958, 1987.

48. H.-P. D. Shieh et al., "Magnetic Properties of Amorphous Tb(Nd,Pr)Co Films," *IEEE Trans. Magn.*, Vol. MAG-23, p. 3208, 1987.

49. M. Masuda et al., "Preparation, Magnetic and Magneto-Optic Properties of Small-Crystalline MnBi Films," *Japan. J. Appl. Phys.*, Vol. 26, p. 707, 1987.

50. M. H. Kryder, "Magneto-Optic Recording Technology," *J. Appl. Phys.*, Vol. 57, p. 3913, 1985.

51. E. Schultheiss et al., "Production Technology for Magnetooptic Data Storage Media," *Solid State Tech.*, p. 107, 1988.

52. S. Middlehoek et al., *Physics of Computer Memory Devices*, Academic, New York, 1976.

53. H.-P. D. Shieh and M. H. Kryder, "Dynamics and Factors Controlling Regularity of Thermomagnetically Written Domains," *J. Appl. Phys.*, Vol. 61, p. 1108, 1987.

54. H.-P. D. Shieh and M. Kryder, "Magneto-Optic Recording Materials with Direct Overwrite Capability," *Appl. Phys. Lett.*, Vol. 49, p. 473, 1986.

55. M. D. Schultz et al., "Performance of Magneto-Optical Recording Media with Direct Overwrite Capability," *J. Appl. Phys.*, Vol. 63, p. 3844, 1988.

56. H.-P. D. Shieh and M. Kryder, "Operating Margins for Magneto-Optic Recording Materials with Direct Overwrite Capability," *IEEE Trans. Magn.*, Vol. MAG-23, p. 171, 1987.

57. M. Masuda et al., "Preparation, Magnetic and Magneto-Optic Properties of Small-Crystalline MnBi Films," *Japan. J. Appl. Phys.*, Vol. 26, p. 707, 1987.

58. M. Kaneko et al., "Optical Operation of a Magnetic Bubble," *IEEE Trans. Magn.*, Vol. MAG-22, p. 2, 1986.

59. K. Balasubramanian and A. Macleod, "Performance Calculations for Multi-Layer Thin-Film Structures Containing a Magnetooptical Film," *Abstracts of the 1986Annual Meeting, Optical Society of America*, p. P28; *J. Opt. Soc. Amer. A*, Vol. 3, No. 13, p. P28, 1986.

60. M. Gomi et al., "RF-Sputtering of Highly Bi-Substituted Garnet Films on Glass Substrates for Magneto-Optic Memory," *IEEE Trans. J. Magn. Jpn.*, Vol. TJMJ-1, p. 75, 1985.

61. M. Gomi et al., "Bi-Substituted Garnet Films Crystallized During RF Sputtering for M-O Memory," *IEEE Trans. Magn.*, Vol. MAG-23, p. 2967, 1987.

62. T. Okuda et al., "Faraday Rotation in Highly Bi-Substituted Yttrium Iron Garnet Films Prepared by Ion Beam Sputtering," *IEEE Trans. Magn.*, Vol. MAG-23, p. 3491, 1987.

63. K. Maeto et al., "Bi-Substituted DyIG Thin Films Prepared by Chemical Deposition for Magneto-Optic Recording Medium," *IEEE Trans. J. Magn. Jpn.*, Vol. TJMJ-2, p. 839, 1987.

64. M. Mansipur, "Coercivity and Its Role in Thermomagnetic Recording," *J. Appl. Phys.*, Vol. 61, p. 3334, 1987.

65. M. Mansipur, "Magnetization Reversal Dynamics in Magneto-Optic Media" (invited), *J. Appl. Phys.*, Vol. 63, p. 3831, 1988.

66. M. Mansipur, "Magnetization Reversal, Coercivity, and the Process of Thermomagnetic Recording in Thin Films of Amorphous Rare Earth–Transition Metal Alloys," *J. Appl. Phys.*, Vol. 61, p. 1580, 1987.

67. M. Kaneko et al., "A Low Loss 0.8 Micron Band Optical Isolator Using Highly Bi-substituted LPE Garnet Film," *IEEE Trans. Magn.*, Vol. MAG-23, p. 3482, 1987.

68. K. Matsuda et al., "Bi-Substituted Rare-Earth Iron Garnet Composite Film with Temperature Independent Faraday Rotation for Optical Isolators," *IEEE Trans. Magn.*, Vol. MAG-23, p. 3479, 1987.

69. H. Hemme et al., "Optical Isolator Based on Mode Conversion in Magnetic Garnet Films," *Appl. Opt.*, Vol. 26, p. 3811, 1987.

70. R. Wolfe et al., "Thin-Film Waveguide Magneto-Optical Isolator," *Appl. Phys. Lett.*, Vol. 46, p. 817, 1985.

71. T. Mizumoto et al., "Measurement of Optical Nonreciprocal Phase Shift in a Bi-Substituted $Gd_3Fe_5O_{12}$ Film and Application to Waveguide-Type Optical Circulator," *J. Lightwave Tech.*, Vol. LT-4, p. 347, 1986.

72. S. Yamamoto and Y. Okamura, "Magneto-Optical Branching Waveguides and Their Applications to Nonreciprocal Optical Devices," *IEEE Trans. J. Magn. Jpn.*, Vol. TJMJ-1, p. 1037, 1985.

73. B. Hill and K. P. Schmidt, "Fast Switchable Magneto-Optic Memory—Display Components," *Philips J. Res.*, Vol. 33, p. 211, 1978.

74. W. E. Ross et al., "Fundamental Characteristics of the Litton Iron Garnet Magneto-Optic Spatial Light Modulator," *Advances in Optical Information Processing*, SPIE Vol. 388, p. 55, 1983.

75. G. R. Pulliam et al., "Large Stable Magnetic Domains," *J. Appl. Phys.*, Vol. 53, p. 2754, 1982.

76. G. L. Nelson et al., "Stripe Domain Light Deflection for Intersattelite Communications," *Communication Networking in Dense Electromagnetic Environments*, SPIE Vol. 876, p. 121, 1988.

77. H. Kogelnik, "Theory of Dielectric Waveguides," in T. Tamir, Ed., *Integrated Optics*, Springer-Verlag, New York, 1975, p. 13.

78. A. D. Fisher, "Optical Signal Processing with Magnetostatic Waves," in J. P. Parekh, Ed., *Circuits, Systems, and Signal Processing*, Vol. 4 (Special Issue on Magnetostatic Waves and Applications to Signal Processing), 1985, p. 265.

79. A. E. Craig et al., "Characterization of Ferrimagnetic Garnets for MSW—Optical Diffraction," *Proc. IEEE Ultrasonics Symp.*, p. 174, 1985.

80. Y. H. Won et al., "Three-Prism Loss Measurements of Optical Waveguides," *Appl. Phys. Lett.*, Vol. 37, p. 269, 1980.

81. A. D. Fisher et al., "Optical Guided-Wave Interactions with Magneto-Static Waves at Microwave Frequencies," *Appl. Phys. Lett.*, Vol. 41, p. 779, 1982.

82. A. D. Fisher et al., "Magnetostatic Wave Devices for Integrated-Optical Signal Processing," *Proc. IEEE Ultrason. Symp.*, p. 226, 1983.

83. A. C.-T. Wey et al., "Enhanced-Bandwidth MSFVW—Optical Interaction Employing an Inhomogeneous Bias Field," *Proc. IEEE Ultrason. Symp.*, p. 173, 1986.

84. C.-T. Wey et al., "Inhomogeneous Field MSFVW—Optical Interaction," *Integrated Optical Circuit Engineering*, SPIE Vol. 704, p. 51, 1986.

85. H. Tamada, "Bi-Substituted LPE Garnet Films with FMR Linewidth as Small as YIG," *IEEE Trans. J. Magn. Jpn.*, Vol. 3, p. 98, 1988.

86. C. S. Tsai et al., "Noncollinear Coplanar Magneto-Optic Interaction of Guided Optical Wave and Magnetostatic Surface Waves in Yttrium Iron Garnet–Gadolinium Gallium Garnet Waveguides," *Appl. Phys. Lett.*, Vol. 47, p. 651, 1985.

87. D. Young et al., "Tunable Wideband Guided Wave Magneto-Optic Modulator Using Magnetostatic Surface Wave," *Acousto-Optic, Electro-Optic and Magneto-Optic Devices and Applications*, SPIE Vol. 753, p. 161, 1987.

88. D. Young and T. S. Tsai, "GHz Bandwidth Magneto-Optic Interaction in Yttrium Iron Garnet–Gadolinium Gallium Garnet Waveguide Using Magnetostatic Forward Volume Waves," *Appl. Phys. Lett.*, Vol. 53, p. 1696, 1988.

89. S. H. Talisa, "The Collinear Interaction between Forward Volume Magnetostatic Waves and Guided Light in YIG Films," *IEEE Trans. Magn.*, Vol. MAG-24, p. 2811, 1988.

90. D. D. Stancil, "Theoretical Investigations for MSW–Optical Interactions," Final Report on U.S. Naval Research Laboratory contract N00173-88-M-X012, September 12, 1988.

91. H. Huahui et al., "Optical Mode Conversion in a Ferrimagnetic Film Containing Magnetostatic Forward Volume Wave," *IEEE Trans. Magn.*, Vol. MAG-23, p. 3500, 1987.

92. G. Srinivasan et al., "Characterization of Magnetostatic Wave Devices by Brillouin Light Scattering," *IEEE Trans. Magn.*, Vol. MAG-23, p. 3718, 1987.

93. H. le Gall, "First and Second-Order Inelastic Scatterings of Light by High Amplitudes Spin-Waves in Ferrimagnetic Crystals," in M. Balkanski, Ed., *Proceedings of the Second International Conference on Light Scattering in Solids*, Flammarion Sciences, New York, 1971, p. 170.

94. A. A. Solomko et al., "Interaction of Laser Radiation with Surface Magnetostatic Waves in Iron–Garnet Films," *Opt. Spectrosc. (USSR)*, Vol. 59, p. 381, 1985.

95. A. A. Solomko et al., "Collinear Interaction of Light with Surface Magnetostatic Waves in Ferrite–Garnet Films," *Opt. Spectrosc. (USSR)*, Vol. 61, p. 804, 1986.

SUGGESTIONS FOR FURTHER READING

6.2 Units

D. J. Craik, *Structure and Properties of Magnetic Materials*, Pion Ltd., London, 1971.

B. D. Cullity, *Introduction to Magnetic Materials*, Addison-Wesley, Reading, MA, 1972.

6.3 Material Physics of Magnetics

S. Chikazumi, *Physics of Magnetism*, Wiley, New York, 1964.

B. D. Cullity, *Introduction to Magnetic Materials*, Addison-Wesley, Reading, MA, 1972.

C. Kittel, in *Introduction to Solid State Physics*, 4th ed., Wiley, New York, 1971, Chaps. 14–17.

B. Lax and K. J. Button, *Microwave Ferrites and Ferrimagnetics*, McGraw-Hill, New York, 1962.

M. McCaig and A. G. Clegg, *Permanent Magnets in Theory and Practice*, Wiley, New York, 1987.

M. S. Sodha and N. C. Srivastava, *Microwave Propagation in Ferrimagnetics*, Plenum, New York, 1981.

6.4 Spin Waves

B. Lax and K. J. Button, *Microwave Ferrites and Ferrimagnetics*, McGraw-Hill, New York, 1962.

J. P. Parekh, Ed., *Circuits, Systems, and Signal Processing* (Special Issue on Magnetostatic Waves and Applications to Signal Processing), Vol. 4, No. 1–2, 1985.

M. S. Sodha and N. C. Srivastava, *Microwave Propagation in Ferrimagnetics*, Plenum, New York, 1981.

D. D. Stancil, *Magnetostatic Waves* (in press).

M. G. Cottam and D. R. Tilley, *Introduction to Surface and Superlattice Excitations*, Cambridge University Press, Cambridge and New York, 1989.

P. E. Tannenwald, "Spin Waves," *Microwave J. for July*, 1959, p. 25.

P. E. Wigen, "Magnetic Excitations," in A. Paoletti, Ed., *Physics of Magnetic Garnets*, North-Holland, New York, 1978, p. 196.

P. E. Wigen, "Microwave Properties of Magnetic Garnet Thin Films," *Thin Solid Films*, Vol. 114, No. 1/2, p. 135, 1984; Special Issue on Magnetic Garnet Films, A. Paoletti, guest editor.

6.5 Material Physics of Optics

D. Clarke and J. F. Grainger, *Polarized Light and Optical Measurement*, Pergamon, New York, 1971.

P. Hlawiczka, *Gyrotropic Waveguides*, Academic, New York, 1981.

E. Hecht and A. Zajac, *Optics*, Addison-Wesley, Reading, MA, 1974.

R. G. Hunsperger, Ed., *Integrated Optics: Theory and Technology*, 2nd ed., Springer-Verlag, New York, 1985.

F. A. Jenkins and H. E. White, *Fundamentals of Optics*, 4th ed., McGraw-Hill, New York, 1976.

G. B. Scott and D. E. Lacklison, "Magnetooptic Properties and Applications of Bismuth Substituted Iron Garnets," *IEEE Trans. Magn.*, Vol. MAG-12, No. 4, p. 292, 1976.

T. Tamir, Ed., *Integrated Optics*, Springer-Verlag, New York, 1975.

W. Wettling, "Magneto-Optics of Ferrites," *J. Magnet. Magn. Mat.*, Vol. 3, p. 147, 1976.

6.6 Magneto-optics

J. F. Dillon, Jr., "Magneto-Optical Properties of Magnetic Crystals," in J. Smit, Ed., *Magnetic Properties of Materials*, McGraw-Hill, New York, 1971, Chap. 5, p. 149.

J. F. Dillon, Jr., "Magneto-Optical Properties of Magnetic Garnets," in A. Paoletti, Ed., *Physics of Magnetic Garnets*, North-Holland, New York, 1978, p. 379.

M. J. Freiser, "A Survey of Magnetooptic Effects," *IEEE Trans. Magn.*, Vol. MAG-4, No. 2, p. 152, 1968.

G. S. Monk, in *Light: Principles and Experiments*, 2nd ed., Dover, New York, 1963, Chap. 16.

G. B. Scott and D. E. Lacklison, "Magnetooptic Properties and Applications of Bismuth Substituted Iron Garnets," *IEEE Trans. Magn.*, Vol. MAG-12, No. 4, p. 292, 1976.

J. C. Suits, "Faraday and Kerr Effects in Magnetic Compounds," *IEEE Trans. Magn.*, Vol. MAG-8, No. 1, p. 95, 1972.

6.7 Real Materials

J. O. Artman, "Magnetic Anisotropy and Structure of Magnetic Recording Media," *J. Appl. Phys.*, Vol. 61, p. 3137, 1987.

G. Bate, "Materials Challenges in Metallic, Reversible, Optical Recording Media: A Review," *IEEE Trans. Magn.*, Vol. MAG-23, p. 156, 1987.

A. H. Eschenfelder, *Magnetic Bubble Technology*, Springer-Verlag, New York, 1981.

P. Hansen and J.-P. Krumme, "Magnetic and Magneto-Optical Properties of Garnet Films," *Thin Solid Films*, Vol. 114 (Special Issue on Magnetic Garnet Films), p. 69, 1984.

P. Hansen and K. Witter, "Growth-Induced Uniaxial Anisotropy of Bismuth-Substituted Iron-Garnet Films," *J. Appl. Phys.*, Vol. 58, p. 454, 1985.

P. Hansen et al., "Optical Switching with Bismuth-Substituted Iron Garnets," *Philips Tech. Rev.*, Vol. 41, p. 33, 1983/84.

P. Hansen et al., "Magnetic and Magneto-Optical Properties of Bismuth-Substituted Lutetium Iron Garnet Films," *Phys. Rev. B*, Vol. 31, p. 5858, 1985.

M. Kaneko et al., "Optical Operation of a Magnetic Bubble," *IEEE Trans. Magn.*, Vol. MAG-22, p. 2, 1986.

B. Knorr and W. Tolksdorf, "Lattice Parameters and Misfits of Gallium Garnets and Iron Garnet Epitaxial Layers at Temperatures between 294 and 1300 K," *Mat. Res. Bull.*, Vol. 19, p. 1507, 1984.

G. Nelson and W. A. Harvey, "Optical Absorption Reduction in $Bi_1Lu_2Fe_5O_{12}$ Garnet Magneto-Optical Crystals," *J. Appl. Phys.*, Vol. 53, p. 1687, 1982.

A. Paoletti, Ed., *Physics of Magnetic Garnets*, North-Holland, New York, 1978.

W. Reim et al., "$Tb_xNd_y(FeCo)_{1-x-y}$: Promising Materials for Magneto-Optical Storage?" *J. Appl. Phys.*, Vol. 61, p. 3349, 1987.

B. Strocka et al., "An Empirical Formula for the Calculation of Lattice Constants of Oxide Garnets Based on Substituted Yttrium- and Gadolinium-Iron Garnets," *Philips J. Res.*, Vol. 33, p. 186, 1978.

W. Tolksdorf and C.-P. Klages, "The Growth of Bismuth Iron Garnet Layers by Liquid Phase Epitaxy," *Thin Solid Films*, Vol. 114 (Special Issue on Magnetic Garnet Films), p. 33, 1984.

W. H. von Aulock, Ed., *Handbook of Microwave Ferrite Materials*, Academic Press, New York, 1965.

6.8 Modern Magneto-optical Devices

G. Bouwhuis et al., *Principles of Optical Disc Systems*, Adam Hilger Ltd., Boston, 1985.

A. D. Fisher, "Optical Signal Processing with Magnetostatic Waves," in J. P. Parekh, Ed., *Circuits, Systems, and Signal Processing* (Special Issue on Magnetostatic Waves and Applications to Signal Processing), Vol. 4, p. 265, 1985.

R. G. Hunsperger, Ed., *Integrated Optics: Theory and Technology*, 2nd ed., Springer-Verlag, New York, 1985.

N. S. Kapany and J. J. Burke, *Optical Waveguides*, Academic, New York, 1972.

N. Koshizuka and T. Okuda, "The Problems in the Development of Magneto-Optical Devices," *IEEE Trans. J. Magn. Jpn.*, Vol. TJMJ-1, p. 1044, 1986.

M. H. Kryder, "Magneto-Optic Recording Technology," *J. Appl. Phys.*, Vol. 57, p. 3913, 1985.

D. Marcuse, *Theory of Dielectric Optical Waveguides*, Academic, New York, 1974.

S. Middlehoek et al., *Physics of Computer Memory Devices*, Academic, New York, 1976.

P. Paroli, "Magneto-Optical Devices Based on Garnet Films," *Thin Solid Films*, Vol. 114 (Special Issue on Magnetic Garnet Films), p. 187, 1984.

W. E. Ross et al., "Two-Dimensional Magneto-Optical Spatial Light Modulator for Signal Processing," *Opt. Eng.*, Vol. 22, p. 485, 1983.

T. Tamir, Ed., *Integrated Optics*, Springer-Verlag, New York, 1975.

K. Tsushima, "Magneto-Optics: Its Past and Present," *IEEE Trans. J. Magn. Jpn.*, Vol. TJMJ-1, p. 920, 1985.

K. Tsushima and N. Koshizuka, "Research Activities on Magneto-Optical Devices in Japan," *IEEE Trans. Magn.*, Vol. MAG-23, p. 3473, 1987.

7

OPTICAL DETECTORS

P. K. L. Yu

Department of Electrical and Computer Engineering
University of California, San Diego
La Jolla, California

H. D. Law

PCO, Inc.
Chatsworth, California

7.1 INTRODUCTION

Any device that converts photon energy into some other form of energy and couples it into a detection circuit can be regarded as an optical detector. For instance, the human eye is an optical detector that excels in its ability to perform parallel signal processing and to adjust itself to the changes in background intensity despite the slow response speed and spectral and sensitivity limitations of its photosensitive cells [1]. Presently, optical detectors play a distinguished role in communication systems that employ photons as information carriers. To this end, devices such as photodiodes, photoconductors, and phototransistors have been developed. A number of basic requirements are common to these devices. They have to be highly sensitive to low-level optical signals; this implies both low noise level and a high responsivity or quantum efficiency to incident photons. In addition, high response speed is an essential requirement for large information capacity in communication systems.

This chapter deals with the basic principles of operation and key performance parameters of optical detectors; it represents an introductory reading for engineers using optical detectors. Consequently, details of most of the relevant equations are omitted, and the reader is referred to other treatises [2–6]. As the needs of optical fiber communication systems have been responsible for the recent development in optical detectors, properties pertinent to these applications will be emphasized in the following sections. Section 7.2 describes the terminology for optical detection used in the literature, including detection parameters and noise parameters. Material issues pertaining to photodetection are also discussed. Section 7.3 describes the characteristics of various types of photodiodes and avalanche photodiodes. Section 7.4 describes photoconductive detectors and some of their applications. Section 7.5 briefly reviews some recent developments of novel optical detectors. In particular, phototransistors and solid-state photomultipliers are described.

7.2 TERMINOLOGY IN OPTICAL DETECTION AND MATERIAL CONSIDERATIONS

Semiconductor materials are employed for optical detection because photons can be absorbed in a quantum-mechanical process in which electrons interacting with such photons make a transition from one state to another. These electrons are subsequently detected in an external circuit. The absorption is dependent on incident photon energy and produces specific absorption spectra for materials. The absorption coefficient α_0 is defined as the incremental decrease in optical intensity per unit length per unit incident power. The photon energy $h\nu$ (where h is Planck's constant and ν is frequency) and the wavelength λ are related by

$$\lambda = \frac{1.24}{h\nu} \tag{7.1}$$

where λ is in micrometers and $h\nu$ is in electron volts.

The coefficient α_0 describes the exponential decay of the optical power as a function of its penetration into a given material. In general, a large α_0 is desirable for getting a large output signal. However, too large an α_0 results in a small penetration depth, and the photogenerated electrons can be lost via recombination at the surface; they will not be collected, therefore, in an external circuit. Optimization of α_0 depends on both material and structural aspects of the detector.

The quantum efficiency η of an optical detector is defined as the ratio of the number of electrons generated upon absorption to the number of photons absorbed before any photogain occurs. In general, the quantum efficiency depends on the absorption coefficient of the material and the dimensions of the absorption region [e.g., see Eqs. (7.7) and (7.25)]. The photogain M can arise from carrier injection in semiconductor materials, as in the case of photoconductive devices [see Eq. (7.22)], or from impact ionization, as in the case of avalanche photodiodes [see Eqs. (7.14) and (7.15)]. The responsivity R of a detector describes the dependence of output response on input optical power and is expressed as

$$R = M\eta \frac{q}{h\nu} \quad \text{(A/W)} \tag{7.2}$$

where q is the electronic charge.

Response speed of a photodetector is measured in terms of the signal decrement of 3 dB at a characteristic 3-dB frequency ($f_{3\mathrm{dB}}$) in the spectral response where the normalized photocurrent (or voltage) is plotted versus the signal frequency. The $f_{3\mathrm{dB}}$ is related to the response time τ of the detector by

$$f_{3\mathrm{dB}} = \frac{1}{2\pi\tau} \tag{7.3}$$

The response time constant can be related to the transit time of electrons and holes in the detector, carrier diffusion process, RC time constant, and carrier multiplication processes in the semiconductor. In practice, τ can be measured by observing the dependence of the output response of the detector on the pulsed optical signal. If the rise time τ_{rise} and fall time τ_{fall} of the optical pulse are so short that it appears as a square wave to the detector, then τ is determined by the larger of the τ_{rise} and τ_{fall} of the signal current. [The rise (fall) time is defined as the time for the photosignal to increase (decrease) from 10% (90%) to 90% (10%).]

For detectors with gain, an additional figure of merit, the so-called gain–bandwidth product (GB) is defined. It is simply the product of the 3-dB bandwidth and the dc gain. However, in many cases, GB is found to be insensitive to device geometry and carrier dynamics and only depends on some of the basic material properties.

As mentioned earlier, noise characteristics are important criteria of detectors. Three commonly used measurements of noise performance are the signal-to-noise (S/N) ratio, noise equivalent power (NEP), and specific detectivity (D^*) [2]. The S/N ratio is simply the ratio of detected electrical signal power to the noise power at the output of the detector (or receiver). The NEP of a detector is the optical power that must be incident on the detector so that S/N equals unity for a given wavelength, detector temperature, and bandwidth. The specific detectivity is the inverse of the NEP normalized to the square root of the detector area A and the system bandwidth B,[†]

$$D^* = \frac{\sqrt{AB}}{\text{NEP}} \tag{7.4}$$

Common noise sources are background optical radiation, the Poisson (shot) noise due to the quantized nature of photons and finite response time of the detector; dark-current noise due to generation-recombination in the depletion region, the neutral regions, as well as at the surface of detectors; and thermal (Johnson) noise due to the resistance of the external load [4].

In selecting a semiconductor material for efficient optical detection, several factors must be considered. First, the band-gap energy E_g of the absorption region must be smaller than the photon energy in order to get a large α_0. Second, a direct band-gap material is to be preferred to an indirect gap material due to the larger absorption coefficient of the former. Third, close matching of crystal lattice constants of different materials within a detector structure is important in order to minimize trapping of photogenerated carriers by interfacial defects and dislocations.

For band-to-band transition of electrons that involves the absorption of a photon—in the so-called intrinsic absorption process—the photon energy must be larger than that of the band gap to achieve a large α_0 [7]. There are other ways that can lead to the absorption of photons with an energy less than the band-gap energy. For example, in extrinsic materials, an impurity-assisted transition occurs when the photon absorption causes a transition from an impurity level within the band gap to the conduction band. Free-carrier absorption can also occur when the electron makes a transition within the same band while a photon is being absorbed. However, free-carrier absorption plays an important role only when the electron (or hole) density is larger ($>10^{17}$–10^{18} cm^{-3}). The Franz–Keldysh effect produced by the electric-field-dependent shift of the absorption edge is also responsible for below-band-gap energy absorption [8,9]. However, the band-to-band absorption process is preferred in most cases as it has the largest number of possible initial and final states for the absorptive transition.

In contrast to the direct band-gap semiconductor materials, indirect band-gap materials rely on a phonon-assisted transition process for absorption near the band-gap energy. This process has a small probability to occur. Consequently, for photon energy near the band-gap energy, a relatively smaller absorption coefficient is observed [10].

[†] While the noise equivalent power is normally used to characterize the sensitivity of a photodetector, the specific detectivity is used when the incident flux is larger in area than the detector.

It is important to have good quality interface between layers of different materials in an optical detector. Any interfacial states or traps may cause degradation of the dark-current characteristics and device reliability. Also, they can reduce the response speed. For optimal designs, other device parameters, such as device dimensions, have to be considered. However, two material-related issues are commonly encountered: the doping profile in the high-field region of the device and the nature of the metal–semiconductor contact. The doping level affects the depletion region width, and in many cases this region coincides with the absorption region. More importantly, the depletion region width affects the capacitance of the photodetector and can degrade the response speed. A low doping level of the depletion region is often desirable for fast carrier sweep out at low bias voltage as well as for small capacitance values.

Digression. The *RC* limitation is due to the time required to charge and discharge the photodiode capacitances. Capacitance C denotes the sum of the junction capacitance (C_j) and other parasitic (stray) capacitance (C_s) related to the contact and the package; R is the resistance of the load. The junction capacitance C_j of the photodiode is given by

$$C_j = \frac{\varepsilon A}{W} \tag{7.5a}$$

where ε is the dielectric constant; A is the junction area of the diode, and W is the depletion region thickness. In many cases, one-sided, abrupt junction photodiodes are used. In the case of a p^+n diode, W can be related to the doping of the n layer, N_D, and the bias voltage V through Poisson's equation:

$$W = \sqrt{\frac{2V\varepsilon}{qN_D}} \tag{7.5b}$$

From Eqs. (7.5a) and (7.5b), one can reduce the diode capacitance by reducing the doping in the active region. For instance, a 50-μm-diameter InGaAs photodiode will have a capacitance of approximately 0.06 pF at a bias of 5 V if the doping of the active layer can be kept below 5×10^{14} cm^{-3}. To further reduce the junction capacitance, one can either reduce the area of the photodiode or increase the depletion region width (say by increasing the intrinsic region thickness of the photodiode). However, the minimal area is determined by the manner in which the optical signal is coupled to the photodiode. Complicated optical coupling schemes may be used to focus the beam; in this way the photodiode area can be reduced to a few micrometers in diameter. The depletion region width is also related to the quantum efficiency and the transit time, so that in general, an optimization process is necessary. It turns out that in practice, after the junction capacitance is reduced to below a certain value, say around 0.05 pF, further reduction can be obtained only if the parasitic capacitance can be reduced.

The nature of the metal–semiconductor contact can affect carrier injection (and thus gain) and the response speed since the contact itself is an energy barrier. Studies of metal–semiconductor contacts on III–V semiconductors and their effects on the performance of detectors are part of current research [11].

In addition to the *RC* time effect, the response speed of detectors can be affected by carrier diffusion and drift. On the one hand, materials with a long minority carrier diffusion length assure the collection of the carriers generated outside the high-field region and thus provide a high quantum efficiency. On the other hand, the detector response time can be degraded by the slow diffusion process. The average diffusion velocity v_{Diff}

is defined as

$$v_{\text{Diff}} = \frac{L_p}{\tau_p} = \frac{D_p}{L_p} \tag{7.6a}$$

for minority carriers in an n-type material, where L_p is the hole diffusion length, D_p is the diffusion coefficient for holes, and τ_p is the minority carrier lifetime for holes. A similar equation can be written for electrons in p-type materials. At low carrier concentration, the diffusion constants are related to the carrier mobility by the Einstein relationship

$$D = \frac{kT}{q}\mu \tag{7.6b}$$

where μ is the corresponding carrier mobility, k is Boltzmann's constant, and T is the temperature measured in kelvin.

The transit time can be minimized by reducing the depletion region width. However, this also increases the capacitance of the photodiode. Therefore, a well-designed diode should have a doping profile that optimizes the operating voltage and achieves the highest speed. This usually occurs when the RC time constant and the transit time are comparable in value.

As noted earlier, the photogenerated carriers outside the depletion region will have to rely on the slow diffusion process to reach the depletion region. Because of this, the photodiode response exhibits a long "diffusion tail" when excited by a short optical pulse.

Most direct band-gap III–V materials have short carrier lifetimes and a large electron mobility, in addition to a large absorption coefficient (10^4–10^5 cm^{-1}). These properties make them attractive for detector applications that require fast response time and high responsivity. For example, $In_{0.53}Ga_{0.47}As$, whose quantum efficiency (because of the difference in photon energy, responsivity at 1.55 μm is higher) is flat over the 1.3–1.55 μm wavelength range, is suitable for optical fiber communication. On the other hand, II–VI materials such as HgCdTe, which can in principle be used for 1.3 μm wavelength detection, are handicapped by the long diffusion length and result in a slow response time. Silicon can be used for 0.8 μm wavelength detection where its absorption coefficient is $\sim 10^3$ cm^{-1} (due to its indirect band gap), and this corresponds to an absorption length of ~ 10 μm. Since its diffusion length is about 25 μm, diffusion effects can severely impede its applications in high-speed optical detection. An alternative approach is to apply an electric field over the entire absorption region so that most of the photogenerated carriers are transported by drift rather than by diffusion [12]. This approach requires a low dopant concentration in silicon, as noted from Eq. (7.5b). Figure 7.1 shows the absorption coefficient versus wavelength for silicon, germanium, GaAs, and InGaAs.

7.3 PHOTODIODES

Photodiodes usually have an absorption region where a strong electric field is applied in such a manner that the photogenerated carriers can be swept out to produce a current or voltage signal in external electronic circuitry. Generic photodiodes are *pn* junction photodiodes, *p*-intrinsic-*n* (PIN) photodiodes, Schottky junction photodiodes, and avalanche photodiodes (APDs).

A *pn* photodiode usually consists of a p^+n (or n^+p) junction in which the optical signal is incident on the top or the bottom surfaces, which are perpendicular to the

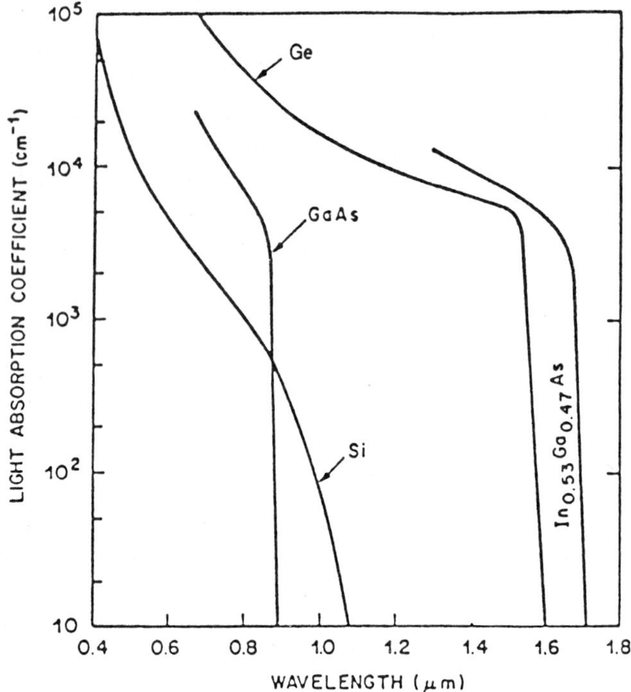

Figure 7.1 Absorption coefficient vs. wavelength for silicon, germanium, gallium arsenide, and indium gallium arsenide.

junction (see Fig. 7.2). For some applications, such as in switching high power, the optical signal can also be incident in a direction parallel to the junction [13]. In normal operation, the diode is reverse biased, and the electron–hole pairs photogenerated in the depletion region and those reached there by diffusion are separated by the electric field and are collected at the opposite electrodes. The resistive load of the external circuit senses the signal current as the photogenerated carriers transverse the depletion region. The induced current stops when the carriers reach the bulk material outside the depletion region. This phenomenon is responsible for the transit time limit of the response speed. Depending on the velocity (v) of the carriers and the width (W) of the depletion region, a different transit time T_{tr} ($T_{tr} = W/v$) can be obtained. Typically, for III–V compounds each micrometer of the depletion region corresponds to approximately 10 ps of transit time.

The quantum efficiency η of the photodiode shown in Fig. 7.2 with a thin p region is given by

$$\eta = (1 - R_0)\left[e^{-\alpha_0 W_1}(1 - e^{-\alpha_0 W}) + (e^{-W_1/L_n} - e^{-\alpha_0 W_1})\left(\frac{\alpha_0}{\alpha_0 - 1/L_n}\right) + e^{-\alpha_0(W_1 + W)}\left(\frac{\alpha_0}{\alpha_0 + 1/L_p}\right)\right] \tag{7.7}$$

where R_0 denotes the reflectivity at the air–semiconductor interface, W_1 is the p region width, and L_n and L_p are the diffusion lengths of electrons in the p region and holes in the n region, respectively. The first term inside the bracket of Eq. (7.7) accounts for the

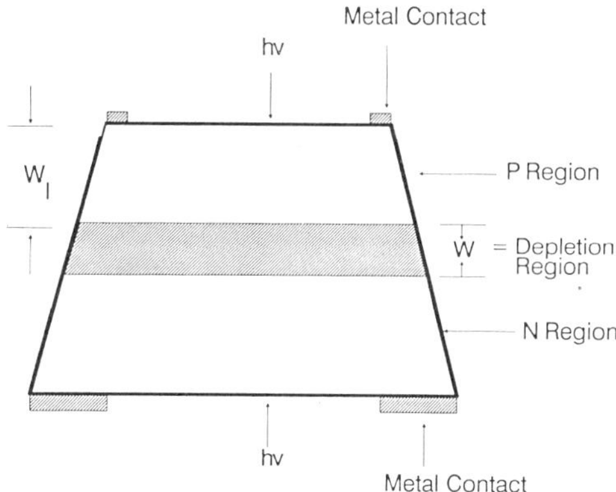

Figure 7.2 A *pn* junction photodiode with a surface reflectivity R.

photogenerated carriers within the depletion region; the second and third terms account for those in the p and n regions, respectively. For most III–V photodiodes in which the absorption coefficient is large, the absorption in the n region can be ignored. In the case of a heterojunction photodiode where the p region is fabricated by a material transparent to incident photons, W_1 can be set to zero. As can be seen from Eq. (7.7), η can be enhanced by reducing R_0; this is achieved by depositing antireflection coating on the incident facet of the diode. The quantum efficiency in some well-designed heterojunction III–V photodiodes can be as high as 95% [14].

There is no photogain in junction photodiodes where carriers are swept by a moderate electric field (10^3–10^4 V/cm). The responsivity can be calculated from Eq. (7.2) by setting M to unity. For instance, *pn* junction photodiodes made on silicon with a resistivity less than 100 Ω-cm have a depletion width of about 10 μm and a responsivity of ~0.4 A/W at 0.8–0.9 μm wavelength [15].

For simple *pn* junction photodiodes, it is thus difficult to optimize the quantum efficiency and the transit time at the same time due to the fact that the depletion region width depends on the applied electric field. This is further complicated by the nonlinear relationship between electron velocity and the electric field.

7.3.1 PIN Photodiodes

The PIN photodiode, which consists of a low-doped (either i or π) region sandwiched between p and n regions, represents a better alternative to the simple *pn* photodiode. This is because at a relatively low reverse bias, the intrinsic region is fully depleted and the total depletion width, including those from the p and n regions, remains almost constant as the peak electric field inside the depletion region is increased. A schematic diagram of a PIN photodiode is shown in Fig. 7.3*a* with light incident from the left; its energy band diagram at reverse bias is depicted in Fig. 7.3*b*. Typically, for a doping concentration of $\sim 10^{14}$ cm^{-3} in the intrinsic region, a bias voltage of 5–10 V is sufficient to deplete several micrometers, and the electron velocity also reaches the saturation value. However, as mentioned earlier, transit time is not the only factor affecting the response speed of the PIN photodiode. Other effects such as carrier diffusion and the RC time constant can also be important. For p and n regions less than one diffusion length, the response time due to diffusion alone is typically 1 ns/μm in p-type silicon and about 100 ps/μm in p-type III–V materials [16]. The corresponding value for n-type III–V materials

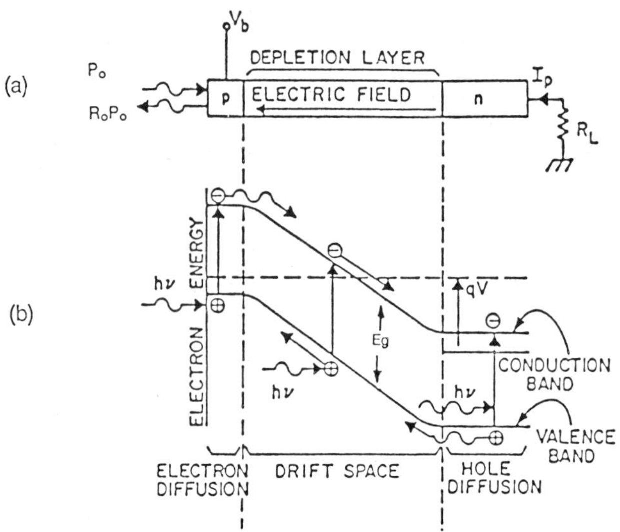

Figure 7.3 A PIN photodiode schematics. (*a*) Structure: P_0, incident power; V_b, bias voltage; R_0, reflectivity; I_p, photocurrent; R_L, load resistor. (*b*) Energy band diagram at reverse bias: E_g, band gap energy. (From H. Melchoir, *J. Lumin.*, Vol. 7, p. 390, 1973, with permission.)

is several nanoseconds per micrometer due to the lower mobility of holes. To minimize the diffusion effect, one can employ very thin *p* and *n* regions; alternatively, one can employ a *p* or *n* region that is transparent to the incident light, as in the case of heterojunction PIN photodiodes; or one can employ a combination of both [14,17].

For wavelengths in the 1.0–1.55 μm range, which is suitable for optical fiber communication, germanium and a few III–V compound semiconductor alloys stand out as candidate materials for PIN photodiodes, primarily because of their large absorption coefficients (α_0). The absorption edge of germanium is near 1.6 μm at room temperature. However, its α_0 is almost flat in the 10^4 cm^{-1} range over the 1.0–1.55 μm wavelength region [18]. In comparison, silicon's absorption edge is near 1.1 μm. There exist some problems, however, in using germanium for making PIN photodiodes. Due to its small band gap, the dark current arising from generation-recombination across the band gap can degrade the signal-to-noise ratio [19]. This is further aggravated by surface recombination [20]. So far, no satisfactory surface passivation technique exists for germanium PIN photodiodes, and the surface leakage current tends to be very high and unstable, especially at high ambient temperature.

Currently, III–V compound semiconductor materials are overtaking germanium as materials for fiber-optical compatible detectors [21–23]. By properly selecting the material composition of III–V materials, they can be lattice matched to each other. As commented earlier on heterostructure PIN photodiodes, one can choose a material composition at the intrinsic region with a band gap smaller than those of the *p* and *n* regions, while at the same time its absorption edge is just above that of the photon wavelength. This assures high quantum efficiency, high response speed, and low dark current [24]. The materials of particular interest are the ternary alloys $In_{0.53}Ga_{0.47}As$ and the quaternary $In_xGa_{1-x}As_yP_{1-x}$ alloys lattice matched to InP. As in the case with germanium, unpassivated InGaAs surfaces also exhibit an unstable dark current. However, this problem can be alleviated by passivating the surface with polyimide [25]. PIN detectors of excellent performances in the 1–1.6 μm wavelength range have been demonstrated in the InGaAs/InP material system.

Two generic InGaAs/InP PIN photodiodes, top-illuminated and back-illuminated photodiodes, have been developed [26,27]. A simple front-illuminated diffused InGaAs/InP homojunction photodiode is shown in Fig. 7.4*a*, where a layer of $In_{0.53}Ga_{0.47}As$ with an absorption edge at 1.65 μm is grown on top of the InP material. The residual doping level for a liquid phase epitaxially (LPE) grown InGaAs layer is usually in the 5×10^{14}–1×10^{16} cm^{-3} range. For similar materials grown by the vapor phase epitaxy (VPE) system, the doping level tends to be higher than 1×10^{15} cm^{-3}. As noted in the preceding, a thick *p*-type top layer will degrade both the response speed and quantum efficiency. A thin layer (on the order of 0.5 μm) of *p*-type material is thus formed by shallow zinc diffusion or beryllium ion implementation [26,28]. Additionally, a mesa is subsequently etched in order to provide electrical isolation and define the contact and light coupling area. Surface passivation material (polyimide) and antireflection coating layer are then deposited. Dark currents as low as a few nanoamperes have been observed for 50-μm-diameter devices. Speeds higher than 10 GHz have been achieved. Lifetimes

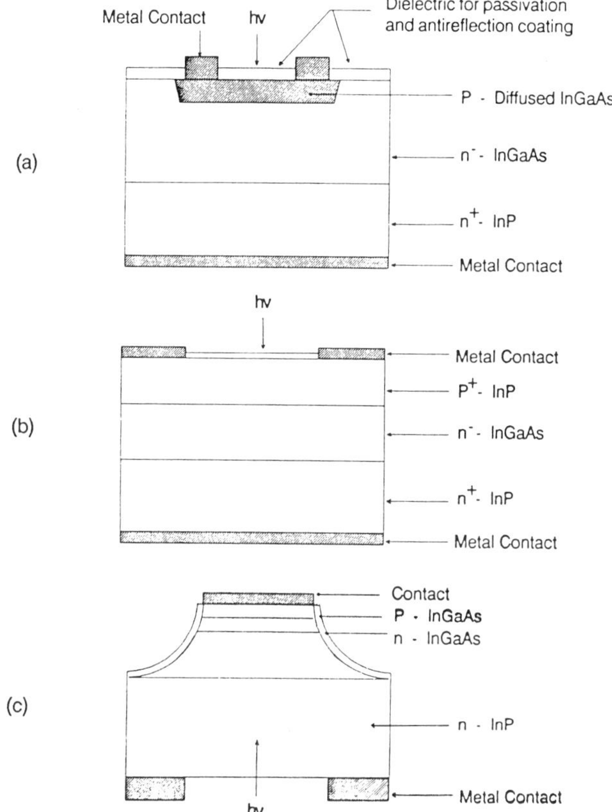

Figure 7.4 Schematic diagram of InGaAs/InP PIN photodiode. (*a*) Single-heterostructure PIN photodiode with front illumination through the thin *p*-InGaAs layer. (*b*) InP/InGaAs/InP PIN photodiode with a window for front illumination. (*c*) Back-illuminated PIN photodiode with side-well passivation.

of 10^8 h have been observed. The disadvantage of this simple structure is that the contact area needed for forming the bonding pad introduces additional capacitance and dark current to the photodiodes [24].

A variation of the preceding structure is the front-illuminated double-heterojunction photodiode shown in Fig. 7.4*b*. The *p*-type transparent InP window layer can be grown by LPE or VPE [29,30]. The light coupled from the top is absorbed directly at the InGaAs active layer. This structure has a quantum efficiency and speed higher than those of the homojunction diode.

Back illumination is possible for material systems whose substrates are transparent to the optical beam. Figure 7.4*c* illustrates such a structure. In this case, the active-layer thickness is selected such that at the operating bias, the electric field punches through to n^+ InP material. In principle, this structure has the same basic advantages as the heterojunction front-illuminated photodiode, although the quantum efficiency can be reduced by the free-carrier absorption in the relatively thick substrate. For this case, a lightly doped or undoped substrate is preferred to a heavily doped substrate. The main advantage of this configuration is that the front-side contact pad is no longer a concern in light coupling to the photodiode. Therefore, for the same InGaAs optical area, the back-illuminated diodes have lower capacitance and lower dark current than those of the front-illuminated diodes.

It is important to note that for back-illuminated diodes in the InGaAs/InP system, the responsivity is low for photon wavelengths below the 0.9-μm region because of the absorption edge of the InP substrate.

7.3.2 Schottky Photodiodes

Schottky photodiodes can consist of an undoped semiconductor layer (usually *n* type) on bulk material and a metal layer deposited on top to form a Schottky barrier. An example is shown in Fig. 7.5 [31]. As a strong electric field is developed near the metal–semiconductor interface due to the surface barrier, photogenerated electrons and holes are separated and collected at the metal contact and in the bulk semiconductor materials, respectively. By increasing the reverse bias to these diodes, both the peak electric field and the depletion region width inside the undoped layer can be increased, and thus the

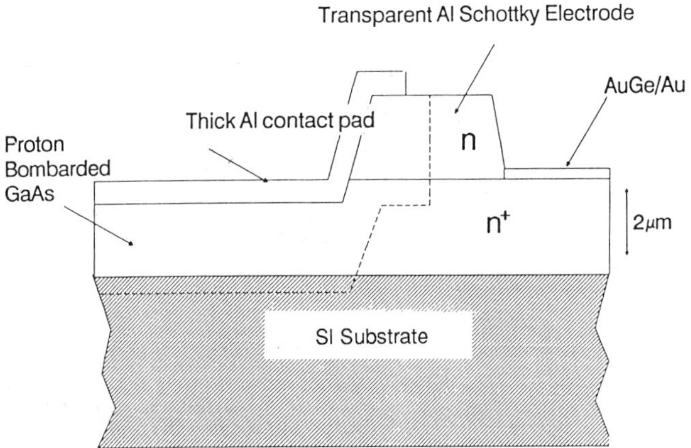

Figure 7.5 Schematic diagram of a GaAs Schottky photodiode. (From Z. Ray-Noy et al., *Electron. Lett.*, Vol. 19, p. 753, 1983, with permission.)

responsivity can be increased. The complete absorption of optical signals within the high-field region ensures high-speed performance, as absorption outside the high-field region generates a diffusion current. Again, the undoped layer should not be too thick, as the photodiode response speed can then become transit time limited.

In comparison with PIN photodiodes, Schottky photodiodes are simpler in structure. Unfortunately, while good Schottky contacts are readily made on GaAs, for long-wavelength InGaAsP/InP materials, stable Schottky barriers remain to be developed [32]. GaAs Schottky photodiodes with a 3-dB bandwidth larger than 100 GHz and operating at less than 4 V reverse bias have been reported [33]. This high speed is attained by restricting the photosensitive area to a very small mesa ($5 \times 5\ \mu m^2$) in order to minimize capacitance and by using a semitransparent platinum Schottky contact such that good responsivity can be achieved with top-side illumination.

7.3.3 Avalanche Photodiodes

Avalanche photodiodes are based on the impact ionization effect. When free carriers created by photoabsorption are accelerated by a strong electric field, they gain sufficient energy to promote more electrons from the valence band to the conduction band, thus giving rise to internal current gain [34–36]. Avalanche photodiodes usually require higher bias voltage than PIN photodiodes to maintain a high electric field. Also, the current gain is not a linear function of the applied voltage and is sensitive to temperature changes [37]. (This is mainly due to the fact that phonon scattering rate increases with temperature, which increases the mean free path between ionization collisions.) Figure 7.6*a* shows a schematic diagram of a silicon avalanche photodetector that consists of an n^+ contact, a *p*-type multiplying region, a π drift region, and a p^+ contact. The corresponding electric field profile is shown in Fig. 7.6*b*. In operation under the large reverse-bias voltage, the depletion region extends completely from the n^+ to the p^+ contact. Avalanche multiplication occurs in the high-field region. Figure 7.6*b* shows that when an electron–hole pair is photogenerated, electrons are injected into the multiplying region if the light is mostly absorbed to the right of the multiplication region. Similarly, hole injection occurs for light absorbed to the left. Mixed injection of both carriers occurs when light is absorbed within the avalanche region. The doping profile and position of the multiplication region as well as the absorption of light must be optimized for each material system to achieve the highest performance. It is important that carriers of higher ionization coefficient are injected to the avalanche region in order to have the minimum multiplication noise. For instance, silicon APDs are generally designed so that the electrons are the primary carriers that undergo multiplication as electrons in silicon have a larger ionization coefficient than holes. Consequently, both the hole injection or mixed injection are to be avoided [14,38,39]. For low-noise silicon APDs, a long drift region is incorporated to ensure high quantum efficiency and pure electron injection. The guard ring in the planar structures, shown in Fig. 7.6*a*, prevents edge breakdown at the perimeter of the multiplying region [16]. Reverse-bias voltage of up to 400 V is common for this kind of diode because of the large voltage drop across the long drift region.

The electric field required for impact ionization depends strongly on the band gap of the material. The minimum energy needed for impact ionization is known as the ionization threshold energy E_i [40,41]. Table 7.1 shows the ionization threshold energy of various materials [41]. The ionization energy influences strongly the ionization rates (or coefficients) for electrons (α) and holes (β). These quantities are defined as the reciprocal of the average distance traveled by an electron or a hole, measured along the direction of the electric field, to create a secondary electron–hole pair. In other words, $1/\alpha$, $1/\beta$

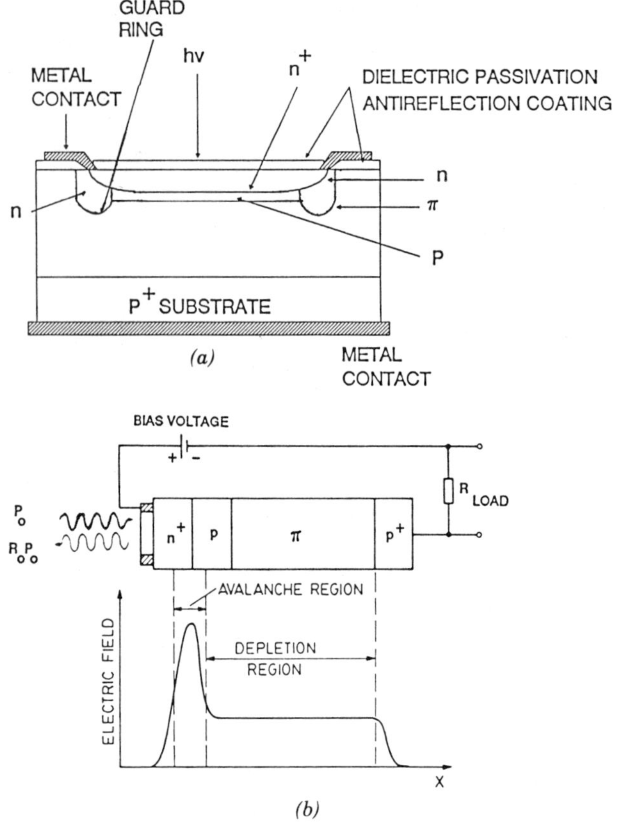

Figure 7.6 Schematic diagram of a silicon APD (*a*) structure and (*b*) electric field distribution. (From Ref. 12 with permission.)

are the mean free paths of the secondary ionization of electrons and holes, respectively. It should be noted that the threshold energy can be estimated by taking into account the conservation of carrier energy, crystal momentum, and group velocity during the ionization process. Avalanche multiplication is the result of many consecutive occurrences of this process.

The multiplication depends on a three-body collision process and is consequently statistical in nature. As a result, it contributes a statistical noise component in excess of the shot noise already present in the diode [39–42]. This excess noise has been studied for the case of an arbitrary α/β ratio for both uniform and arbitrary electric field profiles. For the uniform electric field case, the excess noise factors for electrons, F_n, and holes, F_p, are given as [42]

$$F_n = M_n\left[1 - (1-k)\left(\frac{M_n - 1}{M_n}\right)^2\right] \tag{7.8}$$

and

$$F_P = M_P\left[1 - \left(1 - \frac{1}{k}\right)\left(\frac{M_P - 1}{M_P}\right)^2\right] \tag{7.9}$$

TABLE 7.1 Ionization Threshold Energies

Direction	Initiation	Si, 1.1 eV, Indirect Band Structure	Ge, 0.7 eV Indirect Band Structure	GaAs, 1.4 eV Direct Band Structure	GaP, 2.3 eV Indirect Band Structure	InSb, 0.2 eV Direct Band Structure
⟨100⟩	Electrons	1.1 U	0.9 U, D	2.1 D^a	2.6 U^b	0.2(5) D
		1.5 D	1.0 D^a, U	2.3 U^a	3.0 D^b	1.6 D
		1.6 U^a	1.2 D	2.4 D^a	3.1 D^b	1.7 D^a, D^a
	Holes	1.8 D	1.3 D, D^a	1.7 D^a	2.4 D	0.2 D^a
		2.1 D^a	1.7 D^a	1.9 D	—	0.6 D
		—	4.5 D^a	2.5 D	—	2.2 D^a
⟨111⟩	Electrons	3.1 U^a	0.8 U, U	3.2 D^b, D^b	3.0 D^b	0.2(5) D
		3.3 U^a	2.5 D	3.6 D^b, D^b	3.4 D^b	—
		3.5 U^a	2.6 D^a	3.7 D^b	3.5 U^b, D^b	—
	Holes	2.9 D^a	0.9 D	1.6 D^b	2.9 D^b	0.4 D^a
		4.4 D^a	1.0 D^a	2.3 D^b	3.6 D^b	1.5 D^a
		4.7 D^a	1.4 D^a	2.5 D^b	—	1.8 D^a
⟨110⟩	Electrons	2.1 U	1.1 D^a	1.7 D	2.8 D^b	0.2 D
		4.0 D^a	1.2 D	1.9 D^a	2.9 D^b	1.6 D
		4.2 U^a	1.3 U	2.2 D	3.3 D^b	1.7 D^a, D^a
	Holes	1.8 D	0.9 D^a, D^a	1.4 D^a	2.3 D^b	0.2 D^a, D^a
		4.0 D^a	1.3 D	1.6 D^a	2.6 D^a	0.4 D
		4.1 D^a	1.8 D^a	1.9 D	2.8 D^a	1.7 U^a, D^a

Source: Ref. 41.

Note: D, direct process (no reciprocal-lattice translation vector involved); U, umklapp process.

[a] Initiating carrier comes from other than the normal conduction or valence band;

[b] Normal conduction or valence band cannot supply the initiating carrier (even if momentum conservation is neglected) because the energy range of the band is less than the band gap.

where $k = \beta/\alpha$ and M_n and M_P are the dc multiplication factors for pure electron and pure hole injection, respectively. For cases where k is not constant, F_n and F_p are given as

$$F_n = k_{\text{eff}} M_n + \left(2 - \frac{1}{M_n}\right)(1 - k_{\text{eff}}) \tag{7.10}$$

$$F_P = k'_{\text{eff}} M_p + \left(2 - \frac{1}{M_p}\right)(1 - k'_{\text{eff}}) \tag{7.11}$$

where $k_{\text{eff}} \approx k_2$ and $k'_{\text{eff}} = k_2/k_1^2$ depending on the spatial averages k_1, k_2 of α and β and

$$k_1 = \frac{\int_0^{W_a} \beta(x) M(x)\, dx}{\int_0^{W_a} \alpha(x) M(x)\, dx} \tag{7.12}$$

$$k_2 = \frac{\int_0^{W_a} \beta(x) M^2(x)\, dx}{\int_0^{W_a} \alpha(x) M^2(x)\, dx} \tag{7.13}$$

where W_a denotes the width of the avalanche region.

Thus, from Eqs. (7.8) and (7.9), if low-noise APDs are to be made, the ionization rate for one type of carrier must be much greater than that of the other, and the carrier with the larger ionization coefficient should initiate the avalanche process.

The dc avalanche gain of an APD is expressed as

$$M_n = \frac{1}{1 - \int_0^{W_a} \alpha \exp\left[-\int_0^x (\alpha - \beta)\, dx'\right] dx} \tag{7.14}$$

$$M_P = \frac{1}{1 - \int_0^{W_a} \beta \exp\left[\int_x^{W_a} (\alpha - \beta)\, dx'\right] dx} \tag{7.15}$$

where M_n, M_P denote the current gain for electron and hole injection, respectively. For cases where $\beta = 0$, Eq. (7.14) can be reduced to

$$M_n = \frac{1}{1 - \int_0^{W_a} \alpha \exp(-\alpha x)\, dx} \tag{7.16}$$

For a uniform field, α becomes constant and M_n of Eq. (7.16) increases exponentially with the number of ionizing carriers in the high-field region, and there is no sharp breakdown. Figure 7.7*a* depicts the avalanche buildup process under the $\beta = 0$ condition. A current pulse is induced when the photogenerated electron starts to drift toward the multiplication region. This pulse increases in magnitude during the electron transit time through the high-field region due to electron impact ionization. Then it is reduced to zero when the last hole is swept outside the high-field region. Neglecting the transit time inside the absorption region, the current pulse width is approximately the average of electron and hole transit times and decreases a little as the gain increases. The corresponding excess noise F [see Eqs. (7.8) and (7.9)] is small because when M increases, statistical variation of the impact ionization process only causes a small fluctuation in

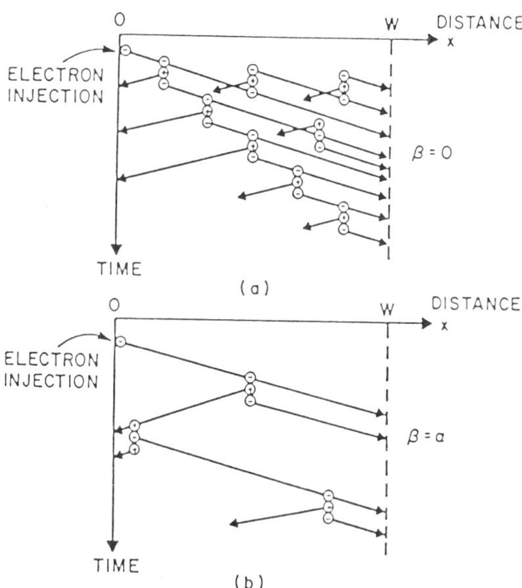

Figure 7.7 Physical representation of the avalanche buildup process with (*a*) $\beta = 0$ (electron multiplication only) and (*b*) $\alpha = \beta$ (both carriers multiply). (From H. Melchoir, in F. T. Arecchi and E. O. Schulz-Dubois, Eds., *Laser Handbook*, Vol. 1, North-Holland, Amsterdam, 1972, with permission.)

the total number of carriers [40]. In fact, when $\beta = 0$ and pure electron injection occurs at the avalanche region, $F_n \approx 2 - 1/M_n$.

When both carriers ionize at the same rate (i.e., $\beta = \alpha$), the APDs become quite noisy. For the sake of simplicity, in the case of a uniform electric field, the current gain becomes

$$M_n = M_p = \frac{1}{1 - \alpha W_a} \tag{7.17}$$

which gives rise to a sharp breakdown ($M \to \infty$) situation when αW_a approaches unity. This corresponds to the situation when, on the average, each injected electron or hole produces one electron–hole pair during its transit through the high-field region, as depicted in Fig. 7.7*b*. The secondary hole generated transverses in the $-x$ direction and generates further electron–hole pairs, and the electron repeats the process. This process persists for a long time when the gain is large. Theoretically speaking, during breakdown ($M \to \infty$), an optical pulse can produce a dc current. It is clear that for a large gain, there are fewer ionizing carriers in the high-field region than the case when $\alpha \gg \beta$ (or $\beta \gg \alpha$). Thus any statistical variation in the impact ionization process will produce large fluctuation in the gain and cause considerable excess noise.

The gain is also a function of signal frequency. The overall low-frequency gain, M_0, of a APD can be empirically expressed as a function of bias voltage [43]:

$$M_0 = \frac{I}{I_p} = \left[1 - \left(\frac{V - IR}{V_B}\right)^n\right]^{-1} \tag{7.18}$$

where I is the multiplied diode current, I_p is the primary photocurrent, V_B is the breakdown voltage, V is the applied voltage, and R is the sum of the diode series resistance and the load resistance. For $V_B \gg IR$, a maximum for M_0 can be obtained:

$$M_0(\text{max}) = \left(\frac{V_B}{nI_pR}\right)^{1/2} \tag{7.19}$$

The high-frequency gain can be approximated by

$$M(\omega) = \frac{M_0}{[1 + (\omega M_0 \tau_{\text{eff}})^2]^{1/2}} \tag{7.20}$$

where ω is the angular frequency and τ_{eff} is approximately equal to $[\beta/\alpha]\tau_{\text{av}}\delta$, where δ is a parameter in the range of 1/3 (at $\beta = \alpha$) to 2 (at $\beta \ll \alpha$) and τ_{av} is the average transit time. Figure 7.8 depicts the theoretical gain–bandwidth product as a function of M_0 for various α/β values [34].

The applications of APDs are very much related to the receiver design. Consider the signal-to-noise ratio of an APD receiver:

$$\frac{\text{S}}{\text{N}} = \frac{\text{signal power}}{\text{receiver noise (Johnson noise + capacitance noise) + APD noise}}$$

$$= \frac{(q\eta P_0 M/hv)^2}{B[[4kT/R_f + i_c^2] + 2q[I_d + q\eta P_0/hv]M^2F]} \tag{7.21}$$

where η is the quantum efficiency at unity gain, P_0 is the optical power incident on the diode, M is the multiplication factor, B is the bandwidth; $4kT/R_f$ is the Johnson noise of the feedback resistor R_f, i_c^2 is the amplifier noise due to the capacitance, I_d is the dark current of the APD when $M = 1$, and F is the excess noise factor. Using Eqs. (7.8) and

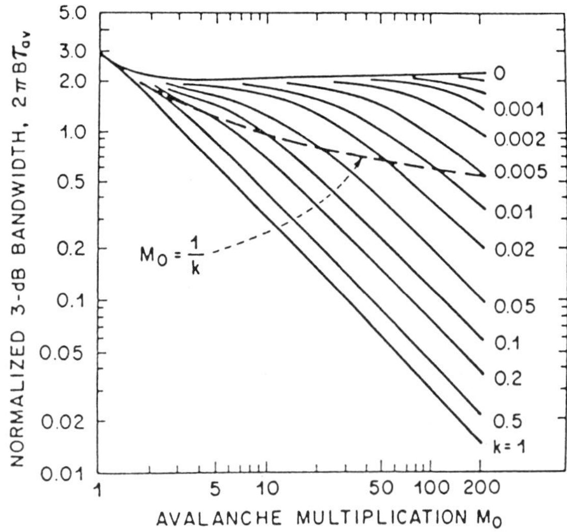

Figure 7.8 Normalized 3-dB bandwidth as a function of average avalanche multiplication M_0 for several values of the ionization rate ratio k. For pure electron injection $k = \beta/\alpha$; for pure hole injection $k = \alpha/\beta$; τ_{av} is the average of the electron and hole transit times in the avalanche region. (From Ref. 34 with permission.)

(7.9), one can find the optimum gain for certain receiver designs from Eq. (7.21) and its derivative with respect to M. As the noise of an APD goes up faster than the signal term (M^{2+x} versus M^2, where x is between 0 and 1), once the APD noise becomes the dominant term in the denominator of Eq. (7.21), the signal-to-noise ratio actually degrades with a further increase in gain. This sets the usable gain of the APD.

Silicon APDs are used in the wavelength range of 0.4–1.0 μm because of the low dark current and a large α/β. The structure shown in Fig. 7.6a was employed in the case of silicon APDs to obtain carrier multiplication with very little excess noise. The high-field n^+p junction served as electron-initiated avalanche region and can be formed by diffusion or ion implantation. At a bias below the breakdown of the n^+p junction, the π region will be depleted. Beyond this point, the voltage drop across the wide π region increases with bias faster than that across the n^+p, thus increasing the operating bias range for multiplication. The full depletion of the π region is essential for high response speed, as the electrons generated by absorption within the π region will then drift to the n^+p region with high velocity. Holes, on the other hand, will drift opposite to the electrons, and thus a pure electron current is injected into the multiplication region. (Although there exist some hole injections due to the absorption in the n^+ region and the avalanche region, they are much smaller than the electron current due to the absorption in the long-drift region.) An excess noise factor of 4 at a gain of 100 has been achieved in this APD [44]. The response speed is limited mostly by the transit time across the π region. To increase the response speed without increasing the operating voltage, a built-in field in the π region can be profiled by grading the doping profile so as to increase the carrier velocities [45]. Another scheme involves the use of an antireflective coating at the back contact to increase the quantum efficiency of a relatively narrow π region [46]. Response times of 150–200 ps have been achieved.

For long-wavelength application, germanium APDs have been used because silicon cannot respond to wavelengths beyond 1.1 μm [43]. As can be seen from Fig. 7.9, β/α ratio for germanium has a maximum value of 2. A relatively high excess noise factor ($F \approx \frac{1}{2}M$), in addition to a relatively high dark-current noise, is present in germanium APDs due to the small band gap and surface leakage. This noise behavior is depicted in Fig. 7.10. For this particular germanium APD receiver, the usable gain is only around 10. Beyond this, the signal-to-noise ratio degrades due to the fact that the signal power increases as M^2 while the excess noise increases as M^3 [see Eq. (7.21)]. By reducing the sensitive area of the APDs, dark current in the nanoampere range can be achieved.

Attractive alternatives to germanium APDs for long-wavelength applications are III–V APDs because the material band gap can be tailored to achieve both high absorption and low dark current. Early APD research in the ternary and quaternary system have shown low usable gain because the difference in α and β in these materials is not as high as that in silicon. For homojunction InGaAs devices, the dark current increases exponentially with reverse bias as the bias approaches avalanche breakdown. This is attributed to the tunneling current, which is more profound in materials with a small band gap. To reduce the tunneling current, a heterostructure that separates the absorption region from the multiplication region was proposed and is depicted in Fig. 7.11 [47]. This is known as separate absorption and multiplication (SAM) APD. With this arrangement, the low-band-gap InGaAs material will not experience a high electric field, and thus the tunneling effect is reduced. The absorption in this structure takes place in the narrow-band-gap material, which, like the silicon APDs, is low doped; while the multiplication takes place in the wide-band-gap pn junction, which usually consists of InP. Consequently, the noise of the SAM APDs reflects much of the ionization process in InP [50]. Reasonably small unmultiplied dark currents are observed for APDs with InGaAsP absorption region (nanoampere range) [48] and InGaAs absorption region (tens of nanoamperes) [49], while the gain is maintained in the range of 10–60. Recently, trapping of holes at the

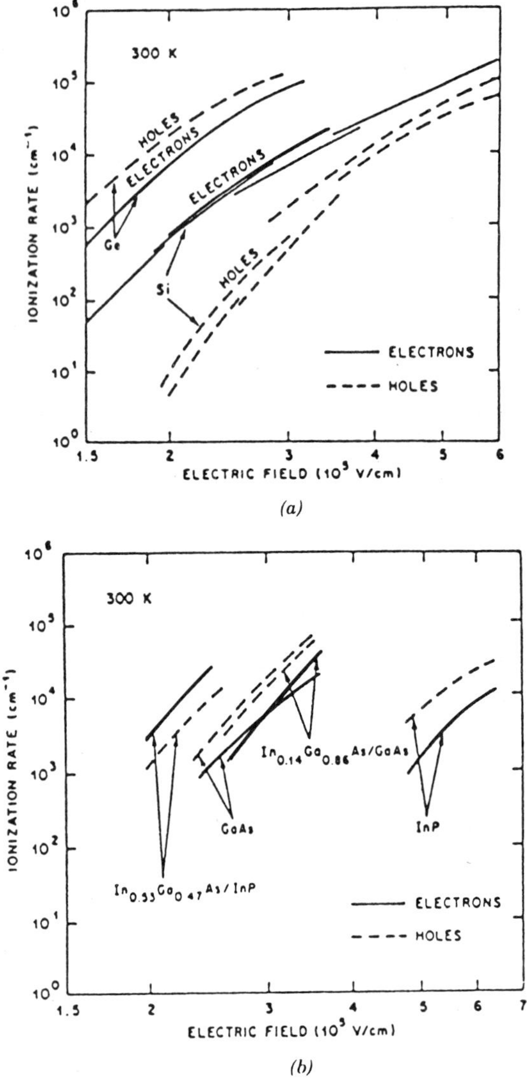

Figure 7.9 Experimentally obtained carrier ionization rates at 300 K for (*a*) silicon and germanium and (*b*) GaAs, InP, $In_{0.14}Ga_{0.86}As/GaAs$, and $In_{0.53}Ga_{0.47}As/InP$. (From H. Melchoir, *Phys. Today*, Vol. 30, p. 32, 1977, with permission.)

InGaAs/InP heterojunction was observed, which greatly degrades the speed of the APD. To overcome this problem, an additional layer of quaternary InGaAsP with intermediate energy band gap (or layers of graded band-gap region) is incorporated. A relatively low-noise (as compared with germanium) APD can be achieved with proper design. For instance, at a gain of 10, APDs with an excess noise figure of 5, a dark current of 20 nA, and a bandwidth in excess of 1 GHz have been demonstrated.

Other material systems have also been investigated for avalanche photodiode applications. For instance, GaAlSb/GaSb materials operating at 1.55 μm wavelength have

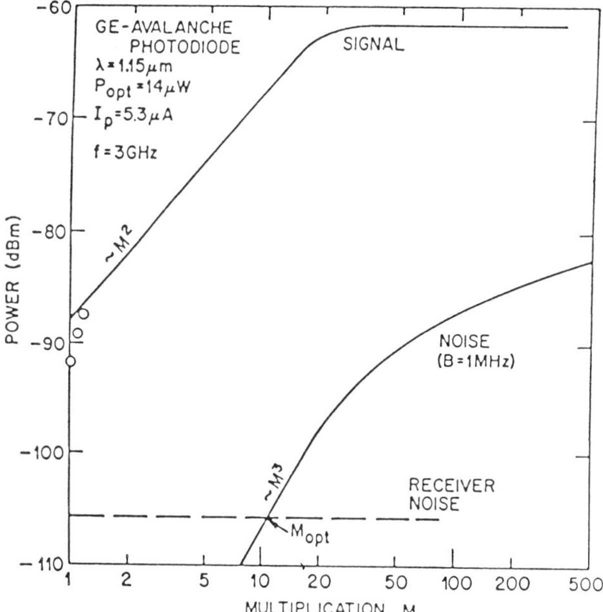

Figure 7.10 Signal and noise power output of Ge avalanche photodiode in a 1-MHz band at 3 GHz. The optimum operating point for the best S/N ratio and sensitivity is indicated. (From H. Melchoir, in F. T. Arecchi and E. O. Schulz-Dubois, Eds., *Laser Handbook*, Vol. 1, (North-Holland, Amsterdam, 1972, with permission.)

shown very high gain. More interestingly, the band gap of this alloy is very close to valence band spin orbit splitting energy [51]. This greatly enhances the hole multiplication process and leads to an extremely high β/α (>20). However, the dark current (mostly due to surface recombination) of this material system is too high at this moment.

Avalanche gain has also been observed in the HgCdTe crystal at 1.3 μm wavelength [52]. However, not enough research has been performed to make this into a commercial product.

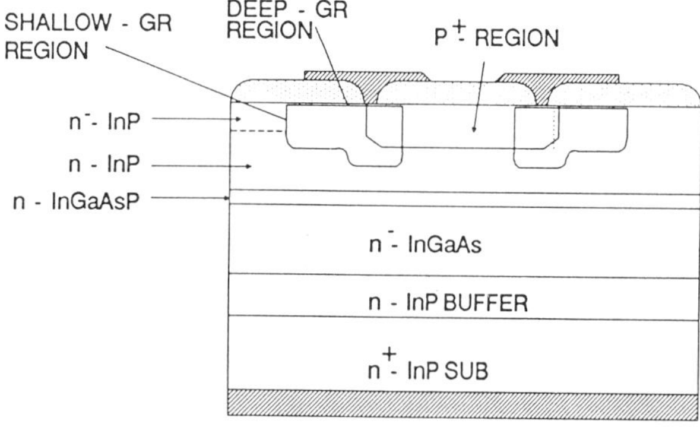

Figure 7.11 InGaAsP/InP APD with separated absorption and multiplication regions. The intermediate InGaAsP layer reduces the hole-trapping effect. GR stands for guard ring.

7.4 PHOTOCONDUCTIVE DEVICES

Photoconductive devices depend on the change in the electrical conductivity of the material upon irradiation [53,54]. Although photoconductive effects have been established for a long time, only recently are they employed for high-sensitivity, high-speed detection. In semiconductors, two basic types of photoconductors, extrinsic and intrinsic, are studied. In the intrinsic case, photoconduction is produced by band-to-band absorption. The absorption coefficient can be very large ($\sim 10^4\ \text{cm}^{-1}$) in this case due to a large density of electronic states associated with the conduction and valence bands. In the extrinsic case, photons are absorbed at the impurity level, and consequently, free electrons are created in the *n*-type semiconductor and free holes are created in the *p*-type semiconductor. Extrinsic photoconduction is characterized by lower sensitivity because the absorption is limited by the smaller number of available states. These devices are usually operated at low temperatures (e.g., 77 K) to freeze out impurity carriers (thus reducing the background noise) so that they become available for optical absorption.

Consider the simple photoconductive device shown in Fig. 7.12. A voltage V is applied across the electrodes and light is illuminated from the top. The photosignal is detected either as a change in voltage across a resistor in series with the photoconductor or as a change in current through the sample. Frequently the signal is detected as a change in voltage across a resistor matched to the dark resistance of the detector.

The photoconductive gain M of these devices can be expressed as [54]

$$M = \frac{\tau_n}{t_n} + \frac{\tau_p}{t_p} \tag{7.22}$$

where τ_n, τ_p are the lifetimes of the excess electrons and holes, respectively, and t_n, t_p are the corresponding transit times across the device. For cases where electron conduction is dominant and the response time of the device is τ, the gain–bandwidth product can

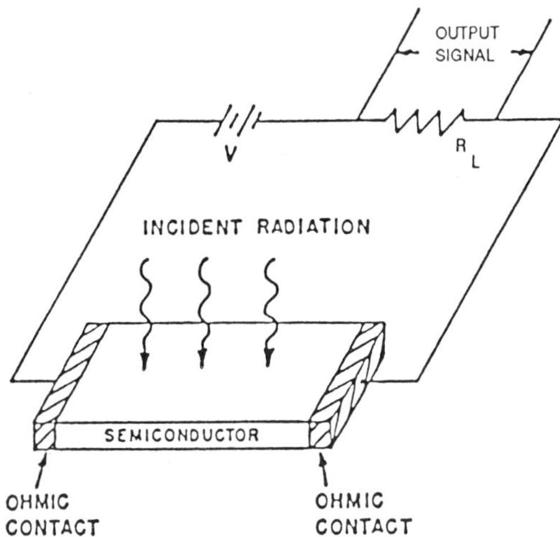

Figure 7.12 Schematic diagram of a simple photoconductor; R_L, load resistor.

be expressed as

$$GB = \frac{\tau_e}{t_n} \frac{1}{\tau} \tag{7.23}$$

For material where carrier trapping is not significant, the response time is determined by carrier lifetimes. However, as the device dimension gets smaller, the response time can be altered by the nature of the metal contact and the method of biasing. When a Schottky barrier is formed at the electrode–semiconductor interface and its depletion region extends across the length of the device to reach the other electrode, the device operates more or less like a photodiode with τ_e replaced by t_n, and its GB value is then equal to $1/\tau$ [55].

When a steady optical beam with power P_0 and wavelength λ is incident on the photoconductor, the optical generation rate of electron–hole pairs versus the distance x into the sample is given by

$$g(x)\,dx = \frac{(1 - R_0)P_0}{h\nu} \alpha_0 e^{-\alpha_0 x}\,dx \tag{7.24}$$

The quantum efficiency η is

$$\eta = (1 - R_0)(1 - e^{-\alpha_0 D}) \tag{7.25}$$

where D is the effective sample thickness. For an intrinsic photoconductor at steady state, the average excess electron and hole concentrations are

$$\Delta n = \Delta p = \frac{P_0 \eta \tau}{WLDh\nu} \tag{7.26}$$

where W, L are the respective width and length dimensions of the device.

The change in conductivity, $\Delta\sigma$, due to P_0 is given by

$$\Delta\sigma = q(\Delta n\, \mu_n + \Delta p\, \mu_p) \tag{7.27}$$

where μ_n and μ_p are the electron and hole mobilities, respectively.

For many signal-processing applications, the ratio of the on-resistance to the off-resistance of the photoconductive device is an important parameter. The off-resistance is primarily limited by the material resistivity in the dark. The on-resistance, as discussed earlier, depends on the carrier mobility and the excess carriers generated.

Due to low surface recombination velocity of InP-related materials, planar photoconductive devices on iron-doped semi-insulating InP have been extensively studied [56]. This material has a room temperature resistivity of 10^5–10^8 Ω-cm and electron mobility of 1500–4500 cm^2/V-s depending on the crystal quality and iron concentration. The carrier lifetime ranges from less than 100 ps to 3 ns [55–57].

A typical photoconductor structure consists of either (i) a single narrow gap made on a metal stripe deposited on semi-insulating InP [58] or (ii) interdigital electrodes deposited on the substrate [57]. Response times as short as 50 ps have been achieved in both devices. As noted earlier, the nature of metal contacts on semiconductors affects much of the device performance. For devices with nonalloyed contacts, lower photoconductive gain is observed. However, the response time, especially the fall time, is shorter than that of devices with alloyed contacts. Such alloyed contact devices show lower contact resistance and a higher photoconductive gain. Various contact materials (such as Au/Sn and Ni/Ge/Au) as well as alloying conditions [59] (such as alloying temperature and duration) have been investigated. Compared to photodiodes, the planar photoconductive detector has advantages such as ease of optoelectronic integration, especially with field-effect transistor (FET) circuits. Also, detector arrays are more readily achieved with planar geometry photoconductors.

A photoconductive device can function as a current switch. However, this requires low on-resistance. For alloyed contact devices, the current–voltage characteristics are linear at low voltage, and the current tends to saturate at increased voltage, in contrast to Eq. (7.26). The current saturation is believed to be caused by electrons transferred from a high-mobility valley to a low-mobility valley, especially at the contact region where the electric field is strong [60]. The photoconductance, on the other hand, is linearly dependent on the light intensity over several orders of magnitude, as shown in Fig. 7.13.

For 1.0–1.6 μm wavelength applications, InGaAs lattice matched to InP has been considered for photoconductive detectors [61]. It is an attractive material due to its high saturation drift velocity. Compared to semi-insulating InP, however, semi-insulating InGaAs materials obtained so far tend to be less resistive ($\sim 10^5$ Ω-cm), the typical dark resistance in the range of kilohms, which compromises some of the device performance.

Due to the fact that InP can be used to detect radiation ranges from near IR to UV region with good sensitivity, it becomes very attractive for broadband detection [62,63].

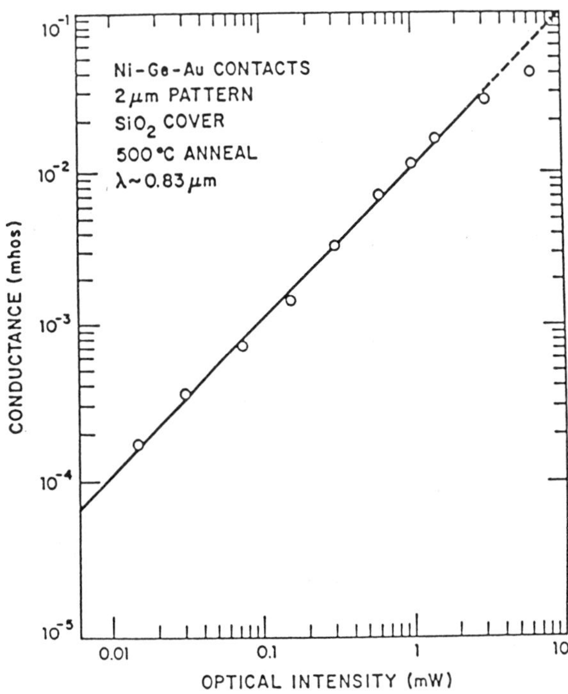

Figure 7.13 On-state conductance versus incident optical power ($\lambda \sim 0.85$ μm) for an InP optoelectronic switch with Ni/Ge/Au contacts. This device had a 2-μm finger pattern and was alloyed at 500°C with a SiO_2 cover. (From A. G. Foyt and F. J. Leonberger, in C. H. Lee, Ed., *Picosecond Optoelectronic Devices*, Academic, Orlando, FL, 1984, with permission.)

Together with its small surface recombination velocity, very fast detection (< 90 ps rise time) has been achieved at the X-ray wavelengths where $\alpha_0 \sim 10^3\ \text{cm}^{-1}$. Recently, such devices have been studied in conjunction with analog-to-digital (A/D) conversion where an analog electrical signal is time sampled and coded into a digital format [56]. In this application, photoconductive devices operate as a time switch triggered by laser pulses and serve for tracking of the analog signal voltage. When compared to electrical A/D converter circuits typically employing bipolar and FETs, the photoconductive switch can lead potentially to less time jitter as well as higher sampling rates. Up to 6-bit resolution with a sampling rate greater than 1×10^9 samples per second has been predicted with InP photoconductive switches.

Another important application of photoconductive device is in the generation of high-power, picosecond duration, electrical pulses. In this case, the photoconductor operates as a switching element connected to a low-impedance transmission line and is terminated in a load resistor. With the switch in the off state, one side of the line is charged to a high voltage. Upon laser illumination, the switch is turned on, and a high-power electrical pulse propagates down the transmission line. The on-impedance of the switching element needs to be much smaller than the line impedance so that a small fraction of the voltage (and also power dissipation) will drop across the switching element. The carrier lifetime determines the maximum time window when the switch is in the on state. Silicon photoconductive switches have been used to generate gigawatt pulse power, and the duration of the electrical pulse is in the nanosecond range, depending on the length of the charging transmission line [64]. On the other hand, III–V semiconductor materials are inferior to silicon materials because of their relative shorter carrier lifetimes which limits the duration of electrical pulse. Recently, a photodiode switch has also been investigated as an alternative to the photoconductive switch for this application [13].

7.5 RECENT DEVELOPMENT OF OPTICAL DETECTORS

Due to advances made in material epitaxy, new ideas on photodetection have been proposed and many novel device structures have been tested. Among these, the most notable are those that have led to the development of phototransistors and the band-gap engineering approach to achieve solid-state photomultiplication. However, many of the devices conceived are still in the embryonic stage. Their potential for widespread application will critically depend on their device performance characteristics and receiver designs.

7.5.1 Phototransistors

Previously, most of the work on phototransistors has been based on silicon and germanium [65,66]. Recent interest in III–V phototransistors is primarily due to the good material quality of heterojunction devices and the higher gain achievable in the heterojunction bipolar transistor. By employing a heterojunction, a wide-band-gap emitter-configured phototransistor is realized, and this greatly improves the emitter efficiency [67]. The heterojunction relieves the restriction on relative dopant concentrations on both sides of the emitter–base junction [68], and a high gain–bandwidth product is possible in principle.

For the *npn* phototransistor shown in Fig. 7.14, the base is unbiased, and voltage is applied between the emitter and collector such that the emitter and collector junctions are forward and reverse biased, respectively. The wide-band-gap emitter is transparent to incident photons, which are mostly absorbed in base and collector regions. The

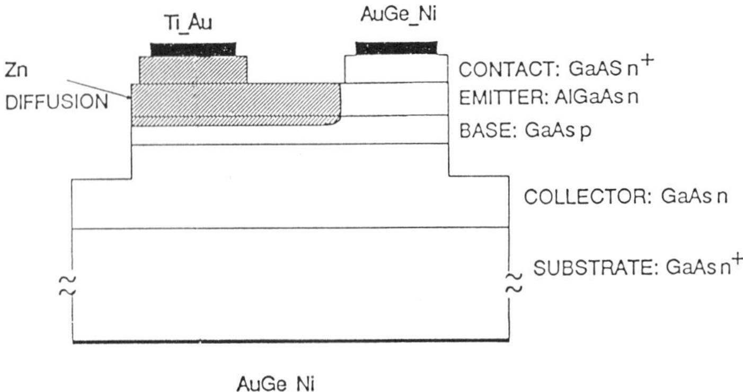

Figure 7.14 Schematic diagram of an *npn* GaAs/AlGaAs phototransistor with diffused base contact region. (From A. Scavennec et al., *Electron. Lett.*, Vol. 19, p. 384, 1983.)

photogenerated holes that accumulate at the base change the potential at the emitter–base junction and cause injection of electrons from the emitter into the base. Current gain is obtained when the electron lifetime in the base is larger than the base transit time. The absence of base contact facilitates the reduction of capacitance. The frequency response of the phototransistor can be improved by employing designs with a lower emitter capacitance (lightly doped emitter) and a reduced base resistance (heavily doped base). This could have compromised the emitter injection efficiency in the case of homostructure *npn* transistors. However, by employing a wider band gap in the emitter than in the base, the valence band discontinuity can greatly enhance the emitter injection efficiency despite the doping effect [69,70].

For the GaAs/AlGaAs material system, phototransistors typically consist of an n^--GaAs collector, a *p*-GaAs base, and an n-$Al_xGa_{1-x}As$ wide-band-gap emitter. The short-wavelength cutoff near 0.65 μm is caused by the emitter absorption, and the long-wavelength cutoff near 0.88 μm is due to the absorption edge of GaAs. For a floating-base configuration, current gain of 5000 has been achieved [69], while high optical gains occur only at high signal levels ($\gg 1\ \mu$W), due to the defect current at the emitter junction. In a configuration with a diffused base, as shown in Fig. 7.14, the emitter–base capacitance is minimized, and a current gain of more than 10^4 has been observed [71].

7.5.2 Novel Material Structures for Photodetection

Graded Band-Gap Avalanche Photodiode. The operation of this device is based on the observation that inside the material with the graded band gap, the extent to which the conduction band is graded can differ from that of the valence band as a result of different band discontinuities in conduction and valence bands from one material to another. As shown in Fig. 7.15, if the conduction band has a larger grading than the valence band, an electron can be accelerated to a higher energy than a hole even when both are subjected to the same electric field (from external bias). Consequently, the input ionization ratio α/β can be enlarged provided the accelerating electrons travel across the different band-gap regions without collision with phonons and reach the threshold energy for impact ionization [72]. In practice, an ionization rate ratio k [which equals $(M_p - 1)/(M_n - 1)$] less than 10 has been obtained for a *pn* junction with a 0.4-μm graded-band-gap $Al_xGa_{1-x}As$ (x ranges from 0 to 0.45) layer [73].

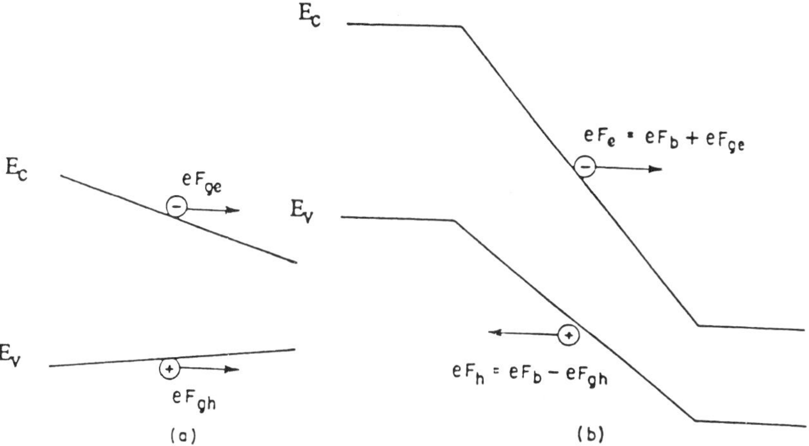

Figure 7.15 (*a*) Effect of quasi-electric field in a graded band gap material. (*b*) Combined effect of applied and quasi-electric field. (From Ref. 73 with permission.)

Superlattice APDs. As in the case of the graded-band-gap APDs, the difference in ΔE_c and ΔE_v values between two adjacent layers can enhance α/β, as the electrons and holes will then suffer different drops in potential energy in the narrow-band-gap layer. As shown in Fig. 7.16, provided the heterojunction is abrupt enough such that the transition region is much shorter than a mean free path for phonon scattering, the potential energies gained can be added to those gained through the electric field to reach the threshold energy [74]. Since ΔE_c is larger than ΔE_v at GaAs/AlGaAs heterojunctions and GaAs has a lower threshold energy than AlGaAs (due to the dependence of the threshold energy

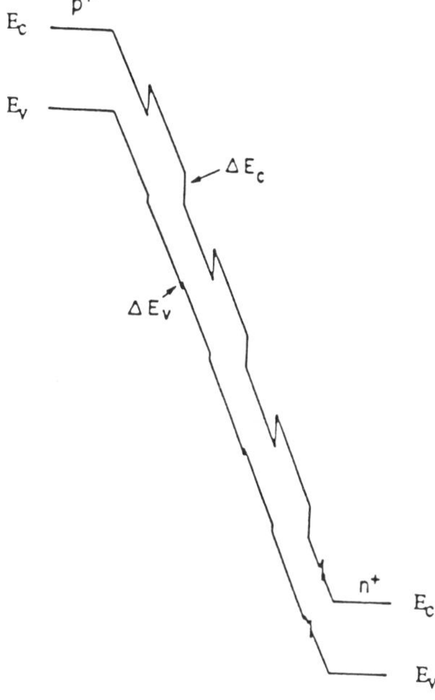

Figure 7.16 Energy band diagram of a superlattice APD with unequal ΔE_c and ΔE_v. (From Ref. 76 with permission.)

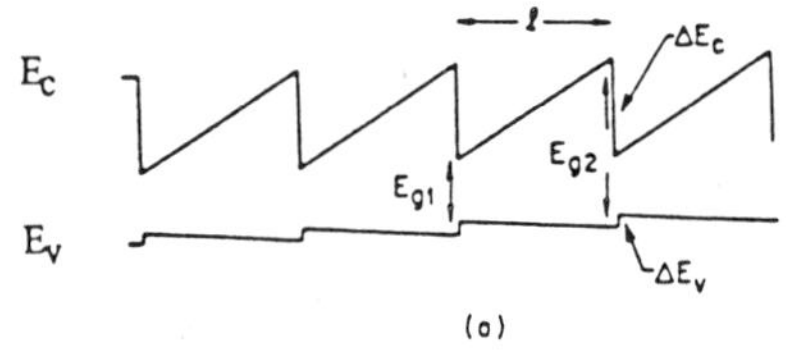

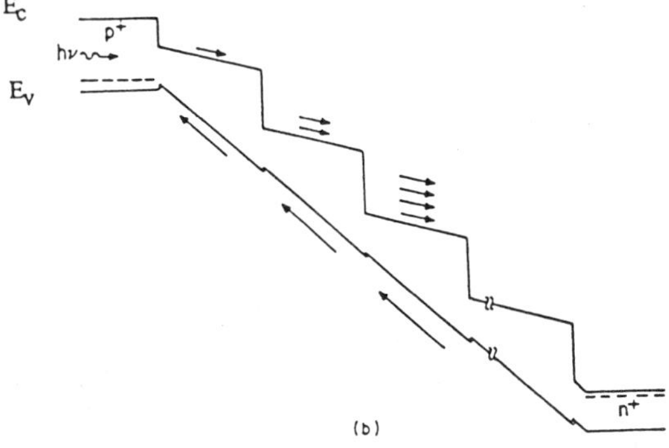

Figure 7.17 Energy band diagram: (*a*) unbiased graded multilayer region of the staircase detector and (*b*) biased detector. (From Ref. 77 with permission.)

on the band gap in the band-to-band impact ionization) [75], electrons have a larger chance to impact ionize in GaAs. With a superlattice structure consisting of pairs of these heterojunctions, the photon can be absorbed in any one of the wells, and thus absorption is not limited to the surface. For the case of an APD consisting of a PIN structure with the intrinsic region consisting of 50 alternating layers of GaAs (~450 Å)/$Al_{0.45}Ga_{0.55}As$ superlattice, an α/β ratio of 8 has been achieved [76–78].

Staircase Solid-State Photomultiplier. This proposed device is a multistage graded-band-gap structure, as shown in Fig. 7.17. Each stage in this photomultiplier structure is linearly graded in composition from a low band gap to a high band gap, with an abrupt transition back to the low-band-gap material [79]. The energy step should be equal to or greater than the electron ionization energy. Since the conduction band discontinuity in GaAs/AlGaAs materials accounts for most of the band-gap difference, the corresponding grading and potential steps for the valence band are considerably smaller.

In this staircase structure, a photogenerated electron at the p^+ contact will drift, under the net effect of the bias field and grading field ($\Delta E_c/e$), toward the first conductive band step without ionization. However, as it falls down the step, the energy gained can cause impact ionization, and the same process repeats at each stage. For the hole, ionization is caused by the applied electric field and the small grading of the valence band only, as the valence band step is of the wrong sign to assist ionization.

7.5.3 Infrared Detector

Another application for superlattice detector structures is in the detection of far-IR (>6-μm) radiation. As shown in Fig. 7.18, when the well region of the superlattice is heavily

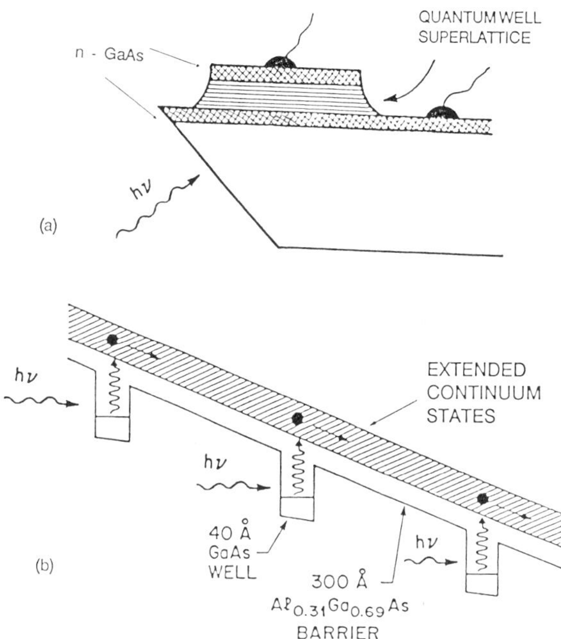

Figure 7.18 A GaAs/AlGaAs superlattice IR detector showing (*a*) the structure and (*b*) the energy band diagram of the excitation to the continuum state. (From Ref. 80 with permission.)

doped, the bound-state electrons within the superlattice can absorb an IR photon and make a transition to the extended continuum states close to the top of the barriers and contribute to the current [80]. In operation, this is similar to an extrinsic photoconductive device, except that electrons are more localized. Good responsivity (6 kA/W) at wavelengths close to 8.3 μm has been achieved with a 50-period 40-Å GaAs/300-Å $Al_{0.31}Ga_{0.69}As$ superlattice where the GaAs well region is doped to 2×10^{18} cm^{-3}.

Acknowledgment

The authors would like to express their appreciation to H. H. Wieder (University of California at San Diego) for his valuable comments.

REFERENCES

1. R. Feymann, *Lecture on Physics*, Vol. 1, Addison Wesley, Reading, HA, 1963, Chaps. 35 and 36.
2. R. J. Keyes, Ed., *Optical and Infrared Detectors, Topics in Applied Physics*, Vol. 19, Springer-Verlag, Berlin, 1980.
3. D. P. Schinke, R. G. Smith, and A. R. Hartman, "Photodetectors," in H. Kressel, Ed., *Semiconductor Devices for Optical Communication*, 2nd ed., *Topics in Applied Physics*, Vol. 39, Springer-Verlag, Heidelberg, 1982.
4. R. G. Smith and S. D. Personick, "Receiver Design for Optical Fiber Communication Systems," in H. Kressel, Ed., *Semiconductor Devices for Optical Communication*, 2nd ed., *Topics in Applied Physics*, Vol. 39, Springer-Verlag, Heidelberg, 1982.

5. R. H. Kingston, *Detection of Optical and Infrared Radiation*, Springer-Verlag, Berlin, 1979.
6. R. K. Willardson and A. C. Beer, Eds., "Infrared Detector," *Semiconductor and Semimetal*, Vols. 5 and 12. Academic, New York, 1970, 1977.
7. J. I. Pankove, *Optical Processes in Semiconductors*, Prentice-Hall, Englewood Cliffs, New Jersey, 1971.
8. W. Franz, *Z. Naturforsch*, Vol. A13, p. 484, 1958.
9. L. V. Keldysh, *Zh. Eksp. Teor. Fiz.*, Vol. 34, p. 1138, 1958. English trans.: *Sov. Phys.*, Vol. JETP 7, p. 788, 1958.
10. K. Seeger, *Semiconductor Physics*, 2nd ed., Springer-Verlag, Berlin, 1982, Chap. 11.
11. W. X. Chen, S. C. Hsueh, P. K. L. Yu, and S. S. Lau, *IEEE Electron. Dev. Lett.*, Vol. EDL-7, p. 471, 1986.
12. H. Melchoir, A. R. Hartman, D. P. Schinke, and T. E. Seidel, *Bell Syst. Tech. J.*, Vol. 57, p. 1791, 1978.
13. D. Giorgi, P. K. L. Yu, J. R. Long, T. Navapanich, and O. S. F. Zucker, *J. Appl. Phys.*, Vol. 63, p. 930, 1988.
14. H. D. Law, K. Nakano, and L. R. Tomasetta, *J. Quant. Electron.*, Vol. QE-15, p. 549, 1979.
15. S. M. Sze, *Physics of Semiconductor Devices*, Wiley, New York, 1981, p. 751.
16. S. M. Sze, *Physics of Semiconductor Devices*, Wiley, New York, 1981, p. 649.
17. H. Melchoir, "Demodulation and Photodetection Techniques," in F. T. Arecchi and E. O. Schulz-Dubois, Eds., *Laser Handbook*, Vol. 1, North-Holland, Amsterdam, 1972, pp. 725–835.
18. H. Ando, H. Kanbe, T. Kimura, T. Yamaoka, and T. Kaneda, *IEEE J. Quant. Electron.*, Vol. QE-14, p. 804, 1978.
19. R. P. Riesz, *Rev. Sci. Instrum.*, Vol. 33, p. 994, 1962.
20. H. Kanbe, G. Grosskopf, O. Mikami, and S. Machinda, *IEEE J. Quant. Electron.*, Vol. QE-17, p. 1534, 1981.
21. M. A. Washington, R. E. Nahory, and E. D. Beebe, *Appl. Phys. Lett.*, Vol. 33, p. 854, 1978.
22. T. P. Lee, C. A. Burrus, and A. G. Dentai, *IEEE J. Quant. Electron.*, Vol. QE-17, p. 232, 1981.
23. A. R. Clawson, W. Y. Lum, G. E. McWilliams, and H. H. Wieder, *Appl. Phys. Lett.*, Vol. 32, p. 549, 1978.
24. C. Fan, P. K. L. Yu, and P. C. Chen, *Electron. Lett.*, Vol. 23, p. 571, 1987.
25. V. Diadink, C. A. Armiento, S. H. Groves, and C. E. Hurvitz, *IEEE Electron Device Lett.*, Vol. EDL-1, p. 177, 1980.
26. T. P. Lee, C. A. Burrus, A. Y. Cho, and K. Y. Cheng, *Appl. Phys. Lett.*, Vol. 37, p. 730, 1980.
27. K. Li, E. Rezek, and H. D. Law, *Electron Lett.*, Vol. 20, p. 196, 1984.
28. H. D. Law, L. R. Tomasetta, and K. Nakano, *Appl. Phys. Lett.*, Vol. 33, p. 920, 1978.
29. G. H. Olsen and T. J. Zamesowski, *IEEE J. Quant. Electron.*, Vol. QE-17, p. 128, 1981.
30. H. Ando, Y. Yamauchi, H. Nakagama, N. Susa, and H. Kanbe, *IEEE J. Quant. Electron.*, Vol. QE-17, p. 250, 1981.
31. M. V. Schneider, *Bell Syst. Tech. J.*, Vol. 45, p. 1611, 1966.
32. H. Schumacher, H. P. Leblanc, J. Soole, and R. Bhat, *IEEE Electron. Dev. Lett.*, Vol. EDL-9, p. 607, 1988.
33. S. Y. Wang and D. M. Bloom, *Electron. Lett.*, Vol. 19, p. 554, 1983.
34. R. B. Emmons, *J. Appl. Phys.*, Vol. 38, p. 3705, 1967.
35. H. W. Ruegg, *IEEE Trans. Electron. Dev.*, Vol. ED-14, p. 238, 1967.
36. H. D. Law, L. R. Tomasetta, K. Nakano, and J. S. Harris, *Appl. Phys. Lett.*, Vol. 33, p. 416, 1978.
37. J. Conradi, *Solid-State Electron.*, Vol. 17, p. 99, 1974.
38. K. Nishida, *Electron. Lett.*, Vol. 13, p. 419, 1977.
39. R. J. McIntyre, *IEEE Trans. Electron. Dev.*, Vol. ED-13, p. 164, 1966.

40. G. E. Stillman and C. M. Wolfe, in P. K. Willardson and A. C. Beers, Eds., *Semiconductor and Semimetals*, Vol. 12, Academic, New York, 1977, p. 291.
41. C. L. Anderson and C. R. Crowell, *Phys. Rev. B.*, Vol. 5, p. 2267, 1972.
42. R. J. McIntyre, *IEEE Trans. Electron. Dev.*, Vol. ED-19, p. 703, 1972.
43. H. Melchoir and W. T. Lynch, *IEEE Trans. Electron. Dev.*, Vol. ED-13, p. 829, 1966.
44. J. Conradi and P. P. Webb, *Proc. 1st Eur. Conf. Opt. Fibre Commun.*, Vol. 1, p. 128, 1975.
45. H. Kanbe, T. Kimura, Y. Mizushima, and K. Kajiyama, *IEEE Trans. Electron. Dev.*, Vol. ED-23, p. 1337, 1976.
46. J. Müller and A. Ataman, *Tech. Dig. Int. Electron. Dev. Meet.*, Washington, DC, 1976, IEEE, New York, p. 416.
47. N. Susa, H. Nakagame, H. Ando, and H. Kanbe, *IEEE J. Quant. Electron.*, Vol. QE-17, p. 243, 1981.
48. R. Yeats and K. Von Dessonneck, *IEEE Electron. Dev. Lett.*, Vol. EDL-2, p. 268, 1981.
49. O. K. Kim, S. R. Forrest, W. A. Bonner, and R. G. Smith, *Appl. Phys. Lett.*, Vol. 39, p. 402, 1981.
50. L. W. Cook, G. E. Bulman, and G. E. Stillman, *Appl. Phys. Lett.*, Vol. 40, p. 589, 1982.
51. O. Hildebrand, W. Kuebart, K. W. Benz, and M. H. Pilkuhn, *IEEE J. Quant. Electron.*, Vol. 17, p. 284, 1981.
52. S. Shin, J. G. Pasko, H. D. Law, and D. T. Cheung, *Appl. Phys. Lett.*, Vol. 40, p. 962, 1982.
53. R. Bube, *Photoconductivity of Solids*, Wiley, New York, 1960.
54. A. Rose, *Concepts in Photoconductivity and Allied Problems*, Wiley-Interscience, New York, 1963.
55. R. B. Hammond, N. G. Paulter, R. S. Wagner, T. E. Springer, and M. MacRoberts, *IEEE Trans. Electron. Dev.*, Vol. ED-30, p. 412, 1983.
56. A. G. Foyt and F. J. Leonberger, "InP Optoelectronic Switches," in C. H. Lee, Ed., *Picosecond Optoelectronic Devices*, Academic, New York, 1984, Chap. 9, p. 271.
57. F. J. Leonberger and P. F. Moulton, *Appl. Phys. Lett.*, Vol. 35, p. 712, 1979.
58. D. H. Auston, *Appl. Phys. Lett.*, Vol. 26, p. 101, 1975.
59. A. G. Foyt, F. J. Leonberger, and R. C. Williamson, *Proc. SPIE*, Vol. 269, p. 109, 1981.
60. J. B. Gunn, *IBM J. Res. Dev.*, Vol. 8, p. 141, 1963.
61. H. J. Klein, R. Kaumanns, and H. Beneking, *Electron. Lett.*, Vol. 17, p. 422, 1981.
62. T. F. Deutsch, F. J. Leonberger, A. G. Foyt, and D. Mills, *Appl. Phys. Lett.*, Vol. 41, p. 403, 1982.
63. R. H. Day, P. Lee, E. B. Solomon, and D. J. Nagel, *Los Alamos Sci. Lab. [REP] LA*, Vol. LA-7941-MS, 1981.
64. G. Mourou, Wilt Know, and S. Williamson, in C. H. Lee, Ed., *Picosecond Optoelectronics*, Academic, New York, 1984, Chap. 7, p. 219.
65. M. A. Schuster and G. Strull, *IEEE Trans. Electron. Dev.*, Vol. ED-13, p. 907, 1966.
66. J. N. Shive, *Phys. Rev.*, Vol. 76, p. 575, 1949.
67. Z. I. Alferov, F. Akhmedov, V. Korol'kov, and V. Niketin, *Sov. Phys. Semiconductor*, Vol. 7, p. 780, 1973.
68. W. Shockley, "Circuit Element Utilizing Semiconductive Material," U.S. Patent Office, Patent No. 2, 569, 347, 1951.
69. M. N. Svilans, N. Crote, and H. Beneking, *IEEE Electron. Dev. Lett.*, Vol. EDL-1, p. 247, 1980.
70. S. Margalit and A. Yariv, in W. T. Tsang, Ed., *Semiconductors and Semimetal*, Vol. 22, Part E, Academic, New York, 1985.
71. S. C. Lee and G. L. Pearson, *J. Appl. Phys.*, Vol. 52, p. 275, 1981.
72. F. Capasso, in W. T. Tsang, Ed., *Semiconductors and Semimetal*, Vol. 22, Part D, Academic, New York, 1985.

73. F. Capasso, *IEEE Trans. Electron. Dev.*, Vol. ED-29, p. 1388, 1982.
74. R. Chin, N. Holonyak, Jr., G. E. Stillman, J. T. Tang, and K. Hess, *Electron. Lett.*, Vol. 16, p. 467, 1980.
75. D. J. Robbins, *Phys. Stat. Sol. (b)*, Vol. 97, Part I, p. 9; Part II, p. 387; Part III, p. 403, 1980.
76. F. Capasso, W. T. Tsang, A. L. Hutchinson, and G. F. Williams, *Appl. Phys. Lett.*, Vol. 40, p. 38, 1982.
77. F. Capasso, W. T. Tsang, and G. F. Williams, *IEEE Trans. Electron. Dev.*, Vol. ED-30, p. 381, 1983.
78. H. Blauvelt, S. Margalit, and A. Yariv, *Electron. Lett.*, Vol. 18, p. 375, 1982.
79. G. F. Williams, F. Capasso, and W. T. Tsang, *IEEE Electron. Dev. Lett.*, Vol. EDL-3, p. 71, 1982.
80. B. F. Levine, C. G. Bethea, G. Hasnain, J. Walker, and R. J. Malik, *Appl. Phys. Lett.*, Vol. 53, p. 296, 1988.

8

LIQUID CRYSTALS: MATERIALS, DEVICES, AND APPLICATIONS

Uzi Efron

Research Laboratories
Hughes Aircraft Company
Malibu, California

8.1 INTRODUCTION

Two important fields in the revolutionary development of technology in this century have been those of information-processing and communications displays. Future advances in both of these areas will require the development of a highly efficient optical modulator technology. In the area of displays the need exists to replace the bulky dim cathode ray tube (CRT) displays with brighter, compact, low-voltage power-efficient devices. In the field of information processing we are faced with the need of performing more complex, higher-throughput computation with smaller, more compact, and lower power-consuming systems. The needs in both of these areas require not an improvement of existing devices, but rather the development of fundamental and revolutionary new technologies. Thus the two extreme ends of miniature displays, on the one hand, and of large-screen projection systems, on the other, are not possible with the existing CRT technology. Similarly, there exists no solution to the problem of real-time robotic function or to the related problem of real-time image understanding. Optical data processing has emerged as a potential technique for high-throughput computation that can solve the need for robotics and image-understanding issues. For this technology to succeed, however, a highly efficient optical modulator would be required. This would allow the conversion of the data to be processed to a (two-dimensional) optical format required by this optical processing technique. One powerful technology that has emerged in the past two decades as a potential solution to those needs in information processing and display fields is that of liquid crystals. As the name implies, this class of material represents an intermediate phase between a solid and a liquid. As a consequence, the physical and optical properties of these liquid crystals represent a "compromise" in some sense between the two phases. This intermediate state of matter is actually a result of breaking the melting transition into two stages. Thus melting, which is normally associated with the breaking of the translational *and* the orientational degrees of freedom of a solid, occurs in two steps. First, a translational disorder occurs in which the lattice structure disappears. The intermolecular forces between the molecules in the resulting liquid are still strong enough, due to the elongated shape of the molecules, so that rotation of the long molecular axis is inhibited. This liquid phase is therefore such that a certain degree of orientational order, or "crystallinity," is still maintained. Hence the name *liquid crystal*. As the temperature is increased further, the thermal agitation of the molecules becomes large enough to tear apart the orientational bonds. The liquid then becomes an "isotropic" liquid such as we normally encounter. This fascinating, intermediate phase phenomenon has been and still is a subject of intense research in physics, aimed at revealing the fundamental laws that govern the behavior of the various states of matter. However, it is exactly this "compromise" between the two phases that is so useful in the employment of liquid crystals as optical modulators. Thus, while solid materials are commonly found to exhibit the electro-optical, or magneto-optical, phenomena required for those applications mentioned in the preceding, the field required to activate those crystals in order to achieve the necessary optical modulation is quite high. This results in a severe limitation of the obtainable resolution as well as in the compactness of the device. Liquids, on the other hand, are more "flexible" in allowing molecules to be rotated by the action of an external field. However, they normally lack the optical anisotropy required for the use of optical modulation. The intermediate liquid crystal state is a very beneficial compromise from this standpoint. This class of material does exhibit the optical anisotropy required and, at the same time, has this "flexibility" in allowing the use of relatively low fields to achieve a sizable change in its optical properties. The resulting compact, high-resolution liquid crystal spatial light modulators may well revolutionize both areas of information processing and displays. In what follows I have tried to give an insight into the fundamental physics of liquid crystals as well as to provide a comprehensive overview of the associated devices and applications. In order to accomplish the first of the two, I have limited the discussion of the physics and optics of these materials to the class of nematic liquid crystals that represents, at least currently, the most widely used one. At the end of Section 8.2, a brief description of other classes of materials, such as cholesteric and smectic materials (in particular the ferroelectric subclass of the latter), is provided. The third section is devoted to the description of spatial light modulators, which represent the most important class of liquid crystal devices for practical applications. Finally, those important applications such as displays, optical data processing, adaptive optics, and wavelength image conversion are presented in Section 8.4.

8.2 PHYSICAL AND OPTICAL PROPERTIES OF LIQUID CRYSTALS

8.2.1 The Liquid Crystal State

As their name implies, liquid crystals represent an intermediate phase involving the properties of both liquids and crystals. From the standpoint of the physics of phases, one can define liquids or solids based on the notion of an "order parameter" that represents the degree of either translational or rotational order of the molecules within the particular phase. This degree of ordering is generally determined by the trade-off between the decrease in the internal energy U and the increase in entropy S in the free energy of the system:

$$F = U - TS$$

As the system becomes more disordered in a "classical" solid–liquid transition, this decrease in internal energy due to increasing the intermolecular distance is offset by the increase in both translational and orientational disorder such that the solid melts. Due to the large coupling that normally exists between the rotational and translational motion of molecules, both degrees of freedom are "released" as the solid melts. However, in the situation where the molecules are either very close to a spherical structure or, conversely, very asymmetrical or oblong, the translational and rotational motions may be sufficiently decoupled, resulting in the separation of the melting transition into two stages. In the first case, with nearly spherical molecules the penalty in losing the rotational coupling energy may be low enough that the solid may give up the rotational order while maintaining the translational one. This will then give rise to a solid–solid transition in which the molecules execute rotational motion in fixed lattice sites. A solid that exhibits a large degree of plasticity due to this rotational order is the outcome of this phase transition. The other extreme structure, rod-shaped molecules, will undergo a melting phase transition that involves only the translational degrees of freedom. In this resulting liquid, molecules will have the freedom to move about, but due to their rod-shaped structure, their rotational motion about an axis perpendicular to the long rod directions will be largely inhibited. We have in this case a liquid that maintains some of its rotational order; hence the name liquid crystal.

The appearance of a liquid crystal (LC) phase may be observed not only within a certain temperature range (thermotropic LCs), as described in the preceding, but also by varying the *concentration* of a polar-head-containing (amphilatrile) molecule in water, for example. A nematiclike formation of the LC phase may then be observed within a certain concentration range. This type of LC is called lyotropic. The most useful family of LC materials so far has been the thermotropic family; in the following we will therefore concentrate on examining the properties of this class.

8.2.2 Thermotropic Liquid Crystal Classes

The three main classes of thermotropic LCs are smectics, cholesterics, and nematics. Figure 8.1 shows the schematic structure of each of these classes. It is observed that the smectic class is characterized not only by possessing an orientational order, common to all LCs, but also by a certain degree of translational order associated with its layered structure. It is also observed from Fig. 8.1 that the smectic class exhibits polymorphism, that is, multiple phases, within the temperature range of liquid crystallinity.

The nematic class of LCs, to which we will devote most of the discussion in the following sections, is completely disordered translationally. The rod-shaped molecules are, however, aligned along a common direction parallel to their long dimension. This alignment is not perfect, as shown in Fig. 8.1*a*, since the thermal energy causes a certain amount of disorder through individual orientational fluctuations of the molecules. The addition of chirallike molecules to the nematic phase will result in the formation of a helical structure, as shown in Fig. 8.1*c*. This type of liquid crystalline class is referred to as *cholesteric* since the same structure exists in pure cholesterol esters. A similar structure of helical shape can also be obtained by imposing different alignment directions on the cell surfaces of a nematic LC. The structures obtained in this case of twisting a purely nematic material are referred to as *twisted nematic LC cells*. These forms of nematic structures are the most commonly used configurations of LCs today.

8.2.3 Physical Properties of Thermotropic Liquid Crystals

Chemical Structure. The building blocks of LC materials are relatively small organic molecules. The molecular structure of *p*-methoxybenzylidene *p*-butylaniline (MBBA) and of *p*-azoxyanisole (PAA) are shown in Figs. 8.2*a* and 8.2*b*, respectively. Both are representative of *nematic* LC molecules. Both of these molecules are relatively small, 20–30 Å, and are typically of an oblong shape. The double-benzene-ring structure with

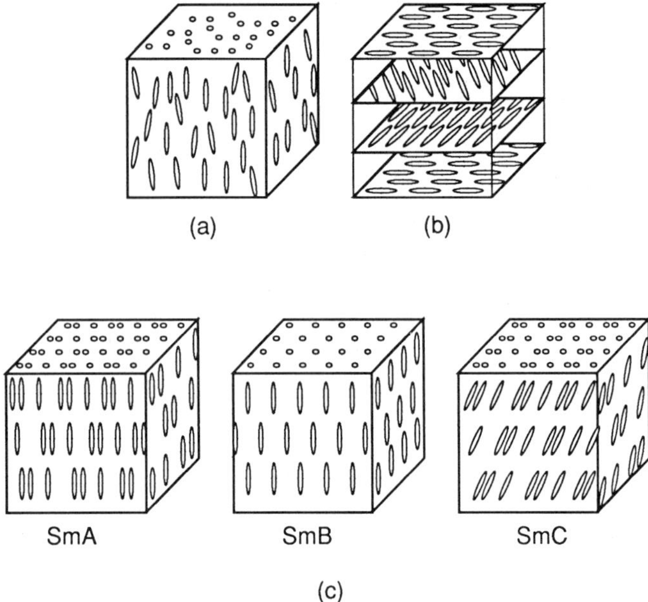

Figure 8.1 Three classes of thermotropic liquid crystals: (*a*) nematic, (*b*) cholesteric, and (*c*) smectics.

the bridging molecules results in a highly anisotropic polarizability that is the origin for both the electrical and optical anisotropy of the LC phases.

The direction of high optical polarizability occurs along the conjugated bonds (alternate single and double bonds), typically between the two benzene rings. The polarizability anisotropy gives rise to the anisotropic refractive index, which, as we shall see later, is the most significant property of LC materials that is used in their numerous optical applications. There exist many chemical classes of LCs, whose detailed description is outside the scope of this chapter. Division into their main classes can be found in Ref. 1. The most important classes of LCs are based on the structure shown in Fig. 8.3. Where the radicals R_1, R_2 and the bridging structures X are either alkoxy alkyl chain or cinnamate types, the bridging structures X are Schiff base azoxy or ester compounds [2].

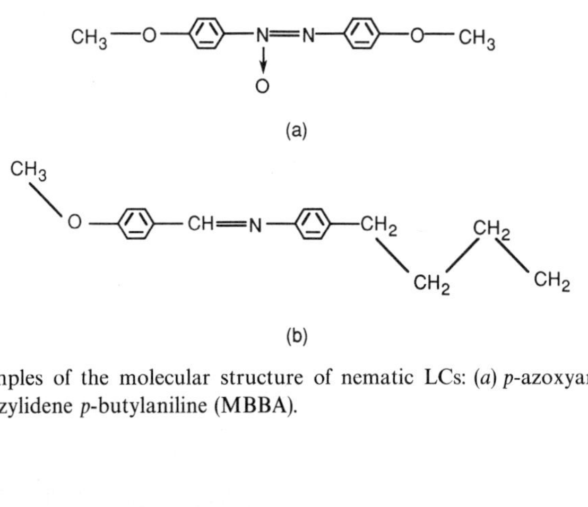

CH_3—O—⟨◯⟩—N═N(→O)—⟨◯⟩—O—CH_3

(a)

CH_3—O—⟨◯⟩—CH═N—⟨◯⟩—CH_2—CH_2—CH_2—CH_2

(b)

Figure 8.2 Examples of the molecular structure of nematic LCs: (*a*) *p*-azoxyanisole (PAA) and (*b*) *p*-methoxybenzylidene *p*-butylaniline (MBBA).

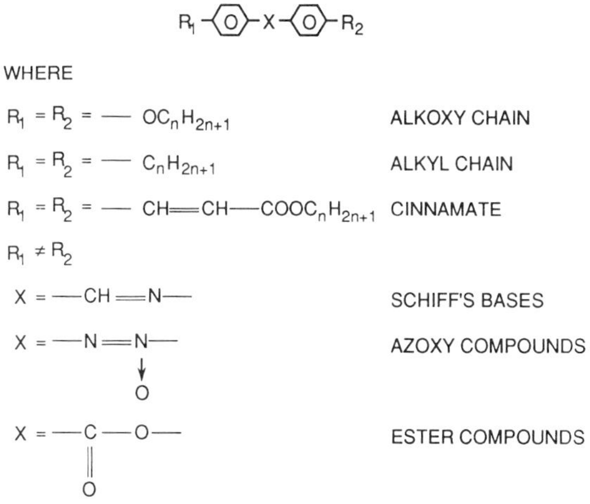

R_1–⟨O⟩–X–⟨O⟩–R_2

WHERE

$R_1 = R_2 = —OC_nH_{2n+1}$	ALKOXY CHAIN
$R_1 = R_2 = —C_nH_{2n+1}$	ALKYL CHAIN
$R_1 = R_2 = —CH{=}CH—COOC_nH_{2n+1}$	CINNAMATE
$R_1 \neq R_2$	
$X = —CH{=}N—$	SCHIFF'S BASES
$X = —N{=}N(\rightarrow O)—$	AZOXY COMPOUNDS
$X = —C({=}O)—O—$	ESTER COMPOUNDS

Figure 8.3 Generalized structure of liquid crystals. (After Ref. 2.)

Thermodynamic Properties. As mentioned earlier, the nematic materials are presently the most useful class of liquid crystals. In the following we shall therefore concentrate mainly on the properties of this class. A brief description of the properties of the other two thermotropic classes will follow.

As explained in Section 8.1, LCs are "ordered liquids" or liquid phases in which a certain degree of orientational order is retained. In order to characterize this order, consider Fig. 8.4, which describes the nematic phase as an aggregate of rigid rods [3]. Since macroscopically the dipole moment of this ensemble vanishes (due to the symmetry of rotating the whole aggregate by π), the directional order of the rods can be expressed as the quadrupole moment:

$$S_L = \langle \tfrac{1}{2}(3\cos^2\theta - 1)\rangle = \tfrac{1}{2}\int d\theta\, f(\theta)(3\cos^2\theta - 1) \tag{8.1}$$

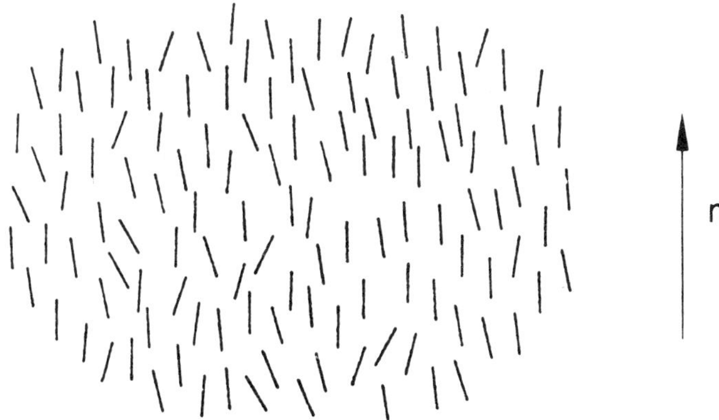

Figure 8.4 Structure of the nematic phase. A long-range order exists in the direction, n, shown.

In this average θ is the angle between the direction of each individual rod and the Z-axis of a specified coordinate system, and $f(\theta)$ is the angular distribution of the rods' orientation. It is observed, of course, that in an isotropic liquid $\cos^2\theta = \frac{1}{3}$ and the order parameter S_L vanishes. The nematic phase transition can be described based on some assumption with regard to the LC molecular interactions. If one starts from the rigid picture of Onsager [3], the molecular interaction is represented simply as an excluded volume, which means that a certain volume surrounding each rod within a certain radius is unavailable to any other rod. The thermodynamics of such a system can be determined, as pointed out in the preceding, by minimizing the free energy:

$$F = U - TS \tag{8.2}$$

The idea is to assume a perturbation from the equilibrium state that is a function of the temperature T. If the perturbation is such that the molecules would tend to establish a more disordered state, then more of them will pay the "penalty" of increased energy, ΔU, due to the constraint of excluded volume. However, by achieving a higher degree of disorder, the *entropy*, ΔS, of the system will also increase. Since the whole expression is temperature dependent, we can see that below a certain temperature T_c, defined as $\Delta U = T_c\,\Delta S$, the ΔU term will "win" over the entropy, $T\Delta S$, term and the system will be ordered. Above T_c, we will have $T\Delta S > \Delta U$, and the disordered state will prevail.

It is quite obvious that the existence of a transition to an ordered state must depend on the volume concentration of the rods. This is so since if this concentration is sufficiently low, the energy penalty associated with the probability of one rod getting into the excluded volume of another one is so low that the term $T\Delta S$ will always dominate for $T > T_m$, where T_m is the melting temperature of the rod-doped solid, and there will be no ordered liquid state. Thus, if one assumes a cylindrical volume of length L and diameter D with a concentration C, the fractional volume occupied by the rods is given by

$$\phi = \tfrac{1}{4}\pi CLD^2 \tag{8.3}$$

Onsager's treatment shows that the critical volume fraction ϕ_c is given by

$$\phi_c = \frac{4.5D}{L} \tag{8.4}$$

where for $\phi > \phi_c$ the rod mixture will be ordered. It is interesting to note that the critical volume fraction is directly proportional to the aspect ratio D/L of the rods. Thus, a smaller critical volume fraction will be required to allow an ordered state with a smaller aspect ratio.

Rather than assuming excluded volumes, one can assume a certain interaction energy between the rods. The theory that describes such interaction, the Meier–Saupe treatment [4], is actually based on earlier treatments. These are equivalent to the *molecular field approximation*, which assumes that (1) the system favors the alignment of the units (rods, molecules, etc.) and (2) each molecule "sees" an average, or molecular field, of all the others as the effective field. This is mathematically expressed by defining the energy of the system, U, as

$$U = -J\sum_{i,j} T_i T_j \tag{8.5}$$

where the T_i, T_j are the "spins," or directions, of the units and J is the attractive potential of aligned units. In 1939 Lennard-Jones and Devonshire [5] showed that melting of a crystalline lattice can occur as a result of balancing the free-energy equation (8.2) when a change in the internal energy of the system, ΔU, resulting from an "alignment error" is offset by the increase in the entropy due to that same error. The authors introduce a free-energy function of the form (8.2) when the internal energy is given by

$$U = -\tfrac{1}{2}\alpha S_L^2 \tag{8.6}$$

where α depends on the pressure and temperature of the system and S_L is the order parameter defined in (8.1). This form of energy represents an interaction term between the molecules. Minimization of the free energy [Eqs. (8.2), (8.5), and (8.6)] results in the appearance of an ordered phase below a temperature T_c determined by $KT_c = 4.55\alpha$. The Meier–Saupe theory [4] treats in an equivalent way the question of rotational order. The theory was fairly successful in describing qualitatively the thermodynamics of the LC phase. The expansion of this theory to allow the separate melting of both *translational* and *rotational* degrees of freedom was carried out by Pople and Karasz [6] in 1961. In this treatment they have defined *two* order parameters for the system: one for translational degrees of freedom and one for rotational degrees of freedom, with a ratio v characterizing the ratio between the energy increment involved in making an "orientation error" and that of a translational error. The authors have shown that for $1.925 > v > 0.325$ we will have a "normal" melting where both the rotational and the translational degrees of freedom are simultaneously melted above a certain (v-dependent) temperature. The occurrence of a combined translational–rotational melting over a large scale of v is due to the coupling between the rotational and translational errors. However, for either $v < 0.325$ or $v > 0.925$, the two melting transitions will separate; for the case of a relatively small rotational barrier ($v < 0.325$), the rotational degrees of freedom will melt first while a "translational lattice" is still maintained. These solids are known as plastic phases, with examples such as CD_4 and CCL_4, whose molecular structure approaches a sphere resulting in a low barrier to rotation. At the other extreme we have, as pointed out earlier, very anisotropic or elongated molecules in which the barrier to rotation of the long axis is much higher than that of translation of these molecules. In this case the melting of the lattice occurs first, with the resulting "liquid crystal" remaining highly anisotropic within a certain temperature range as the oblong molecules are aligned along a certain direction. An obvious extension of this theory is to allow a sequence of phases where the phases become gradually disordered with the increase in temperature.

Dielectric Properties of Liquid Crystals. As we have already seen, LCs are composed of elongated molecules. The distribution of charges in the molecules is anisotropic, giving rise to anisotropic polarizabilities. This in turn results in the appearance of an anisotropic permittivity in the ordered phase. The dielectric anisotropy $\Delta\varepsilon = \varepsilon_{\|} - \varepsilon_{\perp}$ is given by [7]:

$$\Delta\varepsilon = 4\pi NhF\left[(\alpha_e - \alpha_t) - F\left(\frac{\mu^2}{2KT}\right)(1 - 3\cos^2\beta)\right]S_L \tag{8.7}$$

where α_e, α_t are the longitudinal and transverse polarizabilities of the molecule, μ is the permanent molecular dipole moment (which is absent in some LC materials), F depends on the reaction field factor, β is related to the angle between the molecular axis and the direction of the applied field, and S_L is the order parameter. The important features of (8.6) are (a) the linear proportionality of $\Delta\varepsilon$ to the order parameter and (b) the (obvious) proportionality to the polarizability anisotropy $\Delta\alpha = \alpha_e - \alpha_t$.

The existence of an anisotropic dielectric constant, $\Delta\varepsilon$, allows the distortion of the LC medium by an external electric field **E** that will induce a dipole moment $\boldsymbol{\mu} \approx (\Delta\varepsilon/4\pi)\mathbf{E}$ and will exert a torque on this dipole moment proportional to $\boldsymbol{\mu} \times \mathbf{E}$.

Turning now to the dynamic properties of the dielectric anisotropy, the governing factor here is the anisotropy in the response time. Generally, one can describe the frequency dependence of each of the two components $\varepsilon_{||}$ and $\varepsilon_{\perp}$ by the Debye relation:

$$\varepsilon_\alpha = \varepsilon'_\alpha - i\varepsilon''_\alpha = \varepsilon_{\infty,\alpha} + \frac{\varepsilon_\alpha - \varepsilon_{\infty,\alpha}}{1 + i\omega\tau_\alpha} \tag{8.8}$$

where $\varepsilon_{\infty,\alpha}$ refers to the infinite (optical) frequency response of the α component ($\alpha = ||, \perp$) and τ_α is the characteristic relaxation time of the α component of ε. The large, low-frequency component of the polarizability in the LC is of an orientational nature. It is therefore expected that the longitudinal relaxation time $\tau_{||}$ will be much longer than the transverse time $\tau_\perp$. This is so since $\tau_{||}$ is associated with the rotation of the elongated molecules around their short axis, a motion significantly hindered in the ordered phase. Thus, while $\varepsilon_\perp(\omega)$ stays more or less constant up to the megahertz region, $\varepsilon_{||}$ decreases rapidly with the frequency and can drop below $\varepsilon_\perp$ at $\omega \approx 10$ kHz in some materials. In this case we have a sign change in the dielectric anisotropy that can be used to enhance the relaxation time of nematic LC cells [8]. At optical frequencies $\varepsilon \sim n^2$, and we observe an anisotropy of the refractive index: $\Delta n = n_{||} - n_\perp$. The magnitude of Δn is quite large, on the order of 0.1. It is this large refractive index anisotropy that, combined with the low field required to modify the LC orientation (see below), results in an extremely large electro-optical coefficient, thus making LCs one of the most important electro-optical materials. The anisotropy in the refractive index, or birefringence, is of course directly related to the anisotropy in the optical absorption coefficient via the Kramers–Kronig dispersion relationship. Figure 8.5 shows the birefringence of two LC materials in the visible and IR spectral regions.

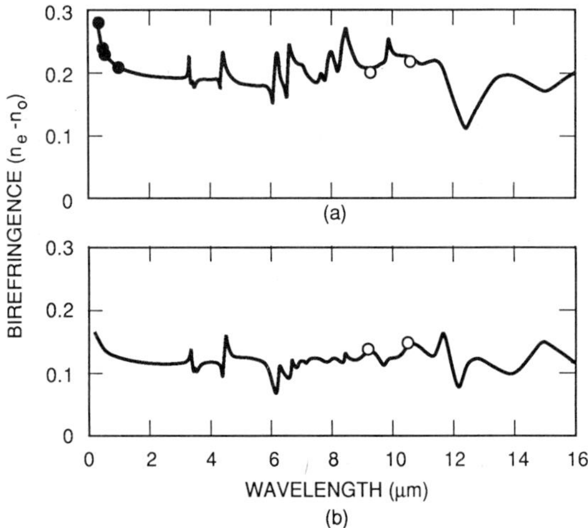

Figure 8.5 Birefringence of nematic LC materials (*a*) BDH E-7 and (*b*) ZLI-1132. Circles represent measurements performed with single laser lines [42].

Viscoelastic Properties and Static Field Effects. The continuum theory, which is used in describing the statics and dynamics of LCs, is based on the empirical observation that the ordering of the LC molecules has a typical region size that is much larger than the molecular dimensions. Thus, in order to describe the LC state and orientation, one uses the concept of a "director," **n(r)**. This vector quantity is of unit length, and its orientation is along the optical axis of the LC at the spatial location **r**. The method of defining the various LC orientations or distortion then consists of defining the spatial (**r**) dependency of **n**. Figure 8.6 shows the three basic configurations of a nematic LC cell: (a) parallel aligned, (b) perpendicular aligned or homeotropic, and (c) a twisted configuration. These basic configurations are the result of free-energy minimization of the LC cell under the particular boundary conditions. Therefore, as we shall see, any attempt to deform the LC from one of these long-range alignments will result in an increase in the elastic energy that must be compensated for by a decrease in the potential energy. This uniform long-range orientation can be determined by proper alignment layers produced at the cell surfaces. These alignment layers are produced by angular deposition of certain insulating materials (e.g., SiO_2) or by mechanical rubbing of the surface. These surface relief patterns interact and anchor the adjacent LC molecules in certain directions determined by these alignment reliefs or "marks." These surface treatments can determine whether the LC molecules will be aligned parallel or perpendicular to the surface as well as the relative orientation of the LC director to the surface. Let us assume, then, that we have treated the cell electrode surfaces to achieve a certain uniform orientation of the LC and examine the possible deformation of the LC layers by some external field. The latter may be electrical, magnetic, or mechanical (acoustic). Three basic elastic deformations can be induced in LC media. In the case of splay (Fig. 8.7*a*) or S-deformation, in which the adjacent layers of parallel-aligned nematic material are sheared angularly, there is a nonvanishing divergence of the director, $\nabla \cdot \mathbf{n} \neq 0$; the associated elastic constant is K_{11}. The second deformation is a bending deformation, or B-deformation (Fig. 8.7*b*), in which, similar to the S-deformation, adjacent layers are angularly sheared for the case of perpendicularly aligned (or homeotropic) LC. The associated elastic constant is K_{33}. Finally, we have the very useful twist deformation (Fig. 8.7*c*) in which the director orientation is slowly rotated, or twisted, between adjacent parallel-aligned layers. As can be seen, we have a nonzero $\nabla \times \mathbf{n}$ with an associated elastic constant, K_{22}. The elastic energy density describing a general contribution from all three distortions is given by [9,10]

$$U_e = \tfrac{1}{2}[K_{11}(\nabla \cdot \mathbf{n})^2 + K_{22}(\mathbf{n} \cdot \nabla \times \mathbf{n})^2 + K_{33}(\mathbf{n} \times \nabla \times \mathbf{n})]^2 \tag{8.9}$$

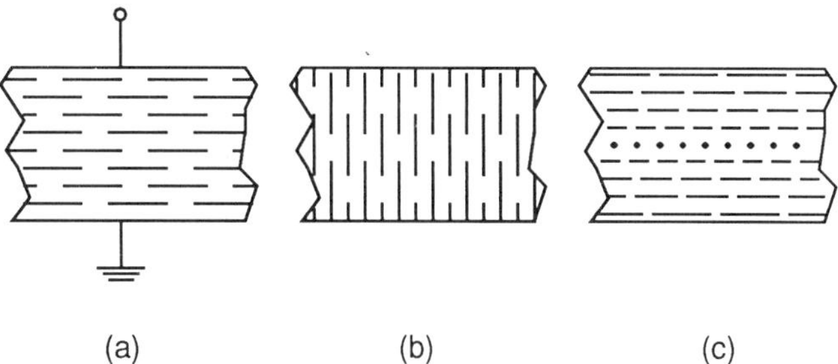

Figure 8.6 Three basic configurations of nematic LCs: (*a*) parallel alignment, (*b*) perpendicular or homeotropic alignment, and (*c*) twisted nematic (TN) configuration.

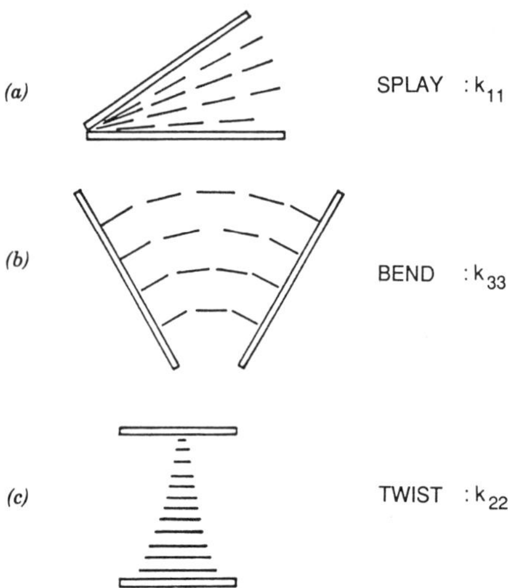

Figure 8.7 Three basic types of distortion in nematic LCs: splay, bend, and twist.

This expression describes the energy density in a volume element under the general distortion of the LC alignment. This distortion will depend on the particular LC class used (nematic, cholesteric, or smectic). While splay (K_{11}) and bend (K_{33}) effects can be observed in all three configurations, twist effects are observed only in the nematic or cholesteric classes.

As indicated earlier, one can use electric [11], magnetic [12], or acoustic [13] fields to distort the LC medium. The latter has not been practically utilized in the operation of LC cells. Magnetic fields were historically the first to be applied but gave way to electrical activation due to the more practical usage of electro-optical effects (as opposed to magneto-optical effects) in LCs. Sufficiently intense *optical* field can also induce a reorientation or deformation of the LC director through the effect of the associated electric field [14]. Within the scope of this chapter we shall consider mainly the electric field effect, which has been the most practical one for device applications.

Let us first examine the static distortion of a LC cell. The torque associated with the application of the electric field is estimated as follows: The application of the field first results in an induced dipole moment in the direction of the LC director $\mathbf{n}_n$, given by

$$\boldsymbol{\mu} = \frac{\Delta\varepsilon}{4\pi}(\mathbf{E} \cdot \mathbf{n}_n)\mathbf{n}_n = \left(\frac{\Delta\varepsilon}{4\pi}|E|\cos\theta\right)\mathbf{n}_n \tag{8.10}$$

where θ is the angle between the direction of the applied field and the director $\mathbf{L}$. Next, the field acts to exert a torque on the induced dipole moment:

$$|\mathbf{T}| = |\boldsymbol{\mu} \times \mathbf{E}| = \frac{\Delta\varepsilon}{4\pi}|E|^2\cos\theta\sin\theta \tag{8.11}$$

In the static (or steady-state) case, this torque is balanced by the restoring elastic forces discussed earlier. Let us assume, as an example, that we are dealing with a parallel-aligned nematic LC cell. The elastic deformation energy per unit volume is given by

$K_{11}(\nabla \mathbf{n})^2$ [Eq. (8.9)]. Differentiating the elastic energy term with respect to Z (the coordinate along the cell thickness) to obtain the elastic torque results in the following equation:

$$K_{11} \frac{\partial^2 \theta}{\partial Z^2} + \frac{\Delta\varepsilon E^2}{4\pi} \sin\theta \cos\theta = 0 \tag{8.12}$$

The solution to Eq. (8.12) can be expressed implicitly as [15,16]

$$E \cdot d = 2\left(\frac{4\pi k}{\Delta\varepsilon}\right)^{1/2} \int_0^{\theta_m} \frac{d\theta}{(\sin^2\theta_m - \sin^2\theta)^{1/2}} \tag{8.13}$$

This solution shows that the LC will be distorted (tilted) so that the director orientation $\mathbf{n}(z)$ [or $\theta(z)$] will be continuously distributed across the cell (z direction). In this equation d is the cell thickness and θ_m is the angle of maximum tilt that occurs at the center of the cell. The important feature of Eq. (8.13) to the torque equation (8.12) is that the LC will distort under the effect of an electric (or magnetic) field above a critical level given by

$$E_c = \frac{\pi}{d}\left(\frac{4\pi k}{\Delta\varepsilon}\right)^{1/2} \tag{8.14}$$

This effect was first discovered by Freedericksz [17] in 1931, and because of its analogy to a second-order phase transition, it is called the *Freedericksz transition*. Figure 8.8 shows the tilt angle distribution in a 4-μm BDH E-7 LC cell with the bias voltage as the parameter [18]. The critical or threshold voltage $V_{TH} = E_c d \approx 0.9$ V for this particular material. There are three interesting points about the critical field worth noting. (1) The origin of the critical field is in the Freedericksz transition, being a second-order phase transition. In this case fluctuation in the director, $\mathbf{n}$, is proportional to δL^2. This fluctuation results in a fluctuation of the two balanced terms in the free-energy density (which is the

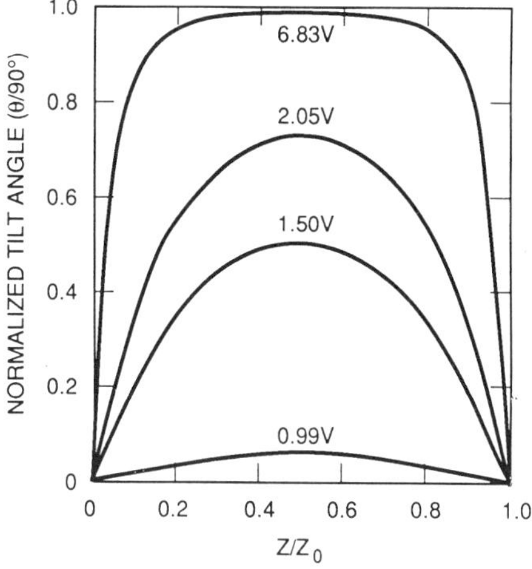

Figure 8.8 Tilt angle distribution in a 4-μm BDH E-7 LC cell with various applied fields [18].

integration over z of Eq. (8.12). For a sufficiently small field the balance is positive and the nondistorted LC is the stable solution. For $E \geq E_c$ the distorted LC becomes the more stable solution [19]. (2) As observed from Eq. (8.14) the threshold voltage (or the product of the critical field and the cell thickness) is basically a balance between the dielectric anisotropy $\Delta\varepsilon$ and the elastic constant. Thus, either a higher anisotropy or a smaller elastic constant will result in a reduced critical field. (3) The third property of the critical field is that it characterizes a coherence or transition length over which a given external field distorts the LC medium. This characteristic length is given by [19]

$$\xi = \frac{V_{\mathrm{TH}}}{E} \tag{8.15}$$

where $V_{\mathrm{TH}} = E_c d$, with E_c given by Eq. (8.14).

Dynamics of Liquid Crystal Distortions. Here again, for the sake of simplicity, we shall use a parallel-aligned nematic LC cell. In this case, assuming that $K_{11} = K_{33} = K$, we simply add the viscosity (γ-) related time-dependent term, $\gamma(d\theta/dt)$:

$$K\frac{\partial^2\theta}{\partial z^2} + \frac{\Delta\varepsilon}{4\pi}E^2 \sin\theta\cos\theta - \gamma\frac{d\theta}{dt} = 0 \tag{8.16}$$

Here we have neglected the inertial term $I(\partial^2\theta/\partial t^2)$ since it can be ignored for transient angular velocities of interest (namely $\omega = \partial\theta/\partial t \leq 10^7$ rad/s).† The solution to Eq. (8.16) can be approximated by [24–26]

$$\theta = \theta_m(t)\cos\frac{\pi z}{d} \tag{8.17}$$

where θ_m, the angle of maximum tilt, is given by

$$\theta_m^2(t) = \frac{\theta_m^2(\infty)}{1 + [\theta_m^2(\infty)/\delta^2 - 1]\exp[-2t/\tau(v)]} \tag{8.18}$$

where δ is equal to $\theta_m(t = 0)$ and the voltage-dependent response time $\tau(v)$ is

$$\tau(v) = \frac{\tau_0}{(V/V_{\mathrm{TH}})^2 - 1} = \frac{\tau^* d^2}{(V/V_{\mathrm{TH}})^2 - 1} \tag{8.19}$$

The characteristic response τ^* is expressed as a function of the material parameters by

$$\tau^* = \frac{\gamma}{\pi^2 k} \tag{8.20}$$

where γ is the rotational viscosity and k is the appropriate elastic constant (K_{11} in our case). Equation (8.18) can be used as a good approximation for describing the LC response time in two cases. The first is the rise, or "turning-on," time associated with the field-

† There are some conflicting reports about the estimates of I, the moment of inertia [20–22]. A recent attempt to measure inertial effects in nematic LCs indicated the possibility of a rather large I [23].

assisted reorientation of the LC director. In this case $\tau(V)$ describes the time required to reach $1/e^2$ of the final orientation value, with V as the activation voltage. The second is the "small-signal" response time. This term refers to the response time associated with small deviations of the director around an equilibrium position attained with the application of a steady-state voltage V. In this case, either positive *or negative* changes in the voltage, $\pm\Delta V$, around the steady-state voltage V will result in a reorientation time given approximately by Eq. (8.19) [27]. The "turning-off," or decay, time of the nematic director, that is, the response time required for the director to relax to within $1/e^2$ of its original undisturbed alignment, is given by

$$\tau_d = \tau_0 = \tau^* d^2 \tag{8.21}$$

and is completely governed by the ratio of the viscosity to the elastic constant.

8.2.4 Electro-optical Properties of Nematic Liquid Crystals

Now that we are equipped with the equations governing the static, field-dependent deformation of the LC medium [Eqs. (8.12)–(8.14)] and the dynamics of the director's motion [Eqs. (8.18)–(8.21)], we can turn to the description of its electro-optical behavior based on those physical models.

Let us first recall that the LC domain is characterized by a rather large refractive index anisotropy, or birefringence, Δn (see the section on the dielectric properties of LCs). This anisotropy $\Delta n = n_{\parallel} - n_{\perp}$ is observed with respect to the director's orientation. Thus, $n_{\parallel}$ refers to the value of n for a beam of light polarized along the director **L**, while $n_{\perp}$ refers to the effective refractive index experienced by a beam polarized perpendicular to **L** (the values of $n_{\perp}$ at both directions orthogonal to **L** are identical as LCs feature uniaxial structure). It follows that by subjecting the LC to the distortion action of an external field, we are effectively changing the apparent refractive index. The relatively large dielectric constant anisotropy $\Delta\varepsilon$ allows a relatively large amount of distortion, say in the tilt direction θ, employing only a small field [Eqs. (8.13) and (8.14)]. For commonly used nematic materials the threshold voltage is on the order of 1 V. A complete distortion (tilt) of the LC medium can be practically accomplished with less than 10 V over a cell a few micrometers thick.

In general, LCs can be used for both phase and amplitude modulation of optical beams. Amplitude or intensity modulation can be accomplished by the following mechanisms:

1. polarization rotation followed by an analyzer,
2. light scattering modulation, and
3. absorption using guest (liquid crystal)–host (absorbing dye) mixtures.

Phase modulation can be accomplished by the same polarization rotation techniques used for amplitude modulation, but with a suitable orientation of the beam polarization vector relative to the LC director. Again, with the limited scope of this chapter, we will confine ourselves mainly to polarization–rotation techniques used mainly with nematic (or smectic–ferroelectric) liquid crystals. A short description of scattering mechanisms will be given at the end of this section. Two basic nematic LC configurations are employed to accomplish polarization rotation. These two modes of operation, controlled birefringence (or birefringent mode) and the twisted nematic configuration, are based on the same physical phenomenon of controlling the effective birefringence of the LC cell by an external (usually electric) field.

Controlled Birefringence. In the controlled birefringence mode (Fig. 8.9) the LC is either parallel or homeotropically aligned. In parallel alignment, which is accomplished by mechanical treatment of the confining surfaces (windows) of the LC cell, the LC molecules are aligned parallel to the surface. The particular alignment direction in the plane of the cell surfaces is determined by the alignment procedure. In this mode of controlled birefringence the alignment on both surfaces is in the *same direction.* An optical beam at an incident angle of ψ to the normal of the LC cell surface and with an angle ϕ between the plane of incidence and the LC alignment will undergo a retardation as it passes through the cell. Let us assume for simplicity that $\psi = 0$, and examine two cases: (a) $\phi = 45^\circ$ and (b) $\phi = 0$.

(a) $\phi = 45^\circ$. For the first case, we have the familiar situation where the polarization is rotated due to the birefringence of the uniaxial medium. In this case, analyzing the exiting beam at 90° to the input polarization, the intensity will be given by

$$I_{\text{out}} = I_{\text{in}} \sin^2(\delta/2) \tag{8.22}$$

where the phase retardation δ is given by [28]

$$\delta = \frac{2\pi\, d\, \Delta n_{\text{eff}}(V)}{\lambda} \tag{8.23}$$

and where the effective, voltage-dependent birefringence $\Delta n_{\text{eff}}(V)$ can be approximated by [29]

$$\Delta n_{\text{eff}} = n_e - n_{\text{eff}} = n_e - \frac{1}{d}\int_0^d \frac{n_e n_o\, dz}{[n_e^2 \sin^2 \bar{\theta}(V) + n_o^2 \cos^2 \bar{\theta}(V)]^{1/2}} \tag{8.24}$$

where n_e, n_o are the extraordinary and ordinary refractive indices; $\bar{\theta}(V)$ is the voltage-dependent tilt angle. As we have previously observed, the tilt angle is actually a distribution of angles across the LC cell (Fig. 8.8). One can, however, approximate this distribution by an *average* voltage-dependent tilt angle $\bar{\theta}(V)$, to which Eq. (8.24) refers. Approximations

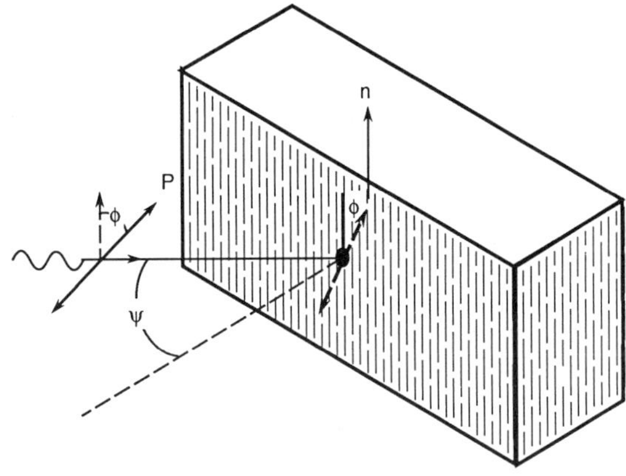

Figure 8.9 Geometry of the incident beam and polarization in the controlled birefringence LC configuration.

for Eqs. (8.23) and (8.24) exist for low and high fields [27]. These are

$$\delta = \alpha\delta_{\max}\left(\frac{V}{V_{\mathrm{TH}}} - 1\right) \quad \text{for } \frac{V - V_{\mathrm{TH}}}{V_{\mathrm{TH}}} \ll 1 \tag{8.24a}$$

$$\delta = \delta_{\max}\left(1 - \frac{\beta}{V}\right) \quad \text{for high fields } \frac{V - V_{\mathrm{TH}}}{V_{\mathrm{TH}}} \gg 1 \tag{8.24b}$$

where α and β are proportionality constants and

$$\delta_{\max} = \frac{2\pi\, d\, \Delta n}{\lambda} \tag{8.24c}$$

is the maximum retardation (phase change) of a LC cell.

The preceding expression assumes negligible absorption and reflection losses from the cell surfaces. Figure 8.10 shows the voltage-dependent transmission of a 10.1-μm-thick BDH/E-7 LC cell at $\lambda = 6328$ Å [30]. The effective squared sine modulation is clearly observed. The voltage-dependent effective birefringence of that material can be retrieved using expressions (8.22)–(8.24) and the experimental data of Fig. 8.10. The result is shown in Fig. 8.5*a*.

(b) $\phi = 0$. In this case the polarization of the incoming beam is parallel to the extraordinary axis of the LC (or to its director). As a consequence, no polarization rotation takes place. The retardation δ or the corresponding effective birefringence $\Delta n_{\mathrm{eff}}(V)$ of Eqs. (8.23) and (8.24) are simply translated into effective *phase shift* and refractive index changes, respectively. This configuration is therefore the one we can use to affect the *phase* rather than the amplitude of the optical beam [31,32]. Alternatively, one can exploit the fact that the same effective $\Delta n_{\mathrm{eff}}(V)$ can be used to modulate either phase shift or retardation. Since polarization rotation is much easier to measure than phase shift (which requires an interferometric measurement), one can evaluate the voltage-dependent phase shift in a given LC cell by measuring the voltage-dependent polarization rotation [18]. Finally, for the general case of an arbitrary angle ψ, we have, for the transmission through a crossed analyzer–polarizer pair [33],

$$T = \sin^2(2\psi)\sin^2(\delta/2) \tag{8.25}$$

which reduces to Eq. (8.22) for $\psi = 45°$.

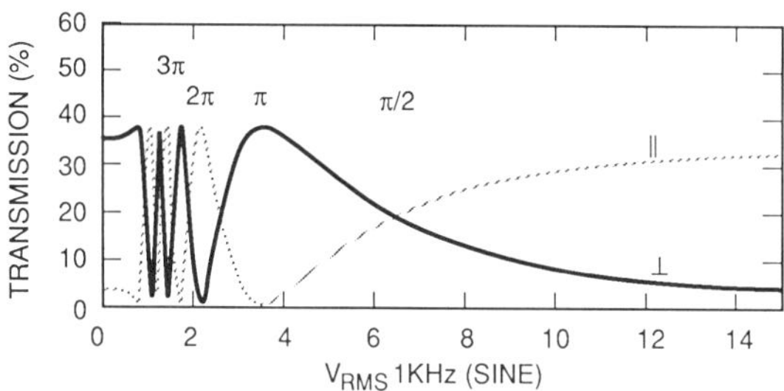

Figure 8.10 Transmission of a 10.1-μm BDH E-7 LC cell in the controlled birefringence configuration (Fig. 8.9); solid curve: crossed analyzer–polarizer; dashed curve: parallel polarizer–analyzer. (After Ref. 30.)

Twisted Nematic Configuration. In this configuration (Fig. 8.11) the nematic LC is twisted (see discussion in the preceding on viscoelastic properties and static field effects) by an angle θ_T determined by the difference between the alignment directions of the front and back surfaces of the cell. Thus as an optical beam propagates through the cell, it sees a LC director that rotates (twists) from one surface of the cell to the other. This twist induces a corresponding rotation of the polarization direction of the beam propagating through the cell. The physical mechanism of the polarization rotation by the twist effect has been extensively studied [34]. The effect is based on the same birefringence-related polarization–rotation as in the controlled birefringence (CB) mode, except that now we are dealing with the effective birefringence that is experienced by the optical beam as it propagates from one layer to the adjacent one in which the director is slightly rotated. If the relative retardation angle ($2\pi\, d\, \Delta n/\lambda$) is large compared to the twist angle θ_T, then the polarization of the beam will be rotated by the twist angle with negligible ellipticity. In other words, the exiting beam will be linearly polarized along the direction of the LC director at the exit surface. Gooch and Tarry [35] developed an expression for the rotatory power for the general case of a twisted nematic (TN) layer. This rotatory power (which is defined as the fraction of light intensity exiting with a polarization parallel to the LC alignment of the exit window) is given by

$$P_R^{(\mathrm{TN})} = 1 - \frac{\sin^2[\theta(1 - U^2)^{1/2}]}{1 + U^2} \tag{8.26}$$

where U is the retardation ratioed by the twist angle:

$$U = \frac{\delta}{\theta_T} = \frac{2\pi\, d\, \Delta n}{\theta_T \lambda} \tag{8.27}$$

A useful approximation to (8.26) that is valid for both $U \gg 1$ and $U \ll 1$ is

$$P_R^{(\mathrm{TN})} \approx \frac{U^2}{1 + U^2} \tag{8.28}$$

showing that for $U \gg 1$ (which is the case discussed in the preceding) $P_R^{(\mathrm{TN})} \to 1$; whereas for $U \ll 1$ the rotatory power approaches zero as U^2. The simplest way of utilizing the

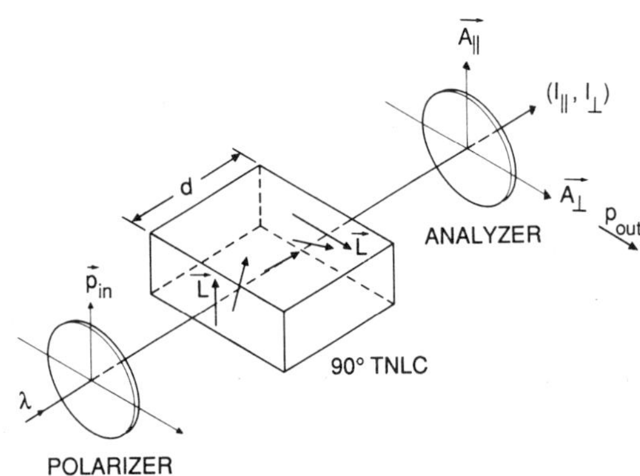

Figure 8.11 Polarization rotation using a 90° twisted nematic LC configuration.

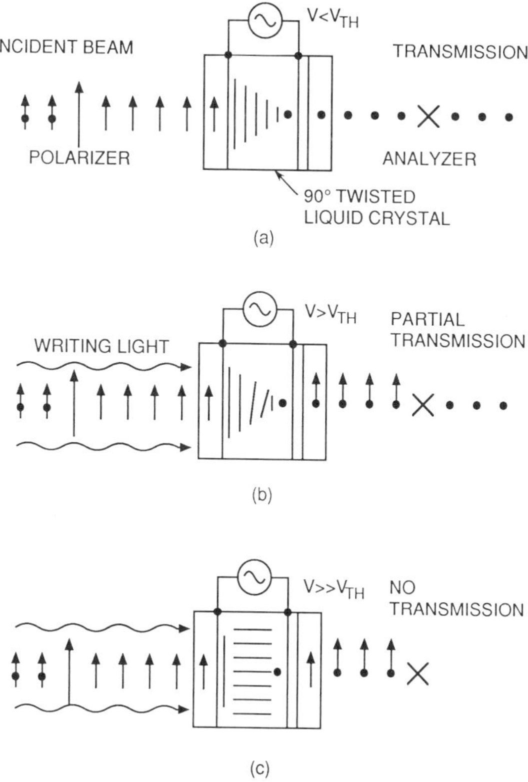

Figure 8.12 A 90° twisted nematic LC cell in a transmissive configuration: (*a*) LC below threshold; (*b*) partially activated state; and (*c*) fully activated state.

TN effect is shown in Fig. 8.12. In this case a 90° TN cell is shown between a crossed polarizer–analyzer pair with bias voltage below threshold (Fig. 8.12*a*). Complete transmission occurs through the cell since the twist configuration acts to rotate the polarization, as previously explained. With an intermediate voltage applied (between threshold and saturation) the LC molecules are partially tilted, resulting in a partial destruction of the twist configuration. The rotatory power of the 90° TN cell decreases, and as a result, only partial transmission occurs through the analyzer. Finally, with a high bias voltage, the molecules are sufficiently tilted to result in a complete destruction of the twist. This in turn renders the cell completely isotropic with zero rotatory power. No transmission of light is observed. It should be noted that in order to utilize this transmissive mode of operation, one needs to sandwich the 90° TN cell between a polarizer–analyzer pair. In many practical applications, a preferred mode is a reflective mode, and as we shall see later, most LC spatial light modulators operate in that mode. Since it is impractical to insert a polarizing sheet between the LC layer and the photodriving substrate† (see

† A polarizing sheet will take on a large fraction of the bias voltage, thus reducing the voltage increment required to activate the LC cell. An exception to that is the use of a wiregrid polarizer within the light valve. A complete description of this concept is found in Ref. 36.

Section 8.3.2), a different LC configuration is needed. Such a configuration, which optimizes the reflected intensity of a LC cell, was invented and demonstrated by Grinberg et al. [37]. In this configuration, which is termed the *hybrid field effect* (HFE) mode, the LC layer is twisted at 45°. A polarizing beam splitter is used as a crossed polarizer–analyzer pair placed in front of the HFE mode LC cell. With no bias voltage (Fig. 8.13) the polarization of the optical beam simply twists and untwists as it traverses in and out (by reflection) of the 45° twisted cell. The crossed analyzer results in a total extinction of the exiting beam as its polarization orientation is unaltered by the LC cell. With a low bias voltage exceeding the threshold, the LC molecules will start to tilt, resulting in an increased birefringence and a partial destruction of the twist arrangement. The polarization will therefore be partially rotated upon exiting the cell, resulting in a nonzero transmission. With increasingly high bias voltage, a point will be reached where full transmission is achieved. Further increase in the voltage beyond this point will result in an increasingly high tilt with decreasing birefringence and hence decreasing rotatory power. This operating regime, in which the output *decreases* with increased input, is called the *backslope* regime. At a sufficiently high bias voltage, the LC cell will appear virtually isotropic with the molecules fully tilted in a direction normal to the surface. With the resulting full bias of the LC, the transmission through the analyzer will vanish. This voltage dependence of the HFE cell transmission is shown in Fig. 8.14 and is compared against the use of a 90° twist in a reflective mode configuration. A computer program calculating the HFE cell transmission [38] showed good agreement with the experimental results shown in the preceding.

The CB mode described can also be used in principle for a reflective mode device. In order to utilize this configuration, one needs to construct a cell with a thickness that will correspond to a π phase shift with no voltage so as to achieve a dark state with no applied voltage. The immediate problem with this choice is that of a wavelength-dependent dark, or "off," state. Thus, the use of a CB mode would be restricted to monochromatic beams. The second problem with the straight use of the CB mode is the high sensitivity to direct cell thickness nonuniformity. Again, this stems from the direct dependence of the transmission on the thickness of the cell [Eqs. (8.22) and (8.23)]. In contrast, the TN effect is quite insensitive to either wavelength variation or cell thickness variation in its off state. Thus, as long as U in Eq. (8.26) satisfies $U \gg 1$, which holds throughout the visible region (400 μm $< \lambda <$ 1000 nm) for a few-micrometer LC cell (with $\Delta n \geq 0.15$), there is no wavelength dependence in the TN transmission [Eq. (8.26) or (8.28)]. The same holds for the cell thickness or d dependence in Eq. (8.26). These problems of using the CB effect can be alleviated to some degree by using a contrast reversal technique [39] and the use of a homeotropic (perpendicularly aligned) LC configuration [40]. Nevertheless, these are strong reasons for using the HFE mode of operation.

Wavelength Dependence Issues. The wavelength dependence of the rotatory power for both the CB and the TN modes becomes an issue for sufficiently long wavelengths, that is, in the IR regimes. It was previously shown that many of the commonly used nematic materials are highly transmissive throughout the IR regime, with the exception of a few, isolated absorption bands [41]. A relatively large birefringence in the IR was also observed. This opened up the possibility of using LCs to modulate IR beams. Studies of both the CB mode [41] and the TN mode [42] demonstrated the capability of LCs in modulating IR beams. The question was then which of the two modes is the more efficient one to use in the IR regime. The answer to this question is given in Fig. 8.15, which compares the two effects [39] in terms of the rotatory power versus the parameter $U = 2\pi d\ \Delta n/\theta_T \lambda$, which for a given cell thickness d, birefringence Δn, and twist angle θ_T is inversely proportional to the wavelength. As observed from this figure, the CB mode

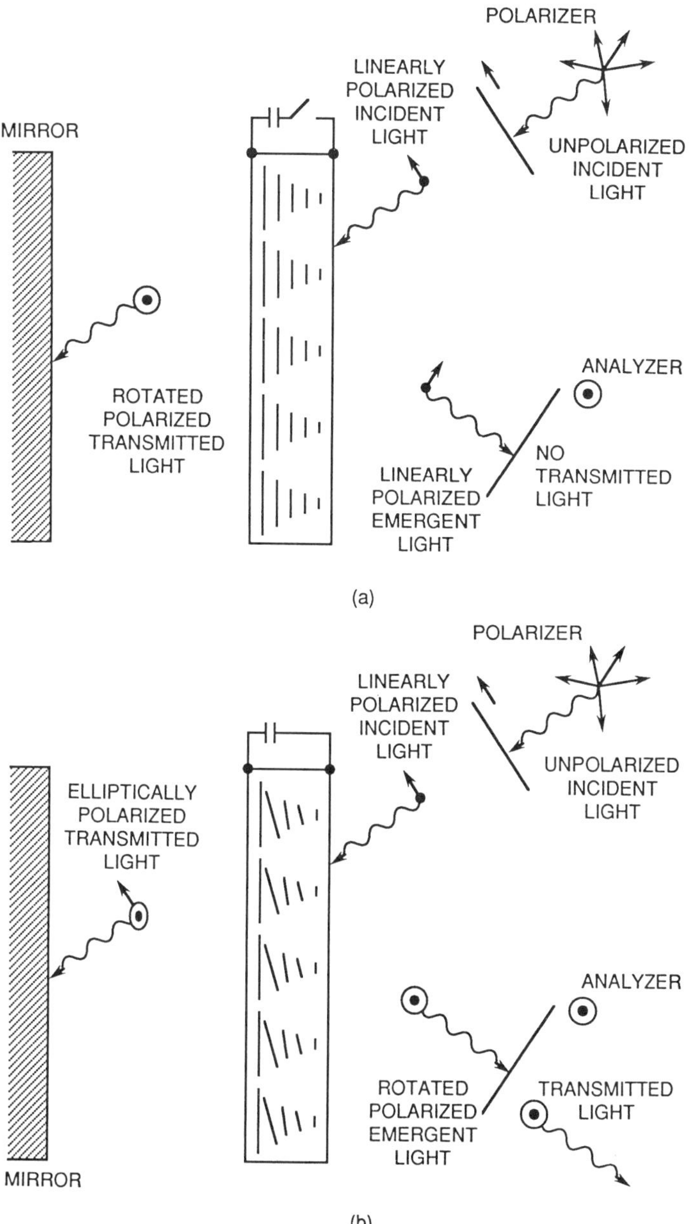

Figure 8.13 Hybrid field effect configuration. (After Ref. 38.)

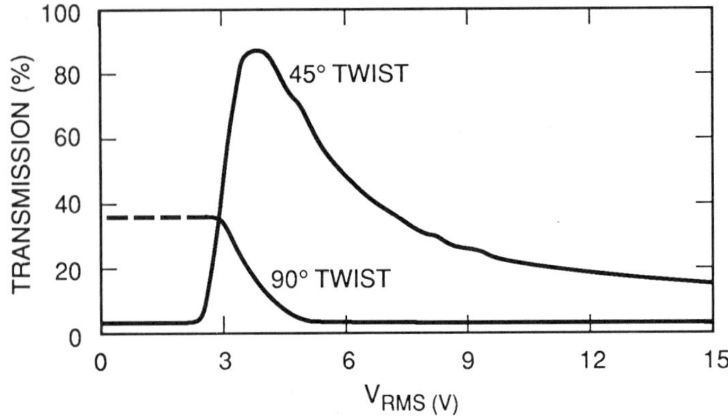

Figure 8.14 Transmission of the 45° twist hybrid field effect LC cell and 90° twist cell, both in the reflective configuration.

[curve (*a*)] is the "winner" in having a higher rotatory power for $U \leq 1.5$. This range of U corresponds to $\lambda > 6\ \mu$m for a 12-μm-thick LC cell with $\Delta n \approx 0.2$.

8.2.5 Dynamic Range/Response Time Trade-off

From the static response of the LC medium to electric field distortion [Eq. (8.24c)], we observe that the maximum dynamic range $\Delta\delta$ or $\Delta\phi$ (phase shift) for a LC is proportional to its thickness d. On the other hand, we observe from Eq. (8.19) that the response time is a quadratic function of the cell thickness. It is therefore clear that if one wishes to maximize the dynamic range while minimizing the response time, one would like to maximize the quantity $\Delta\phi/\Delta t$ [27]. In order to gain some feeling as to the behavior of

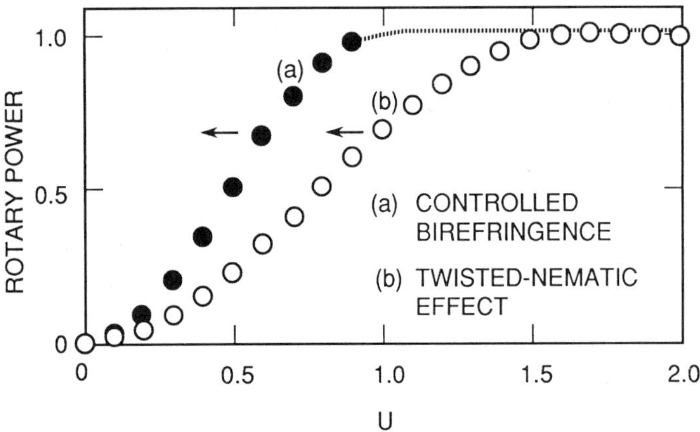

Figure 8.15 Comparison of the rotatory power between the controlled birefringence and the twisted nematic configuration for BDH/E-7 LC. (After Ref. 39.)

$\Delta\phi/\Delta t$, we can make the following approximations [43,44]:

$$\frac{\Delta\phi}{\Delta t} \approx \frac{\Delta\phi}{\Delta V}\frac{\Delta V}{\Delta t} \approx \frac{\partial\phi(V,d)}{\partial V}\frac{\Delta V}{\tau} \tag{8.29}$$

where $\Delta\phi(V,d)/\Delta V$ can be determined from Eqs. (8.24a) and (8.24b). Here ΔV is the voltage swing applied to the LC cell and τ is its response for the bias voltage V, given by Eqs. (8.19) and (8.20). For the low-field approximation [18,23,27], one gets

$$\frac{\Delta\delta}{\Delta t_{(\mathrm{LF})}} = \frac{\alpha\delta_{\max}\,\Delta V[(V/V_{\mathrm{TH}}^2)^2 - 1]}{V_{\mathrm{TH}}\tau^* d^2} \tag{8.30}$$

where $\delta_{\max}$ is the maximum phase shift for the cell, V_0 is the threshold voltage, α is the low-field constant, and τ^* is the characteristic response time given by Eq. (8.19).

Similarly, for the high field, one gets

$$\frac{\Delta\phi}{\Delta t_{\mathrm{HF}}} = \frac{\beta\delta_{\max}\,\Delta V[(V/V_{\mathrm{TH}})^2 - 1]}{V^2\tau^* d^2} \tag{8.31}$$

where β is the high-field constant [Eq. (8.24a)]. We observe that in both cases $\Delta\delta/\Delta t \sim \delta_{\max}/\tau^* d^2$, and since $\delta_{\max} \sim d\,\Delta n$, we see that $\Delta\delta/\Delta t \approx \Delta n/\tau^* d$. We can turn this expression into a "figure of merit" [18] that depends only on the material properties by requiring that the maximum phase shift $\delta_{\max}$ of the cell be equal to a certain desirable level, say, 2π. Thus for $\delta_{\max} = 2\pi d\,\Delta n/\lambda = 2\pi$, it follows that $d = \lambda/\Delta n$, and hence

$$\mathrm{FM} = \frac{\Delta n^2}{\tau^*} = \Delta n^2 \frac{K}{\gamma} \tag{8.32}$$

where this equation expresses the dependence of the phase rate of change as a function of material parameters. Selecting materials or physical conditions that optimize the preceding figure of merit will result in an optimized operation of the LC cell with regard to the speed–dynamic range trade-off. An optimized temperature operation based on a maximized figure of merit was recently analyzed and demonstrated in a LC light valve [45,46]. Not only does the quantity $\Delta\phi/\Delta t$ represent essentially the figure of merit of LC materials with respect to electro-optical performance, as it turns out this same quantity is directly related to the *gain–bandwidth* product of a LC based modulator system [18] as will be shown (see Section 8.3.6).

8.2.6 Other Liquid Crystal Classes and Electro-optical Effects

Nematic Supertwist Configuration. In the twisted-cell configuration (either 90° or 45°), the electro-optical modulation is achieved by the induction of a tilt via the external field. The presence of tilt results in a partial destruction of the twist configuration, depending on the degree of tilt. It is recalled that it is the twisted configuration that allows the polarization rotation to occur. The partial destruction of the twist results, therefore, in a partial loss of the rotatory power, and with it an intensity modulation using a suitable polarizer–analyzer arrangement. This effect can be significantly enhanced by overtwisting or "supertwisting" of nematic material beyond the conventional 90° or 45° twist angles. The main motivation for this enhancement is that it tends to sharpen or increase the slope of the transfer curve of the twisted nematic configuration (see Fig. 8.14). This slope,

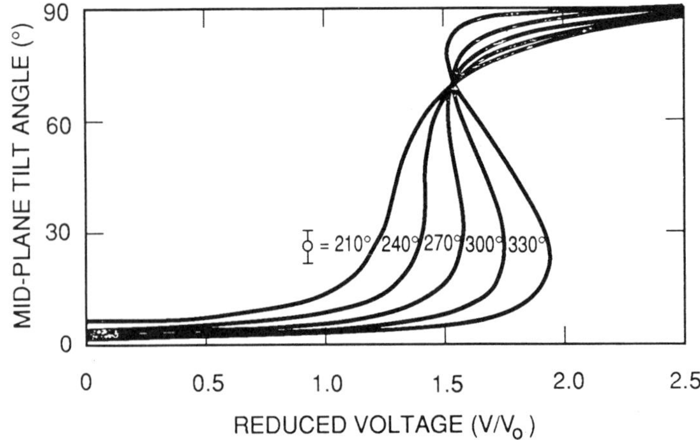

Figure 8.16 Theoretical curves of the tilt angle of the local optical axis in the layer midplane versus the reduced voltage V/V_{TH}, with total twist angle as the parameter. (After Ref. 47.)

as we shall later see, is critical in determining the usefulness (or the degree of multiplexity) of the particular LC configuration for multiple-element display application. The "superbirefringence" effect discovered in 1984 by Scheffer and Nehring [47] was based on the use of a 270° twist angle; in other words, a 180° "overtwist" was added to the 90° twist configuration. The ability to attain a sharp slope in the transmission slope can be inferred from Fig. 8.16, in which the midplane tilt angle for 270° twist undergoes a short transition from approximately 0° to approximately 90° at a normalized voltage ratio of about 1.5. This transition would correspond, for a *parallel* analyzer–polarizer pair, to an optical transmission change from a low to a high transmission state. This arrangement, while allowing a sharp activation slope to be attained, suffers from background color effects and the need to attain a fairly large pretilt angle, thereby complicating the cell fabrication procedure. The discovery of the optical mode interference (OMI) effect by

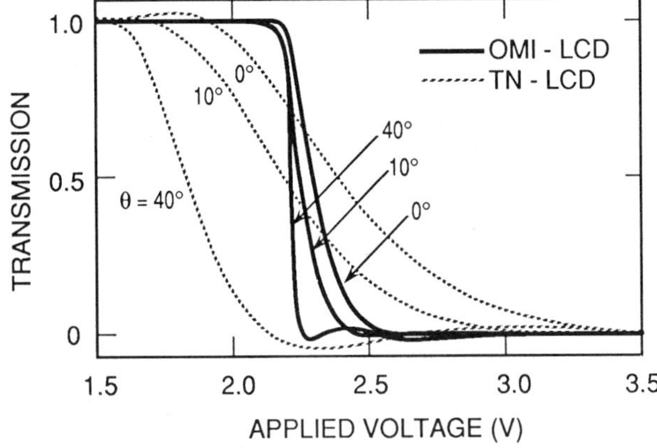

Figure 8.17 Transmission-versus-voltage characteristics of the OMI (solid curve) and the TN configuration (dashed curve). (After Schadt et al, *Mol. Cryst.*, Vol. 165, p. 405, 1988.)

Schadt and Leenhouts [48] and later by Kawasaki et al. [49] resolved the issue of colored background using a π or 180° twist cell. The sharpening of the activation (transmission-versus-voltage curve) is shown in Fig. 8.17. The main issue with the OMI configuration is the relatively low transmission. Another method of overcoming the color background problem that avoids the sacrifice of transmission is the use of a dual supertwist cell to compensate for the color-dependent birefringence [50]. This concept can be made cost-effective by the introduction of a polymer film compensator [51].

Nematics: Dynamic Scattering/Variable Grating Mode. Some negative dielectric anisotropy LCs ($\Delta\varepsilon < 0$) exhibit a peculiar phenomenon of forming a periodic domain structure where the director's orientation varies spatially in a periodic manner. These domains, called William domains, were extensively investigated in the 1960s and early 1970s [52]. The treatment of this phenomenon is again beyond the scope of this chapter. In general, those materials will exhibit a weak dependence of the effect upon the voltage applied up to a certain threshold voltage, beyond which electrohydrodynamic effects take place, leading to a dynamic scattering of light. The dynamic scattering effect was one of the first effects to be utilized in LC devices. However, due to the relatively low contrast and speed of response, this approach has now been abandoned. A treatment of both the subthreshold regimes (William domains) and the electrohydrodynamic phenomenon (dynamic scattering effects) can be found in several textbooks on LCs [53–55]. Here we concentrate on one related effect in which the formation of the grating structure previously mentioned is sharply dependent on the applied voltage, an effect called the *variable grating mode* (VGM). In this effect, which was studied by several researchers [56–61], a phase grating is formed with a period that depends upon the voltage applied across the cell. This phase grating originates from a variation of the optical path length due to a periodic orientational perturbation of the LC uniaxial index ellipsoid. The direction of periodicity is perpendicular to the quiescent state alignment of the LC molecules, that is, the domains are parallel to the LC alignment in the off state. Typical spatial frequency variation is from 100 to 600 cycles/mm. Figure 8.18 shows typical voltage-induced behavior of a VGM cell as seen through a polarizing microscope. The period of the phase grating can be seen to decrease as the applied voltage increases.

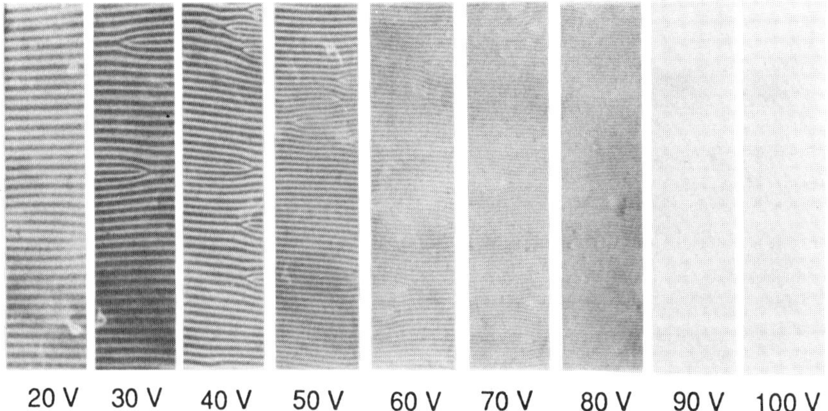

Figure 8.18 Variable grating mode (VGM): dependence of the VGM period on the applied voltage. The grating period variation is between approximately 100 and 600 cycles/mm. (After Ref. 59.)

The tunability of the grating period with the applied voltage renders this effect a potential tool for optical data processing, as we shall later see.

Cholesteric Liquid Crystals. Cholesteric LCs [62–64] can be synthesized by dissolving a chiral (helical) molecule in a nematic material. The helical distortion resulting in this structure can also be found in some pure cholesterol esters, hence the name cholesteric LCs. The helical structure of the cholesteric phase (Fig. 8.19) is the same as that occurring in a nematic having a twisted configuration. The only difference is the pitch (length of the helical period) of the helix, which is much shorter in the cholesterics (a few thousand angstroms). In every layer of the cholesteric structure the local arrangement of the molecules is similar to that of the nematics, namely long-range order in the orientation of the molecules but no ordering of their center of gravity. The structure of these substances generally consists of long hydrocarbon chains having protruding side groups that alter the rodlike nematic molecule shape. There are three possible configurations of the cholesteric phase when confined in a planar cell (Fig. 8.20): (a) the planar configuration in which the helical axis is perpendicular to the cell walls, (b) the "fingerprint" texture in which the helical axis is parallel to the walls, and (c) the focal-conic structure in which the helical axis is randomly aligned in analogy to a polycrystalline material.

Optical Properties [65]. Similar to the nematic phase, the cholesteric materials exhibit polarization rotation (rotatory power). Dichroic reflectivity is also observed in these materials. Both the optical rotatory power and the dichroic reflectivity exhibit significant enhancement near a characteristic wavelength $\lambda_0 \approx \bar{n}P$ for an optical beam propagating normal to the helical axis where $\bar{n}$ is the mean refractive index of the material and P is the pitch length. This behavior is shown schematically in Fig. 8.21. It is observed that the dichroic reflectivity has a flat peak of 50–100% within $\pm$10–20% of the characteristic wavelength λ_0. The rotatory power peaks to typical levels of a few hundred cycles per millimeter and *changes sign* as the wavelength is increased past λ_0. With a beam of light

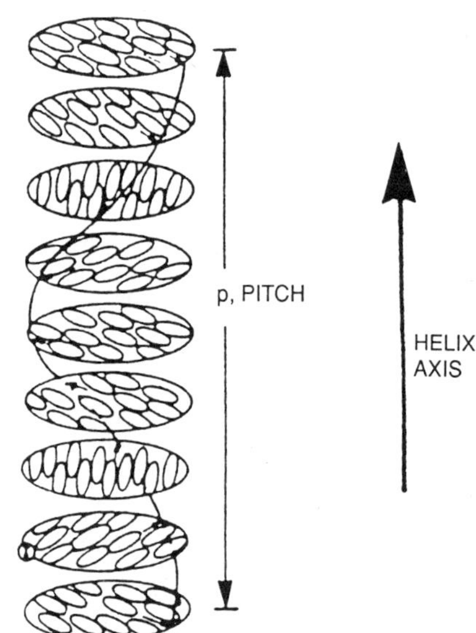

Figure 8.19 Helical structure of a cholesteric phase.

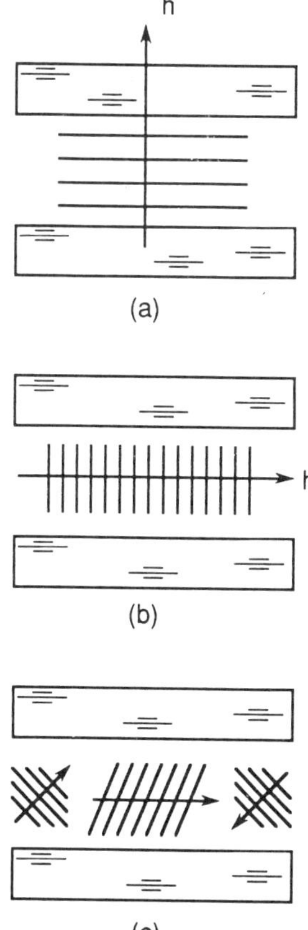

Figure 8.20 Three basic textures of cholesterics: (*a*) planar, (*b*) "fingerprint," and (*c*) focal comic. (After Ref. 1, Chap. 6.)

propagating *normal* to the helical axis, several diffraction peaks occur. The selective dichroic reflectivity discussed in the preceding will, in general, have a cosine law dependence on the angle between the beam propagation direction and the helical axis.

Electro-optical Properties. Similar to the nematic case, the cholesteric materials also feature anisotropies in the dielectric constant ($\Delta\varepsilon$), the refractive index (Δn), and the conductivity ($\Delta\sigma$). As pointed out before, the cholesteric phase can be regarded as a fine-pitch twisted nematic phase. Thus, the local director **n** rotates or twists in the helical sense, as shown in Fig. 8.19. As a result, the macroscopic $\Delta\varepsilon$ or Δn differ from the local material constants, which were defined earlier for the (homogeneous) nematic case. For a positive LC material where locally $\Delta\varepsilon = \varepsilon_{||} - \varepsilon_{\perp} > 0$, the "macroscopic" $\Delta\varepsilon$ for the cholesteric phase, averaged over many helical cycles, will be given by

$$\Delta\varepsilon_h = \varepsilon_{||h} - \varepsilon_{\perp h} = \varepsilon_{\perp} - \tfrac{1}{2}(\varepsilon_{||} + \varepsilon_{\perp}) = \tfrac{1}{2}(\varepsilon_{\perp} - \varepsilon_{||}) < 0 \tag{8.33}$$

where the subscript h refers to the orientation with respect to the helical axis. Under the

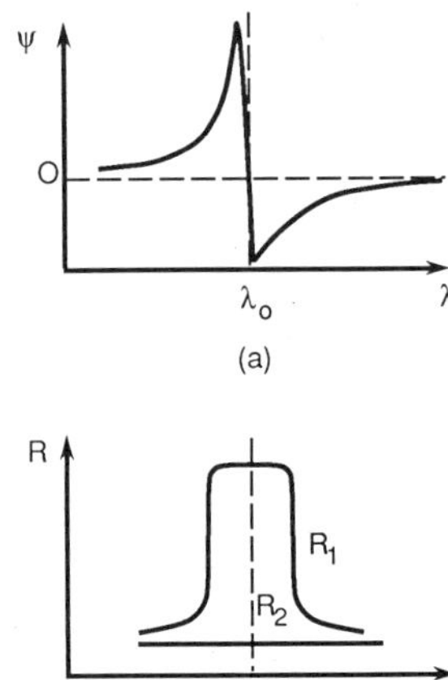

Figure 8.21 Wavelength dependence of (*a*) optical rotatory power and (*b*) dichroic reflectivity of cholesterics. (After Ref. 1, Chap. 6.)

influence of an electric field the cholesteric phase will orient so as to minimize the free energy whose local electrical contribution is proportional to $\Delta\varepsilon\, E^2$. Several electro-optical effects are observed depending on (a) the static configuration of the phase (Fig. 8.20) and the sign of the dieletric anisotropy. In general, the applied electric field will tend to orient the helical structure parallel to the field lines for $\Delta\varepsilon_h > 0$ and perpendicular to the field lines for $\Delta\varepsilon_h < 0$, as expected. This includes the multiple domain or focal-conic configuration as well. An important electro-optical effect observed in the cholesteric phase occurs when a field is applied *perpendicular* to the helical axis in a *positive-anistropy* material. In this case the field will tend to "untwist" the helical structure, resulting in the complete destruction of the helix and the appearance of a nematic like aligned phase. Optically this results in a change from a scattering structure with resonant dichroic reflection (as previously discussed) to a transparent medium.

Smectic Materials. Smectic materials (the name derived from the Greek word for soap) are LC materials possessing layered structure with soap like mechanical properties. While there exist several subclasses of smectic materials, the two important types of this class of materials are smectics A and C (SmA, SmC), shown in Fig. 8.22, which differ by the angle in which the molecules tilt with respect to the layers, or "shelves." As observed from this figure, a *crystalline* order, namely a well-defined period, exists in both smectic phases between the layers. Clearly, the degree of ordering is higher in this case than in the nematic case where there exists no long-range translational order.

In terms of the optical properties, SmA is a uniaxial material. The extraordinary index is along the long molecular axis, or perpendicular to the smectic layers in Fig. 8.22. The refractive indices along the two orthogonal directions in the smectic layer planes are both equal to the ordinary index value. Due to the tilt of the molecules in the SmC layers, the symmetry around the molecular axis is lost. This phase is therefore biaxial.

In some classes of materials, the smectics *and* nematics constitute phases of the same material. Furthermore, the SmA and SmC may also constitute different phases of the

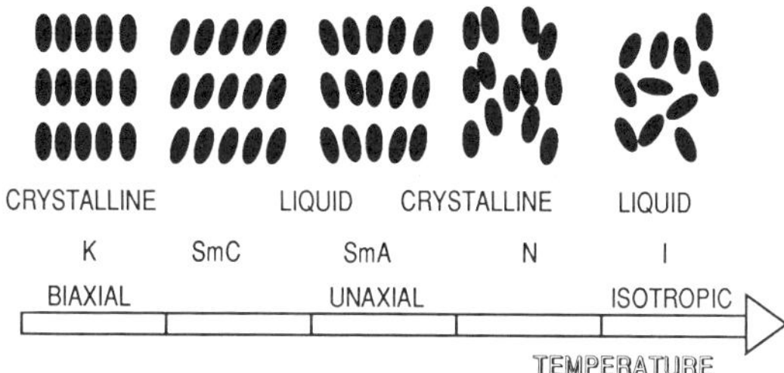

Figure 8.22 Sequence of phases from crystalline through SmC to SmA to isotropic liquid. (After B. S. Scheuble, *Soc. Inf. Disp.*, 1989, Seminar 12.)

same material. As shown in Fig. 8.22, the sequence of phases as a function of temperature is expected on the basis of the degree of ordering of the various LC phases, the highest being in SmC and the lowest in the nematic phase.

In the following we will concentrate on the two main classes, SmA and SmC. Both of these classes exhibit the electro-optical effect are switchable under the external field (e.g., electrical) at extremely short response times.

Ferroelectric Liquid Crystals. As discussed in the previous section, SmC materials comprise the most ordered liquid crystalline phase. The next level of order is full (solid-state) crystallinity as the LC freezes (see Fig. 8.22). As observed from the figure, the SmC is arranged in layers with the director (or the molecular axis) tilted with respect to the normal. Note that this phase possesses a twofold symmetry axis (normal to the director) as well as a mirror symmetry. If helicity is introduced into the SmC layers by having the director rotate around an imaginary axis perpendicular to the smectic planes between consecutive smectic layers, the mirror symmetry is lost (Fig. 8.23). The loss of the mirror symmetry, however, implies that a nonzero dipole moment, or ferroelectricity, is possible.

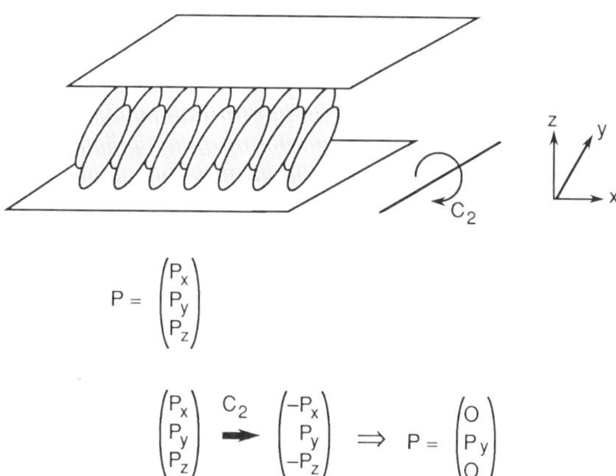

Figure 8.23 The C_2 symmetry operation applied to a single layer of SmC material. (After B. S. Scheuble, *Soc. Inf. Disp.*, 1989, Seminar 12.)

This argument, first derived and demonstrated by Meyer [66,67] and his group, is based on the fact that the C_2 symmetry around the Y-axis in the tilted structure (Fig. 8.23) allows a dipole moment along the C_2 (Y) axis. Note, however, that while a dipole moment perpendicular to the long molecular axis can exist in *each layer*, it will be averaged out to zero on a macroscopic scale due to the helical structure. Suppression of the helical pitch is therefore necessary to actually realize a dipole moment, or ferroelectricity, in a macroscopic sample. The suppression of helical pitch using the surface alignment method was first demonstrated by Clark and Lagerwall [68] in 1980. Consider Fig. 8.24. The figure shows the concept of a surface-stabilized ferroelectric LC (FLC) in which the molecules are forced to align to one of the two directions determined by the conic cut, as shown. The cone represents the rotation of the molecular axis with the helical pitch. This rotation is suppressed by strong anchoring forces of the surface alignment, resulting in a homogeneous structure where the alignment of the long axis of the molecules is in one of the two directions shown. The application of an electric field can, through coupling to the dipole moment P, result in flipping of the dipole moment. This will result in rotation along the cone surface of the molecular axis to the other allowed orientation. The result is a reorientation of the optical axis by 2θ, θ being the tilt angle as shown.

The transmission through a crossed polarizer–analyzer pair of the FLC cell following the reorientation of the optical axis will be given by

$$I_{\text{out}} = I_{\text{in}} \sin^2(4\theta) \sin^2(\delta/2) \tag{8.34}$$

[see Eq. (8.25)], where the retardation δ is given [see Eq. (8.23)] by:

$$\delta = \frac{2\pi d \, \Delta n_{\text{eff}}}{\lambda} \tag{8.35}$$

where it is assumed that the optical axis was initially aligned with the polarizer orientation.

Since θ is close to 22°, the reorientation of the optical axis results in an efficient polarization rotation for a suitable choice of the cell thickness d ($\sin 4\theta \approx 1$). Similar to the nematic case Δn_{eff} is around 0.2. The main advantage of the FLC cell compared to the nematic case is the speed of response associated with the field-assisted optical axis reorientation described in the preceding. Since a permanent dipole moment exists in the

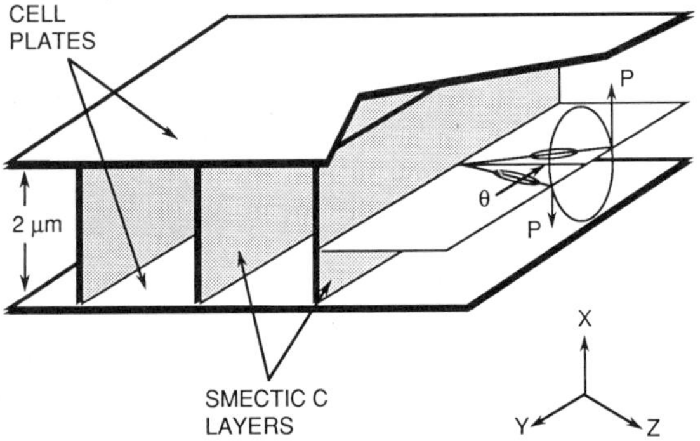

Figure 8.24 Smectic C ferroelectric liquid crystal in the surface-stabilized configuration. (After B. S. Scheuble, *Soc. Inf. Disp.*, 1989, Seminar 12.)

FLC case, the torque exerted by the electric field, $\mathbf{T} = \boldsymbol{\mu} \times \mathbf{E}$, is now proportional to $\mathbf{E}$ and not to E^2 [compare Eq. (8.11)]. The significance of the linear E dependence is that upon reversal of the $\mathbf{E}$ direction to $-\mathbf{E}$ by reversing the voltage across the cell electrodes, we can orient the optical axis with the aid of the electric field. It should be recalled that in the nematic LC case, while the turning-on time for the LC reorientation (Freedericksz transition) is field assisted, the turning-off time relies entirely on the natural relaxation and is independent of the field applied [see Eq. (8.21)]. The equation of motion for the director in the FLC case is given by [69]

$$\mathbf{T}(\mathbf{E}) - \mathbf{T}(\gamma) = 0 \tag{8.36}$$

where $\mathbf{T}(\mathbf{E})$ is the applied E field torque

$$\mathbf{T}(\mathbf{E}) = \boldsymbol{\mu} \times \mathbf{E} \tag{8.36a}$$

and $\mathbf{T}(\gamma)$ is the viscosity reaction torque given, as in the nematic case, by

$$\mathbf{T}(\gamma) = -\gamma \frac{\partial \theta}{dt} \mathbf{e}_{\perp} \tag{8.36b}$$

where θ is the orientation angle of the director and $\mathbf{e}_{\perp}$ is a unit vector in the direction of the axis of the director's rotation. The change in the director's tilt angle, $\delta\theta$, is the same as the change in the orientation of the polarization, $\delta\phi$ (which is perpendicular to the director), as it moves on the conic surface. Thus, Eq. (8.36) is given by

$$\mu E \sin\theta - \gamma \frac{d\theta}{dt} = 0 \tag{8.37}$$

Using a small angle approximation ($\theta \leq 22^\circ$), $\sin\theta \approx \theta$, the solution can be approximated by

$$\theta \approx \theta_0 e^{-t/\tau(E)} \tag{8.38}$$

where the response time $\tau(E)$ is given by

$$\tau(E) = \frac{\gamma}{\mu E} \tag{8.39}$$

We note that in contrast to the nematic case, the response time behaves as $1/E$ and not as $1/E^2$. It should be borne in mind that the preceding derivations are approximations only. In reality, FLC materials display a varying dependence on the response time, as shown in Fig. 8.25. The fast response, which is in the *microsecond range* for applied voltage of ≈ 10 V, should be noted.

In summarizing the properties of ferroelectric SmC* materials in the surface-stabilized configuration discussed in the preceding, the main advantage of this material is its extremely fast response in electro-optical switching. The main issues are the lack of gray scale (there are only two stable orientations of the director corresponding to positions "up" and "down" of the dipole moment) and the microscopic defects associated with the tendency to form a helical structure.

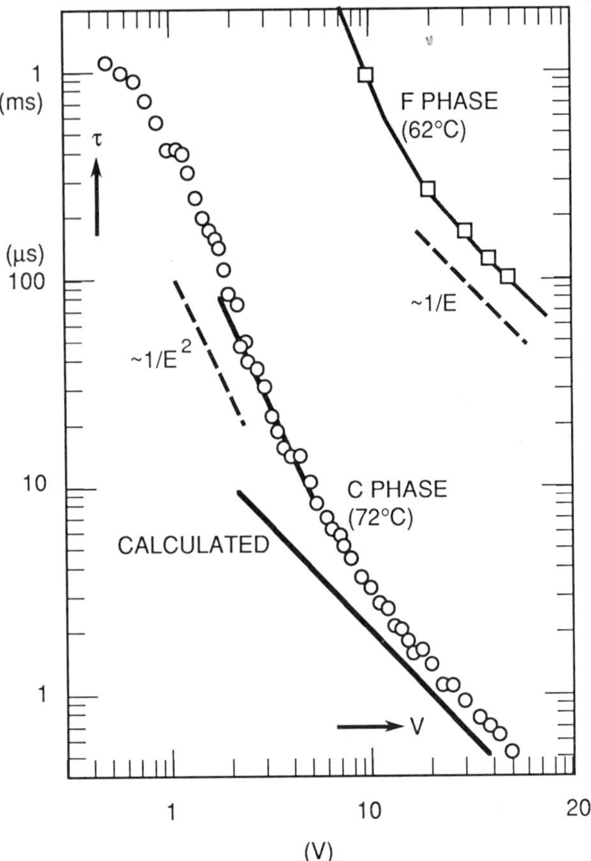

Figure 8.25 Response time versus bias voltage of an SSFLC (surface stabilized ferroelectric liquid crystal) device employing 1-μm-thick Sm.*C HOBACPC at $T = 68°C$. (After N. A. Clark et al., *Mol. Cryst. Liq. Cryst.*, Vol. 94, p. 213, 1983.)

Electroclinic (Soft Mode) Effect in Smectic A. As pointed out, in the SmA layers the average molecular tilt with respect to the layer normal is zero. As the temperature is decreased toward the SmA* → SmC* transition, tilt fluctuations with increasing amplitude and spatial extent occur. As the temperature reaches the transition point, the extent of fluctuation becomes infinitely long and they "freeze" in the new phase state in which the tilt extends (with a helical configuration) throughout the sample. In a phase transition language, one speaks of a tilt fluctuation mode that becomes increasingly of long extent and "softens," or becomes infinitely slow, in terms of its frequency dependence at the transition. If an electric field is applied close to the SmA* → SmC* transition point, this field will bias the tilt fluctuation in such a way as to favor the particular fluctuation that tends to align the dipole moment with the applied field. As a result, a tilt angle will be induced in SmA* due to the application of an electric field, with the tilt angle proportional to the applied field [70,71]. Again we have a *linear* coupling of the tilt angle to the electric field, and hence the possibility to switch or reorient the polarization and with it the optical axis, by reversing the sign of the applied field. The effect can be used in the same way as the surface-stabilized FLC configuration, as shown in Fig. 8.26. In comparing

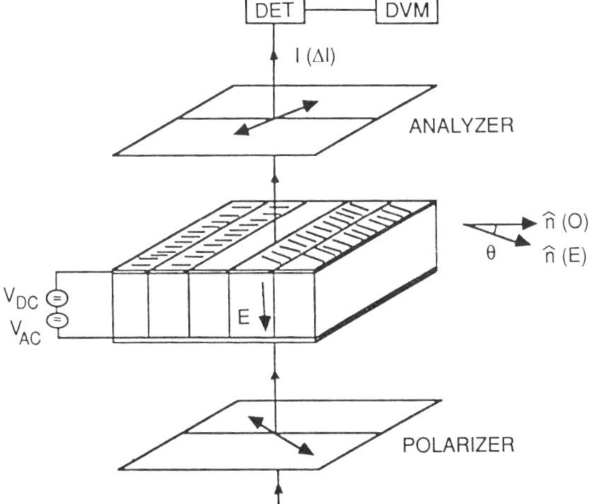

Figure 8.26 Optical modulation using field-induced tilt angle (θ) change in SmA LC. (After S. T. Lagerwall et al., *Mol. Cryst. Liq. Cryst.*, Vol. 152, p. 503, 1987.)

the electroclinic effect in SmA* to the surface-stabilized mode of the SmC* phase, we find that while the electroclinic effect is faster and allows a continuous variation in the tilt angle with the applied field, the tilt angle itself is limited to ≈ 10–12°, thereby reducing the dynamic range as compared to SmC*.

Polymer-dispersed Liquid Crystals. Polymer-dispersed liquid crystal (PDLC) structures are thin films containing LC droplets of micrometer size embedded in a polymer matrix. The main advantages of this structure, first suggested by Fergason in 1985 [72], over other LC configurations are (1) the fact that the PDLC film is self-supported, being based on a polymer matrix (and not a LC), and (2) that this technology is amenable to large-area production. The preparation of the PDLC film is based on the phase separation of the LC phase from the polymer material to form dispersed LC droplets. There exist several fabrication methods for producing PDLC films [73]. In *temperature-induced phase separation*, the polymer is heated up above its melting point. The LC material is then added in the desired volume ratio. The homogenized mixture is then cooled down, and phase separation occurs below a certain temperature.

In *solvent-induced phase separation*, a solvent common to both the polymer and the LC is used in conjunction with heat. Separation of LC droplets is achieved by extracting the solvent from the homogenized mixture. Finally, *polymerization-induced phase separation* is based on the formation of LC droplets as a result of the sharp reduction of the solubility of the LC in the polymer matrix due to curing of the polymer. The polymer material and the LC are chosen such that the ordinary refractive index value of the LC matches the polymer's index. The electro-optical use of the PDLC film is shown in Fig. 8.27. Due to the large surface-to-volume ratio resulting in strong surface forces, the LC molecules tend to align along the surface of the droplets, forming a well-defined alignment direction in each droplet. However, the alignment orientation of each of the droplets is randomly distributed in the nonactivated state. Since the optical axis of the LC droplets is randomly distributed, a light beam incident on the film will undergo heavy scattering due to the refractive index variations that occur over spatial dimensions close to that of a visible wavelength (typical radii of the droplet are approximately 1 μm).

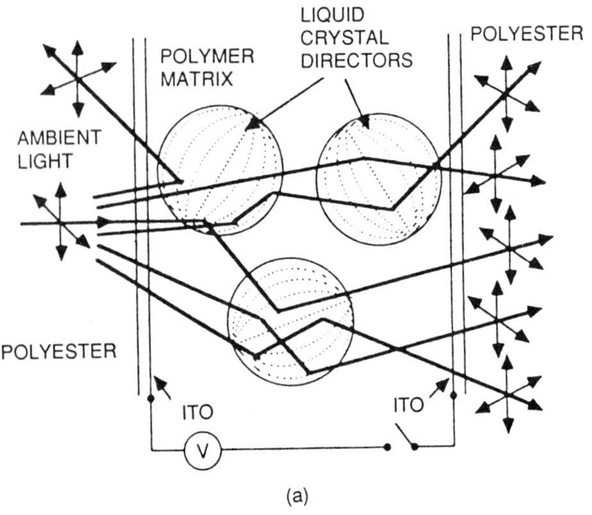

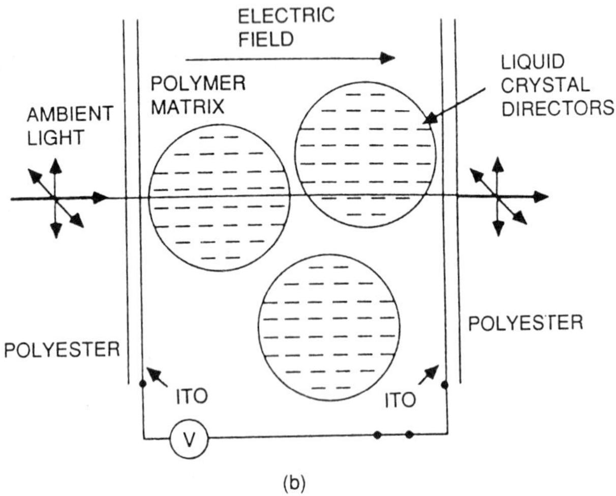

Figure 8.27 Polymer-dispersed LC or nematic curvilinear aligned phase (NCAP) structure used for optical modulation: (*a*) scattering state (randomly aligned droplets); (*b*) transparent state alignment due to *E* field. (After P. Drzaic and K. Werner, *J. Inf. Display*, 2–89, p. 7.)

Upon application of an electric field, the LC molecules in each of the droplets will tend to align with the field. The optical beam will now sense a uniform refractive index distribution equal to the ordinary index of the LC matched to that of the polymer. This will therefore constitute a high transmission state. The static and dynamics of the nematic LC droplets dispersed in the polymer basically follow the same behavior discussed earlier (see Section 8.23). As pointed out, the surface anchoring forces are relatively large due to the large surface-to-volume ratio. This, coupled with the small size of the droplet, will result in a faster decay time compared to the regular nematic configuration.

The decay time of a nematic droplet in a PDLC film can be approximated by [74]

$$\tau_{\text{off}} = \frac{\gamma_1 a^2}{K(l^2 - 1)} \tag{8.40}$$

where γ_1 is the (rotational) viscosity, a is the long axis of the ellipselike droplet, and l is the aspect ratio (< 1). This is similar to Eq. (8.21), with $a^2/l^2 - 1$ replacing the cell thickness dependence (d^2) of the regular nematic configuration. This same factor also replaces the d^2 dependence in the voltage-driven case [similar to Eq. (8.19)].

Typical relaxation times in PDLC films will be in the millisecond range (compared to tens of milliseconds in mematic cells that are a few micrometers thick).

The large anchoring forces in the droplets and the loss of voltage due to the inert polymer matrix require a larger voltage for switching the PDLC films. This technology is especially promising for large-area shutter applications such as solar radiation control or "electric window" panels. As the fabrication of large-area PDLC, or "NCAP," films is expected to be much cheaper than the cost of large-area nematic LC cells, the elimination of the need to polarize the optical beam is another important advantage of this technology.

Guest–Host Liquid Crystal Mixtures. This concept is based on the introduction of dye molecules as a "guest" substance dissolved in the "host" LC mixture. The oblong dye molecules will follow the host LC molecules as the latter realign themselves in an electric field. Due to their dichroic properties, namely an extremely anisotropic absorption, these molecules will heavily absorb in one orientation while acting as a transmissive medium in the other. The effect, first reported by Heilmeier et al. [75], can be used in twisted, planar, or homeotropic (perpendicularly aligned) configurations. In these early configurations analyzers were needed to obtain high contrast. White and Taylor [76] pointed out in 1974 that the use of dye molecules in a cholesteric host can result in a high-contrast display without the need for polarizers. The guest-host effect is still used in circumstances requiring the enhancement of contrast or a wide viewing angle. The use of particular dye molecules can lead to a variety of color backgrounds. The reader is referred to general articles about the effect and its uses that appeared in recent publications [77,78].

8.3 LIQUID CRYSTAL SPATIAL LIGHT MODULATORS

8.3.1 Electrically Addressed Devices

Alphanumeric Displays. We have already seen how the various LC modes of operations can be used for optical modulation. It is quite obvious then that such configurations as twisted nematic or controlled birefringence can be used directly as optical shutters for wavelengths ranging from UV through visible all the way to the IR region. All that is needed is to coat the glass substrates of the LC cell with conductive transparent electrodes (typically of indium tin oxide), then to from the LC alignment structure on the electrodes, and finally to assemble the cell with the appropriate LC material. However, a considerable amount of design and fabrication is required if a *spatial* addressing of the LC cell is desired. The simplest case is that of an alphanumeric display where the number of elements, or *pixels*, is relatively small. Thus, less then 200 pixels are typically needed for a digital watch. In this case one can use the simple *brute-force* method of accessing *individually* each element with its own electrode. Figure 8.28 (taken from one of the original patents in this field by Berreman et al. [79]) shows a cross section of such a LC alphanumeric display using conductive transparent leads of tin oxide and a fixed seven-segment conductive electrode mask to form two eight numerals. The device operates as follows: an optical beam enters the cell from the front side (top), becomes polarized following the polarizing layer (1), and is polarization modulated by the twisted-nematic LC layer (2). As previously explained (Section 8.24), the spatial field pattern imposed on the LC layer by the patterned top electrodes (3) and the bottom uniform electrode (4) will result in a partial destruction of the twist pattern. This in turn will result in the rotation of the beam polarization under those activated portions of the patterned electrode. The portions of the beam, the polarization of which was rotated, will be absorbed by the analyzer (5). Since the entire optical beam is retroreflected by means of the mirror (6), the regions under the activated portions of the mirror will appear dark to the viewer.

Multiplexed Addressing. Using a fixed electrode pattern and an individual feeding line for each is clearly practical for only a small number of elements, for example, in a digital watch. However, the fabrication of such a display becomes totally impractical if, for example, one needs to address a two-dimensional display with 400 × 500 elements, as in a TV application. One solution for such an application is to use a matrixlike connection scheme where each of the pixel electrodes is connected to a matrix of lines and columns. The idea is to perform the addressing in a sequential process. A whole line of elements is partially activated, and each element on that line is fully activated by supplying the additional voltage to each of the columns sequentially, one at a time. There are two generic configurations based on this technique, which is called *multiplexing*. The simplest structure is that in which the crossed line column elements are made to exhibit a "nonlinear" electrical behavior. Thus, the element will only turn on if both line and column

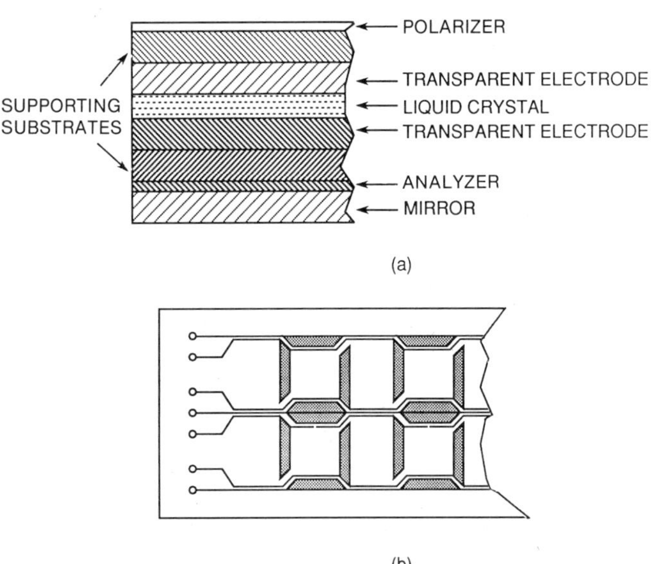

Figure 8.28 Seven-segment LC display device. (After Ref. 80.)

are activated. It will remain in an off state if either the line or the column is not activated. The sequential addressing described in the preceding can be carried out activating each element one at a time. The other method is based on the use of active elements at each node. This "active matrix" approach will be discussed in the next section.

There are two basic types of nonlinear elements that one can use for the first type of passive element multiplexed addressing. The first and oldest concept was to use materials exhibiting a large nonlinearity in their current–voltage characteristics, such as Ta_2O_5 [80]. These materials are deposited in between the two transparent metallic electrode layers. These structures are referred to as metal–insulator–metal (MIM) devices. The other more recent approach is to fabricate a multiple-diode structure, such as a back-to-back configuration with either *pn* or Schottky effect diodes, commonly using amorphous silicon. The diode configuration is advantageous over the MIM configuration in showing sharper nonlinearity. However, the more complex fabrication involved reduces the difference in fabrication between this structure and the much more powerful thin-film transistor (TFT) active matrix technology (see next section).

In this form of addressing we have therefore replaced the $N \times M$ lines needed for individual pixel addressing by only $N + M$ lines. The savings for $M \sim N$ is by a factor of N. This is a significant simplification of the device fabrication. There are two constraints that we have in this configuration. One is that the partial activation of each pixel by a line (or column) voltage will not be sufficient without the activation of the corresponding column (or line) to drive it into the on state. Note that this can be quite demanding for large N (say, 1000) since the incremental complementary column voltage sufficient to drive a line-activating pixel into the on state will be on the order of V column/N, assuming the voltage drop across the column to be linear. It is also clear that one would not be capable of using this method without the *threshold* voltage effect in LCs. This allows a partially activated pixel due to, say, a line activation to still be in an off position. Figure 8.29 shows the optical characteristics of a nematic LC and the threshold effect. We observe that the optical curve must be sufficiently "sharp" such that the ΔV needed to carry the LC pixel across the threshold value will be small enough. If we define as V_{on} and V_{off} the two values that determine the state of the LC as being above or below the threshold, it is quite clear that $V_{off}/\Delta V$ will be a rough measure of the number N of elements on a line (column) that can be used in such a device. (This capability of allowing N elements to be multiplexed is often referred to as the *degree of multiplexing*.) The second serious constraint that must be satisfied using this scheme is that of timing. As previously discussed, we have a typical rise time τ_r [Eq. (8.19)] for a given LC material in a given configuration (say, twisted nematic) and cell thickness. A typical decay time

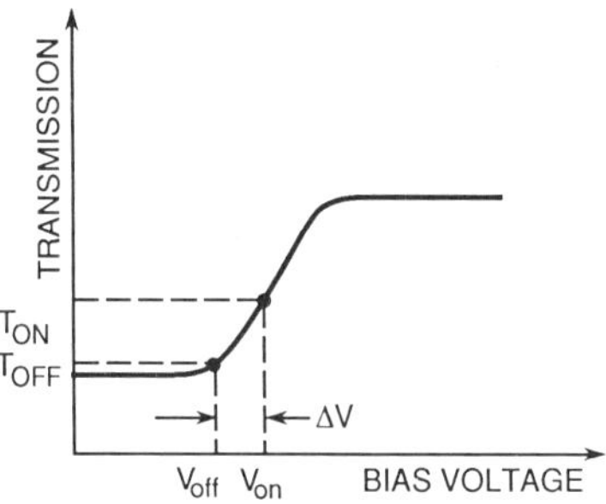

Figure 8.29 Typical transfer curve of a 90° twisted nematic configuration. A voltage increment ΔV is required to switch the device from the off to the on state.

τ_d is also associated [Eq. (8.20)] with that same LC configuration. This means that if one activates a certain line of elements by driving sequentially each of the associated columns, one must allot τ_r for each element at a time before passing to the next one. However, the addressing of the whole frame (array) must be completed within a frame time $T < \tau_d$ to avoid significant decay of the activation. Since $\tau_d \approx 1/V_{\text{off}}^2$ and $\tau_r \approx 1/V_{\text{on}}^2$, we must have $\tau_d \sim T_f \gg \tau_r$, to allow this scheme to be used. However, it is clear that this constraint limits, in a similar manner to the previous one, the total number of elements in the display. A serious effort has been launched since the early 1980s to improve the multiplexing capabilities of LCs. The effort has concentrated on the following:

1. improvement of the response time of nematic LCs by chemical engineering and formulation of new mixtures,
2. development of SmC-based ultrafast FLCs (Section 8.2.6),
3. development of supertwist nematic LC configurations (Section 8.2.6), and
4. development of laser-addressed smectic LC displays (Section 8.3.3).

To this one can also add the development of guest–host mixtures for improved contrast (see Section 8.2.6).

Active Matrix Displays. As pointed out in the preceding, there are practical limitations in the use of multiplexed addressing schemes. Due to the finite slope of the transfer curve, only a limited number of pixels can be addressed ($\approx 100 \times 100$ with nematic LCs). There are also limitations on the achievable gray scale, since the signals at the line and column of each node (pixel) are used both for the activation of the LC element above threshold and for delivering the gray level information. An obvious solution to this problem is to construct a driving circuitry with *active* elements at each node (pixel). For simplicity if we consider a regular field-effect transistor (FET) device at each node, then the address lines can be used to "open" the gates of each of the FETs while the columns can be used to transmit the data to each of the pixels by modulating the source voltage for each device. This immediately relieves the device design from the multiplexing issue discussed in the preceding, allowing much larger array sizes to be constructed. However, other practical issues are raised with this concept of active matrix addressing. In order to take advantage of the smaller depth (thickness) of the active matrix liquid crystal display (LCD) and at the same time provide ample brightness, one is driven into a back-lighting design. This in turn means a transmissive mode LCD with the obvious need for a *transparent substrate*. Also, in order for such a device to compete with existing TV monitors, *large-area* active devices of a few tens of centimeters on the side are required. These two requirements necessitate the fabrication of a thin-film-based circuitry, or what is popularly called TFT technology. When this concept of LCD was initiated about 10 years ago, the TFT technology was in a very primitive stage. It is fair to say that the LCD application is perhaps one of the most significant drives for this technology that has achieved to date a very respectable level of performance. The main issues in this technology are (1) a sufficient mobility of the thin-film semiconductor substrate to drive the required currents, (2) sufficiently low leakage current in the off state, and (3) uniformity and reproducibility. To these one should add the obvious constraint of cost competition versus color TV monitors, which means that the production cost of such a device should be on the order of \$100 or less. A schematic of a typical active matrix circuitry for a LCD is shown in Fig. 8.30 [81]. The operation of this driving circuitry, as already pointed out, is based on (a) selecting the line to be addressed and (b) applying the data voltages to the columns, charging each pixel's capacitor to the voltage level required and scanning

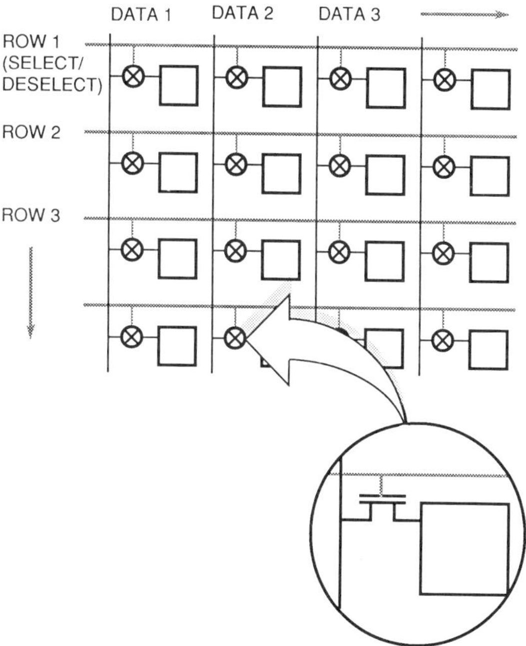

Figure 8.30 Basic thin-film transistor (TFT) active matrix-driving circuitry. (After Ref. 83.)

the lines and columns until the whole two-dimensional array has been updated. As to the semiconductor selection for the TFT technology, there are currently three basic approaches. The first two are based on silicon. Amorphous silicon (α-Si) technology is the most commonly used. It is the lowest in cost due to the simplicity of the low-temperature deposition and the glass substrate that can be used without particular (thermal) requirements. The amorphous silicon is usually deposited in a hydrogen atmosphere to reduce trapping of charge in dangling bonds to which the hydrogen atoms get attached, thus improving its mobility. The symbol α-Si:H is therefore commonly used to describe this material. The main issues with this technology are (1) low mobility (0.5–1 cm^2/V-s) and (2) some degree of threshold voltage instability. The second TFT technology under development is based on polycrystalline silicon, (poly-Si). The main advantage of this technology [81] is the relatively high mobility (10–50 cm^2/V-s) of the polycrystalline phase. The main disadvantage is (1) the higher cost (compared to α-Si:H) of producing such devices. This increased cost is associated with (1) the need to anneal an α-Si substrate at high temperature ($\approx$600°C) in order to form the polycrystalline phase and (2) the need for a high-temperature glass substrate. Another disadvantage of this technology is the relatively high leakage currents in the TFT's off state. The third technology is based on CdSe, with some ongoing research still continuing on the earlier CdS version. Both of these materials, which are depositable at low temperature ($\approx$350°C) enjoy a relatively high mobility of up to 100 cm^2/V-s [82]. The main disadvantages are (1) the relatively high off-state leakage current, similar to the poly-Si case, and (2) the instability of the depositable film. The structure of typical TFT circuitry for each of these three main technologies is shown in Fig. 8.31 [83]. The specific requirements of driving

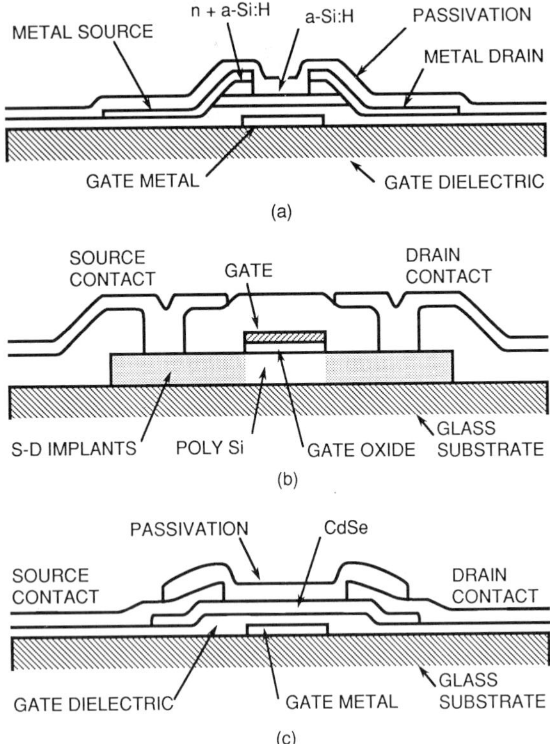

Figure 8.31 Three classes of TFT devices: (*a*) α-Si:H, (*b*) poly-Si; (*c*) CdSe-based TFT. (After Ref. 83.)

a LC modulator are associated with the need to (1) eliminate a net dc current to avoid electrolytic effects and (2) accommodate the finite impedance of the LC, which could vary with temperature and in time. Switching the data polarity in each frame is commonly used to solve the first problem, while an external pixel capacitor (in addition to the LC capacitor) is sometimes used to avoid voltage drop due to leakage current. The requirements of an active matrix LCD vary according to the applications in terms of the need for a certain array size, gray scale, and color. A partial list of recent LCD developments available in the 1988–1990 literature is provided in Table 8.1.

The CCD-addressed Device

Structure and Operation of the CCD Liquid Crystal Light Valve (LCLV): The CCD LCLV is an electrically addressed spatial light modulator. It is based on the silicon LCLV technology developed at Hughes Research Laboratories [96]. It features a serial electrical input that is converted to a truly parallel optical output and has been developed for use as an input device in coherent optical data processing systems (see Section 8.4.2). The device can utilize both coherent and incoherent readout light sources with a spectrum extending from the visible to the IR.

A schematic of the structure of the CCD LCLV is shown in Fig. 8.32. A CCD integrated circuit is fabricated on one side of a silicon wafer and is used to supply a spatially resolved

TABLE 8.1 Recent (1988–1990) LCD Developments

Asahi Glass	360 × 240 poly-Si TFT [84]
David Sarnoff Center	poly-Si LCD [85]
General Electric	10^6 pixel α-Si TFT LCD [86]
Hitachi	5-in. diagonal LCD [87]
IBM	10.4-in. diagonal 640 × 480 LCD [88]
Mitsubishi	10-in. diagonal full-color LCD [89]
NEC	12-in. diagonal 960 × 240 α-Si TFT LCD [90]
Ovonics	11.3-in. diagonal 1296 × 1296 diode array LCD [91]
Seiko-Epson	9.5-in. diagonal poly-Si TFT LCD [92]
Sharp	10.4 in. diagonal 1920 × 480 α-Si TFT LCD [93]
UK-JOERS/ALVEY	720 × 400 FLC LCD [94]
University of Stuttgart/Raychem	288 × CdSe TFT/PDLC for projection TV [95]

signal to the light valve structure on the opposite side. The essential device components are:

- CCD circuits
- Readout structure
- Dielectric mirror
- Liquid crystal

The CCD circuits convert the serial input voltage signal into sampled charge packets and distribute them into a regular two-dimensional array, which is presently configured to be 256 × 256 pixels. The readout structure transports the charge information from the epitaxial layer, upon which the CCDs are built, to the opposite side of the silicon wafer, while retaining the spatial resolution of the charge packets. The dielectric mirror reflects a high percentage of the readout light while greatly attenuating the nonreflected

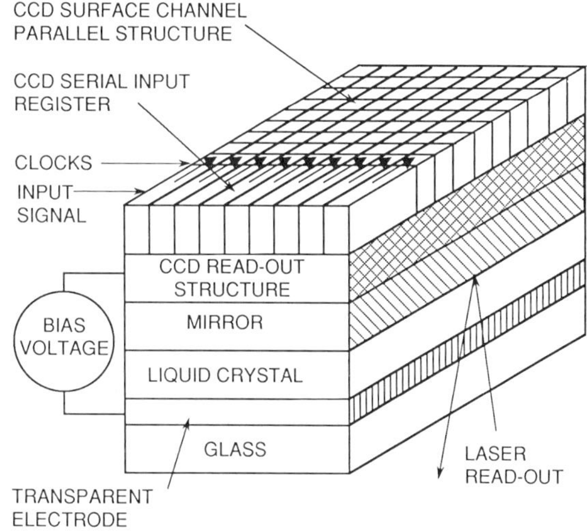

Figure 8.32 Schematics of the CCD-addressed LCLV. (After Ref. 97.)

portion in order to prevent activation of the silicon substrate, which is photosensitive. The nematic LC used converts the amount of charge in each of the packets into either an amplitude modulation using the HFE configuration or a phase modulation using the CB effect.

CCD CIRCUITS. The CCD circuits convert the serial electrical input voltage signal into a two-dimensional parallel array of charge packets using a four-phase clocking design. They are composed of three elements:

- Serial input CCD register
- Serial-to-parallel buffer transfer structure
- Parallel array CCD

The serial electrical input is sampled in a standard Tomsett scheme (charge presetting). As shown in Fig. 8.33, the charge samples are shifted across the serial input CCD until the register is full. When the serial input register is full, shifting (clocking) of charge stops and each of the charge samples is simultaneously shifted into the serial-to-parallel transfer structure. Upon completion of the transfer, the serial input CCD begins shifting a new line of information into the serial register. While this new line is being shifted into place, the lines of charge packets already in place in the transfer structure and the parallel array are shifted down one line in order to accept the next line of charge packet information. This process continues until the entire parallel array CCD is filled.

When all shifting of charge is completed, each line of charge is held under each of the phase 4 gates of the parallel array. The voltage on these gates is slowly reduced to release all of the charge packets. They diffuse through the epitaxial layer to be transported to the opposite side of the silicon chip by the readout structure.

READOUT STRUCTURE. The readout structure is identical to that of the photoactivated silicon LCLV, a detailed description of which will be given in Section 8.3.2. As shown in Fig. 8.34, it is composed of:

- High-resistivity silicon substrate
- Microdiode focusing grid

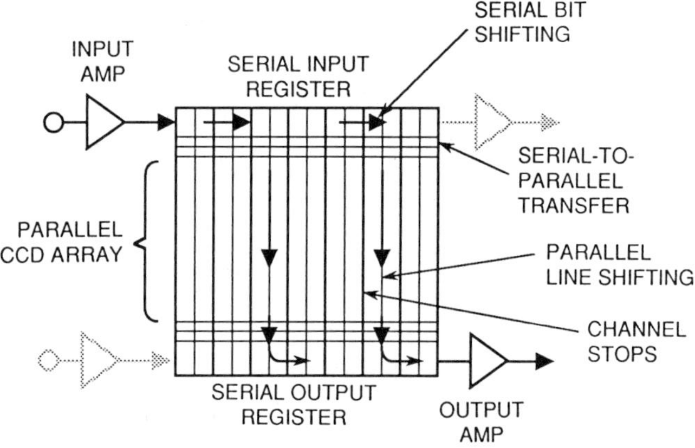

Figure 8.33 Block diagram of the CCD array operation. (After Ref. 97.)

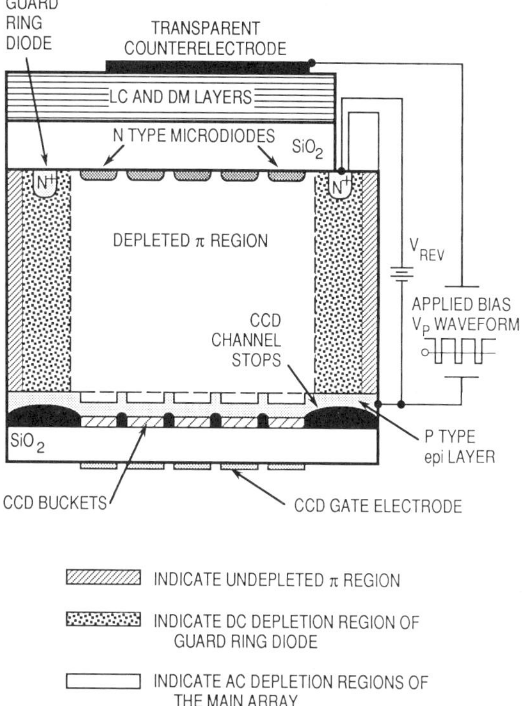

Figure 8.34 Cross section of the CCD LCLV. (After Ref. 97.)

- Guard ring diode
- Metal–oxide–semiconductor (MOS) gate oxide (SiO_2)

During transport of the charge information from the epitaxial–high resistivity silicon interface to the gate oxide, the readout MOS structure is in the depletion phase. The positive voltage applied to the structure via the transparent counterelectrode (and its associated electric field) repels the majority carriers and attracts the signal charge packets, which are minority carriers. The microdiodes produce a focusing effect to maintain the spatial resolution of the information at the Si/SiO_2 interface. The guard ring serves to prevent peripheral minority carrier injection into the active area of the device, thus maintaining the ratio of signal to noise.

DIELECTRIC MIRROR. The dielectric mirror is composed of multiple alternating pairs of $\lambda/4$ Si and SiO_2 tuned to the wavelength of the incident readout light. The mirror is designed to provide better than 98% reflectivity for monochromatic or narrow-band radiation. It will attenuate the nonreflected portion of the readout light sufficiently to prevent front-surface photoactivation of the silicon substrate.

LIQUID CRYSTAL. The electro-optical modulation is performed by a LC layer that presently consists of a positive-anisotropy twisted nematic LC operating in the hybrid field-effect mode (45° twist) discussed in a previous section. A continuous, phase-only modulation using parallel-aligned LC is also possible.

Present Status. The present CCD LCLV is operated in the reflective mode and has an active area of 5 × 5 mm holding an array of 256 × 256 pixels. The expected resolution, based on the design of the elemental CCD cell, is 20 μm. This resolution, which has been demonstrated, as shown in Fig. 8.35, corresponds to one CCD cell period, where the width of the cell is 17 μm and the spacing is 3 μm. The complete array has been activated at serial data rates exceeding 6.5 MHz, corresponding to 100-Hz frame rates, using external drive electronics. Depending on the device used, contrast ratios in bistable applications have been measured at up to 50:1; real-time video has been demonstrated with gray scale capability (Fig. 8.36).

8.3.2 Photoactivated LCLVs

Due to the complexities involved in the development of active driving circuitry (either an "active matrix" or a CCD-addressed device), the photoactivated LCLVs were the first LC spatial light modulators developed. The main motivation for the development of these has been their use for large-screen projection systems and optical data-processing applications. These two applications are detailed in Section 8.4.

CdS-based LCLV. Cadmium sulfide (CdS) and cadmium selenide (CdSe) were the available photoconductors in the early seventies when LC device development began. The CdS-based LCLV developed by the Hughes Research Laboratories group was the first commercially available LCLV technology [97].

The general configuration of the device is shown in Fig. 8.37. The ac light valve consists of a number of thin-film layers sandwiched between two glass substrates. A low-voltage (5–10 V_{rms}) audio frequency power supply is connected to the two outer, thin-film indium tin oxide (ITO) transparent electrodes. Thus it is connected across the entire thin-film sandwich. The photoconductor (cadmium sulfide) and the light-blocking layer

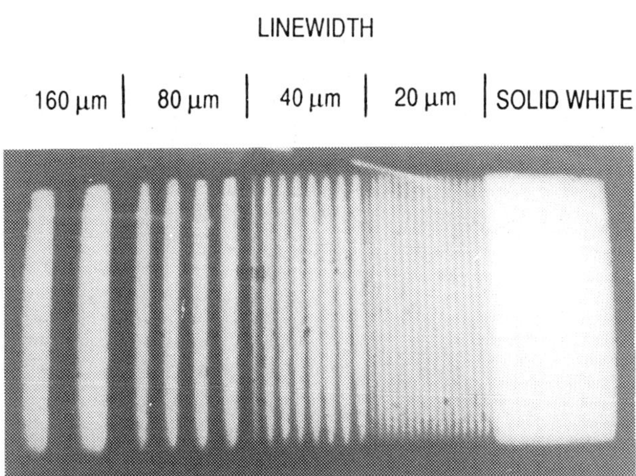

Figure 8.35 Resolution test pattern on the CCD LCLV showing resolution down to the CCD/microdiode size of 20 μm. (After Ref. 97.)

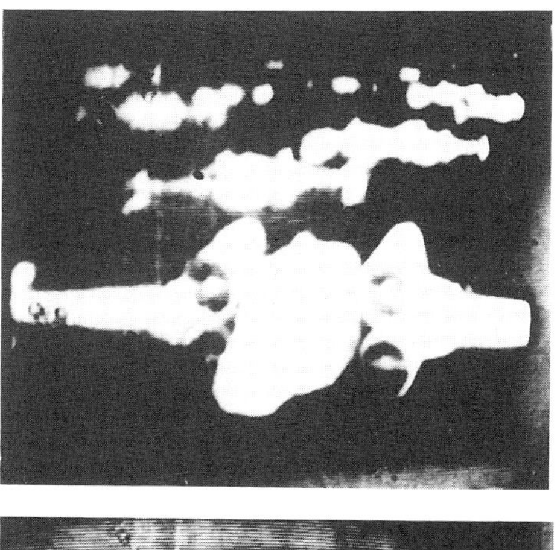

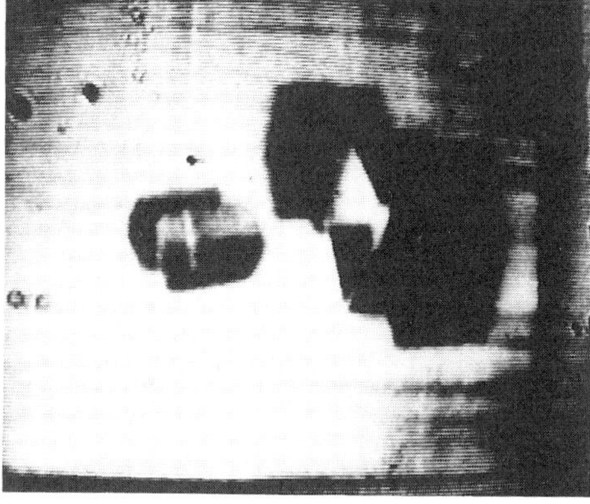

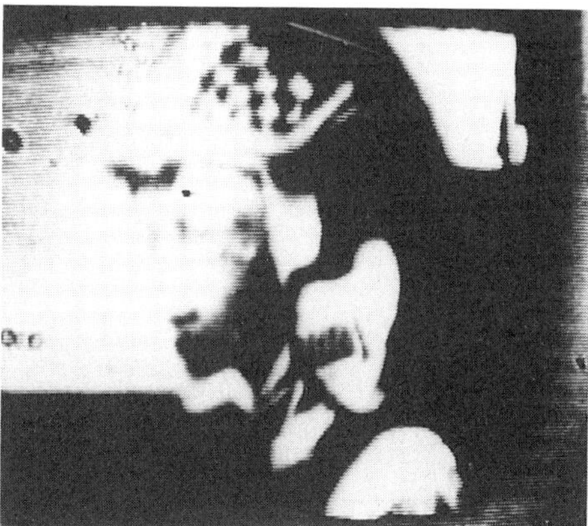

Figure 8.36 Live video frame taken from the CCD showing a scene from "Star Wars." (After M. S. Welkowsky and W. Byles, internal report, Hughes Research Laboratories, 1987.)

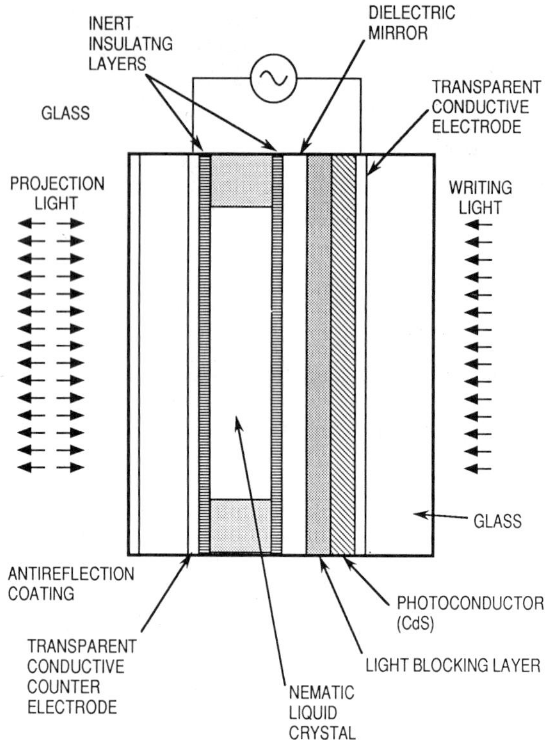

Figure 8.37 Structure of the photoactivated CdS-based LCLV. (After Ref. 98.)

(cadmium telluride) combine to create a rectifying heterojunction. The dielectric mirror and the blocking layer separate the photoconductor from the readout light beam. This is a major design feature of the ac light valve. It enables simultaneous writing and reading of the device without regard to the spectral composition of the two light beams. Furthermore, the dielectric mirror prevents the flow of dc current through the liquid crystal. This enhances the lifetime of the device. Finally, the mirror can be "tuned" to reflect any portion of the visible spectrum, thereby optimizing the separation efficacy of the mirror and, at the same time, maximizing the reflectivity of the device. This, together with the chemically inert insulting layer, SiO_2, that bounds the LC layer ensures a very long lifetime for the device. The LC that is used in this device is typically a biphenyl nematic material.

The operation of the device can be explained using an equivalent circuit of an ideal light valve substrate (Fig. 8.38). The diode represents the CdS/CdTe heterojunction diode, and the capacitor represents the capacitance of the dielectric mirror. If an ac voltage power supply is connected in the dark state (no input illumination) to the circuit, the capacitor will be charged to the negative peak voltage ($-V_p$) of the power supply during the first cycle. This voltage will then serve as a back bias voltage on the diode. Assuming infinite back resistance for the diode, the steady-state current flow in this circuit will be zero. Thus, there will be no current flow in the nonilluminated resolution element of the ideal LCLV.

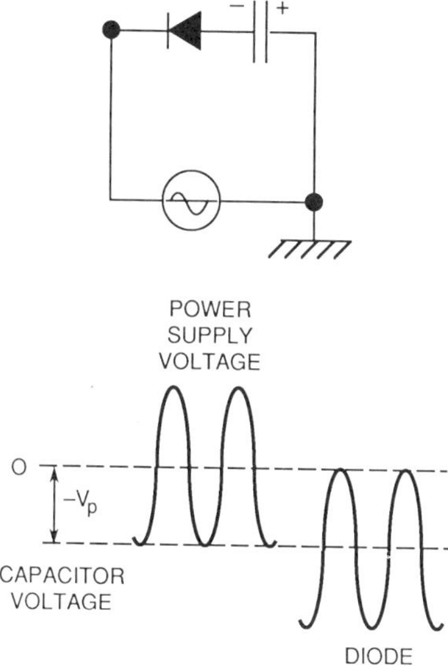

Figure 8.38 Equivalent circuit of the CdS LCLV under dark conditions. (After Ref. 98.)

If the element is illuminated by an input beam (Fig. 8.39*a*), the incident photons introduce a leakage resistance across the diode; this resistance discharges the capacitor during the back bias cycle of the diode. The approximate waveforms are shown in Fig. 8.39*b*. If the current is sufficiently high, the LC in that pixel will be driven above its threshold by the field developed across it. This effect can then be read out by the projection beam.

The device uses a nematic LC in the hybrid field-effect configuration [97]. As previously discussed (Section 8.2.4), this mode is more efficient in terms of maximum transmittance when used in a reflection configuration. A typical modulation transfer curve is shown in Fig. 8.40. A summary of the device's performance is given in Table 8.2.

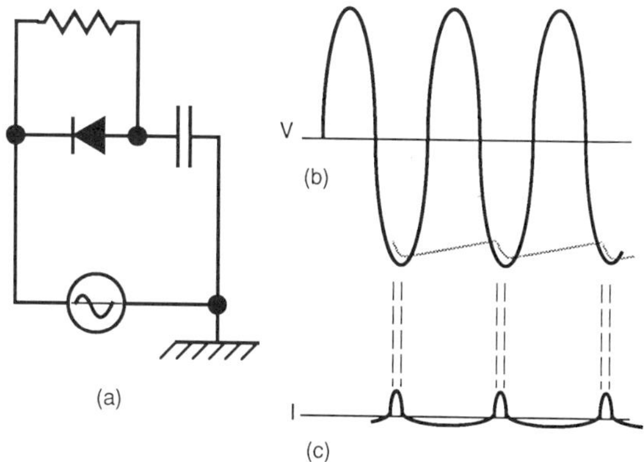

Figure 8.39 Equivalent circuit of the CdS LCLV under illumination.

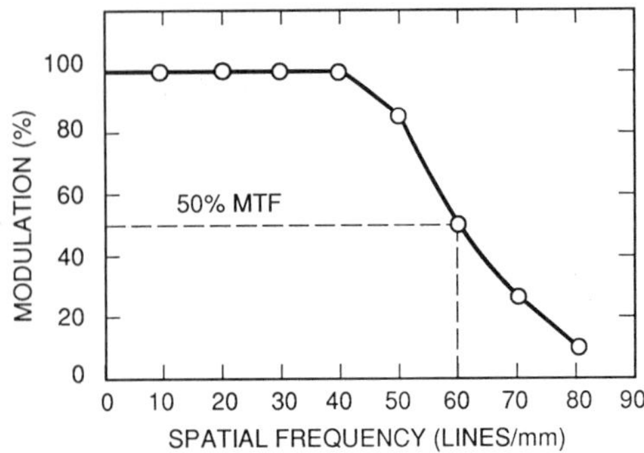

Figure 8.40 Modulation transfer function (MTF) curve of the CdS LCLV. (After Ref. 98.)

Silicon, GaAs, and Other Single-Crystal Photodriver Devices

Silicon (Single-Crystal) LCLV. As pointed out in the preceding, the CdS-based LCLV was perhaps the first successful demonstration of an LCLV technology. However, the device suffered from relatively slow response due to the CdS photoconductor. Trapping of photoactivated charges in defect levels whose density is quite high in polycrystalline substrate and the slow release of that trapped charge due to the relatively large band gap of the CdS (2.4 eV) were the main causes for the slower-than-video-rate response. The single-crystalline silicon-based photosubstrate with its low defect levels and low energy band gap (1.1 eV) was selected as a second-generation development to overcome these problems. It should be pointed out that a similar work on single-crystal silicon (and GaAs-) based metal–insulator–semiconductor configurations as photosubstrates for LCLVs was published by Soviet scientists [98,99].

Structure and Operation of the Silicon Liquid-Crystal Light Valve. The device [100] (Fig. 8.41) consists of a silicon photoconductor coupled with an oxide layer to form an MOS structure. A unified thin-film structure consisting of a dielectric mirror and a light-blocking layer provides the high (>90%) broadband reflectivity required as well as an optical isolation of the photoconductor from the high-intensity readout beam. The read-

TABLE 8.2 The CdS–LCLV: Summary of Performance

Aperture size	1 in.^{2}
Sensitivity (full contrast)	160 μW/cm^2 at 525 nm
Resolution	60 lines/mm at 50% MTF
Contrast	>100:1
Gray levels	9
Speed	
Excitation (0–90%)	10 ms
Extinction (100–10%)	15 ms
Projection light	
Throughput	>100 mW/cm^2
Reflectivity	>90%
Optical quality	<2 wavelength curvature at 6328 Å
Voltage	6 V_{rms} at 10 kHz

Source: Ref. 97.

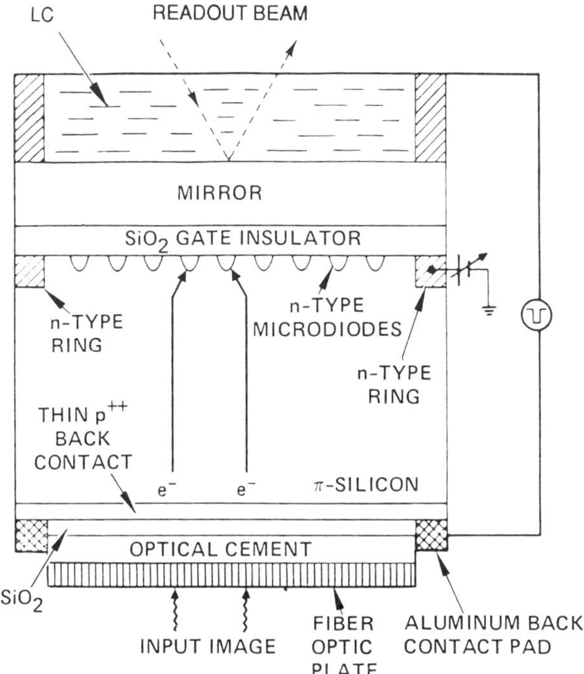

Figure 8.41 Schematic cross section of the Si LCLV [100].

out beam is retroreflected by the dielectric mirror through the LC. The latter is operated in a hybrid field-effect mode. The MOS mode of operation consists of periodic depletion and accumulation phases. In the depletion (active) phase, the high-resistivity π silicon is depleted completely, and electron–hole pairs generated by the input light are swept by the electric field, thereby producing the signal current that activates the LC. The electric field existing in the depletion region acts to focus the signal charges and, hence, to preserve the spatial resolution of the input image. The grid acts to focus the incoming electrons into the resolution cell defined by it as well as to form charge buckets of the electrons already residing at the Si/SiO_2. This prevents lateral "spillover" and consequent smearing of the charge pattern. As a result, higher resolution and contrast can be achieved. A dc-biased diode guard ring is implanted at the silicon wafer edge to prevent peripheral minority carrier injection into the active region; a thin layer of fast-response, positive-anisotropy, 45°-twisted nematic LC is employed as the light modulator. The input electrode consists of a fiber-optical faceplate to allow a direct coupling to a CRT. The fiber-optical faceplate is coated with indium tin oxide. A LC spacer as well as an indium tin oxide conductive layer are deposited on the front glass electrode. The whole structure is assembled within an airtight anodized aluminum holder.

Detailed Description of the MOS Structure and Operation. The high-resistivity MOS structure (Fig. 8.41) forms the essential part of the silicon light valve.

There are two basic phases in the operation of the MOS device:

1. *Accumulation.* The gate is charged with a voltage having the opposite sign to the type of dopant used (i.e., negative gate voltage for p-type material). This causes the

majority carriers (holes for p type) to accumulate at the Si/SiO_2 interface. The minority carriers (electrons) residing in the interface (from the previous frame) will be pushed toward the back contact and will be recombined with the majority carriers in the bulk p region. This phase, when used as a "refresh" phase, is relatively short (10–20 μs, or about 1–10% of the total cycle time).

2. *Depletion.* In the depletion phase (which constitutes the main, active phase) the gate is charged with voltage of the same polarity as that of the carriers (i.e., positive gate voltage for π-type silicon). The majority carriers (holes) are pushed away from the π region, while negative immobile ions remain. The gate voltage may be expressed as

$$U_G^{(\text{dep})} = \frac{Q_s(t)}{C_{\text{ox}}} + q\frac{N_a^{(\pi)}X_\pi^2}{2\varepsilon_s} + q\frac{N_a^{(p^+)}X_{p^+}^2(t)}{2\varepsilon_s} + q\frac{N_a^{(p^+)}X_{p^+}(t)X_\pi}{\varepsilon_s} \tag{8.41}$$

where $N_a(\pi)$ and $N_a(p^+)$ are the doping levels of the π region and the p^+ back contact, respectively; X_π and X_{p+} are the depletion width of the π region and the p^+ region, respectively; q is the electron charge; ε_s is the silicon dielectric constant; and Q_s is the total surface charge at the SiO_2 interface. The first term represents the voltage dropped across the oxide, the second term represents the voltage dropped on the high-sensitivity π region, and the third term represents the voltage invested in partially depleting the p^+ back contact. The last term represents the integration of the p^+ charge field along the π region. The surface charge may be expressed as

$$Q_s(t) = qN_a^{(\pi)}X_\pi + QN_a^{(p^+)}X_{p^+}(t) + Q_\lambda(t) + Q_{\text{th}}(t) \tag{8.42}$$

with the first and second term representing the charge resulting from the depletion of the π and p^+ regions, respectively. The third term represents the minority carriers (electrons) generated by the photo (input) signal, and the last term represents the thermally generated minority carrier charge, which is approximated by

$$Q_{\text{th}}(t) \approx \frac{n_i}{2\tau}X_\pi t \tag{8.43}$$

where n_i is the intrinsic carrier concentration ($\approx 2 \times 10^{10}$ cm^{-2}), τ is the minority carrier (generation) lifetime (typically 10–100 μs), and t is the time from the onset of depletion. Equation (8.41) implies, of course, that the π region is fully depleted. According to Eq. (8.41), this requires a fairly low doping level $N_a^{(\pi)}$. Thus, in order to fully deplete 125 μm (5 mils) of Si using $V_{\text{dep}} \leqslant 25$ V, the doping level must be $N_a \leqslant 2 \times 10^{12}$ cm^{-3}.

As may be observed from Eqs. (8.38)–(8.40), a "deep depletion" state is first established at the transition from accumulation to depletion where $Q_\lambda(0) = Q_{\text{th}}(0) = 0$. However, due to an externally injected charge (photocharge Q_λ) and a thermally generated charge Q_{th}, the accumulated surface charge will finally result in the voltage being dropped, mostly across the oxide [the first term in Eq. 8.41). Thus, a collapse of the depletion region is expected if the depletion voltage is left unchanged. This collapse will start at the p^+ back contact and move into the π region. It may be shown that the collapse, or "fill," time T_f is proportional to the minority carrier lifetime τ assuming thermal generation to be the sole mechanism for minority carrier charge generation. Thus,

$$T_f \approx \left(\frac{2C_{\text{ox}}}{C_{\text{dep}}}\right)\left(\frac{N_a^{(\pi)}}{N_i}\right)\tau \tag{8.44}$$

where C_{dep} is the depletion capacitance determined mainly by the wide π region. However, the depletion region cannot collapse entirely since the minority carriers causing this collapse are generated inside the depletion region. When equilibrium depletion width has been reached, the MOS is said to be in an "inversion" state. The width of the depletion region at inversion is independent of the gate voltage, while the surface potential at the Si/SiO_2 interface is of the order of the band-gap energy.

In using this configuration within the silicon light valve, a periodic asymmetric voltage waveform is applied, so that the device alternates between a long (active) depletion phase (≈ 1 ms) and a short (refreshing) accumulation phase. The design is such that the π region should not collapse during depletion. This is indicated by a constant, time-independent X_π term in Eq. (8.42). The time periods in each of these phases are determined as follows: The ratio of the active state to the "idle" state of the refreshing mode should be large to ensure a high signal-to-noise ratio. The duration of the active depletion phase, which largely determines the total cycle time, is limited by the TV frame rate to avoid a flickering effect and by the possible collapse of the depletion region due to thermal generation, as explained. The latter may be estimated, assuming $\tau \approx 10\ \mu$s, as a lower limit to the minority carrier lifetime. Using Eq. (8.43), $T_f \approx 30$ ms. Thus, from the preceding arguments, it follows that T_{dep} should be a few milliseconds. The accumulation time must be on the order of the time required for the recombination. Other constraints on the accumulation time are that it must be a small fraction of the depletion time, as noted, and that the transition current spikes caused by going to and from accumulation must be small. These constraints limit the accumulation time to a few microseconds.

THIN-FILM STRUCTURE. This structure provides the critical functions of readout light reflection and light blocking. While the first determines the degree of output brightness attainable on the screen, the latter will affect the contrast ratio, or in fact, the dark-state activation of the device. This activation is determined by the dark current generated in the photoconductor and by the degree of the photoconductor activation by the readout light that leaks through the mirror. Typical output brightness required for large-screen displays is on the order of 1000 lm, or about 500 mW/cm^2 of equivalent light intensity incident on the output surface of the light valve. The threshold sensitivity of the silicon photoconductor is on the order of 1 μW/cm^2. Thus, a readout light blocking efficiency of about 10^6 is required to prevent activation of the device by the readout light. A high reflectivity is required in order to (a) obtain the high degree of brightness required and (b) minimize absorption of the high readout light intensity involved. A high sheet resistivity is also required. This is essential in avoiding signal charge spread in the thin-film layers.

It can be shown that the required sheet resistivity of the thin film, $\rho_\square^{(\text{TF})}$, may be approximated by

$$\rho_\square^{(\text{TF})} \geqslant \frac{2\rho_{\text{LC}} T_{\text{LC}}}{a^2} \tag{8.45}$$

where ρ_{LC} and T_{LC} are the liquid crystal resistivity and thickness, respectively, and a is the resolution element size. Taking $\rho_{\text{LC}} = 2 \times 10^9\ \Omega\,\text{cm}$, $T_{\text{LC}} = 4\ \mu$m, and $a = 6\ \mu$m, we have

$$\rho_\square^{(\text{TF})} > 5 \times 10^{12}\ \Omega/\square \tag{8.46}$$

Additional requirements of the thin-film structure are those of low through-impedance. Also, inertness with respect to LC, high thermal stability to allow proper annealing, and compatibility with silicon-processing technology are needed. A Si/SiO_2 mirror structure was implemented in the device structure. The large difference in the refractive indices ($n_{Si} = 4.0$, $n_{SiO_2} = 1.55$) results in a high reflectivity. The use of 12–14 $\lambda/4$ pairs of Si/SiO_2 optically tuned to 430, 540, and 650 nm proved adequate in providing high (>90%) broadband reflectivity as well as moderate light-blocking capability.

LIQUID CRYSTAL OPERATION. The hybrid field effect mode was primarily used. Controlled birefringence was used for the IR modulation "VIDIC" device and for phase conjugation (see the following).

COMPUTER SIMULATION OF THE Si LCLV. A one-dimensional computer simulation model of the silicon–MOS LCLV was developed in order to provide a practical design tool for the Si–MOS LCLV [23, 101, 102]. The simulation confirmed most of the theoretical predictions discussed in the preceding. The combination of this Si–MOS modeling with the electro-optical behavior of the nematic LC layer [103] is expected to provide a complete electro-optical description of the Si–MOS LCLV.

Experimental Data

DEVICE FABRICATION. Fifty millimeter *p*-type ⟨100⟩ and ⟨111⟩ silicon ingots are used as the starting material. The resistivity of these ingots varies between 5 and 8 kΩ cm. Minority carrier lifetime specifications are typically between 1000 and 3000 μs. Following the slicing of 10–15-mil-thick wafers, final chemopolishing using both wax and waxless mounting techniques are used to produce 125-μm-(5-mil-) thick substrates.

The silicon-processing sequence of the thinned-down wafers is shown in Table 8.3.

Following the silicon processing the substrate is subjected to thin-film evaporation to form the light blocking and the dielectric mirror layers required. The previously described Si/SiO_2 structure consisting of 12–14 pairs optically tuned in thickness to 430-, 540-, and 660-nm spectral ranges is deposited. This results in broadband (visible) reflectivity of above 90% as well as an optical isolation factor of about 10^5 in the visible. Following the mirror deposition the wafer is mounted using optical cement to the fiber-optical base electrode using a special uniform press-down fixture. Both the mounted wafer and the counter glass electrode (indium tin oxide coated) are treated by SiO shallow-angle deposition for LC alignment using the medium and shallow-angle evaporation

TABLE 8.3 Major Processing Steps Used in the Fabrication of a *p*-type Substrate

1. Thorough wafer cleaning in organic and acid solvents.
2. Diffusion field mask and pyroelectric oxidation.
3. Backside oxide etch and boron diffusion.
4. Guard ring diode implant region: photoresist and etch.
5. Guard ring diode: phosphorous implant.
6. Microdiode array pattern: photoresist and etch.
7. Microdiode array: phosphorous implant.
8. Oxide strip and wafer clean.
9. Final gate oxidation.
10. Guard ring diode and ohmic back contacts: photoresist and oxide etch.
11. Si–Al guard ring diode metallization.
12. Ohmic back contact: metallization.
13. Anneal.
14. 14 pair Si/SiO_2 dielectric mirror evaporation.
15. SiO_x shallow-angle LC alignment evaporation.

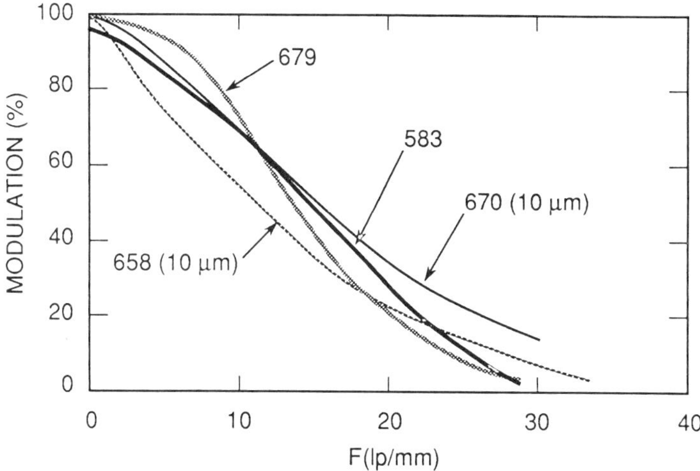

Figure 8.42 Modulation transfer function (MTF) curves of Si LCLVs; experimental results of several devices are shown [100].

techniques. Alignment is designed to result either in a 45° twisted nematic configuration utilized in the hybrid field-effect mode or in a parallel-aligned (0° twist) configuration. The LCs used were BDH E-7, E-18, and E-44.

LIGHT VALVE OPERATION. The voltage waveform employed is typically 1–5-kHz frequency with a 10–20-μs accumulation or refresh phase. Typical voltages are 30 V of depletion and $-$ 20 to $-$ 30 V of accumulation. The guard ring is reverse biased by 20–40 V with a typical leakage current of <1 mA.

SILICON LIGHT VALVE PERFORMANCE

1. *Contrast Ratio.* Typical values are 20:1–30:1; the best results are 100:1. The difference is due mainly to the fact that the contrast ratio is highly sensitive to the LC alignment layers, which are degraded by repeated use of the electrodes.

2. *Resolution.* This parameter is affected by two main factors: the retention of full depletion throughout the active (depletion) phase and focusing grid geometry and efficiency. The degree of depletion retention determines the behavior of the modulation transfer function throughout the whole spatial frequency range. The grid geometry determines the high spatial frequency behavior (limiting resolution).[†] Modulation transfer function curves of 50 and 100 gridlines/mm are shown in Fig. 8.42. Typical resolution for the 2-in. device is 15–16 line pairs per millimeter (lp/mm) at 50% modulation and a limiting resolution (~5% modulation) at 35 lp/mm [70 TV lines per millimeter (TVL/mm)] using a 100-line/mm focusing grid.

3. *Optical Transfer Characteristics.* As explained in the preceding, retaining the high-resistivity π silicon in full depletion results in a linear conversion of optical input intensity into current activating LC. As may be observed from Fig. 8.43, the main portion of the

[†] Recent calculations indicate that fringing fields constitute an additional factor. Also, the limited focusing capability of the microdiode grid [100] may be the dominant cause for the loss of resolution at medium spatial frequencies.

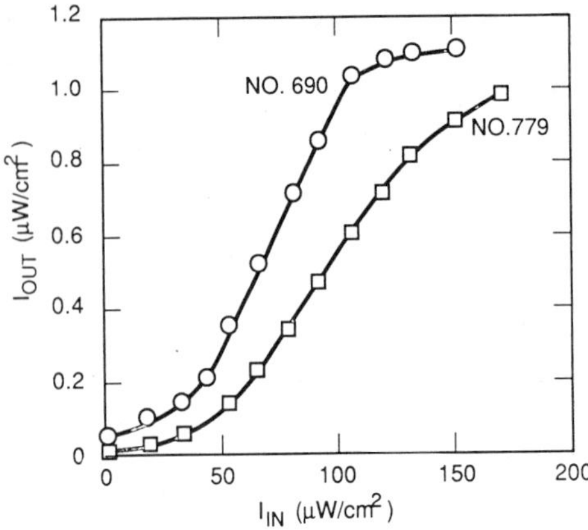

Figure 8.43 Transfer (input/output) curves of the Si LCLV [100].

transfer curves is quite linear. It should be noted that although these measurements were conducted under low output levels, high contrast was also demonstrated with devices operating at output levels of about 100 mW/cm^2 with the same input levels.

4. *Response Time.* As previously discussed, the time response in the silicon-photoconductor-based device is limited by the LC. The LCs commonly used are BDH E-7 and E-18, with thicknesses of 4 μm. A typical curve is shown in Fig. 8.44. Rise times of ~4 ms (to 90%) and decay times of 16 ms (to 10%) are observed. The best results obtained (with a 2-μm LC layer) are 2 and 5 ms for the rise and decay times, respectively. Results on the Si–MOS LCLV, including comparison with the computer simulation model, were recently published [104].

***pn* Diode Si LCLV.** A *pn* diode array Si LCLV was recently reported by the General Electric Corporation (U.K.) group [105]; its structure is shown in Fig. 8.45. The structure

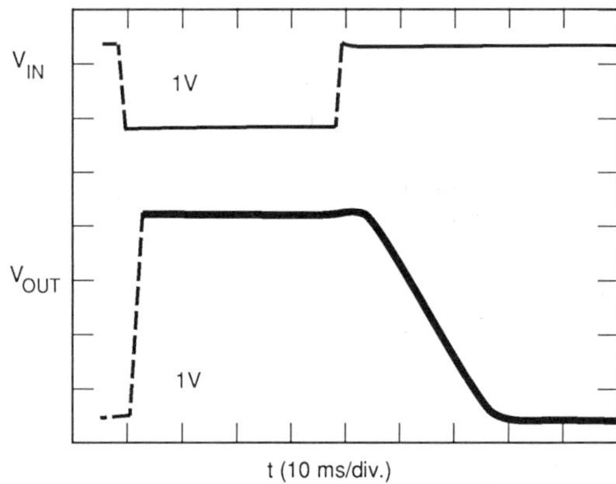

Figure 8.44 Typical response time of the Si LCLV. Top: LED activation pulse (negative means on). Bottom: photo response curve (10 ms/div) [100].

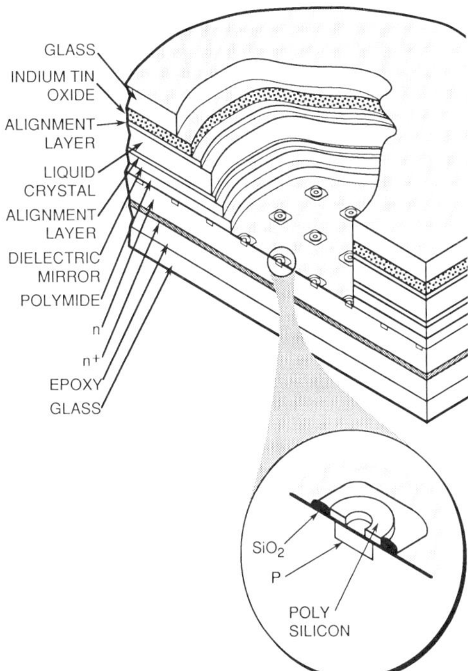

Figure 8.45 Cross section of the silicon *pn* diode array LCLV structure. (After Ref. 105.)

is quite similar to the Hughes device described earlier. However, while the device still uses a *pn* microdiode array similar to the Hughes design, no oxide layer is used. It is assumed that the developers rely on the application of a true ac bias voltage and/or the use of the dielectric mirror to avoid the electrolytic deterioration of the LC layer due to the effect of a dc current.

Schottky Diode Silicon LCLV. A device based on the use of a single-crystal silicon and a metal matrix mirror serving also as a Schottky mirror was recently published [106]. The concept is to use a Schottky metal–semiconductor interface that allows depletion through the silicon to be attained and thus maintain the impedance match between the photoconductor and the LC. This match is critical in achieving the operation of a photoactivated spatial light modulator (SLM). Since a high Schottky barrier-forming metal such as aluminum or platinum can be deposited on the *n*-type silicon substrate at low temperature, the silicon substrate can be mounted on a flat glass substrate, polished, and then processed. This allows a high degree of output uniformity to be attained. Also, by eliminating the SiO_2 insulator present in the MOS LCLV version previously discussed, one can reduce the sheet conductivity at the Si/LC interface and achieve higher resolution and dynamic range.

Two Schottky LCLV configurations have been designed: a double-Schottky-layer structure (Fig. 8.46) and a single-Schottky configuration. The double-Schottky configuration is designed to allow a balanced ac current to flow through the device, thus minimizing the LC decomposition that occurs due to dc current flow. The back (input) side Schottky diode layer (for a *n*-silicon-based device) is fabricated using either a platinum grid or a thin platinum layer (100 Å), followed by deposition of indium tin oxide to

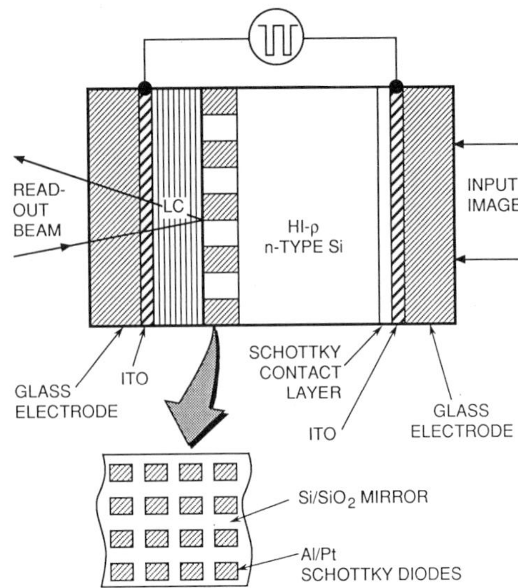

Figure 8.46 Double-Schottky Si LCLV configuration [106].

enhance the sheet conductivity of the back contact. Both structures have been fabricated and tested. The front Schottky contact consists of a metal matrix of platinum islands (17×17 μm) separated by chemical vapor deposited oxide (3-μm channels). The operation of a Pt–Si Schottky diode LCLV in the conversion of a visible image into an IR image (see Section 8.4.5) is shown in Fig. 8.47. A thermal bar chart of the LCLV using a blackbody source of 780°C is shown.

GaAs and Other Semiconductors. As mentioned previously, single-crystalline GaAs-based metal–insulator–semiconductor (MIS) LCLVs were first reported by Soviet scientists in 1979 [107] and later [108]. Reasonable resolution (10–15 lp/mm) and relatively high sensitivity ($\cong 0.5$ μW/cm^2) were observed in the early work. It appears that no dielectric mirror or light-blocking layer were incorporated into the device. These devices can therefore be read out using relatively low light levels in order to avoid activation of the photosubstrate.

Recently a GaAs-based Schottky diode LCLV development was published by the British Aerospace Group [109]. The device (Fig. 8.48) is based on a single-crystal GaAs substrate. The substrates used were 20 or 70 μm thick, and a microdiode array of 2×2 μm Au–Ti pads formed the Schottky contacts. Good overall performance was achieved (Table 8.4). The Lockheed group reported the development of both GaAs and silicon single-crystal-based *ferroelectric* LCLVs [110,111]. Performance results of these devices are still sketchy at the time of writing this report. Early work on *bismuth germanate* (BGO) and *silicate* (BSO) was reported by the Lebedev Physics Institute group in 1980 [112].

The use of BSO as a photoconductor was also reported by the Thomson CSF group [113]. Finally, one of the earlier operating LCLV devices based on selenium photosubstrate was demonstrated by the Xerox Research group [114].

Wire Grid Mirror-based LCLV (WGLV). This concept, which was recently introduced [115], is based on the use of a polarization-sensitive mirror in the silicon LCLV. The

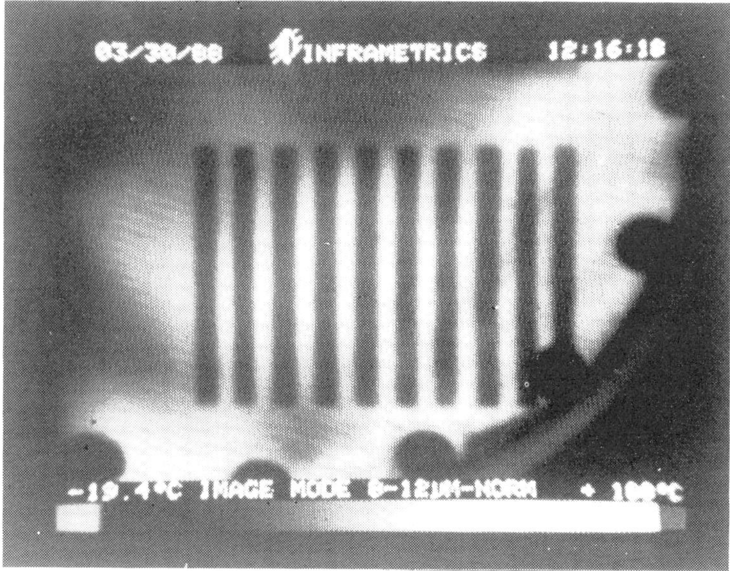

(a)

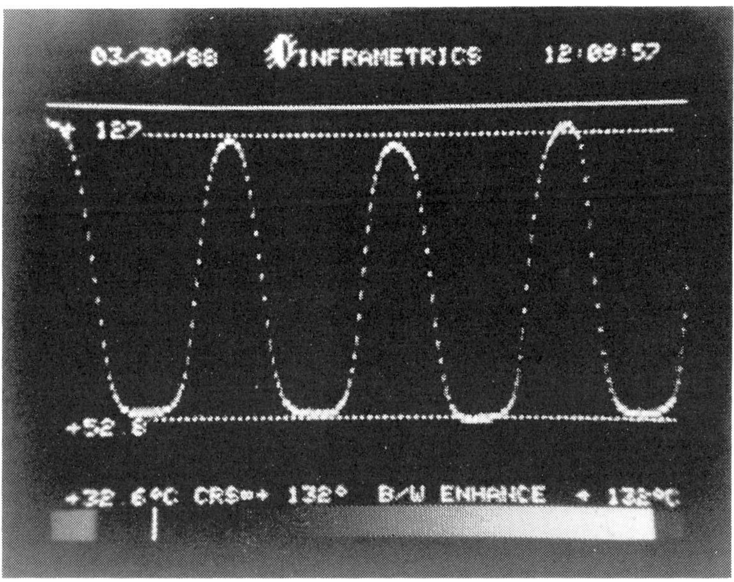

(b)

Figure 8.47 Schottky diode Si LCLV used for IR modulation: (*a*) image of a thermal bar pattern; (*b*) temperature profile across the bar pattern [106].

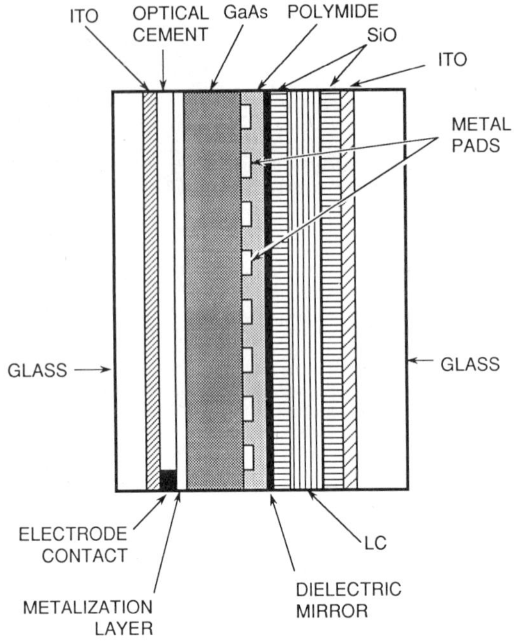

Figure 8.48 A GaAs Schottky diode array LCLV. (After Ref. 109.)

structure of a WGLV is shown in Fig. 8.49 in conjunction with a 90° twisted-nematic (TN) LC configuration. The wire grid polarizer simply acts as a polarization-sensitive mirror, allowing one component of the polarization to be transmitted while blocking with a given extinction ratio the other component. In this application of a polarizing mirror within a LCLV, the submicrometer wire grid structure is cross-etched in order to avoid the shortening of the voltage along the grid lines, which would "shorten" or smear the spatial image information transmitted to the LC modulator.

Various configurations of the LC cell in conjunction with the wire grid mirror are possible. In the one shown (90° TN), the device will act as a non linear bistable device. The use of a parallel-aligned LC configuration can lead to vastly different behavior that

TABLE 8.4 Optimum Overall Performance of the GaAs LCLV

Sensitivity	50 μW/cm^2
Resolution	16 lp/mm 50% MTF
Contrast ratio	15:1
Switching speed	100 ms off/on 90% 70 ms on/off 90%
Drive frequency	218 Hz (13% duty cycle)
Accumulation voltage	+12 V
Depletion voltage	−10 V

Source: Ref. 109.

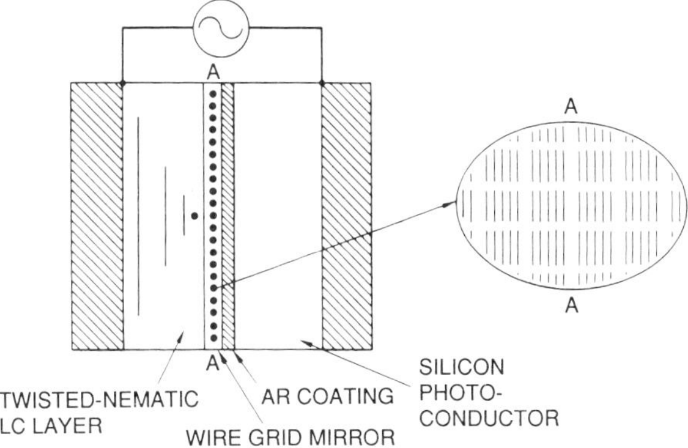

Figure 8.49 A wire grid mirror-based Si LCLV. (After Ref. 115.)

can, in fact, be used to enhance the phase conjugation capabilities of the LCLV (see Section 8.4.4).

To illustrate the use of a WGLV, let us concentrate on the 90° TN configuration shown in Fig. 8.49 which, as pointed out, leads to highly nonlinear transfer characteristics in the LCLV. The optical configuration in this case is shown in Fig. 8.50. An optical readout beam transmitted by the polarizing beam splitter is incident on the the WGLV. If the intensity of the beam is not too high, the polarized ray will get rotated by the 90° TN LC cell and will be incident on the wire grid mirror in such an orientation as to be reflected from it. Upon exiting the LC cell, the reflected beam will have the same polarization orientation as the incident beam and hence will not appear at the output. If a high-intensity spot or pixel is present in the input pattern, then upon reflection from the wire grid mirror, a certain level of light will leak through the WGLV mirror at that pixel location, causing the silicon photosubstrate to be partially activated. This partial activation will cause an increased voltage drop to appear across the LC layer. This, in turn, will result in an increased tilt angle of the LC molecules and a partial local destruction of the twist, causing more light at that spot to leak through the polarizing mirror and a further increase in the activation of the photosubstrate. We therefore see that a positive-feedback loop has been set, resulting ultimately in the transition of the optical output

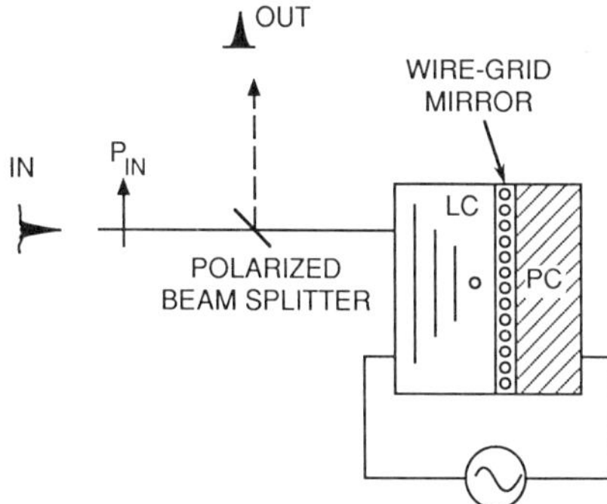

Figure 8.50 Optical set-up for the use of the wire grid LCLV for nonlinear (thresholding) optical operation [115].

at the bright spot location from a low level to a high level, as shown in the figure. Preliminary results with a 0.5-μm-period wire grid grating demonstrated the large optical nonlinearity expected with the use of this device. We therefore see that the insertion of a polarization-sensitive mirror inside a LCLV results in a significant change of its optical–optical transfer characteristics. The preliminary results of the first device assembled (Fig. 8.51) show a sharp threshold with a peak-to-threshold ratio of 3:1. The threshold intensity level was tunable from 2.5 to 50 $\mu W/cm^2$ by changing the bias voltage from about 5 to 25 V. These properties allow the polarization-sensitive mirror-based LCLV to be used as a nonlinear SLM with applications in optical computing (bistable device), adaptive image thresholding, as well as phase conjugation and as an interface device for an optical co-processor (see Sections 8.4.2 and 8.4.4).

Amorphous Silicon Devices. An early publication on the use of a gadolinium/amorphous-silicon Schottky diode as a photosubstrate for an LCLV was demonstrated by the IBM/San Jose group in 1979 [116]. The group has demonstrated photoactivation sensitivities on the order of a few microjoules per square centimeter, which is comparable to that of the single-crystalline silicon discussed previously. Although no resolution tests were reported, laser spot sizes on the order of 30 μm were resolved. These studies were also conducted on molybdenum–amorphous silicon Schottky diode photosubstrate. More recent efforts on photoactivated LCLVs based on amorphous silicon were published by the GEC group in the United Kingdom [117] and by the University of Colorado group [118]. The GEC device (Fig. 8.52) is based on hydrogenated α-Si as the photoconductor and a chiral smectic (ferroelectric) liquid crystal. The group has reported quite high resolution over the 40×40–mm active area of the device (28 lp/mm at 50% modulation), reasonable sensitivity (50–250 $\mu W/cm^2$ for $\approx$90% of the full-activation level), and a response time of 250 μs.

The α-Si:H-based ferroelectric LCLV reported by the University of Colorado [119] operates at 30 V peak to peak and features total response time of about 300 μs, resolution of 83 lp/mm, and switching energy of $12 \rho\tau\ cm^{-2}$. If one assumes 100 μs for the rise time, this energy sensitivity corresponds to a switching *power* of about 0.1 $\mu W/cm^2$.

Finally, an α-Si:H LCLV using nematic LC was recently reported by Hughes [119a].

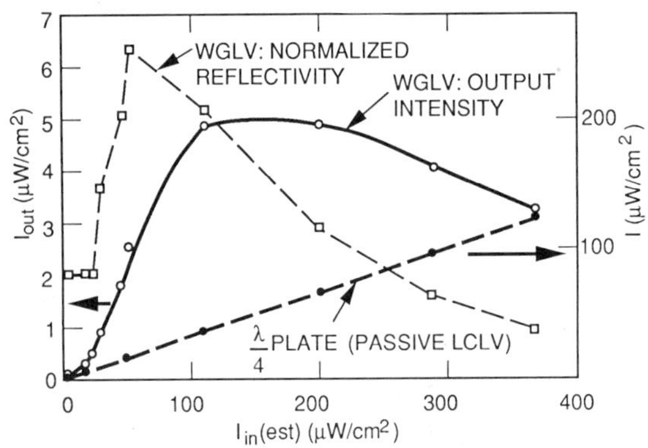

Figure 8.51 Wire grid mirror LCLV. Open circles: input/output characteristic. Closed circles: input/output behavior of a $\lambda/4$ plate. Open squares: normalized reflectivity of the WGLV [115].

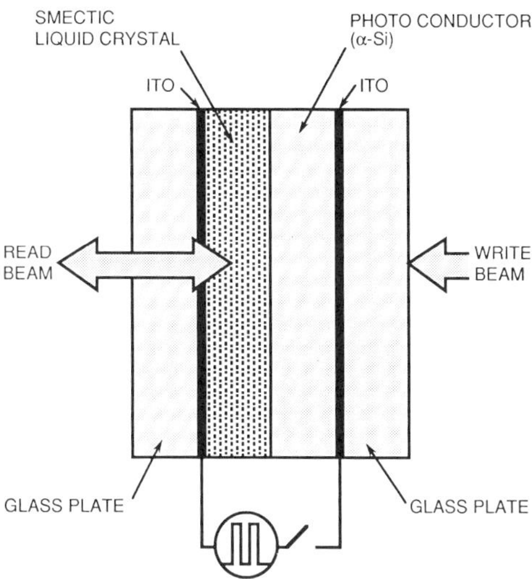

Figure 8.52 An α-Si-based LCLV using smectic C (ferroelectric) LC as the modulator. (After Ref. 117.)

8.3.3 Laser-addressed Liquid Crystal Light Valves

Here the idea is to use the heating effect of a laser beam to trigger a direct phase change in the LC material. The concept is shown in Fig. 8.53. The capability of a laser-addressed LC system to perform partial and total erasure of the image was reported by Kahn in 1973 [119]. This was achieved using SmA materials. The thermally induced smectic–cholesteric phase transition was used as a basis for a direct laser-addressed LCLV by

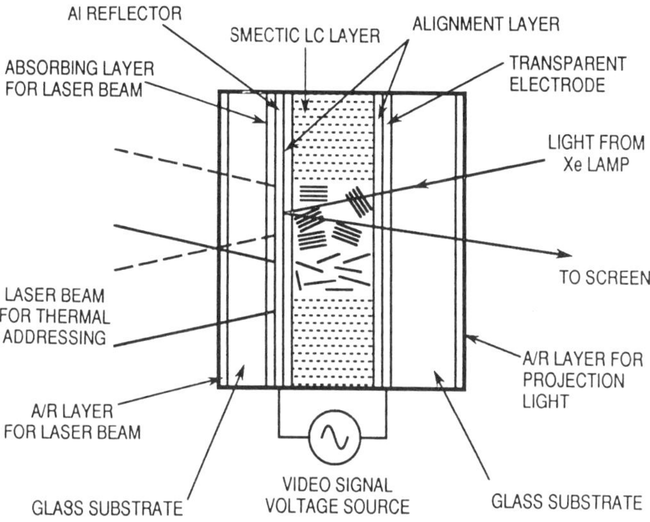

Figure 8.53 Thermally induced phase transition using laser beam addressing in a smectic LC cell. (After Ref. 122.)

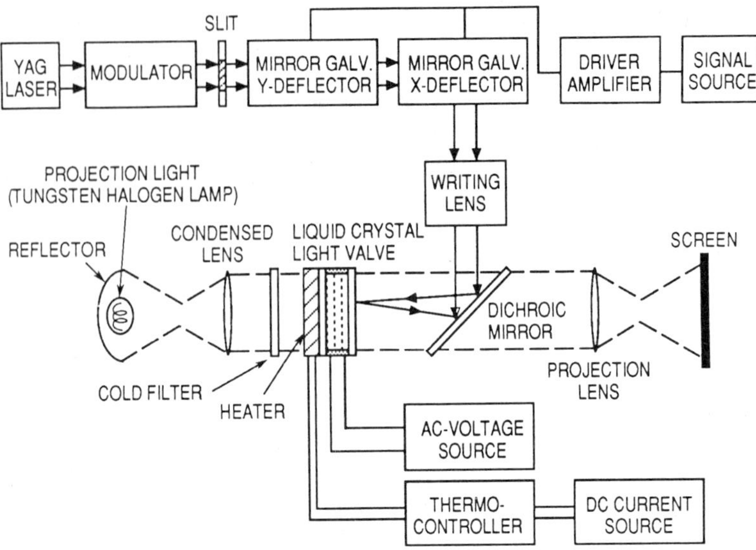

Figure 8.54 Projection system based on a laser-addressed LCLV. (After Ref. 120.)

the group at Tokyo University [120]. Figure 8.54 shows a schematic of the LCLV projection system with a thermally controlled device. The change in the optical transmission due to the phase transition is shown in Fig. 8.55. The figure shows (a) a pure optical–optical, or thermo-optical, effect; (b) the effect of combined electrical and thermal fields; and (c) the effect of a purely electrical field near the phase transition point. A 6×10^6-pixel display based on the phase change in smectics was demonstrated in 1983 [121]. More recently, a full-color laser-addressed LC projection system was reported by the Hitachi Research Group [122] with a 2000 × 2000 array display size (Table 8.5). A very high resolution was recently reported by Greyhawk Systems [123], allowing an up to 5000 × 7500-pixel display to be attained over 50 × 75 mm of LCLV active area with a contrast ratio of 20:1 or better and a writing time of about 150 s for full update. Finally,

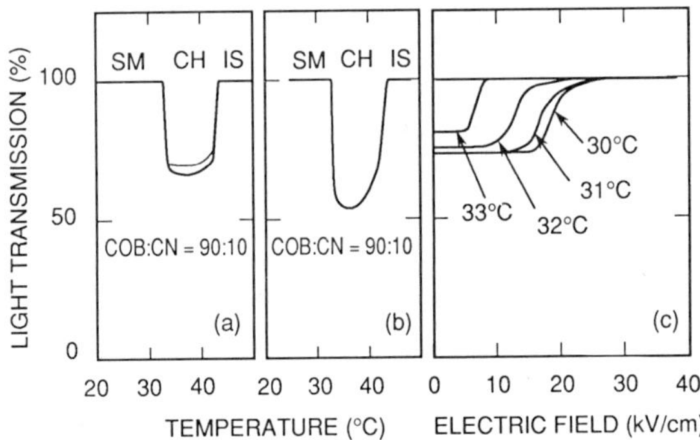

Figure 8.55 Thermal and thermoelectric effects in smectic materials: (*a*) pure thermal effect, (*b*) thermoelectric activation, and (*c*) electro-optical effect. Here SM, CH, and IS denote the smectic, cholesteric, and isotropic phases, respectively. (After Ref. 120.)

TABLE 8.5 Specifications of Laser-addressed Display [122]

Screen size	2 × 2 m
Number of addressable pixels	1000 × 1000,[a] 2000 × 2000[b]
LC Cell size	20 × 20 mm
Laser power	12 mW at each LC cell
Writing speed	1 min/screen,[a] 30 min/screen[b]
Brightness	30 lm (screen gain 3.5)

[a] Full color with raster scan.
[b] Multicolor with vector scan.

the elimination of the CRT as an input device for the LCLV by laser addressing of the photoconductor to attain higher resolution was recently demonstrated by the Naval Ocean Systems Center at San Diego, California [124].

8.3.4 Direct *e*-Beam Addressed Liquid Crystal Light Valves

Direct addressing of a LC cell by an electron beam was reported by the Tektronix group [125]. In this device the electron beam induces a field across the LC (Fig. 8.56) to spatially modulate its effective birefringence, as we have seen in previous configurations. The idea is to eliminate the need for a CRT coupling (similar to the laser-addressed cases) and to simplify the image input mechanism. The 36 × 48-mm device demonstrated about 600 × 500 resolution elements at 30:1 contrast using 60-Hz frame rates.

8.3.5 Photoactivated Variable Grating Mode Device

The VGM effect previously discussed was employed in conjunction with a ZnS photoconductor [59,60]. The device (Fig. 8.57) was operated in a transmissive mode using a passband between 410 and 550 nm for the activation of the ZnS photoconductor and a readout beam at 632.8 nm to which the ZnS is insensitive. In this configuration the device basically operates as an intensity-to-spatial-frequency image converter. Thus, since the spatial intensity pattern is translated into a spatial voltage pattern across the LC, the resultant grating period will be locally dependent upon the spatial voltage pattern [126]. The use of this device for nonlinear optical processing applications is discussed in Section 8.4.

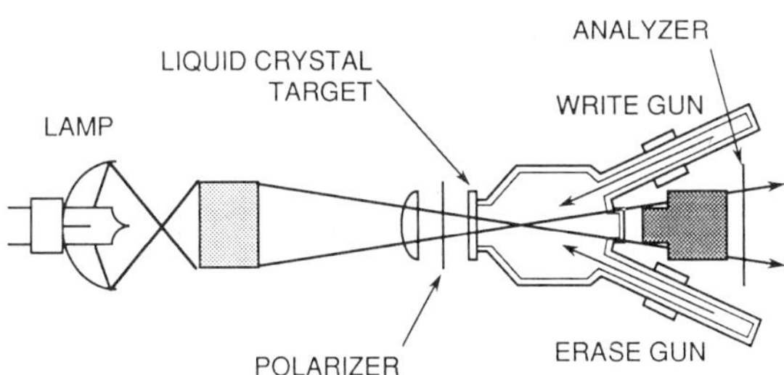

Figure 8.56 Direct *e*-beam addressed LCLV. (After T. S. Buzak et al., *Soc. Inf. Displays Tech. Dig.*, 1987, p. 64.)

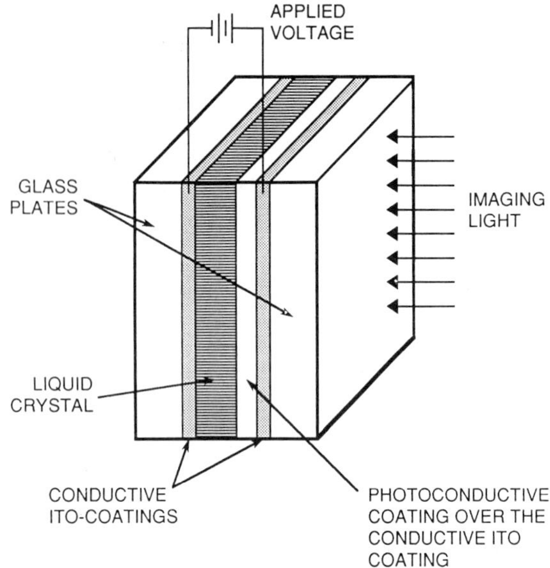

Figure 8.57 Photoactivated VGM LCLV; negative-anisotropy LC is used in its VGM mode. (After Ref. 59.)

8.3.6 Optical Feedback Configurations in LC SLMs

Optical feedback schemes in LC devices can be used for optical computing as well as for adaptive optical applications (see Sections 8.4.2 and 8.4.4). In order to implement binary operation in LC devices, one generally needs some form of a positive feedback to transform from the linear transfer curve characteristic of a nematic-based LCLV to a bistable one. As we shall see, while optical logic operation can be achieved using a standard LCLV (discussed in a previous section), the 0 and 1 states generated by the LCLV are not pure states in the sense that an intermediate state will exist if the input intensity levels slightly deviate from the maximum (high) or the minimum (low). This problem can be remedied by incorporating a positive-feedback loop in the system such that intermediate states are driven into either the low or the high levels with a sufficiently high (>1) loop gain similar to electronic feedback amplifiers [127]. The initial attempts at implementing a feedback loop in a LCLV-based system were made using external elements. However, changing the internal structure of an LCLV can result, as will be shown later, in the effective implementation of an internal feedback mechanism.

An optical flip-flop system using external feedback is shown in Fig. 8.58 [128]. The reinsertion of the optical output of $LCLV_2$ back into the input port of $LCLV_1$ results in the desirable bistability and the capability of performing the "flip-flop" operation. In another method of implementing external optical feedback the CdS-based LCLV [129] is configured with each pixel having a photosensitive area and a transparent area. Again the feedback is achieved by redirecting a portion of the readout beam back to the input photosensitive portion of the device.

Several schemes for implementing *internal* feedback in a LC device were proposed in the late 1970s [130]. A device implementing an internal feedback was built by Kompanets et al. [131]. The device (Fig. 8.59) is based on a semitransparent photoconductor (or

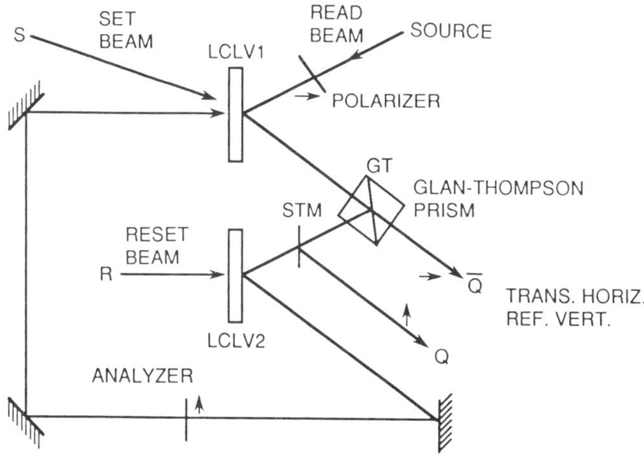

Figure 8.58 Optical flip-flop using two LCLVs (STM, semitransparent mirror). (After Ref. 128.)

"photoelectric") substrate with a LC layer containing absorbed dye in a "guest–host" configuration. The increased voltage across the LC layer as a result of photoactivation at a certain spatial location results in the tilting of the LC molecules, including the aligned dye (guest) molecules, and in the subsequently reduced absorption that reduces the absorption of the constant readout light reflected from the polished photoconductor surface. This in turn results in a further increase in the activation. The insertion of a polarizing mirror into an LCLV structure as previously discussed constitutes another solution to the implementation of an internal feedback loop within the LCLV. The achievement of bistability using this configuration is evident from Fig. 8.51.

Gain–Bandwidth Product in Closed-Loop LC SLM Systems. We have previously discussed the relationship between the dynamic range of a LC cell and its speed of response (see Section 8.2.5). In the case of a LC SLM we may also have to take into account the response time and dynamic range limitation imposed by the other components of the SLM (in particular the photosubstrate). As it turns out, the relationship between the

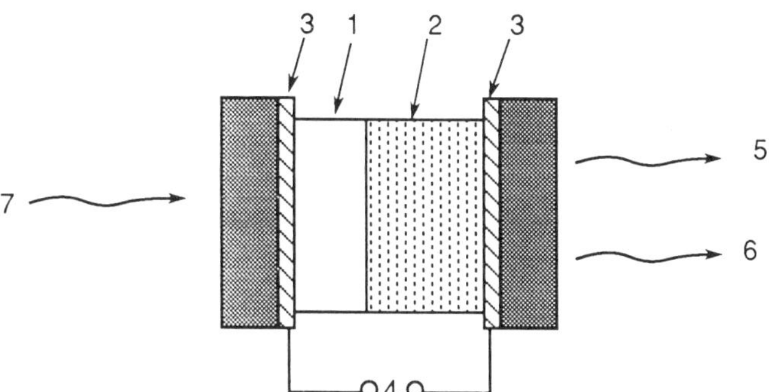

Figure 8.59 Bistable LCLV based on a guest–host LC layer: (1) photoconductor, (2) guest–host LC layer, (3) conductive transparent electrodes, (4) voltage source, (5) constant-bias red beam, (6) reflected beam, and (7) writing beam. (After Ref. 131.)

dynamic range and the speed of a LC SLM is, in fact, also preserved (with some restrictions) in the case where a photoactivated LC SLM is placed in a closed feedback loop, as we shall now show.

In order to show that, assume that we are using a LC cell within a photoactivated SLM configuration (Section 8.3.2). For the present purpose let us simply assume that we have a photoconductor–mirror–LC configuration in which a change in the spatial pattern of the input intensity will result in a proportional change in its output phase modulation (we shall assume that the LC is in a *phase modulation* configuration). The operational definition of the phase rate of change, $\Delta\phi/\Delta t$ [18,23,27], is that we first change the bias voltage across the LC cell by an amount ΔV. This results in a change of the phase $\Delta\phi$, which for small ΔV excursion will be linear with the applied voltage. This phase change will build up exponentially with a certain time constant τ, with V (or $V + \Delta V$) as the bias. Assume now that we split a portion of the output and feed it back to the input (Fig. 8.60), where we also interfere this fed-back beam with a phased (coherent) uniform beam. In this case the change in output phase will translate into a change in the input intensity through the interference with the reference beam, which, depending on the phase difference between the beams and the intensity–phase transfer characteristics of the LC device, may result in either a positive or negative contribution to the input

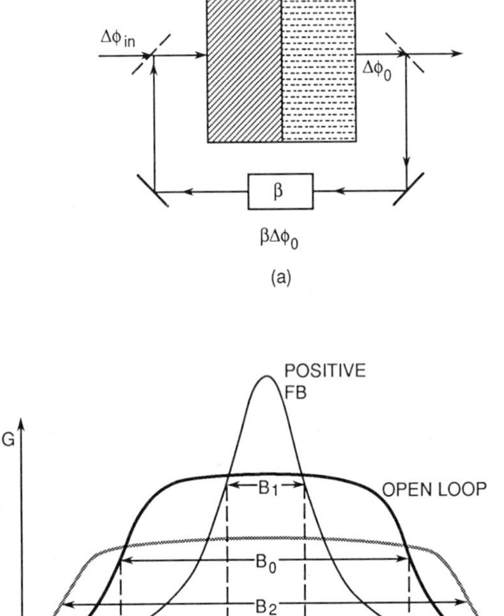

Figure 8.60 An LCLV in a closed-loop configuration. (*a*) Schematics of the closed loop. (*b*) Various gain-versus-bandwidth (ω) curves illustrating a fixed gain–bandwidth product. The peaked curve shows a positive-feedback case where the high gain is compromised by a narrow bandwidth.

intensity. Let us assume that we have established a *positive*-feedback configuration. We first define the gain, G_0, of a phase-modulating LCLV in an open-loop configuration as

$$\Delta\phi_{out} = G_0 \, \Delta\phi_{in} \tag{8.47}$$

where the change in the input intensity ΔI_{in} to the LCLV at a particular pixel location is proportional to the change in the *phase* of that particular pixel, $\Delta\phi_{in}$, through the interference of the phase front with a reference beam. Then for a given equivalent input phase change $\Delta\phi_{in}$ we will have for the output of the closed-loop system

$$\Delta\phi_{out} = G_{CL} \, \Delta\phi_{in} \tag{8.48}$$

where the closed-loop gain is given by

$$G_{CL} = \frac{G_0}{1 - \beta G_0} \tag{8.49}$$

and where G_0 is the open-loop gain of the LC SLM and β is the fraction of the output beam that is fed back to the input. For $\beta G_0 \approx 1$ quite a large overall gain G_{CL} can be achieved. However, while the gain of the system can be quite large allowing a large $\Delta\phi_{out}$ to be attained with a small $\Delta\phi_{in}$, the speed of response will be correspondingly low. To see that, let us assume that the change in bias voltage ΔV is proportional to the change in the *input intensity* $\Delta\phi_{in}$. Let us also recall that the response time τ will depend on the bias voltage V as $\tau \approx \tau^*/[V/V_0^2 - 1]$ [Eq. (8.19)]. In the absence of the feedback we will have an open-loop gain G_0 for the device, which means that an incident intensity equivalent to a phase change $\Delta\phi_{in}$ will translate into a bias voltage $V_b^{(0)}$ and an associated response time τ_0. The change in the modulated phase, which will be set fixed, $\Delta\phi_{out}$ (e.g., π), will be given by Eq. (8.47). As we now "turn on" the positive feedback, a much higher gain G_{CL}, given by (8.48), will be present. Thus, for the same output phase change $\Delta\phi_{out}$, we will have a much *smaller* input signal: $\Delta\phi_{in} \approx \Delta\phi_{out}/G_{CL}$. Since it is $\Delta\phi_{in}$ that determines the response time through the bias voltage dependence, we see that the response time τ increases. In fact, we can show again in analogy to the theory of feedback amplifiers [18,127] that the closed-loop response time Δt_{CL} is proportional to $\tau_0/G_{CL} = \tau_0/1 - \beta G_0$. We therefore have

$$\left(\frac{\Delta\phi}{\Delta t}\right)_{CL} = \frac{\Delta\phi_{out}}{\Delta t_{CL}} \approx \frac{G_0/1 - \beta G_0}{\tau_0/1 - \beta G_0} = \frac{G_0}{\tau_0} \tag{8.50}$$

In other words, the rate of change of the phase $\Delta\phi/\Delta t$ is proportional to the gain–bandwidth product of the LC cell. Furthermore, to the extent that the dynamics of the LC determine those of the entire system, this product is constant and does not depend on the configuration of the LC cell within the LC SLM system, provided that the response time of the photoconductor is negligibly small compared to that of the LC.

8.4 APPLICATIONS

8.4.1 Displays

As pointed out in Section 8.1, LCs are ideally suited for display applications since they can respond in approximately 10–20 ms, which is fast enough to avoid "smearing" sensation to the human observer. The extremely large electro-optical coefficient allows the use of compact, low-voltage and low-power solid-state driver arrays. These driver arrays are also producible in large density using microelectronic technology, to allow production of large-scale, high-resolution displays, as we have seen in the previous section on active matrix LCDs.

We have already discussed in some detail several devices used in displays, namely, alphanumeric displays and matrix-addressed schemes (Section 8.3). This section is organized as follows. I will first describe briefly the main display parameters. Next the wide range of display applications will be discussed. In order to gain some perspective on the relative role of liquid crystal displays (LCDs), other existing display technologies will be briefly outlined. The next section will cover a particularly important class of LC-based display systems, namely, large-screen projection displays. This section will be concluded by providing a brief historical overview of the LCD development, finally detailing one of the more exotic future uses of this display technology, namely, high-definition television (HDTV).

Display Parameters

Contrast. The contrast ratio is defined as $(I_{max} - I_{min})/I_{min}$, where I_{max}, I_{min} are the maximum and minimum intensity levels in the display used. The above contrast level for LCD systems depends on the mode of operation of the LC used. Very high contrast ratios are attainable with nematic LCs used in the controlled-birefringence mode (Section 8.2.4). However, as already discussed, due to the sensitivity of this mode to the cell thickness, only the homeotropically (perpendicularly) aligned configurations can be practically used. These, however, have not yet been realized in practice due to the difficulty of surface treatment for LC alignment. Quite decent contrast ratios (over 100:1) can be obtained using the HFE mode if used in a narrow spectral band.

Color. Another significant advantage of using LCs for displays is their large operational spectral region, which extends from near UV to mid IR. Liquid crystals are particularly suited for operation in the full visible range (400–650 Å). Four principal schemes are used in order to implement color information in LCDs. The first and most widely used method is to space-multiplex the color information such that each pixel is formed of three subpixels that are activated with the red, green, and blue (RGB) information in analogy with the TV addressing scheme [132]. In order to generate the proper color for each of the pixels, two configurations are possible. The first one is based on fabrication of an RGB shadow mask filter that, when aligned with the LCD array, provides a fixed color pattern that matches the color subsection of the pixels [133]. The second configuration utilizes the thickness dependence of a birefringent LC cell, so that with the same voltage range used, each subpixel will activate a different LC subpixel that will have a different color (RGB) depending on its thickness [134]. The second principal scheme also uses space multiplexing but uses different spatial regions on the active array for the RGB data. Thus for an $N \times N$ array, physically separated red, green, and blue $N \times N$ regions are formed on the active array. With the use of RGB filters and optical elements the three-color images are superposed to result in a full-color image [135]. The third principal scheme is based on the use of three separate devices to produce red, green, and blue images. Finally, it is possible, in principle, to use a single device with *sequential color* addressing. However, due to the complex readout scheme required and the slow response of present liquid crystal materials, this concept has not been implemented so far.

Resolution is defined in either the number of cycles or line pairs per millimeter or in terms of TV lines per millimeter, which is double the figure expressed in line pairs per millimeter. Resolution is usually stated at a certain modulation transfer level. The modulation transfer function (MTF) is defined as the dependence of the spatially (sine-) modulated intensity on the spatial frequency where the transfer ratio at zero frequency (uniform illumination) is defined as 1.0. The resolution is often stated at 50% MTF and as a "limiting resolution," which is typically at 5–10% MTF (see, e.g., Table 8.2 for the CdS LCLV).

Speed of Response. We have dealt quite extensively with this parameter in the context of LC and LC device performance. On the (display) system level we usually speak about frame time. Video frame rates are typically 30–60 Hz depending on the particular application and whether or not one deals with field interlacing. The other aspect of speed of response is the human response time. "Typical" human response is on the order of 100 ms. One usually needs frame rates greater than 20 Hz to avoid the sensation of "flickering."

Display Types and Applications. The main display applications are given in Table 8.6.

Current (Non-LCD) Display Technologies. *Gas discharge* (or *plasma*) *displays* are one of the older display technologies using excitation of neon–argon gases under high voltage. Relatively large display segments, as in "neon signs," are produced in this technology. Small plasma displays were and still are also used in small ($\approx$1-mm) segments in seven-segment displays (e.g., calculators).

Electroluminescent (EL) displays are based on the luminescent properties of some solids upon the application of an electric field. The brightness is usually low, and colors are limited to red and yellow.

Light-emitting diodes (LEDs) are based on direct recombination of electron–hole pairs in certain semiconductors (e.g., GaAs). Relatively small "pixels" ($\approx$100 μm) can be produced in this way. Color is again limited to red-orange.

Cathode ray tubes are the work horses of the video display industry and are based on a scanned electron beam that excites phosphor screens. They feature relatively high resolution and medium brightness (acceptable in subdued conditions). Power consumption, high voltage, and size (depth) are the main drawbacks of this technology.

Light Valves. Two main types of projection systems based on non-LC technology are available today. The *thin-film* (eidephor-) based system manufactured by General Electric is based on the phase modulation of an optical beam by a recycling thin fluid film whose refractive index is spatially modulated by the deposition of charge with an electron gun. A sophisticated method of modulating the *e*-beam information coupled with a schlieren/spatial filtering system provides full-color video imagery. A *KDP*-(potassium–dihydrogen–phosphate) based spatial light modulator is produced by Sodern (France) that is a newer version of the "Titus" light valve in which the refractive index of an electro-optical KDP crystal is spatially modulated, again with the use of an electron gun. The spatial birefringence produced is converted, similar to the LCLV case, by the use of crossed analyzer–polarizer pairs.

Finally, one should mention two types of light valves or spatial light modulators (SLM) [136]. The electrically driven *magneto-optical SLM* manufactured by Semetex is currently produced with a 256 × 256 array size. The low transmission (brightness) makes it of little use for display applications. The *deformable membrane SLM* produced by Texas Instruments with a current array size of 128 × 128 uses micromembranes to spatially deflect the optical beam which, in conjunction with a schlieren readout system, is capable of producing an intensity image. Relatively low output efficiency and limited gray scale

TABLE 8.6 Main Display Types and Applications

Type	Applications	Array size (Typical)	Physical Dimension (cm)	Binary/Gray Scale	Color	Current Technology
Alphanumeric						
Miniature	Digital clocks, calculators	10 × 10	1–10	Binary	No	LCD, ELs, LEDs
Instrument monitors	Dashboard, avionics	100 × 100	10–30	Binary	Desirable	CRT, LCD, plasma, LEDs
Video						
Pocket TV	Personal TV,	500 × 500	5–10	Gray	Yes	LCD
TV monitor	Home TV, instrument display	500 × 500[a]	30–150	Gray	Yes	CRT, LCD
Large screens, subdued illumination	Overhead projectors/ teleconferencing	500 × 500[a]	1.5–3 m	Gray	Yes	Projection CRTs, LCLV projectors
	Command and control	1000 × 1000	3–10 m	Gray	Yes	Thin-film (GE) projectors, LCLV
	Flight simulators	≥1000 × 1000	3–20 m	Gray	Yes	Thin-film (GE) projectors, LCLV, lasers
	Movie theaters	5000 × 5000	10–30 m	Gray	Yes	Arc lamp–film projection systems
	Sports stadiums	200 × 200	3–10 m	Gray desirable	Desirable	Electric bulbs, discharge lamps
Large area	Highway programmable signs	100 × 5	Meters	Binary	No	Electric bulbs, discharge
Large area, transmission control	"Window shades," buildings, cars	—	0.5–10 m	Binary	No	None. potential: PDLC, electrochromics

[a] Expected to scale up to ~1000 × 1000 with introduction of HDTV.

have been its main drawbacks for display applications. Recent developments of a "dark field" mode of operation may allow this device to be used for displays. Information about those and other non-LC SLMs can be found in recent publications on the subject [137,138].

LCLV-based Large-Screen Projection Systems. Having acquainted ourselves with the basics and main technologies of displays, we can now return and revisit the use and applications of LCs in displays. Before embarking on an overview of the role LCs played in this application field, we need to acquaint ourselves with one of the most important system applications of LCs in displays, namely, large-screen projection displays. The main "workhorse" of this technology is the LCLV, which was described in Section 8.32. As pointed out earlier, the photoactivated LCLV was one of the first LC devices to be used for image and video displays. A monochrome projection system employing a CdS-based LCLV is shown in Fig. 8.61. A thin-film Mac-Neille polarizing beam splitter is used in the prepolarization of the beam followed by an analyzèr operation. Following Ledebuhr [139], a full-color projection display system using three (red, green, and blue) LCLVs is shown in Fig. 8.62. A high-efficiency arc lamp reflector is used to focus the light output from the arc lamp to a 1.4-in. square aperture. This light is then relayed to the LCLVs via a magnification telecentric relay lens system. After passing through the first relay lens, the light encounters a broadband polarizing prism that is used to pre-polarize the light. The *s*-polarized light is rejected from the system at this point and the *p*-polarized light is transmitted. A second identical polarizing prism oriented at 90° with respect to the first prism sees this incoming light as *s*-polarized and therefore reflects it toward the light valves. This arrangement of two crossed polarizing prisms increases the contrast ratio of the system and is especially useful in providing a very dark off-state. The *s*-polarized light coming from the second polarizer is then directed through a second relay lens element and encounters the 45° blue-reflecting dichroic filter. This blue filter is on the front surface of the substrate so the light undergoes this reflection without encountering any glass. The green and red wavelengths are transmitted by this filter and next encounter a 45° red reflecting dichroic filter oriented in the same plane as the blue filter but at 90° from it. This filter reflects the red and transmits the green wavelengths.

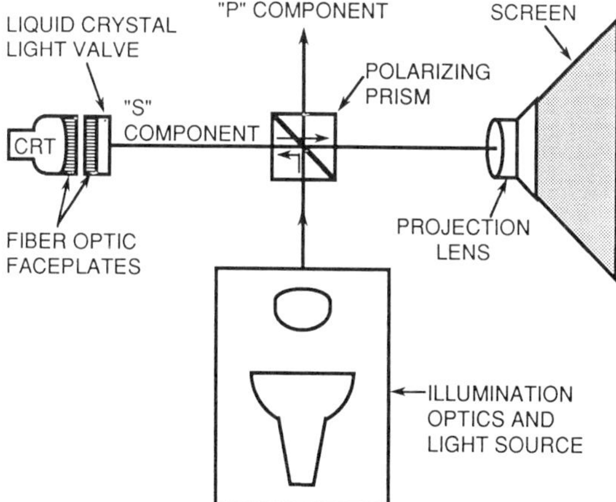

Figure 8.61 Basic reflective LCLV monochrome projector design. (After Ref. 139.)

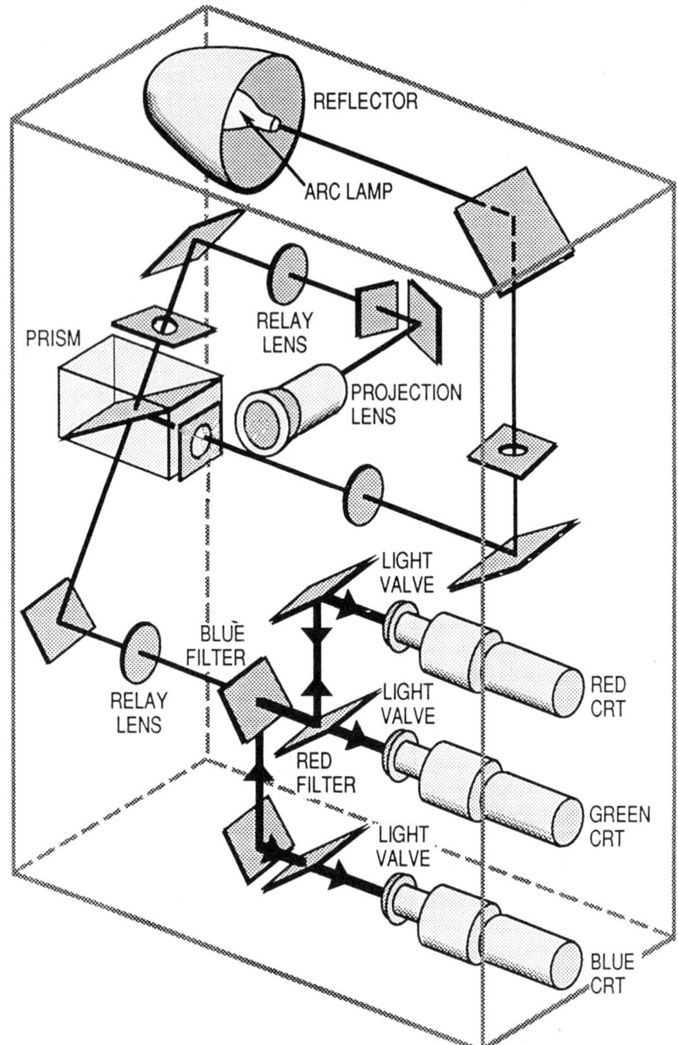

Figure 8.62 Three-color LCLV projection system. (After Ref. 139.)

The green light is then transmitted directly to the green light valve without undergoing any further deviation. The incoming red and blue wavelengths are each reflected by second-surface broadband reflectors so the optical axes of these channels are thus made parallel to the green channel optic axis. These second-surface mirrors are half the thickness of the blue and red reflecting filters so that reflection from these mirrors is equivalent optically to transmission through the thicker dichroic filters. An additional compensating plate is placed in the blue channel so that this light encounters the same amount of glass (with the same orientation) as in the green and red channels. The incoming *s*-polarized light enters a separate LCLV for each primary color channel. Upon reflection by the LCLVs, the light contains a polarization-modulated image that is a replica of the image on the CRTs. This light then retraces its way back through the color filter assembly.

The red and green wavelengths are recombined at the red reflecting filter, and these are then recombined with the blue channel at the blue reflecting filter. The light from the three channels now shares the same common optical axis as it did before polarization modulation. This light passes back through the second illumination relay lens element, which is now analyzed by the second polarizing prism where only *p*-polarized output light is transmitted and *s*-polarized light is reflected back to the arc lamp illumination system. The output now consists of intensity-modulated light, which is directed through a third relay lens element that makes up the second half of the projection relay. The projection relay images the three primary colored light valve images with 1 × magnification to the focal plane of a short back-focus wide-angle projection lens. Table 8.7 summarizes the system's performance.

Recently TFT-based LCDs have also been demonstrated for this application.

An active matrix-(TFT-LCD-) based projection system is shown in Fig. 8.63. The red, green, and blue beams selected by dichroic mirrors read out three TFT LCD devices (transmissive mode); the modulated beams are recombined using two dichroic prisms. Full color at NTSC standards was recently demonstrated [140]. The use of a projection system based on a smectic LC, laser-addressed LCLV was reported by Hitachi [141]. Up to 2000 × 2000 resolution elements on a 2 × 2-m screen with writing speed of 1 min/frame was reported. Finally, the use of the *e*-beam-addressed LCLV previously described [125] in a projection system was reported in 1986 [142]. These types of large-screen projection display systems are already being used in overhead projection and command and control display systems with increased resolution and brightness. It is expected that in a few years this technology will be in a position to invade even the exclusive area of movie projection systems.

Liquid Crystal Displays, Past and Present. While work on LC materials and electro-optical effect was quite actively pursued in Europe in the 1920s and 1930s, actual effort toward practical LCDs did not really start until the 1960s. In these years, the early concepts of dynamic scattering [143] and dichroic dye LCDs [144] were first introduced. The development was enhanced by the introduction of eutectic mixtures that allowed the use of stable LC mixtures over a relatively wide temperature range. Liquid-crystal-based moving advertising displays were introduced by the RCA group in the early 1970s.

TABLE 8.7 Performance Data for Full-Color Projector

Characteristic	Performance
Color range	Full color
Light output (1.0-kW lamp)	>1000 lm (open gate), >650 lm (square aperture)
Light output (1.6-kW lamp)	>1600 lm (open gate), >1000 lm (square aperture)
Contrast ratio (white)	>30:1
Screen size	1–5 m square
Throw distance	1–5 m
Frame rate	30 Hz (interlaced)
Raster format	1075, 625, 525
Vertical resolution	>1024 scan lines
Horizontal resolution	>1400 TVLines limiting (1800 TVLines limiting with upgrade)
Image registration of a 1024 × 1024-pixel display	0.25 pixel in white for all pixels

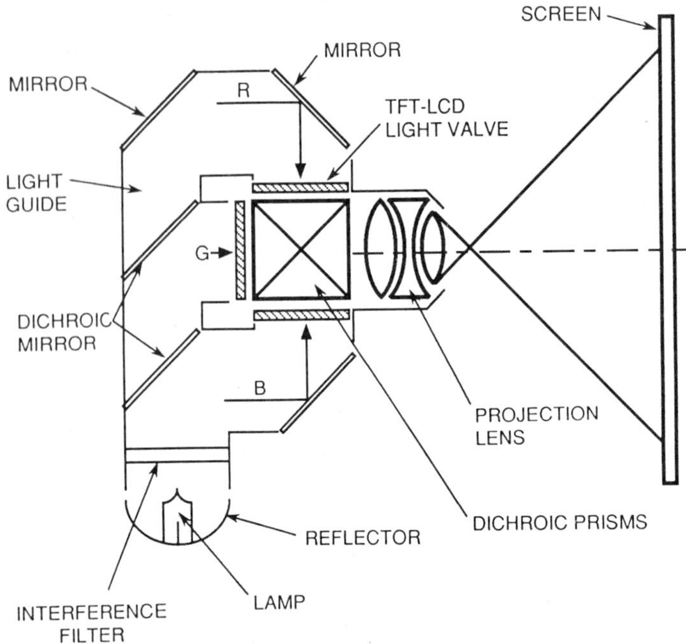

Figure 8.63 Three-color projection system based on three TFT LCD devices. (After T. Uchida et al., *Mol. Cryst. Liq. Cryst.*, Vol. 165, p. 533, 1988.)

These years also saw the first prototypes of LCLVs developed at Xerox [145] and at Hughes [146].

In these years, while reasonably well developed LC materials already existed, the problem was that of driving large arrays. Thus only simple alphanumeric displays (calculators, watches) or LCLV projection systems were under development; as both of these technologies avoided the need for a high-density, large-array-size, active matrix driver. The development of active matrix CdSe TFT arrays then started in the early 1970s [147]. This was followed by the development of MOS-based active matrix devices in the mid-'70s [148]. While these developments were being made in the driving circuitry, a significant breakthrough was made in LC development, namely the invention of the twisted nematic effect by Fergason [149] and by Schadt and Helfrich [150]. A later important invention in the LC domain, which opened the possibility of the use of reflective-mode devices, was that of the hybrid field effect mode by Grinberg et al. [151]. The development in integrated circuits, amorphous silicon materials, and the twisted nematic LC mode resulted in the first successful appearance of an active matrix LCD TV in the early 1980s [152]. Presently we are witnessing the production of full-color LCD TVs with sizes approaching the standard home TV (see the section on active matrix displays). Contrast and color display continue to improve with the latest development in supertwist (STN), superbirefringence (SBE), and the OMI effects described in Section 8.2.6. The LCLV projection systems are taking a strong lead in this type of large-screen projection applications, while small LC displays are slowly penetrating the automotive dashboard display area.

High-Definition Television and the Future of LCDs. The last paragraph was already leading into what appears to be the future: LCDs replacing most of the display applications as we know them today, from alphanumeric miniature displays, calculators, wrist watches, and digital thermometers to medium-size alphanumeric displays such as lap-top calculators, dashboard, and avionic monitors to video displays of various kinds—pocket TV, home flat-panel TV, overhead video projection systems, large-screen projection systems for command and control, flight simulators, and teleconferencing, possibly invading even the movie theater arena. Polymer-dispersed liquid crystals (PDLCs) or nematic curvilinear aligned polymers (NCAPs) will most likely provide an efficient solar luminance control in the form of "electric window shades" in cars and buildings. One particular development in the display arena that appears to be in a rapid state of development is that of HDTV. Japanese and European companies are currently leading this technology of 1125-TVL standard, which will produce approximately five times the number of picture elements of existing 525-TVL standards. The major Japanese companies, including Panasonic, Sanyo, Sharp, Sony, and Toshiba, have already demonstrated the "Hi-Vision" TV. Japan's National Broadcasting Services Network is currently experimenting with HDTV standard broadcasts. There is still a debate in the United States as to the real market behind this concept. While proponents compare it to the successful introduction of Hi-Fi and digital audio systems, opponents compare it to the miserable failure of the quadrophonic systems. Video studios in Japan, Europe, and the United States are experimenting with the production of HDTV standard shows and programs. Equipment requirements are certainly quite demanding, with video signal bandwidth up to 30 MHz from the current 4.2-MHz requirements. Some market surveys predict that between 1995 and the year 2000 HDTV set sales will take the lead over the conventional TV sets of sizes 25-in. and above. The reason I believe that this revolution will indeed happen is because the entire civilization is already in the midst of a formidable change in communication technology and has been since the advent of the telegraph and telephone. High-bandwidth communication allows exchange of information and decisions to be made across the globe (e.g., via teleconferencing) without the need of physical travel. In this era of increasing concern about the future of our planet due to man-made pollution by transportation vehicles, there is clearly something to be said for the reduction of travel needs using high-bandwidth communication as a substitute.

8.4.2 Optical Data Processing

General Concept of Optical Processing. Optical data processing is the title of the general field of data processing and computation using optical means. This technology is based on the fact that a number of mathematical operations and transformations can be performed using optical beams as the carriers of the information to be processed. There are several significant advantages in the use of optical computation methods. First, the natural carrier of information, namely, the optical beams, and the operations involved are one or two-dimensional. This implies a potential advantage in throughput as compared to the serially operated digital computer. Second, optical beams are guided in free space and hence do not require point-to-point conductors such as are needed by their electronic counterparts. Optical beams are also immune to electromagnetic interference and therefore provide for a more rugged system. Third, optical images often represent the raw data to be processed. The direct optical processing of imagery is obviously a very efficient method of data processing compared to the alternative of converting the imagery into electronic information and back to optical image to be relayed to a human operator.

The main issues in the application of optical processing are:

1. The existence of powerful and ever-growing digital electronic computations. This means that in order to justify this new technology, one must show at least two orders of magnitude of improved performance (say in throughput).

2. A related issue in the comparative performances of the optical versus the digital electronic systems is that of accuracy. In optical processing systems, one can typically expect 1–0.1% of accuracy (corresponding to ~7–9 bits) as compared to the digital electronic world in which 32 bits of accuracy are quite common practice. There is a general consensus that in some applications that do not require large dynamic range or accuracy but require a very high throughput of computation, an optical computing module performing a specific operation may provide a competitive solution.

One is still left with two important issues:

3. Efficient interface between the digital host computer and the optical "coprocessor" and
4. A spatial light modulator device that can accept electric or optical data and convert it to the spatial optical pattern required by the optical processor.

It is this last issue that is perhaps the most difficult one in terms of cost and practical development, which has been the main stumbling block in the development of this fascinating new technology. It is also this very issue in which LCs become involved.

The field of optical information processing can be divided into the following areas:

Image processing
Signal processing
Optical computing
Optical implementation of neural networks and optical interconnects

These are not exclusive classes in the sense that there are some overlap among the various categories. Thus one usually refers to signal processing as the handling of a temporal, sequentially transmitted signal. However, this signal may in fact contain image information. Likewise, optical computing may contain two-dimensional data arrays, the processing of which may be quite similar to the processing of images. Optical implementation of neural networks is really a subclass of optical computing with the particular case of artificial neural network techniques. Finally, the area of optical interconnects is only loosely related to optical processing and is often considered a separate topic. In the following I shall detail examples for each of the preceding categories, emphasizing the applications of LC spatial light modulators.

Image Processing. As pointed out earlier, the strength of optical processing is in the inherent parallelism of optics—allowing multiple channels of data to be handled in parallel. One of the most important sets of multiple-channel data is two-dimensional images. Image processing is therefore the area where optical processing has the most to offer. One of the most striking examples of the strength of this technology is in the capability of a simple lens to perform a two-dimensional Fourier transform on an optical beam [153]. This is done naturally, with the only precondition being that the beam must be coherent. This is the basis for the very powerful operation of image correlation, shown in Fig. 8.64. The purpose of the system is to determine the presence of a specific pattern in the field of view of the imaging system. This field of view, $A(x, y)$, is imaged onto a photoactivated SLM that converts this incoherent image into a coherent replica. The Fourier-transformed image is generated by the lens at its back focal plane. An electrically addressed SLM at this plane is programmed to present the Fourier transform of the

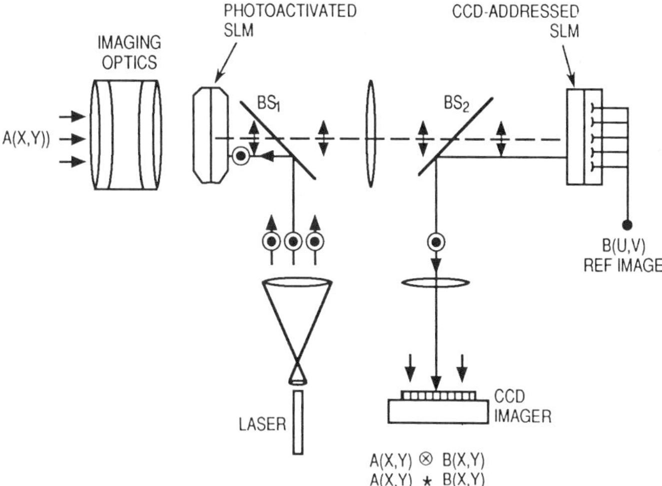

Figure 8.64 Real-time optical correlation system. The photoactivated LCLV converts the (visible) real-time scene into a coherent replica. The Fourier-transformed version (following the first lens) is multiplied by a Fourier-transformed pattern of the reference object. Inverse transformation of the product with the third lens results in the desired correlation.

complex conjugate of the *object*, $B^*(x, y)$, to be detected in the field of view.† As described previously, an electrically addressed SLM such as the (reflective-mode) CCD LCLV will spatially modulate the incoming beam by an effective spatial modulation of its reflectivity via polarization–rotation. The incoming beam in this case, which is already modulated as

$$\iint e^{i(k_x x + k_y y)} A(x, y)\, dx\, dy = F_{\text{qv}}\{A(x, y)\}$$

due the first lens, now gets an additional modulation of $F_{\text{qv}}\{B^*(x, y)\}$. The overall spatial amplitude distribution of this beam following the reflection from the CCD SLM is therefore

$$F_{\text{qv}}\{A(x, y)\} F_{\text{qv}}\{B^*(x, y)\} \tag{8.51}$$

We also note that while the initial coherent beam propagated with its polarization in the plate of incidence (p-wave), the polarization plane of the modulated part of the beam reflected from the CCD SLM will be rotated by 90°. The polarizing beam splitter BS_2 will therefore reflect this portion of the beam into the second lens. This second lens will perform effectively an inverse Fourier transform on the modulated beam. The Fourier transform of a product of two functions is the convolution of the two [154]. It will also produce the *correlation* of the two functions in the case where one of the product terms is the complex conjugate of the original function—as in this example. A more elegant way of producing the cross-correlation of two images, namely, a joint-transform correlator, is shown in Fig. 8.65. In this case the two images A and B are input to two electrically (CCD-) addressed SLMs or to the dual inputs of a single CCD LCLV.

† In practice, the SLM will be used to present *either* the amplitude *or* the phase of the Fourier-transformed image.

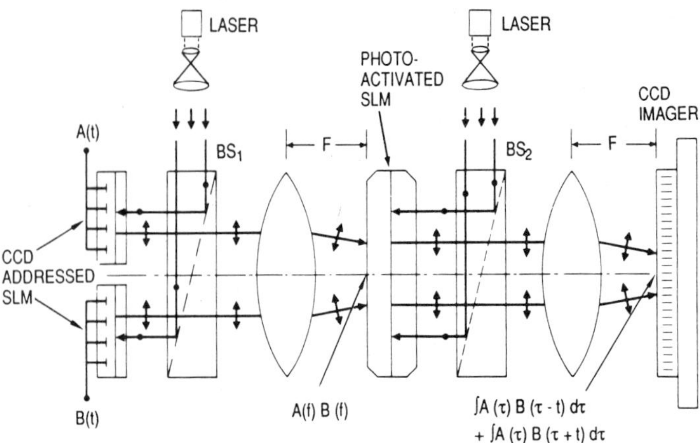

Figure 8.65 Joint transform correlation system based on CCD-addressed LCLVs and a photoactivated LC SLM.

In this case [155] the amplitudes $A(x, y)$ and $B(x, y)$ of the two images are superposed and undergo a Fourier transformation by the first lens. The intensity incident on the photoconductor part of the photoactivated SLM is therefore proportional to $F\{A\}^2 + F\{B\}^2 + 2F\{A\}F\{B\}$. The complex-conjugate terms of this product will also be present. Due to the angular separation of the two images with respect to the optical axis of the systems, the separation of the conjugate terms from the original one is possible using spatial filtering techniques. This will, however, result in some loss of the total available space–bandwidth (resolution) product. Similar to the former example, the second SLM will perform a second, inverse Fourier transform on the product, resulting (among other terms) in the correlation of A and B. In this case, the requirement of pretransforming one of the images is therefore eliminated. The correlation operation performed by either of the preceding optical systems is a critically important function of pattern recognition systems. This particular operation would require $\approx(N \log N)^2$ multiplication operations, where N is the *linear* dimension of the two-dimensional array. Thus, using a photoactivated LCLV with ≈ 1 ms response, which was recently demonstrated [156], the use of two 512×512 array CCD LCLVs for the input of the two arrays $A(x, y)$ and $B(x, y)$ would result in an effective throughput of

$$P = \frac{N_{\text{tot}}}{\tau} \approx \frac{(N \log N)^2}{\tau} \approx 10^{10} \text{ multiplications/s}$$

a result that is beyond the reach of present digital processors.

The use of ferroelectric LC combined with suitable high-speed CCD substrates could result in a further increase of 2 orders of magnitude in the attainable throughput of such correlators.

While image correlation is perhaps the most powerful operation achievable with the use of optical processing techniques, numerous other image-processing operations are also achievable. *Edge enhancement* can be performed very naturally using coherent optical processing by subjecting the (coherent) image to a Fourier transform operation using a lens. A properly designed mask (such as a central disk shape) placed in the Fourier plane will block the transmission of the low-spatial-frequency components, leaving only the higher-frequency terms. Such a Fourier plane filter will effectively act as an edge en-

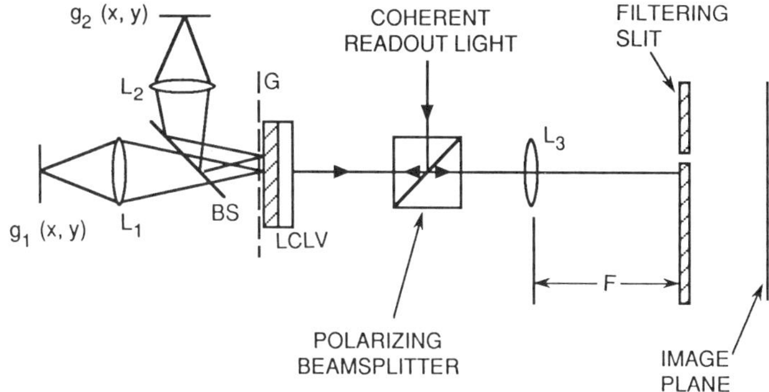

Figure 8.66 Image subtraction system using a LCLV. (After Ref. 157.)

hancement since the edges contain the high spatial frequency terms of an image spectrum. In a similar way, the introduction of an on-axis pinhole in the Fourier plane will enhance the low frequency of the "dc term" of an image.

Image subtraction can be achieved as shown by Marom [157]. In this case (Fig. 8.66) a grid G was placed adjacent to the photoconductor part of a LCLV, resulting in the encoding of a carrier frequency on both images $g_1(x, y)$ and $g_2(x, y)$. These images are superposed with a relative displacement of one-half of the grid period. This displacement results in an effective 180° phase shift at the Fourier plane, following the spatial filtering of the carrier frequency with the subsequent appearance of the difference image at the output.

A general nonlinear operation can effectively be accomplished with the use of half-tone screens [158]. The removal of a multiplicative noise from an image was demonstrated with this combination of a half-tone screen and a VGM LCLV [159]. A powerful example of the previously described VGM for image processing is in the level slicing operation [160]. As discussed in Section 8.2.6, this device can effectively be used in conjunction with a photoactivating substrate to perform local intensity-to-spatial-frequency conversion by the formation of a local phase grating whose spatial frequency is linearly proportional to the local intensity. The concept [160] is shown in Fig. 8.67. In this configuration, a phase grating pattern is formed in the LC whose spatial distribution of

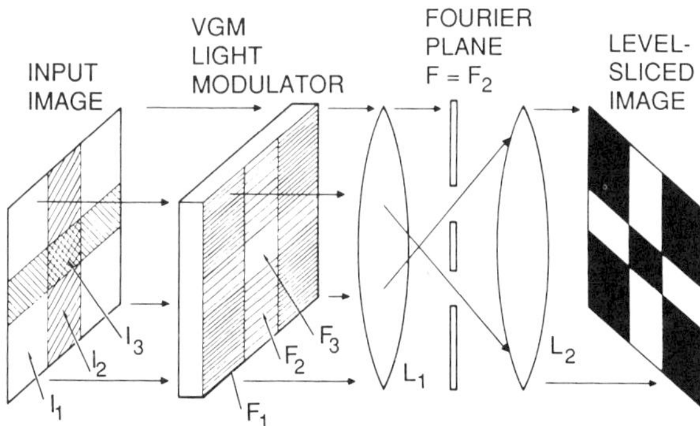

Figure 8.67 Intensity level slicing using a VGM LCLV. (After Refs. 160 and 161.)

frequencies is proportional to the intensity of the input image. The output image, therefore, appears as a spatial distribution of frequencies. A suitable spatial frequency filtering in the Fourier plane will result in the selection of certain spatial frequencies corresponding to certain intensity levels in the input image. The figure shows the selection of I_2 input level.

Image division can also be performed using the forementioned half-tone screen concept [161].

A bispectral imaging–image division system based on this principle is shown in Fig. 8.68. The purpose of this system is to obtain the (logarithmic) image of the intensity ratio of the scene at two wavelengths in the 400–1100-nm spectral range. This operation results in the enhancement of specific textures in the scene. Thus, it has applications in texture recognition, such as the remote-earth-features identification system under development by NASA [162]. The operation is as follows: The scene imaged by the input optics is split into two channels that are each wavelength filtered in the two spectral regions (λ_1, λ_2) required (400 nm $< \lambda_1, \lambda_2 <$ 1100 nm). The filtered images are then spatially modulated by logarithmic half-tone screens with different spatial frequency for each channel, $\lambda_1 - F_1$ and $\lambda_2 - F_2$. A variable attenuation compensator placed at one of the channels acts to compensate for intensity imbalance between the two channels. The two images, each modulated by a different spatial carrier, are then recombined at the input to the silicon LCLV. Thus, each of the two images at the two different wavelengths are "tagged" with a different spatial frequency modulation. The photoactivated silicon LCLV acts as a sensitive, broadband, incoherent-to-coherent image converter. A spatial Fourier transform is then performed on the data read out by the laser beam. The diffractions of the two wavelength images will now appear separately in the Fourier plane, due to the different spatial carriers for each of those images. Spatial filters corresponding to each of the two half-tone screens are placed at the appropriate locations in the Fourier plane. This results in the formation of logarithmic intensity images following a retransforming lens. A 180° phase retardation plate placed at one of the filter locations will result in one of the logarithmic images (λ_2) having a reversed phase with respect to the other. Thus, the amplitude of this image formed at the video detector plane will be proportional to

$$A_{\text{out}} = A_1(x, y) + A_2(x, y) \propto \log I_1(x, y) - \log I_2(x, y) = \log \frac{I_1}{I_2}$$

where $I_1(x, y)$ and $I_2(x, y)$ are the intensities of the input images at λ_1 and λ_2, respectively. The image amplitude following reconstruction at the vidicon input will be proportional to $\log[I(\lambda_1)/I(\lambda_2)]$, that is, to the (logarithmic) ratio of the images at λ_1 and λ_2. Due to the high sensitivity of the silicon photoconductor in the silicon light valve configuration (about 50 $\mu W/cm^2$), the imaging system is expected to have sufficient sensitivity for direct imaging of sun-illuminated scenes.

The *spectral range* of the bispectral imaging–image division system is limited in the case of the silicon LCLV to 400–1100 nm. The use of other types of LCLVs with other photoconductors will allow shifting of the operational spectrum to other regions.

The *dynamic range* of this system is limited by that of the Silicon LCLV, which is typically 100:1. An important advantage of this optical processing system is that the output ratio is presented by a coherent light. This enables a straightforward use of optical postprocessing (e.g., ratio image correlation).

The *spatial resolution* of this system depends on the spatial frequencies employed as well as on the silicon LCLV performance. Taking $F_0 = 25$ cycles/mm at 30% modulation as the current performance of the silicon LCLV and using the two carrier frequencies

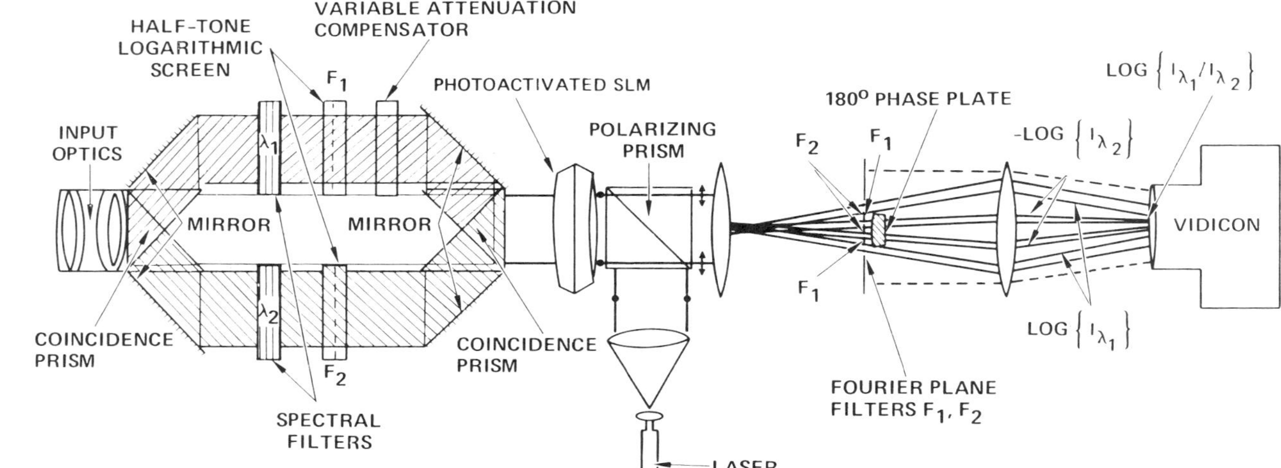

Figure 8.68 Bispectral image division system based on a photoactivated Si LCLV. (After Ref. 163.)

as $F_0/4$ and $3F_0/4$, over 500 pixels of resolution are available using the 43-mm aperture device, with $\Delta F = F_0/2$.

Finally, the WGLV described earlier can be used to perform a variety of image processing functions. The use of the device for adaptive image thresholding was recently described [115].

Optical Signal Processing

Spectrum Analysis: A real-time RF spectrum analyzing system demonstrated several years ago is shown in Fig. 8.69 [163]. The RF signals were amplified and displayed in raster form on a CRT. The signals were obviously asynchronous with the raster scan at a frame time of 7×10^{-2} s. The incoherent optical display was focused on the photoconductive input of a CdS LCLV that acted as a coherent-to-incoherent transformer as the output of the LCLV was illuminated with a coherent He–Ne laser. This transformation permitted an optical Fourier transformation to be performed. It can be easily shown that the Fourier transform of a raster pattern in time is a raster pattern in frequency, as shown in Fig. 8.69. Low-frequency, Morse-coded, tone-modulated RF signals from oil field transmitters displayed a simple textbook spectral pattern of a carrier and two pulsating sidebands. The theoretical resolution is given by the frame time, which was about 14 Hz. Because of the fall-off in resolution of the LCLV and associated optics, the resolution achieved was lower (80 Hz). An obvious improvement of this system is the replacement of the CRT imaging lens with a CCD-addressed LCLV.

In this case the ultimate, 1000-array CCD LCLV would provide 10^6 point resolution over 100 MHz bandwidth at (real-time) frame rates of 100 Hz. The demonstration of a direct spectrum analysis using a CCD-addressed LCLV was recently carried out with the Hughes 256×256 array device [164]. Both amplitude and frequency modulation signals were used with an input bandwidth of 1 MHz. A resolution of up to 100 Hz was demonstrated. The direct image of the signal modulation of the CCD LCLV as well as its Fourier transform pattern are shown in Fig. 8.70.

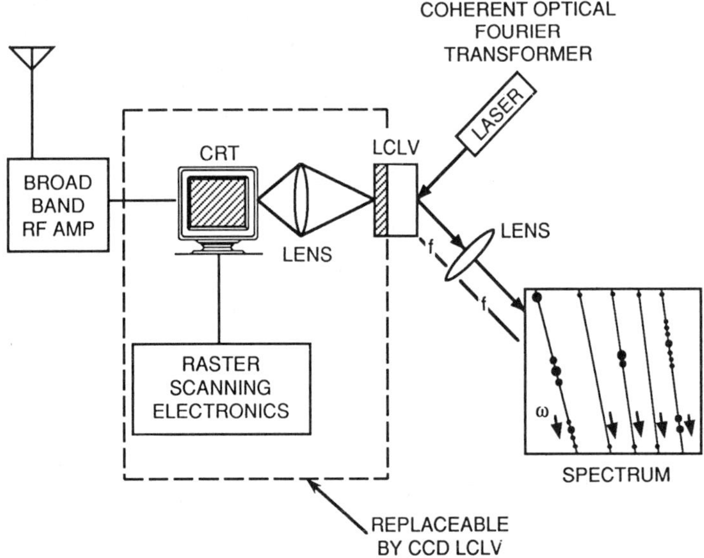

Figure 8.69 An RF spectrum analysis system based on a photoactivated LCLV. (After Ref. 163.)

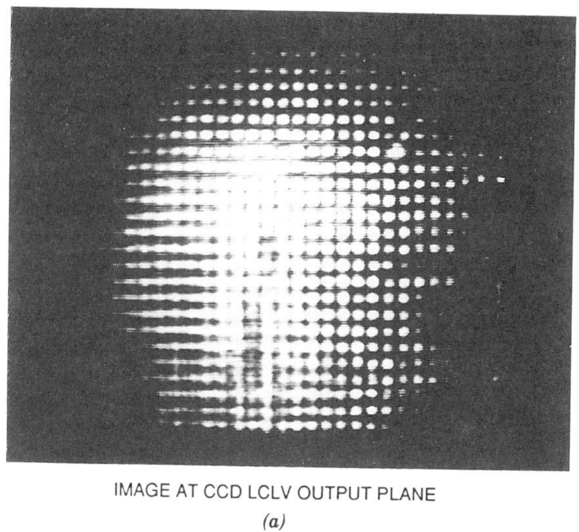

IMAGE AT CCD LCLV OUTPUT PLANE

(a)

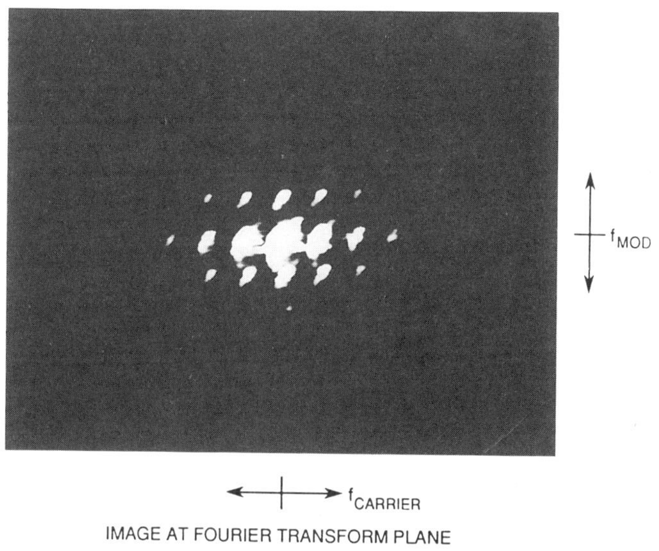

IMAGE AT FOURIER TRANSFORM PLANE

(b)

Figure 8.70 An RF spectrum analysis using a CCD-addressed LCLV: (*a*) direct image of the CCD LCLV; (*b*) Fourier-transformed spectrum showing a carrier (center dots) and sideband frequencies. (After Ref. 164.)

Another useful application is real-time spectrum analysis of a *given scene*. A silicon-light-valve-based system that can perform this operation is shown in Fig. 8.71. The operation of this system is described next.

The radiation from the scene to be analyzed, $I(W)$, is split by the beam splitter in a Michelson interferometer configuration. Two mirrors, a standard one and a staircase one, are used. The interference pattern at the output of the interferometer (i.e., at the input to the LCLV) is the (spatial) Fourier transform of the input spectrum. This is analogous to a conventional Fourier transform spectrometer (FTS) [165], in that each

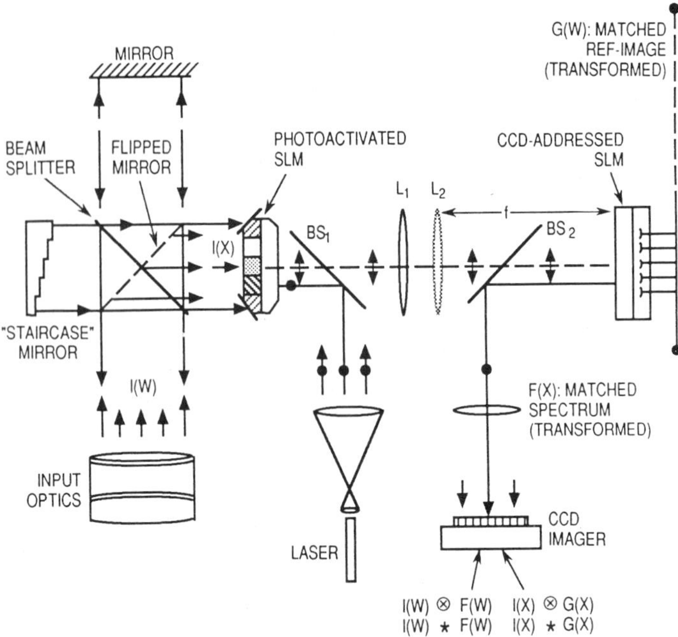

Figure 8.71 Spectral/scene correlation system based on a photoactivated LCLV. Solid curve indicates the positions of the optical elements in the use of this system for spectral correlation. Dashed lines indicate position of elements for scene correlation. (After Ref. 163.)

of the staircase steps represents one mirror location in a moving-mirror spectrometer. The subsequent spatial Fourier transform of the output of the light valve results in the spectral analysis of the input beam at the imaging array. The readout laser beam is spatially modulated by the Fourier-transformed reference spectrum using the CCD light valve. This modulated beam is then used as a readout light for the photoactivated light valve. At the input of the photoactivated light valve, the spatial interferogram of the input beam is present. The emerging output beam consists of a product of the input and the reference, Fourier-transformed spectra presented by the CCD LCLV. The subsequent inverse Fourier transformation carried out by the lens results in the appearance of correlation and convolution terms of the two spectra at the imaging array. This system, which is based on the FTS principle, benefits from two important advantages of the FTS system: the multiplexing, or Felgett, advantage in signal-to-noise ratio and the throughput, or Jaquinot, advantage [165].

An attractive feature of this system is that it can be used for pattern recognition purposes with a flip of a mirror. In this way, shown by the dashed elements in Fig. 8.71, the pattern of the incoming beam, rather than its spectral content, can be analyzed and correlated with a suitable reference image presented by the CCD light valve, as in Fig. 8.64. The system can thus perform both spectral and *pattern* correlations of the scene.

The *spectral range* of this system is limited by the photoactivated light valve, since it must be sensitive in the spectral range used. The existing silicon light valve enables us to use the 400–1160 nm range. Since the detection of longer wavelengths may require cooling of the narrow-band-gap photoconductor light valve, the LC will be the limiting component for such a longer-wavelength light modulator.

The *spectral resolution* largely depends on the manufacturing of the staircase mirror. One could achieve more than 10,000 elements of resolution. It should be pointed out that for the photoactivated and CCD-addressed light valves, a resolution on the order of 10^6 elements is possible.

One obvious limitation for the preceding application is the intensity of the input beam, or the radiation level, from the scene analyzed. Using the silicon light valve, a rough estimate for the input illumination level required is 100 $\mu W/cm^2$ in the visible. Projected performance of such a correlator for two spectral regions is presented in Table 8.8. Finally, it should be pointed out that other, possibly more efficient methods of self-interference of the incoming analyzed beam have been previously suggested [166].

A particularly important type of signal processing where the CCD LCLV may be used is radar signal processing. This field encompasses ambiguity-function generation and synthetic aperture radar (SAR) processing.

An ambiguity-function generation system using two LCLVs was previously described [167]. The replacement of the photoactivated LCLV by a CCD-addressed LCLV will significantly improve the system, eliminating the CRT and the acousto-optical units required. The feasibility of using a CCD-addressed LCLV for SAR signal processing was established by a detailed analysis performed in 1982 [168].

Optical Computing. As pointed out earlier, optical computing is really not a subject on its own within optical information processing, since computations are carried out in optical image and signal processing as well. This section will be limited to numerical operations that can be performed on arrayed data. Such operations include multiplication, division, logic operations, and several other types of nonlinear operations.

Analog multiplication is done very naturally with optics. If an optical beam carrying the spatial information of an array of data is made to pass through a transparency, the spatial intensity modulation of the outgoing beam will be proportional to the product of the two intensities. This statement can be extended from the real-to the complex-number domain by modulating the spatial complex *amplitude* of a beam with a complex mask (or spatial light modulator) with a spatial distribution of its (complex) transmission.

TABLE 8.8 Projected Specifications of an LCLV-based Fourier Transform Spectrophotometer/Correlator

Visible Range: 400 < λ < 1100 nm (Si LCLV)	
Bandwidth	$\Delta f = 16{,}700\ cm^{-1}$
Number of resolution elements	$N = 100 \times 100$
Spectral resolution	$\delta f = 1.67\ cm^{-1}$
Maximum "stroke"	$\delta D_{max} \cong 1/\delta f = 0.6$ cm
"Rough" steps	$\delta D_X = 0.6\ cm/100 = 60\ \mu m$
"Fine" steps	$\delta D_Y = 60\ \mu m/100 = 0.6\ \mu m$
1.5–4.5 μm Region Photoconductor with $E_{g,eff} = 0.27$ eV	
Bandwidth	$\Delta f = 4440\ cm^{-1}$
Number of resolution elements	$N = 100 \times 100$
Spectral resolution	$\delta f = 0.44\ cm^{-1}$
Maximum "stroke"	$\delta D_{max} = 1/\delta f = 2.27$ cm
"Rough" steps	$\delta D_X = 227\ \mu m$
"Fine" steps	$\delta D_Y = 2.27\ \mu m$

Note: Steps dimension (both cases), $\cong 0.5 \times 0.5$ mm for 50-mm aperture.

Array Division. To carry out *division*, we need a more complicated optical system. In one experiment two LCLVs were used in a feedback configuration [169] (Fig. 8.72). The output of the system was shown to be proportional to

$$I_0 = \frac{I_{RO_1}\alpha_1(I_{in} - I_A)}{1 - r\alpha_1 I_{RO_1}(R_0 + \alpha_2 I_B) + r\alpha_1 I_{RO_1} I_2 \alpha_2} \tag{8.52}$$

where I_{RO_1} is the intensity of the readout beam reflected from LCLV 1 operated in its normal mode; α_1 is the slope of the transfer curve (I_{out} versus I_{in}) of LCLV 1, I_{in} is the input intensity proportional to the two-dimensional array data of the numerator, I_A is the threshold intensity of LCLV 1, r is the feedback fraction of the light returning back to the loop, R_0 is the maximum zero-input reflectivity of LCLV 2, α_2 and I_B are the slope and threshold intensity level of LCLV 2, which is addressed by the two-dimensional data proportional to the denominator array with an intensity I_2. This LCLV, however, operates in the "back-slope" mode; that is, in the tail portion of the transfer curve of a hybrid field-effect mode (Fig. 8.14). In this part of the transfer curve, further increase in the input light intensity results in a *decrease* in the output intensity. With a proper selection of the variables r, α, I_{RO_1}, andα_2, one can satisfy

$$r\alpha_1 I_{RO_1}(R_0 + \alpha_2 I_B) = 1 \tag{8.53}$$

in which case we have

$$I_0 = \frac{(I_{in} - I_A)}{r\alpha_2 I_2} \tag{8.54}$$

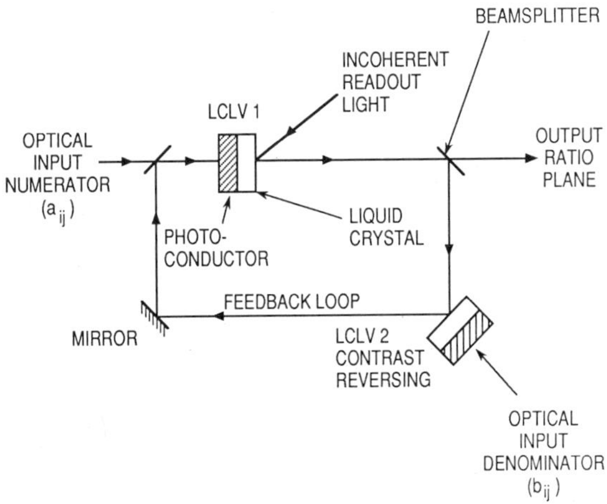

Figure 8.72 Array divisions using two LCLVs in a feedback loop configuration: LCLV1, operated in the normal HFE mode, accepts the numerator array. LCLV2, which operates in the "back-slope" mode, is addressed by the denominator array (see the section on twisted nematic configuration). (After Ref. 169.)

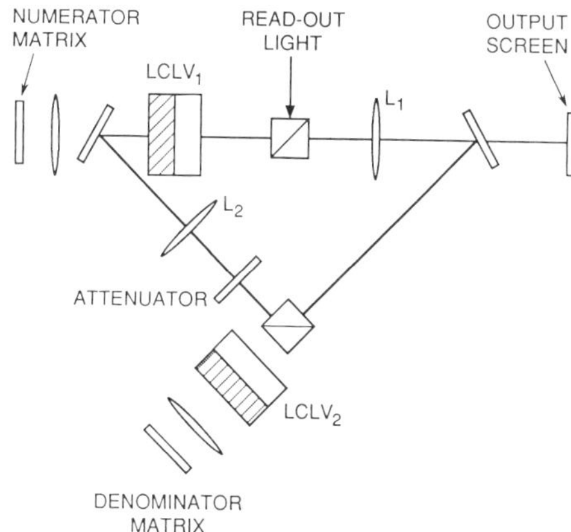

Figure 8.73 Experimental arrangement for array division using LCLVs [169].

In other words, the output intensity is proportional to the ratio of the nominator data (I_{in}) to that of the denominator (I_2).

The optical setup indicated schematically in Fig. 8.73 was used to test the validity of the concepts just described. Lens L_1 images the output of LCLV_1 onto the output of LCLV_2, thus acting as its own readout beam. On the other hand, L_2 images the output of LCLV_2 onto the input of LCLV_1, where the feedback signal is combined with the input pattern I_{in}. A triangular loop, rather than a rectangular one, was used since this was the simplest configuration that preserved the orientation of the fed-back image. Lenses L_1 and L_2 were chosen to provide a system with unity magnification.

An argon laser ($\lambda = 544$ nm) was used as a light source for the readout beam (I_{RO}) as well as for I_{in}. Input I_2 was illuminated by an incandescent microscope source. The condition required by Eq. (8.55) was achieved by balancing the I_{in} and I_f signal levels as displayed on the photoconductive input layer of LCLV_1.

In the experiment, the input I_{in} (numerator) consisted of two vertical stripes; the intensity of the left-hand stripe was 0.63 that of the right-hand stripe. The second input, I_2 (denominator), consisted of two horizontal stripes, the lower one having only 25% of the intensity of the upper one. The resultant matrix contained four picture elements; the expected intensity levels are indicated in Fig. 8.74*a*. The experimental results are shown in Fig. 8.74*b*.

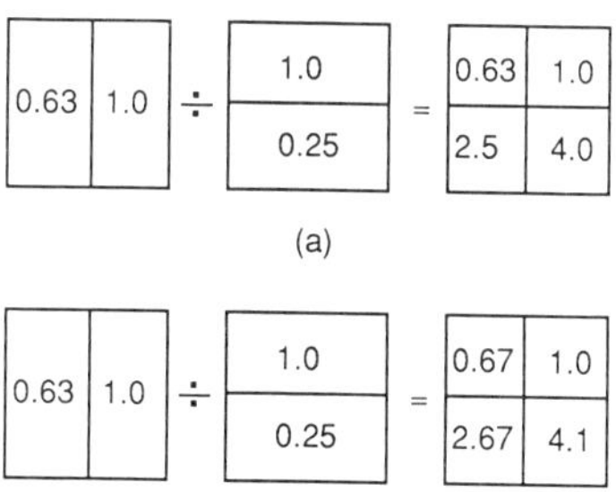

Figure 8.74 Array division using LCLVs: (*a*) theoretical expected intensity levels; (*b*) experimental results [169].

An accuracy of about 5% has been obtained in this experiment. The data were represented here by various intensity levels obtained by illuminating varying density-template arrays. In a dynamic system the data could be introduced electronically through a charge-coupled LCLV device or through a CRT display coupled to an LCLV.

Logic Operations. The pioneering work in the area of logic gates using LC devices was carried out by Collins and his group at Ohio State University. A schematic illustration of a Boolean logic implementation using a LC light valve is shown in Fig. 8.75 [170]. The two input beams *P* and *R* (on the left side) are imaged onto the LCLV, each of the input elements having binary levels only 1 and 0, where those intensity levels are defined according to their being above or below the threshold of the LCLV operated in the HFE mode previously discussed. The polarized readout beam reflected from the LCLV is analyzed and reflected back into the LCLV as shown and is finally bounced into the end polarizer. This configuration is capable of providing all the Boolean logic functions of two variables. Thus, as an example, to attain the operation of $P\bar{R}$ (*P* and not *R*), the readout light is polarized perpendicular to the plane of the figure, whereas all of the analyzers are polarized in an in-plane orientation. Thus, if we have a bright spot at the *P* input and another one at *R*, the one at *P* will result in a 90° rotation of the readout beam. This will allow the beam to pass through the first analyzer. However, since *R* is also positive, this readout beam will rotate again by 90° and will be blocked by the final analyzer. Only if *R* was at a zero level would a bright (1) output corresponding to $P = 1$, $R = 0$, and $P\bar{R} = 1$ appear as required.

The use of optical feedback schemes to achieve a true binary (bistable) operation of a LCLV was discussed in Section 8.3.6. Another possibility of attaining a bistable operation in a LC device is the use of ferroelectric LC materials. As previously discussed, the transfer curve of such devices is inherently bistable. A recent implementation of logic operations using LC devices was performed by the group at the University of Colorado. In this case an electrically addressed ferroelectric LC device was used with polarization encoding of the binary states, as shown in Fig. 8.76 [171]. Light from a He–Ne laser is spatially filtered and then collimated by lens L_1. Polarizer P_1 transmits vertically polarized incident light to array *A*. If a pixel in array *A* has a driving waveform applied such that the average optical axis within that pixel is also vertical, then the light is transmitted unrotated and the pixel is considered *nonswitched*. Another pixel in the same array considered *switched* may have a driving waveform applied such that the average optical axis within that pixel is rotated approximately 45°, thus inducing a rotation of the incident

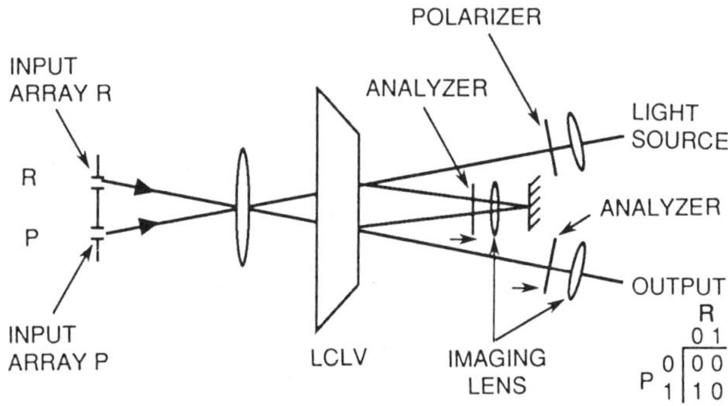

Figure 8.75 Use of the LCLV for logical operations. (After Ref. 170.)

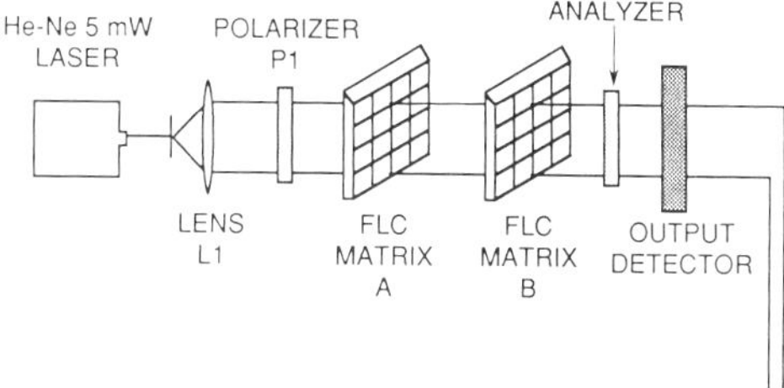

Figure 8.76 Polarization-based optical logic gate implementation using a surface-stabilized ferroelectric LCLV. (After Ref. 171.)

polarization by nearly 90°. With the light now incident on array *B*, a nonswitched pixel transmits the light unrotated, while a switched pixel rotates the polarization of the transmitting light. An analyzer or Wollaston prism can be used as the optical controller at the output to separate the polarization components. A horizontal analyzer yields the XOR output, while a vertical analyzer yields the XNOR output.

An elegant way of implementing optical logic gates is the use of the theta modulation [172]. In this case the encoding of different states is achieved via encoding of grating in various orientations. If those gratings are encoded in two orthogonal directions, subsequent spatial filtering in cascade will allow the AND operation, for example, to be achieved. The VGM LCLV [126, 173] (Section 8.3.5) can be used to implement this function. In this case the VGM is used as a device performing point nonlinear operations. Since the four basic logical functions, AND, OR, NOT, and XOR, are all characterized by a certain form of nonlinearity (Fig. 8.77), a proper tailoring of a spatial filtering at the Fourier plane will produce the desired function. Thus as an example, the operation NOT can be implemented by allowing the Fourier plane filter to transmit only the lowest spatial frequency and reject the higher ones above a certain level. The authors used two separate input sources to demonstrate the truth table, as shown in Fig. 8.78. The operation of analog-to-digital conversion on a two-dimensional array was demonstrated a few years ago using an LCLV [174]. In this case the sinusoidal transfer curve of a controlled-birefringence-based LCLV is used to convert analog information (proportional to the

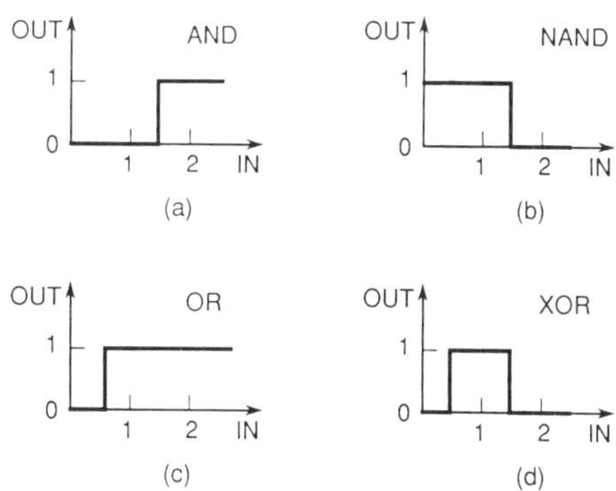

Figure 8.77 Logic operation as a simple nonlinear function of two incoherently superposed inputs *A* and *B*: (*a*) AND; (*b*) NAND; (*c*) OR; (*d*) XOR. (After Ref. 59.)

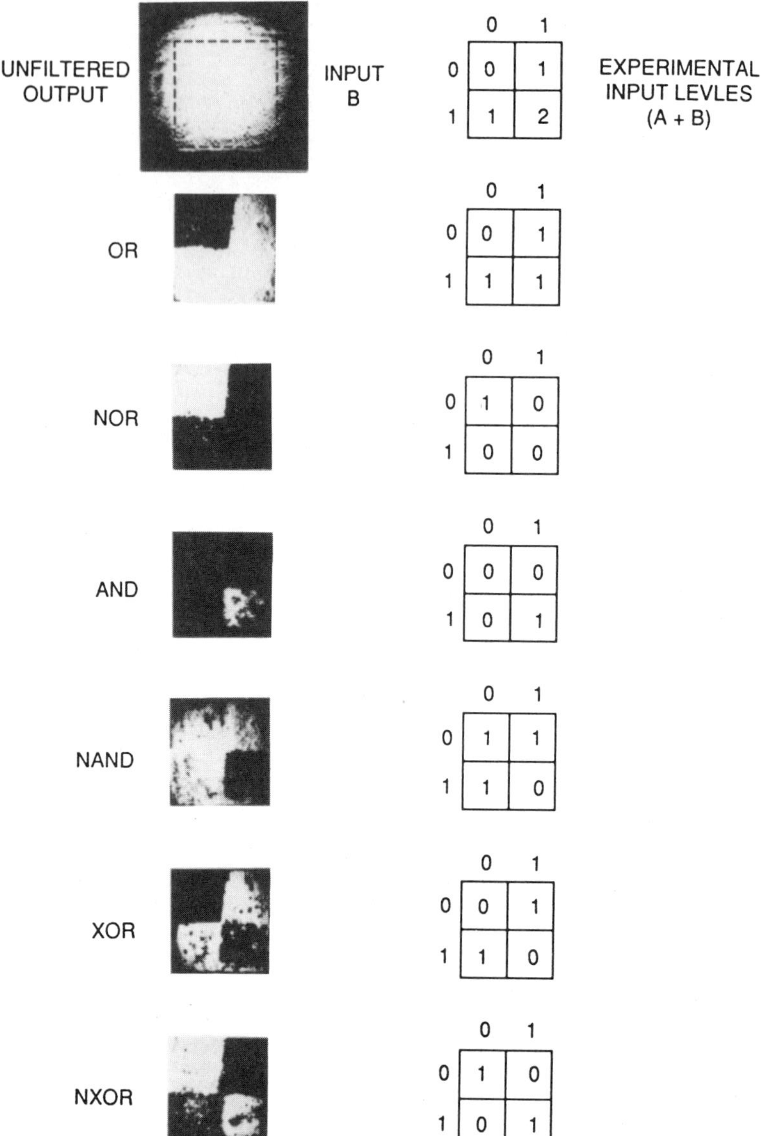

Figure 8.78 Experimental (left column) and theoretical (right column) results of the VGM LCLV used for logical operations. (After Soffer Ref. 59.)

voltage across the LC cell) to digital data. The latter is done by counting the cycles (or high/low levels) generated by the sinusoidal transmission of the LC cell using attenuators to "rescale" the horizontal axis for each bit level, as shown in Fig. 8.79.

Vector–Matrix Multiplication. Following the first concept of an optical implementation of vector–matrix multiplication [175], a LCLV implementation of the concept was proposed by the Ohio State Group [170]. The original concept (Fig. 8.80) is based on

DEVICE RESPONSE CURVES FOR 3-BIT A/D

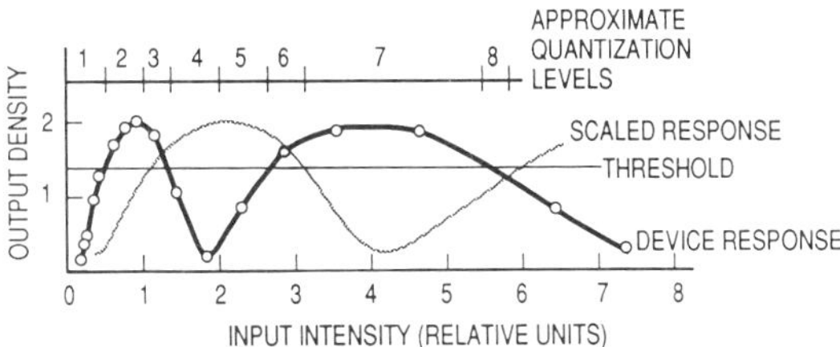

Figure 8.79 Transfer curve of a LCLV operating in the birefringence mode (solid line) and with an attenuated input (dotted line), showing the use of the device as an analog-to-digital converter. (After Ref. 174.)

expanding the input vector V_i into stripes that are projected (using a cylindrical lens) on a two-dimensional matrix mask. The optical output is then integrated in the crossed direction again using a cylindrical lens onto a linear imaging detector. In the implementation proposed by Clymer and Collins [170] a LCLV is used as the two-dimensional matrix mask (Fig. 8.81). It is interesting to note that this same concept of vector–matrix multiplication is the basis for an optical line-to-line interconnection commonly referred to as a "crossbar-switch" (Section 8.4.3).

Systolic Array Processing. Optical numerical processing offers a rather unique application for the CCD-addressed LCLVs described in the preceding. For high speed, an optical numerical processor must utilize spatial parallelism. A two-dimensional data array offers a high degree of parallelism but can entail a significant addressing problem. If, however, data could be entered a line at a time and be made to march across the LCLV at the chosen clock rate, a single $N \times 1$ CCD line could address a full $N \times N$ data array. The use of moving electronic data in a plane for such numerical operations was popularized as "systolic array processing" by Kung [176]. The first extension of systolic array processing to the optical domain used one-dimensional transducers (acousto-optical delay lines and CCD detectors) in direct analogy with VLSI transducers [177]. Recently Bocker et al. [178] proposed the use of optics for systolic array processing in three dimensions, which electronics cannot do. Their Rapid Unbiased Incoherent Calculator cube (or RUBIC cube) uses two electronically addressed spatial light modulators to move components of matrices A and B across the spatial light modulator at certain clock rates. One possible configuration is shown in Fig. 8.82. Because two pixels are needed for real-number representation, one can multiply the two $(N/2) \times (n/2)$ matrices together with the RUBIC cube in $N - 1$ clock periods. To use the CCD-addressed LCLV for the RUBIC cube, one must use an external buffer memory that will feed the CCD

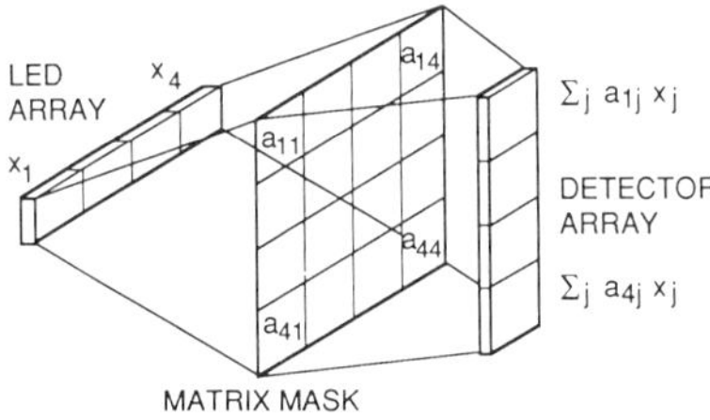

Figure 8.80 Vector–matrix multiplication using a matrix mask. For *programmable* matrix a LCLV can be used. (After Ref. 175.)

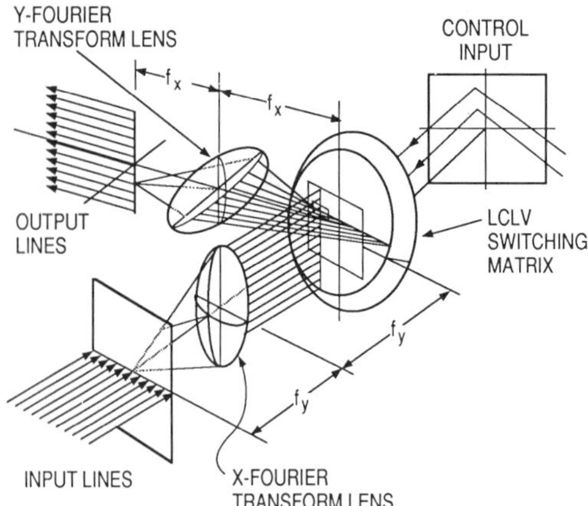

Figure 8.81 Vector–matrix multiplication using a LCLV as a programmable matrix mask. (After Ref. 170.)

LCLV with one line/column displacement in each frame. Alternatively, it may be possible to modify the structure of the CCD LCLV to incorporate an internal buffer memory. This will enable the line/column clocking operation required.

Optical Coprocessor Interface. Since optical computing systems are not expected to be as flexible or "programmable" as the electronic digital processor, it is generally agreed that an electronic computer will provide the host function in a hybrid optoelectronic, high-throughput processor. One could therefore imagine a hybrid processor where the electronic host unit will control the various functions to be executed, activating, when

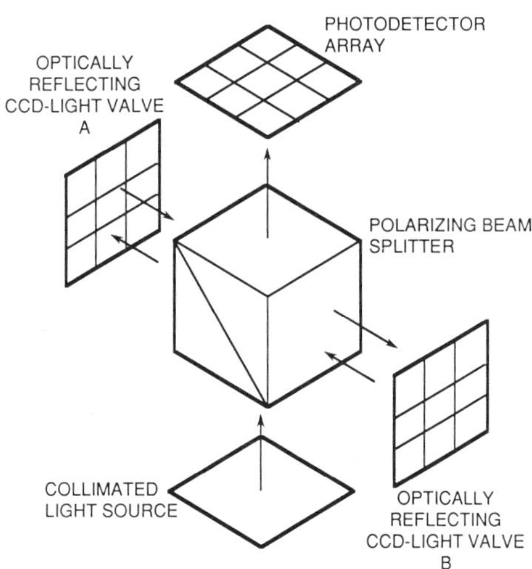

Figure 8.82 Use of a CCD-addressed LCLV for systolic array processing. (After Refs. 178 and 163.)

appropriate, the various optical processing modules that will provide an ultrafast processing of special algorithms or numerical computing. One problem, however, in the way of realizing such an ultrahigh throughput hybrid computer is that of interfacing the optical modules to the electronic host.

The problem is that of interfacing inherently parallel processors (optical modules) to an inherently serial machine (host electronic computer). The need to convert the serial information to parallel and back to serial again can significantly reduce the high-throughput advantages of the parallel optical processors. It therefore seems that a better approach is the use of a parallel host computer to control and interface the various optical modules. If one uses such a machine, one can input and output the necessary information to and from the optical processing modules in the most efficient way. Since the parallel electronic computing architectures have an inherently low degree of interconnectivity, the optical coprocessor can enhance their computation capabilities for operations that require a high degree of interconnectivity such as Fourier transformation, correlation/convolution, as well as shuffling and sorting operations.

An interfacing device is needed for such a system to bridge between the parallel electronic host computer and the various optical processing modules, providing a parallel link of information. This function can be fulfilled with the use of an SLM. The particular implementation of a modified WGLV for this application was recently suggested [115].

Optical Implementation of Artificial Neural Networks

Potential Applications of Neural Network Technology. The basic application areas for neural network (NN) technology [179] are

1. pattern (image, signal) recognition and associative memories and
2. optimized solution under multivariable constraints (decision making), a specific example of which is adaptive motor-sensor operation (robotics).

Even though one usually finds long lists of applications for NN technology, from the technology standpoint these can be grouped under one of the above two categories:

1. The pattern recognition/associative memory category covers both the fields of object or pattern identification and/or its classification. The latter could be defined both as a recognition or as an association operation. Pattern recognition operations cover primitive low-level operations or image processing and preprocessing, mid-level operations (feature extraction), and high-level, ultimate pattern recognition. Associative memories or content-addressable memories cover association as well as image/pattern restoration functions.
2. The area of optimized solution under constraints, or "constraint satisfying systems," covers, as its title implies, the general category of computing an optimized solution to a problem containing a large number of constraints; a typical well-known example for this category is the traveling salesman [180] problem, where the required solution is an optimal (minimum) traveling path under the given constraints of the intercity distances.

Optical Implementation of Neural Networks. The main reasons to use optical techniques for NN applications are as follows:

1. A "natural" massively parallel operation on all elements (image pixels) can be achieved.

2. Three-dimensional interconnectivity is naturally provided by optical beam propagation.
3. Optical interconnectivity: the important advantage of immunity to interference from electromagnetic fields as compared to the electronic counterpart.
4. Images are often themselves the data to be processed. Optical NN processing takes advantage of the natural parallelism of the images.
5. A particularly powerful implementation of optical neural nets in image processing is that of associative memories. In this implementation we can enjoy the added advantages of holographic techniques in that (a) holograms are natural, high-density, interconnected arrays of associated elements and (b) there is a multiplexing advantage in that information is globally stored in the hologram, making the system robust.

In the following we describe three systems in which artificial neural nets are optically implemented. The first two describe holographic implementations; the third is a proposed method for the implementation of the Anderson–Hopfield model for NNs using hybrid electro-optical implementation.

Holographic Implementation of Associative Memory. The pioneering work in this area was performed in 1985 at the Hughes Research Laboratories [181]. This implementation of associative memory (Fig. 8.83) is accomplished by placing a hologram that contains stored images with angularly coded references. Upon illumination of the hologram by a probe image that represents an incomplete version of one of the stored images, the reference beams of the images most closely related to the interrogating image are excited, retroreflected by the phase-conjugating mirrors (PCM), and thresholded to result in a single image–reference pair that represents a stable mode in the cavity. This image is the closest one of the stored group to the interrogating image. Liquid crystal light valves are used for real-time input of the image to be associated.

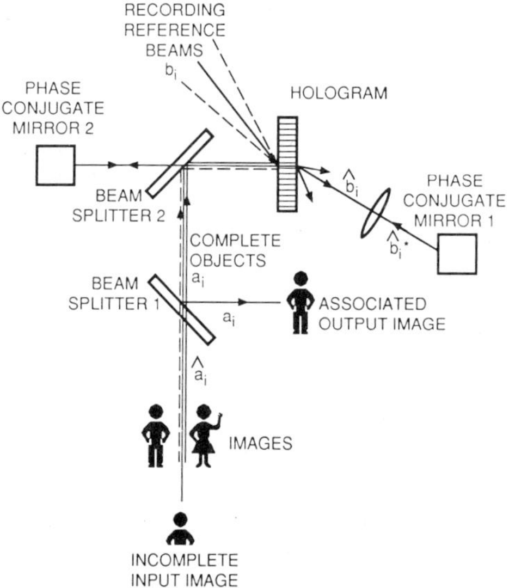

Figure 8.83 Holographic implementation of associative memories. (After Ref. 181.)

Ring Resonator Configuration for Associative Memories. In work performed by a group from the University of Colorado [182] the associative memory function was accomplished by controlling a photorefractive crystal in a ring resonator cavity (Fig. 8.84). In this configuration, the eigenmodes of the resonator are the stored, exemplar images. Similar to the holographic approach described in the preceding, upon probing the system with an interrogating image, the latter will enhance that eigenmode (stored image) that is closest to the input interrogating image while the other modes will be inhibited. Again LC spatial light modulators will be used to enter the interrogating images into this associative memory system.

Self-pumped Optical Neural Networks. Work still in progress [183] is based on the formation of interconnection weights between each pair of optical neurons by the formation of a continuum of angularly and spatially distributed gratings. This approach

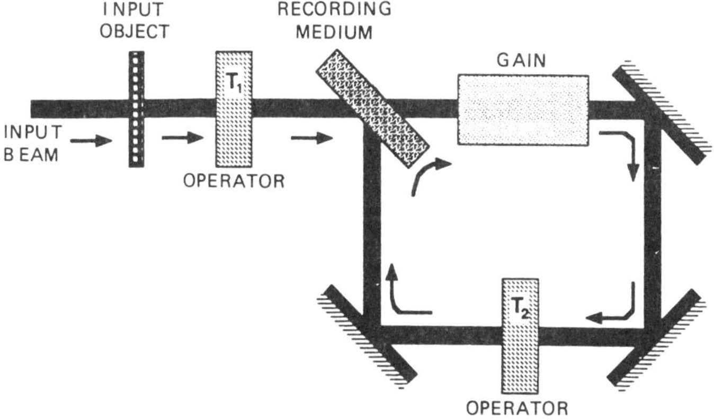

(a) RING RESONATOR MEMORY

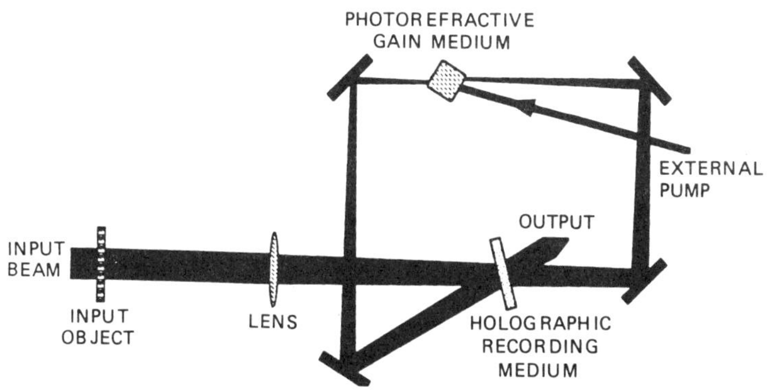

(b) SETUP FOR RECORDING PATTERNS AS EIGENMODES OF A RING RESONATOR.

Figure 8.84 Ring resonator configuration for associative memories. Similar to the holographic approach in Fig. 8.83; as the system is probed with an interrogating image, the latter enhances the eigenmode (stored image) that is closest to the input interrogating image. (After Ref. 182.)

eliminates crosstalk between neurons and allows full utilization of the SLM space–bandwidth product. A diagram of an experimental system used to demonstrate these concepts is shown in Fig. 8.85. The "object plane" corresponds to the plane of neurons represented by pixels on an LCLV. Activation patterns displayed on the LCLV are impressed on a light beam focused into the self-pumped phase-conjugating mirror (SPPCM). Connections between the pixels are formed, and the phase conjugate return is detected by a video camera. The return is processed on a point-by-point basis by the frame grabber/image processor before being displayed again on the LCLV. In NN models such as backpropagation, an error signal would be formed electronically and displayed on the LCLV to adjust the weights between neurons. The error signals are formed on a point-by-point basis (local operations) and so are not computationally intensive.

An experimental demonstration of optical connectivity using the apparatus of Fig. 8.85 is shown in Fig. 8.86. Fig. 8.86*a* shows the phase conjugate return for an input consisting of a complete resolution pattern. The input was then switched to the region enclosed by the dashed ellipse in Fig. 8.86*b*. The return consisted of the complete resolution pattern, as shown in Fig. 8.86*b*, verifying that connection weights were formed globally among all the pixels. Crosstalk suppression was also demonstrated in this experiment. This approach is potentially capable of implementing NNs consisting of up to 10^6 neurons with as many as 10^{10} interconnections.

Optical Implementation of Association and Learning Based on PRIMO/Light Valve Devices. The Hopfield–Anderson model [184,185] shows that a recall or association operation can be implemented by constructing a T_{ij} matrix that is the outer product of the sample "state vectors" V_i, V_j. The recall operation is carried out by multiplying the interrogating vector V_j with the T_{ij} matrix, thresholding the resulting vector, and feeding in the thresholded result again. A state vector closest to the interrogating one is shown to appear as the stable output following several iterations. An electro-optical implementation of this method using microchannel spatial light modulators was previously suggested [186]. A proposal to implement this mechanism electro-optically using LCLVs has recently been presented [187]. The main approach is shown in Fig. 8.87. The system consists of two one-dimensional modulators (MOD1, MOD2) based on PLZT technology [188]. These two layers are used to construct the interconnect matrix T_{ij} stored in an

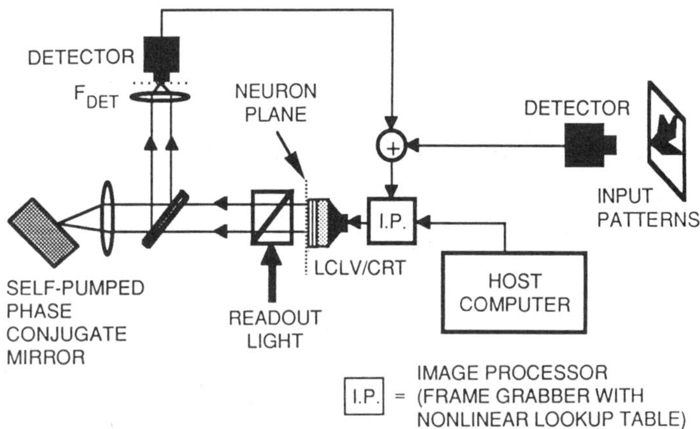

Figure 8.85 Self-pumped optical neural network system. (After Ref. 183.)

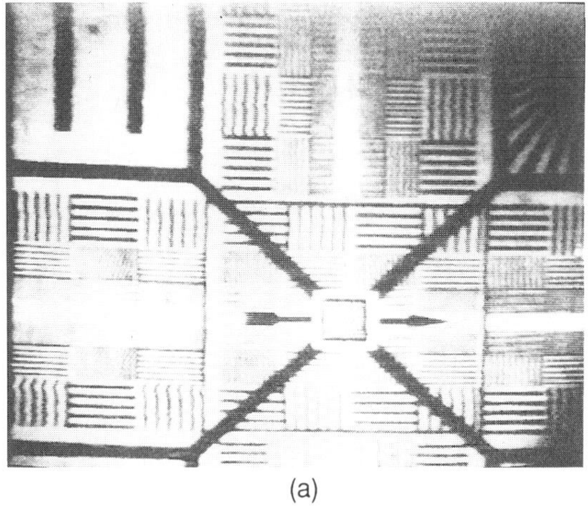

(a)

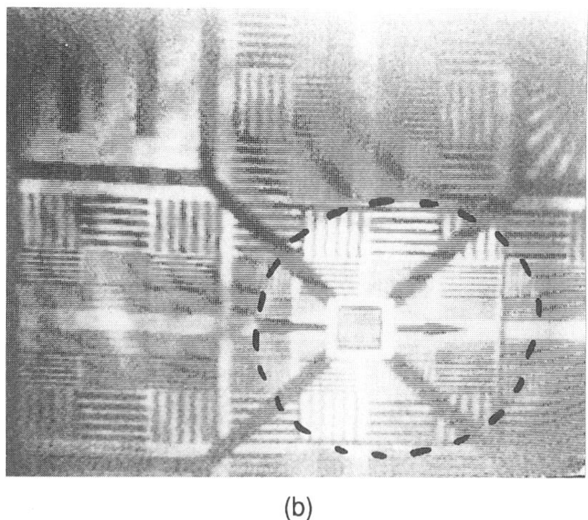

(b)

Figure 8.86 Experimental demonstration of the self-pumped optical neural network (Fig. 8.85). Inputting a partial image [enclosed within the ellipse in (*b*)] resulted in a complete reconstruction of the resolution pattern as shown in (*a*). (After Ref. 183.)

LCLV. Each of the one-dimensional modulators consists of striped-electrode patterns on PLZT. The two layers are oriented so that their electrodes are crossed. Thus, by modulating one with a set of m vectors $U_i^{(m)}$ and the other with a set of m vectors $V_j^{(m)}$ supplied by the microprocessor, one optically forms the T_{ij} matrix (stored in the LCLV) as an outer product, where

$$T_{ij} = \sum_m U_i^{(m)} V_j^{(m)} \tag{8.55}$$

The vector elements are assumed to be ± 1. The vectors $U_i^{(m)}$ and $V_j^{(m)}$ can be supplied at a relatively fast rate ($\cong 10$ μs/vector) by the PLZT modulators. Since the LCLV has a response time of about 10 ms, one will be able to integrate up to a few hundred outer products or vectors in the LCLV-based T_{ij} matrix. Having completed the learning phase, the LC will be modulated with the T_{ij} information for a duration of about 10 ms. During

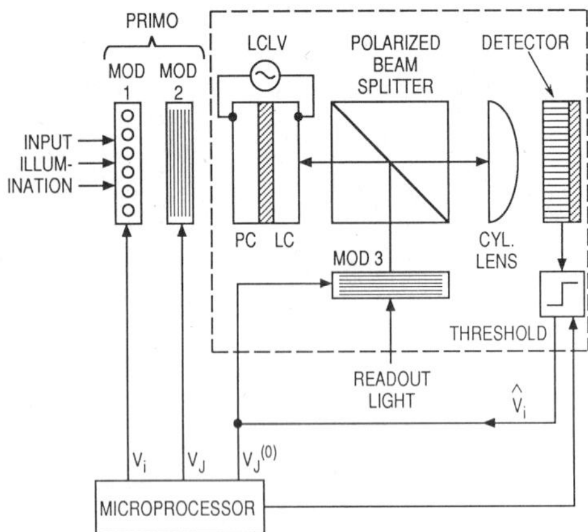

Figure 8.87 An LCLV/PRIMO associative memory system based on the Anderson–Hopfield algorithm. (After Ref. 187.)

this period one can proceed with the interrogation or the association operation. A third one-dimensional PLZT layer (MOD3) will then input the vector $V_j^{(0)}$ to be associated. This one-dimensional vector, whose components are spread in the vertical dimension, will be optically multiplied by the T_{ij} (LCLV) matrix by illuminating the MOD3 modulator using the polarizing beam splitter as shown. Thus, in each line i of the T_{ij} matrix, the columns (running j) are multiplied by the (same) $V_j^{(0)}$ information. By using a cylindrical lens at the output of the beam splitter, as shown, we effectively sum:

$$\hat{V}_i = \sum_j T_{ij} V_j^{(0)} \tag{8.56}$$

for each i-line. Thus each of the i-pixels formed will correspond to the desired ith component $\hat{V}_i$ of the matrix–vector product to be compared against a threshold level according to the outer-product model. This operation can be carried out by detecting the resultant vector $\hat{V}_i$ using the linear detector and an electronic threshold controlled by the microprocessor. To complete the association, the thresholded $\hat{V}_i$ is fed back into MOD3, and the matrix–vector multiplication operation is repeated. The resultant sequence of $V_i^{(n)}$ $(V_i^{(1)} V_i^{(2)}, \ldots, V_i^{(n)})$ is then tested for convergence, which, once reached, will yield the closest association with the interrogating input vector $V_i^{(0)}$. One type of learning that this system can perform is statistical learning, as suggested by Anderson [187]. He showed that for a neuron system coupled in an auto association scheme the multiplication of the weight matrix W_{ij} by the interrogating vector $V_i^{(0)}$ will result in the output vector being one of the stable state, with a weight proportional to the frequency in which this vector appeared during the learning phase.

The system can therefore learn to enhance common features that appear in different patterns during the teaching (learning) phase.

8.4.3 Optical Interconnects

The preceding section was devoted to the description of an important application of optics for computation, namely, optical computing and data processing. A second critically important application of optics for computation is that of optical interconnects.

It is widely recognized by circuit and computer architecture experts that inter- and intra-chip interconnections pose a serious limitation to future increase in the throughput or speed of the electronic processing elements. Interconnection between microprocessors, work stations, and mainframe computers is also a significant challenge. In this case a relatively small number of channels (≈ 100) carry a very high bandwidth data stream (hundreds of gigabytes per second) and require efficient interconnection. Since at least in the latter case the information may be carried by fiber-optical cables, it makes sense intuitively for optics to be used in the interconnection process to avoid a back-and-forth conversion from optical to electronic signals. Here again, similar to the task of high-throughput computation, optics may provide an efficient solution with spatial light modulators again being the critical elements of such a function. It is important to note that the two seemingly different concepts of interconnects and data processing or computing are in fact very much linked together. Consider the operation of vector–matrix multiplication discussed previously (Fig. 8.80). This operation in its optical implementation can in fact be viewed as interconnecting vector $V_j^{(1)}$ through a matrix A_{ij} to a vector $V_j^{(2)}$ by

$$V_j^{(2)} = \sum_j A_{ij} V_i^{(1)} \tag{8.57}$$

This operation is a predetermined connection of each of the $V_j^{(1)}$ elements to the $V_i^{(2)}$ elements, a function known popularly as crossbar-switching [189]. This is precisely the function needed to connect N high-bandwidth channels coming from one computer cluster to another set of N channels of a second cluster. The use of a spatial light modulator for the function of the matrix A_{ij} allows the interconnections to be *reconfigurable*. Free-space interconnections can also be constructed for interchip connections [189] (e.g., clock signal distribution). Such a system is shown in Fig. 8.88. Again, spatial light modulators can replace the fixed holographic routing element shown in Fig. 8.88 to allow for reconfigurability of the interconnections. A detailed design of an optical computer switching network based on a LCLV has been performed by the Ohio State University group in 1985 [190]. They have combined several functions such as crossbar switching, optical threshold detector, and master–slave flip-flop arrays to form a control system for the computer switching. The switching system is designed to connect 80 minicomputers coupled to the switching system via optical fibers. Another concept of programmable optical interconnects proposed by Marom et al., [191] allowing fanning out, is shown

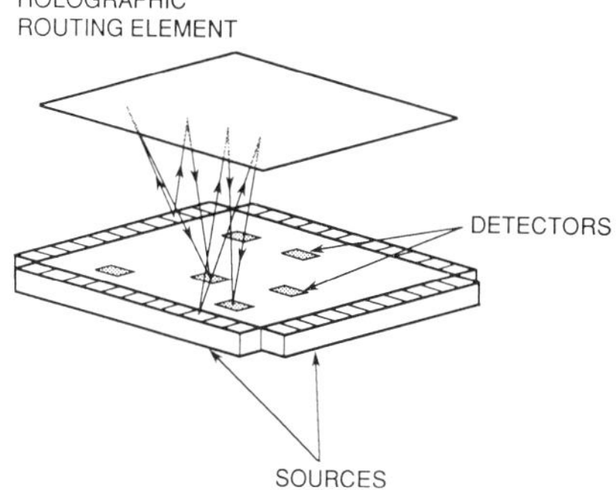

Figure 8.88 Interchip optical interconnection configuration. (After Ref. 189.)

in Fig. 8.89. The authors demonstrate the capability of such a system to interconnect intrachip elements by simulating the LEDs on a chip by a group of fiber-optical light sources and then imaging the resulting light distribution in the same plane. Another treatment of optical interconnections, including reconfigurable connection using LCLVs and other SLMs, was published by the group from the University of Southern California and IBM [192].

While the possibility of using LCLVs was demonstrated, the use of nematic materials for this application is not very attractive due to the long reconfiguration times. Also, it is desirable to have a built-in memory in such reconfigurable devices (SLMs) so that the interconnections in a certain configuration will be maintained unless a change is required. Since only binary operation is required of the modulator, ferroelectric LCs seem to be ideal for such an application [193], offering fast, submicrosecond switching combined with memory.

Finally, the use of an SLM as a storage/parallel addressing element in a hybrid electronic–optical scheme was recently proposed [194].

8.4.4 Adaptive Optics Applications

The objective of adaptive optics technology is to correct for phase aberrations which occur in optical systems due to thermal distortion of components or due to atmospheric turbulence. Such correction can be achieved using a combination of wavefront sensor and spatial phase modulator elements. An example of such a need is in a laser fusion system where the capability of focusing the laser beam down to the diffraction-limited spot size is critical. Aberrations due to thermal stresses oftentimes limit the focusing capability which can be regained by using adaptive optical methods.

The Use of Spatial Light Modulators in a Closed-Loop Arrangement. Following Fisher and Warde [195] Fig. 8.90 is a conceptual diagram of an all-optical phase-compensated communications receiver. The distorted received wavefront passes through an optically addressed monolithic spatial phase modulator, which adds the appropriate phase to present a compensated, approximately planar wavefront to the communications receiver. Part of the compensated wavefront is fed back to a phase estimator whose output is an optical intensity proportional to the phase of the compensated beam. This optical control signal drives the spatial light modulator.

Figure 8.91 illustrates specific all-optical implementation of the adaptive phase compensator employing a homodyne interferometer and a reflection readout monolithic phase modulator, a microchannel spatial light modulator (MSLM). In the MSLM the electron image from the photocathode is amplified by the microchannel array plate and proximity

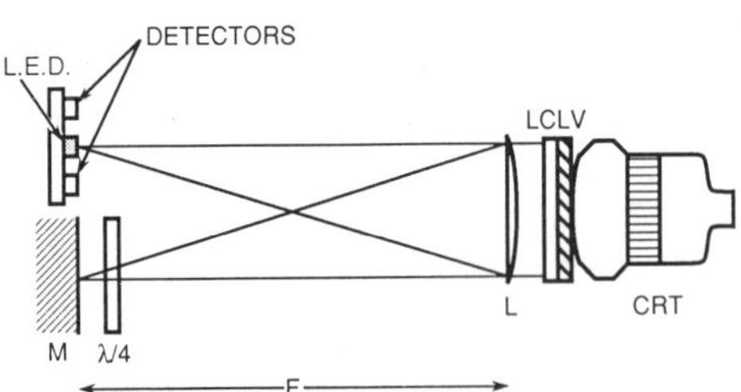

Figure 8.89 Optical interconnection system based on a photoactivated LCLV. (After Ref. 191.)

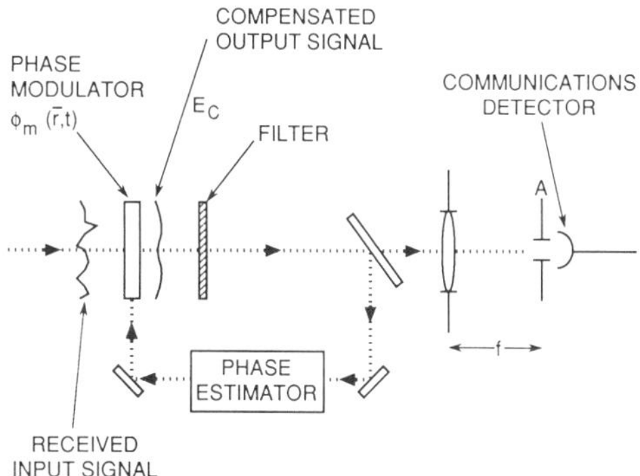

Figure 8.90 All-optical adaptive phase compensator for communications applications. (After Ref. 195.)

focused onto the dielectric mirror. The spatially varying electric field induced by the charge distribution modifies the refractive index of the electro-optical crystal (e.g., $LiNbO_3$), which phase modulates the reflected readout beam. This function can obviously be fulfilled by the use of a photoactivated LCLV.

Direct Phase Conjugation Using the Liquid Crystal Light Valve. Optical phase conjugation [196] has been investigated extensively because it is capable of generating a beam whose wavefront has an envelope proportional to the complex conjugate of another beam and that propagates in a direction opposite to the first one. Such beams are generated primarily by four-wave mixing processes [197] or by using deformable mirrors. In the latter method, some combination of a wavefront sensor and a deformable mirror is used to imprint the spatial phase modulation obtained from the wavefront sensor onto the deformable mirror in a closed-loop operation. The deformable mirror action can be achieved either by mechanical means [198] (piezoelectric activation) or by refractive index modulation [199].

It has already been reported [200] that the four-wave mixing process can be interpreted as a holographic mechanism: two of the incoming beams (one of the pump beams and

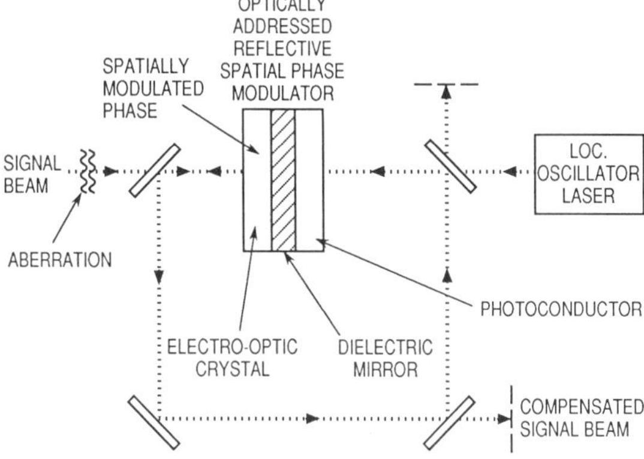

Figure 8.91 Photoactivated SLM used for adaptive phase compensation. (After Ref. 195.)

the signal beam) form a hologram in the nonlinear medium while the third beam (the second pump) reads it out, thus "reconstructing" a conjugate wave. A method of generating a phase conjugate wavefront via a LCLV with a leaky mirror between the LC and photoconductive layers was recently published [201,202].

Following Marom and Efron [201], the leaky mirror allows the two beams addressing the light valve from the readout side to penetrate to the photoconductive layer, thus forming an interference pattern that is transferred as a spatial voltage modulation on the LC layer. The LC then acts as a phase hologram, modulating the same beams used to form it. To see the real-time holographic implementation, let R and S be two beams (reference and signal) impinging on a recording material H (Fig. 8.92*a*) whose transmittance T is then proportional to the irradiation intensity:

$$T = \alpha|R + S|^2 = \alpha(|R|^2 + |S|^2 + RS^* + R^*S) \tag{8.58}$$

If the hologram is now illuminated by the same R and S beams, the transmitted optical field is given by

$$\alpha(R + S)|R + S|^2 = \alpha R(|R|^2 + 2|S|^2) + \alpha S^2R^* + \alpha S(2|R|^2 + |S|^2) + \alpha R^2S^* \tag{8.59}$$

One can schematically display the angular distribution of these beams as shown in Fig. 8.92*a*. It is now evident that if R is impinging on the hologram from a direction normal to its plane and if a mirror is placed just behind the hologram, then the term R^2S^* represents a beam propagation in an opposite direction to S. Since it is a uniform plane wave normally incident on the hologram plane, R does not contribute either spatial or phase information to R^2S^*, so that the latter is a perfect representation of a phase conjugate wavefront.

The preceding requirements are satisfied through the use of a leaky-mirror LCLV with a silicon photoconductive layer [100]. We use a light valve with a metal matrix structure [203] acting as the leaky mirror; the pads, 17-μm square with 3-μm spacing on each side, provide enough leakage for the activation of the photoconductor. The pad periodicity, however, restricts the spatial frequency of the interferometric pattern forming

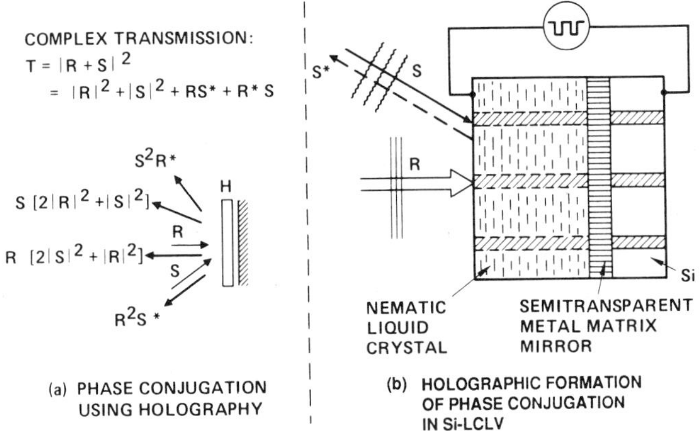

Figure 8.92 Schematics of a direct-phase conjugation system using a leaky-mirror LCLV. (After Ref. 201.)

the hologram to less than 25 cycles/mm, thus limiting the angular separation of the two beams to less than 0.9°.

The 4-μm-thick nematic LC layer (E-7, BDH) is planarly oriented to act as a phase modulator for optical beams with a polarization state parallel to the long axis of the LC molecules. Alignment of the LC was achieved using shallow- and medium-angle deposition of SiO_2.

The optical set-up sketched in Fig. 8.93 was used. The two beams R and S ($\lambda = 632.8$ nm) separated by an angle of 0.8° impinge on the readout (LC) side of the LCLV. The hologram generated by the leaked irradiance in the photoconductive layer is transferred, by virtue of the applied field, as an electric field distribution across the LC. In turn, the LC can now modulate the phase of optical beams traversing that layer, provided that their polarization state coincides with the long-axis orientation of the LC molecules.

It can be shown that, based on the configuration of Fig. 8.93, a phase grating [the last term in Eq. (8.59)] is generated in the LC layer. The object in the experiment performed using this concept was a triangular aperture 1.8 mm on each side placed along the object beam. An aberrator positioned along the object beam was imaged by a lens onto the light valve.

The reconstructed triangle along the back-traveling path was detected by a TV camera and displayed on the monitor (Fig. 8.93). Various aberrators were positioned in the object beam: wedges producing tilt errors of up to 0.25°, a lens ($f = 90$ cm) producing focusing errors, and spatially random phase aberrators (etched glass slides) with various degrees of aberration. All aberrators were successfully corrected by the phase conjugator. Figure 8.94 shows the image of the triangular object as observed along the conjugated path by the TV camera: Part *a*–*d* shows the effect of increasing the level of aberration. The right-hand side of the figure shows the effect with *no* correction using a mirror to replace the LCLV. The left-hand side of the figure shows the compensated phase-conjugated image using the LCLV. As observed, even in the case of an extreme level of aberration (Strehl ratio < 0.1) the center portion of the triangle is still retained in the compensated image. The measured time response of the light valve was 10 ms.

It should be pointed out that the LCLV structure employed did not permit gain; the intensity of the reference beam could not be made much higher than that of the signal beam without producing photoconductor saturation. This problem can be alleviated with the use of a wire grid mirror LCLV [115].

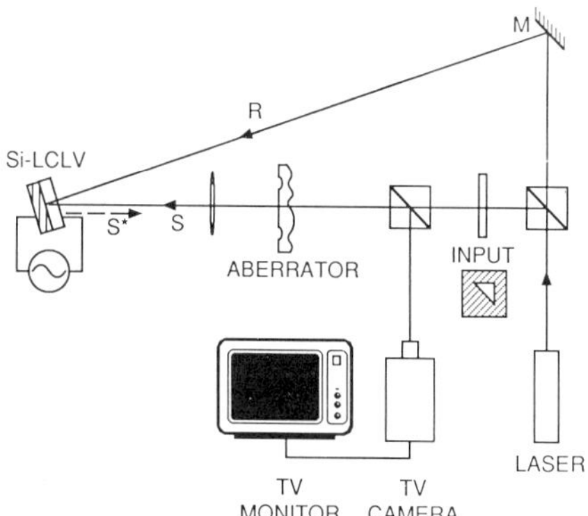

Figure 8.93 Schematic arrangement of phase conjugation using a LCLV. (After Ref. 201.)

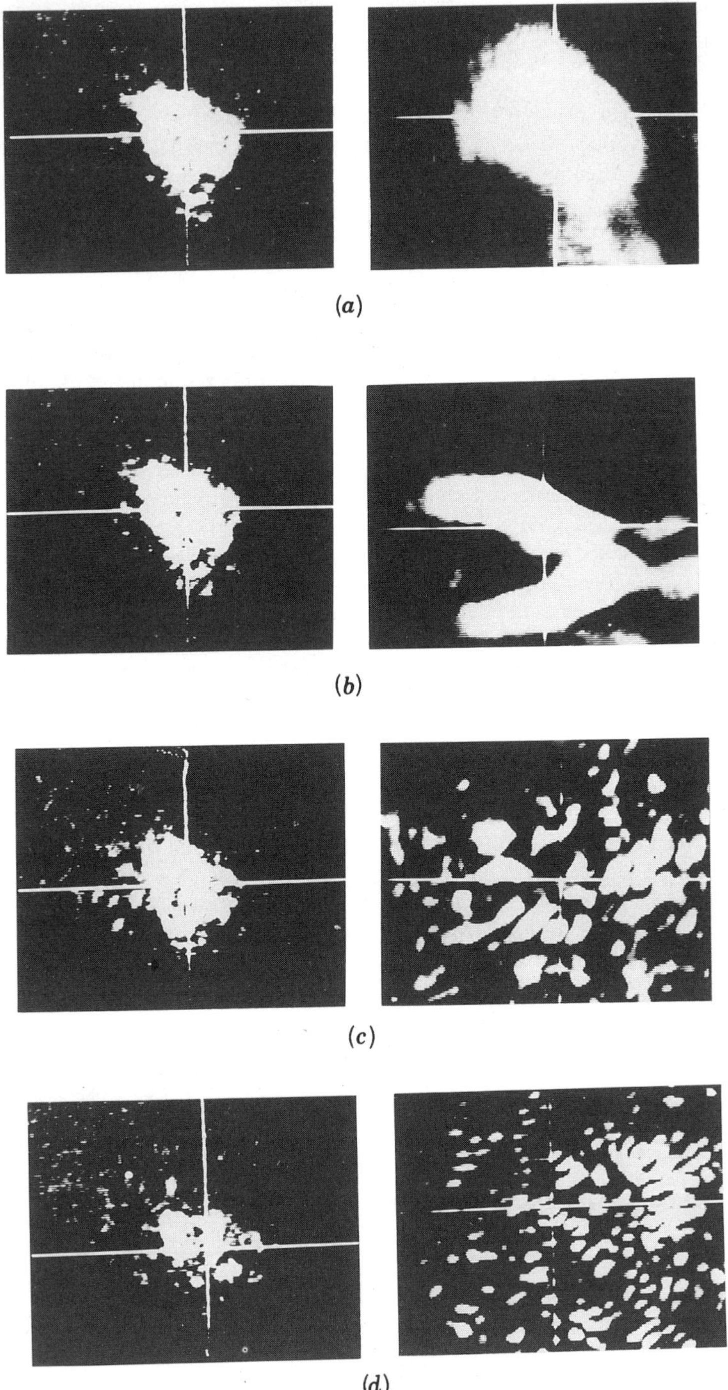

Figure 8.94 Experimental results of direct-phase conjugation using a LCLV. Left side shows the corrected image using the LCLV as a conjugator; the right side shows the image reflected from the LV plane (using mirror) without correction: (*a*) original image; (*b*) with random aberration of 0.17 Strehl ratio; (*c*) with random aberration; $S = 0.1$; (*d*) with random aberration; $S < 0.1$. (After Ref. 201.)

A significant reduction in the response time (down to microseconds) can be expected from replacement of the nematic LC with a ferroelectric material.

8.4.5 Image Wavelength Converters

Liquid-crystal-based spatial light modulators have been successfully utilized to demonstrate visible-to-IR [204, 205] and near-IR-to-visible conversion [206]. Proofs of concept of these two conversion schemes are described in the following.

LC-based Visible-to-IR Dynamic Image Converter. Existing IR simulation systems are thermally based and consequently suffer from low resolution, slow response, and limited dynamic range. A new simulation system is needed that can (1) provide complex, fast-changing IR schemes with a high degree of dynamic range and resolution and (2) be capable of real-time interaction with a computer image generation (CIG) system.

A solution to these limitations is offered by the LCLV-based visible-to-IR dynamic image converter (VIDIC). The two main advantages of the LCLV-based VIDIC over existing systems such as the Bly cell are (1) the generation of a nonthermal image at the focal (target) plane and (2) the separation of the output source from the input image. Optical modulation of the IR beam caused by polarization–rotation effects frees the LCLV VIDIC system from the severe trade-offs associated with a thermal target projection system. Separation of the output image from the input source allows a compact, low-power input image source (e.g., a CRT) to be used without sacrificing the high-IR power required at the output. Thus, much higher performance in resolution, dynamic range, and speed is expected from a LCLV VIDIC system, with the additional advantages of high-temperature simulation and compatibility with a CIG system. It is estimated that more than 1000 × 1000 elements of resolution and over 500K of dynamic range at video (60-Hz) frame rates may ultimately be achieved with the LCLV VIDIC system.

A reflection-type VIDIC system is shown in Fig. 8.95. A CRT coupled through a fiber-optical plate activates the visible-sensitive photoconductor. The latter in turn activates the LC, creating a spatial pattern of birefringence. The IR beam emanating from

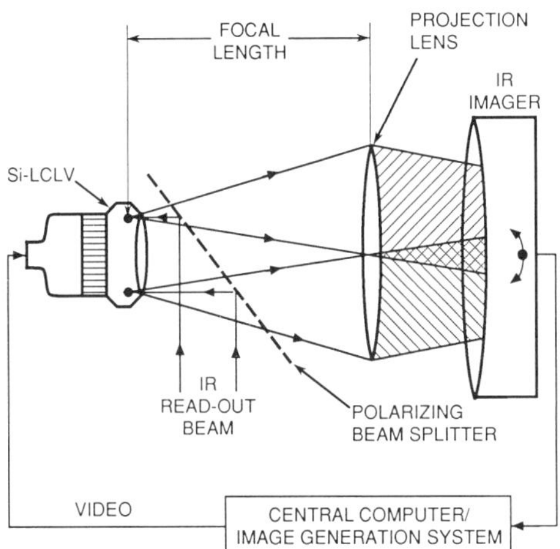

Figure 8.95 Schematic diagram of a visible-to-IR dynamic image converter (VIDIC) based on a Si LCLV. (After Ref. 205.)

the IR source is reflected by a polarizing beam splitter (wire grid) onto the light valve. The IR beam picks up the modulation (in a form of polarization–rotation) from the LC and is retroreflected by the mirror. The beam retraverses through the liquid crystal, picks up additional modulation of the polarization rotation, and passes through the analyzer, which transmits only the cross-polarized component. A projection lens placed one focal length from the plane of the light valve relays the IR image as coming from infinity. Two points should be made here: (1) The extended IR source shown in the figure can be replaced by a point source or a laser, depending on the spectral range desired or the spectral match to a blackbody spectrum desired. (2) The modulation of the IR beam at the focal plane (LC layer) is a spatial distribution of retardation angles (ellipticity). This spatial distribution is converted into an *intensity* or *thermal* image only following the passage of the beam through the analyzer. Thus, the image present at the focal plane is nonthermal. This is an important advantage of the LC-based VIDIC, distinguishing it from thermal target simulators. The latter suffer from a harsh trade-off between resolution and modulation (temperature) level due to thermal image spreading.

The VIDIC development issues that have been addressed are (1) studies of transmission, birefringence, and rotatory power of LCs in the IR; (2) development of an IR mirror and transparent conductive window coating; (3) fabrication of a silicon LC IR modulator; (4) design and construction of a preliminary test-bed system; (5) demonstration of a VIDIC device operating at 10.6 μm; and (6) operation of VIDIC devices with blackbody sources at the spectral regions of 3–5 and 8–12 μm. As pointed out earlier, the conclusions of rotatory power studies in the IR pointed out the controlled birefringence mode as the most efficient one to be used in the IR regime. Silicon-based VIDIC LCLVs were therefore constructed with controlled birefringence (parallel alignment configuration operating in the reverse-contrast mode to attain maximum contrast ratio). The device is based on the silicon LCLV described in Section 8.3.2 with the only modification being the replacement of the dielectric mirror with a metal matrix one [207]. The metal matrix mirror consists of a metallic layer, typically aluminum, that is etched to form a grid pattern, resulting in an array of metallic islands isolated from each other by SiO_2. A plain metal mirror cannot be used because the charge pattern generated in the silicon would be shorted. A dielectric mirror tuned to IR wavelengths would be too thick, resulting in a decreased voltage across the LC layer. Operation of a 10.6-μm-based VIDIC system using a CO_2 laser is shown in Fig. 8.96. An unexpanded 4-mm-diameter laser beam was used in these preliminary experiments. The laser beam was attenuated using a crossed polarizer–analyzer arrangement. The beam was then reflected at an angle of approximately 15° from the light valve and was passed through the output analyzer

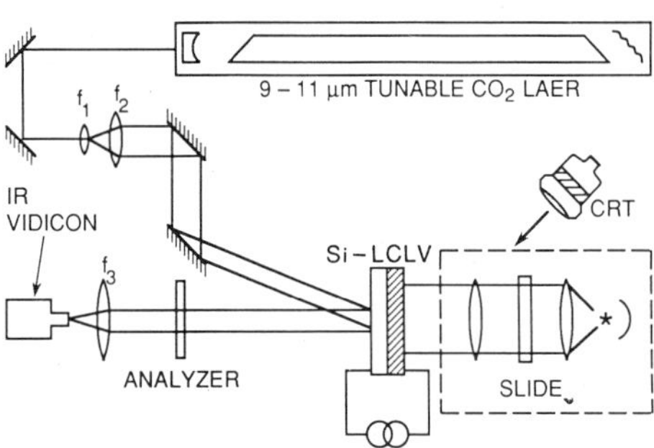

Figure 8.96 Schematics of the 10.6-μm VIDIC test bed system. (After Ref. 205.)

and imaged by a 38-mm ZnSe lens. A pyroelectric vidicon was used in imaging the modulated IR beam. The input channel consisted of a beam from a tungsten iodine lamp (IR filtered out) that was used to illuminate an input pattern. Light levels of a few tens of microwatts per square centimeter provided full activation of the light valve. The light valve was operated with the standard depletion–accumulation cycle used for the silicon device [101]. This consisted of a 25-V, 1-ms depletion phase followed by a 15-V, 30-μs accumulation pulse. A *resolution* of up to 10 lines/mm was observed; this resolution was limited by the imaging system. It is estimated that the actual resolution of the present device is over 20 lines/mm. A *contrast ratio* of over 50:1 was observed with a response time of 200 ms for the 12-μm LC employed. A *response time* of less than 50 ms was achieved with a 6-μm-thick LC layer with similar performance of resolution, contrast ratio, and sensitivity. The transfer curve of the 12-μm LC VIDIC is shown in Fig. 8.97. A *sensitivity* of 50 μW/cm^2 (for 10:1 contrast ratio) was observed. The VIDIC device was also tested at $\lambda = 9.27$ μm. A contrast ratio of over 20:1 was observed with the same operating conditions as used for testing at $\lambda = 10.6$ μm. The full *spectral bandwidth* of this device (at $\pm 10\%$ intensity deviation) is estimated to be 8–12 μm.

Recently, blackbody source-based systems were designed and built as shown in Fig. 8.98. The two systems designed and built are for the 3–5 and the 8–12-μm regions [208]. Dynamic scenes generated from video sources have been simulated in both bands, with characteristics as listed in Table 8.9. The individual optical systems were designed to meet the requirements of the seekers to be tested, for parameters such as field of view and entrance pupil diameter. These requirements do impact the design of various simulator components.

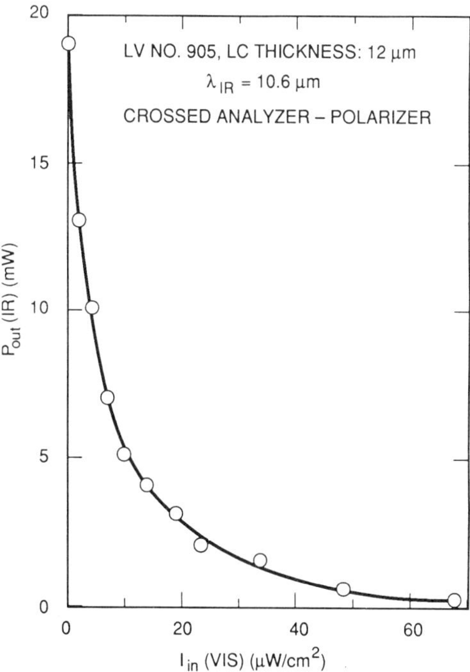

Figure 8.97 Transfer curve (IR output level versus visible input level) of the LCLV–VIDIC system at $\lambda = 10.6$ μm. (After Ref. 205.)

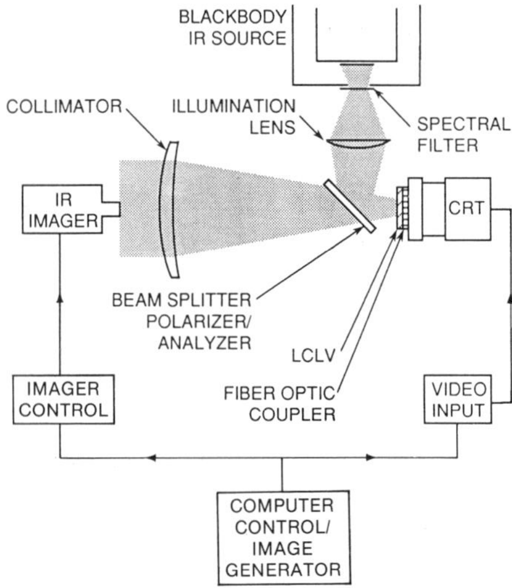

Figure 8.98 Blackbody source-based VIDIC system. (After Ref. 205.)

The main issues in the future development of these systems are (1) the well-known dynamic range–response time trade-off and (2) spectral matching the IR (e.g., blackbody) source. These issues present a problem since (a) the LC exhibits its own wavelength dispersion of birefringence and (b) the efficiency of the VIDIC system is relatively low ($\approx 10\%$); the blackbody has to be heated up to a higher temperature in order to compensate for the losses resulting in a spectral mismatch to the blackbody at the desirable (apparent) temperature. Finally, for some IR simulation it is desirable to remove the "flicker" caused by the periodic activation of the LCLV by the CRT.

Near-IR-to-Visible Image Conversion. The silicon-based LCLV described earlier can be used to convert near-IR imagery to visible based on the fact that the silicon band gap of approximately 1.1 eV allows wavelengths of up to about 1.1 μm to be detected. This allows in particular the detection of the important Nd–YAG line at $\lambda = 1.06$ μm to be imaged. Such an experiment was recently conducted at Hughes Research Laboratories [206]. The experimental set-up for converting an IR image to a visible wavelength is shown in Fig. 8.99. The Si LCLV consists of three major parts: (1) a photoconductor [a 125-μm-thick, high-resistivity ($\sim$5000-Ω-cm) π-silicon], (2) a dielectric mirror (for reflective mode operation), and (3) a nematic LC layer (for modulating the readout beam).

TABLE 8.9 IR LCLV Characteristics

Bandwidth, μm	3–5	8–12
Aperture, mm	43	43
Frame rate, Hz	60	30
Contrast ratio	30:1	20:1
Maximum simulated temperature,[a] °C	500	100
Spatial resolution, lines/mm	>20	>20

[a] Assumes blackbody operation at 1000°C.

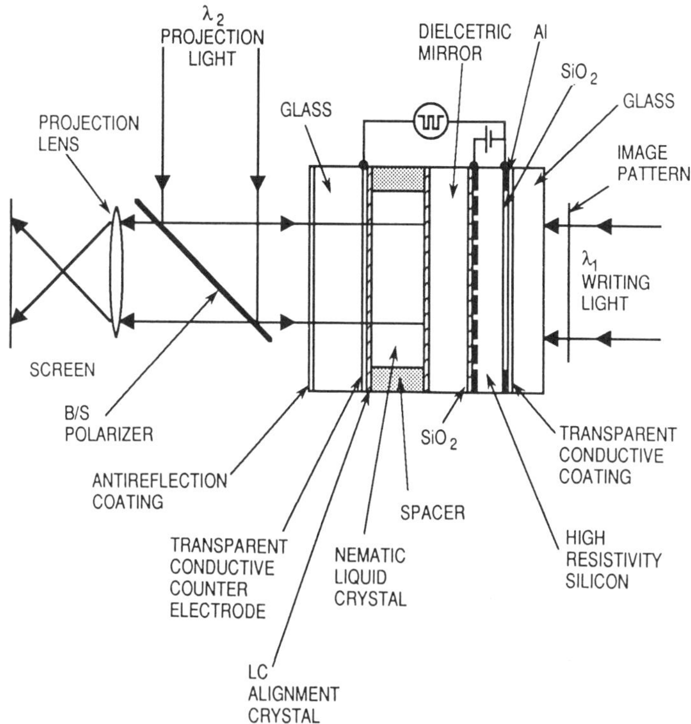

Figure 8.99 Near-IR-to-visible image converter (NIRVIC) system based on the Si LCLV. (After Ref. 206.)

An expanded and collimated ($\sim$40 mm in diameter) single-mode cw Nd–YAG laser beam ($\lambda_1 = 1.06$ μm) was used to illuminate an image pattern onto the photoconductor side of the LCLV. This image was then read out by a He–Ne laser beam ($\lambda_2 = 0.633$ μm). In normal operation of a Si LCLV, the required 1.06-μm input intensity for observing good resolution was measured to be about 100 μW/cm^2. A typical input-intensity-dependent output transmission (which is normalized to the maximum measured output intensity) curve for a Si LCLV operated under normal driving conditions ($V_{\text{acc}} = -30$ V, $V_{\text{dep}} = +20$ V, $\lambda_1 = 1.06$ μm, $\lambda_2 = 0.633$ μm)) is shown in Fig. 8.100. A 4-μm parallel-aligned LC (cyano-biphenyl mixture E-7) layer was used in the reflective mode light valve (Fig. 8.101*a*). The birefringence of E-7 at $\lambda = 0.633$ μm and $T = 24°$C was previously measured to be approximately 0.22. A writing light intensity of a few microwatts per square centimeter is able to cause significant transmission change in the readout light. As $I_{\text{in}}(\lambda_1)$ increases, the output intensity goes through extrema and finally approaches zero at sufficiently high $I_{\text{in}}(\lambda_1)$. This oscillation results from the electro-optical effect of the birefringent parallel-aligned LC layer.

The resolution of the NIRVIC is governed by the imaging and display systems and the Si LCLV. The typical resolution of a Si LCLV is about 18 lp/mm at 50% modulation transfer function or about 35 lp/mm limiting resolution. Figure 8.101 shows the visible image obtained by the NIRVIC system. Due to the coherent readout source used in the experiment, the visually measured resolution is lower ($\sim$10 lp/mm) than that observed using incoherent readout light. Speckle effects as well as reduced modulation transfer of coherent light are probably responsible for this resolution loss. Despite this resolution

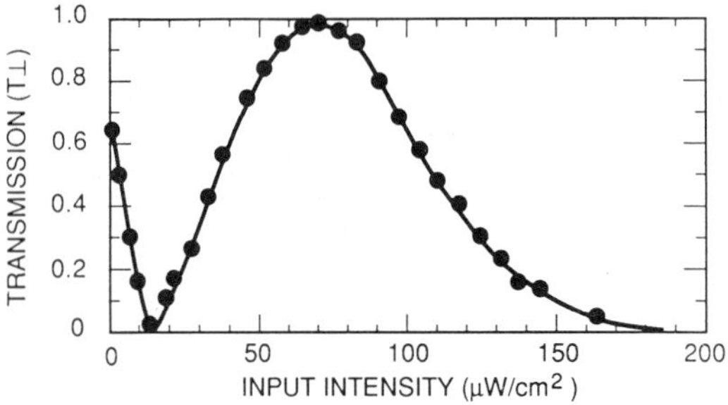

Figure 8.100 Transfer characteristics of an LCLV-based NIRVIC system: transmission at $\lambda = 0.6328$ versus input intensity at $\lambda = 1.06$ μm. (After Ref. 206.)

loss, a total of about 10^5 or more resolution elements is achievable with a 43-mm-diameter Si LCLV activated by a 1.06-μm laser and read out by a He–Ne laser beam.

Both writing and projection beams can be monochromatic or broadband light sources. The bandwidth $\Delta\lambda_1$ of the writing light is determined by the band edge (λ_g) of the photoconductor used in the light valve. For silicon at room temperature, $\lambda_g \cong 1.1$ μm. On the other hand, the bandwidth $\Delta\lambda l_2$ of the readout light is remarkably wide and is determined by the dielectric mirror and the LC.

Direct IR-to-Visible Converter Using LCs. A method using the optical nonlinear properties of LCs was recently demonstrated in which a phase grating pattern was formed in the LC layer using a Nd–YAG laser [209]. A He–Ne beam was then used to read out the grating, which was modulated by transparency. Energies on the order of approximately 1 mJ/cm^2 at 1.06 μm were required for acceptable visible reproduction of the Nd–YAG image.

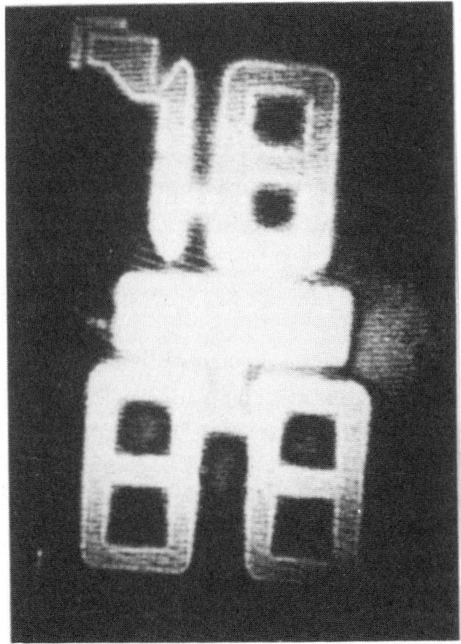

Figure 8.101 Visible image (at $\lambda = 6.328$ Å) obtained by addressing the NIRVIC with that image at $\lambda = 1.06$ μm. (After Ref. 206.)

8.5 CONCLUDING COMMENTS

In this rather concise work I have tried to cover the physics and applications of this unique class of materials called liquid crystals. What has started as a curiosity of an optically anisotropic liquid medium has turned out to be an electro-optical material whose use has already revolutionized the field of displays and is expected to have a significant impact on the optical processing technology.

If there is one lesson to be learned from this amazing development for those of us responsible for technology development (whether scientists or corporate managers), it is that *basic research* allowing "curious" phenomena to be probed and studied is an essential and indispensible part of a successful technology development.

Acknowledgments

The author is grateful to A. M. Lackner for her invaluable help in the studies of LC materials through the years. Thanks is also due to J. Grinberg, whose innovations in the area of liquid crystal devices resulted in many concepts and devices introduced in this work. I wish to acknowledge the fruitful collaboration with B. H. Soffer and S. T. Wu as well as with Y. Owechko, M. S. Welkowsky, J. D. Margerum, D. M. Pepper, W. P. Brown, T. R. O'Meara, and R. C. Lind of the Hughes Research Laboratories and W. P. Bleha and P. C. Baron of the Industrial Product Division and the Ground Systems Group of the Hughes Aircraft Co. I would also like to acknowledge helpful discussions with A. R. Tanguay, Jr. and A. A. Sawchuk of the University of Southern California, S. A. Collins, Jr. of the Ohio State University, D. Casasent of Carnegie Mellon University, S. H. Lee of the University of California at San Diego, and J. S. Patel of Bell Communication Research. Finally, many thanks are due to Y. Owechko, B. H. Soffer, and A. M. Lackner for their careful proofreading of the manuscript and to B. Dover for her skillful help in the preparation of the manuscript.

REFERENCES

1. L. M. Blinov, *Electro-optical and Magneto-optical Properties of Liquid Crystals*, Wiley, New York, 1983, Chap. 1.
2. P. Keller and L. Liebert, *Solid State Physics*, Academic, New York, 1978, Sup. 1.14, Chap. 2.
3. L. Onsager, *Ann. N. Y. Acad. Sci.*, Vol. 51, p. 627, 1969.
4. W. Meier and A. Saupe, *Z. Naturforsch*, Vol. 149, p. 882, 1959.
5. J. E. Lennard-Jones and A. F. Devonshire, *Proc. Roy. Soc. (London)*, Vol. A169, p. 317, 1939.
6. J. A. Pople and F. E. Karasz, *J. Phys. Chem. Solids*.
7. W. Meier and G. Meier, *Z. Naturforsch*, Vol. A16, p. 262, 1961.
8. C. Bak et al., *J. Appl. Phys.*, Vol. 46, p. 1, 1975.
9. F. C. Frank, *Discuss. Faraday Soc.*, Vol. 25, p. 19, 1958.
10. J. L. Ericksen, *Arch. Rat. Mech. Anal.*, Vol. 10, p. 189, 1962.
11. L. M. Blinov, *Electro-optical and Magneto-optical Properties of Liquid Crystals*, Wiley, New York, 1983, p. 107ff.
12. L. Liebert, Ed., *Solid State Physics*, Academic, New York, 1978, Chap. 3.
13. (a) W. Helfrich, *Phys. Rev. Lett.*, Vol. 29, p. 1583, 1972. (b) E. Moritz, *Mol. Cryst. Liq. Cryst.*, Vol. 82 (letters), p. 81, 1982. (c) O. A. Kapustina and V. N. Lupanov, *Sov. Phys. Acoust.*, Vol. 23, p. 218, 1977.

14. I. C. Khoo, *Opt. Eng.*, Vol. 25, p. 198, 1985.

15. H. J. Deuling, "Elasticity of Nematic Liquid Crystals," in L. Liebert, Ed., *Solid State Physics Suppl. 14*, Academic, New York, 1978, p. 81.

16. L. M. Blinov, *Electro-Optical and Magneto-Optical Properties of Liquid Crystals*, Wiley, New York, 1983, p. 109.

17. V. Freedericksz and V. Z. Zolina, *Kristallogr., Kristallogeom., Kristallophys., Kristallochem.*, Vol. 79, p. 225, 1931.

18. U. Efron and T. D. Bates, "Development of a Modified LCLV for Adaptive Optics Applications," Research Report, Hughes Research Laboratories, 1984, p. 9.

19. P. G. DeGennes, *The Physics of Liquid Crystals*, Clarendon Press, Oxford, 1974, Chap. 3.

20. J. S. Fergason and G. H. Brown, *J. Amer. Oil Chem.*, Vol. 45, p. 120, 1968.

21. G. Zhu, *Phys. Rev. Lett.*, Vol. 49, p. 1332, 1982.

22. L. Lin et al., *Phys. Rev. Lett.*, Vol. 49, p. 1335, 1982.

23. U. Efron et al., *Ferroelectrics*, Vol. 73, p. 315, 1987.

24. L. Liebert, Ed., *Solid State Physics*, Academic Press, New York, 1978, p. 117.

25. L. M. Blinov, *Electro-optical and Magneto-optical Properties of Liquid Crystals*, Wiley, New York, 1983, p. 105.

26. P. Pieranski et al., *J. Phys. (PARIS)*, Vol. 34, p. 35, 1973.

27. U. Efron et al., *J. Opt. Soc. Amer.*, Vol. B3, p. 247, 1986.

28. M. Born and E. Wolf, *Principles of Optics*, Pergamon, New York, 1980, Chap. 14.

29. H. J. Deuling, "Elasticity of Nematic Liquid Crystals," in L. Liebert, Ed., *Solid State Physics, Suppl. 14*, Academic, New York, 1978, p. 101.

30. S. T. Wu, U. Efron, and L. D. Hess, *Appl. Opt.*, Vol. 23, p. 3911, 1984.

31. F. J. Kahn, *Appl. Phys. Lett.*, Vol. 20, p. 199, 1972.

32. U. Efron and T. D. Bates, "Development of a Modified LCLV for Adaptive Optics Applications," Research Report, Hughes Research Laboratories, 1984.

33. M. Born and E. Wolf, *Principles of Optics*, Pergamon, New York, 1980, p. 695.

34. J. Grinberg and A. Jacobson, *J. Opt. Soc. Amer.* Vol. 66, p. 1003, 1976.

35. C. H. Gooch and H. A. Tarry, *J. Phys.*, Vol. D8, p. 1575, 1975.

36. U. Efron et al., "A Wire Grid-based Liquid Crystal Light Valve," *Proc. SPIE*, Vol. 1151, 1989.

37. J. Grinberg et al., *Opt. Eng.*, Vol. 14, p. 217, 1975.

38. R. Gagnon, *J. Opt. Soc. Amer.*, Vol. 71, p. 548, 1981. Also R. J. Gagnon, "Computer Simulation of a Twisted Nematic Cell," Internal Report, Hughes Aircraft Co., 1982.

39. U. Efron et al., *Opt. Eng.*, Vol. 24, p. 111, 1985.

40. (a) J. Grinberg et al., *IEEE Trans. El. Dev.*, Vol. ED-22, p. 775, 1975; (b) Lackner et al., U.S. Patent 4,030,997, 1984; Miller et al., U.S. Patent 4,464,134, 1981; and Miller et al., Int. Patent G02F1/133, 1988; 21, April; (c) J. F. Clerc et al., *Proc. 9th Int. Display Res. Conf., Kyoto*, Oct. 16–18, 1989, paper 7–5; (d) T. Sonehara et al., *Proc. 9th Int. Display Res. Conf., Kyoto*, Oct. 16–18, 1989, paper 7–6.

41. S. T. Wu, U. Efron, and L. D. Hess, *Appl. Opt.*, Vol. 21, p. 3911, 1984.

42. S. T. Wu, U. Efron, and L. D. Hess, *Appl. Phys. Lett.*, Vol. 44f, p. 829, 1984.

43. U. Efron, S. T. Wu, and T. D. Bates, *J. Opt. Soc. Amer.* Vol. B3, p. 247, 1986.

44. U. Efron et al., *Ferroelectrics*, Vol. 73, p. 315, 1987.

45. S. T. Wu et al., *Appl. Opt.*, Vol. 26, p. 3441, 1987.

46. S. T. Wu and U. Efron, *Opt. Soc. Amer. Topical Meeting on Spatial Light Modulators, Lake Tahoe, Nevada, Tech. Digest*, Vol. 8, p. 24, 1988.

47. T. J. Scheffer and J. Nehring, *Appl. Phys. Lett.*, Vol. 45, p. 1021, 1984.

48. M. Schadt and F. Leenhouts, *Appl. Phys. Lett.*, Vol. 50, p. 236, 1987.

49. K. Kawasaki, K. Yamada, and K. Mizunoya, *SID Dig.*, p. 391, 1987.
50. Katoh et al., *Jap. J. Appl. Phys.*, Vol. 26, p. L1784, 1987.
51. H. Odai et al., *IEEE Int. Display Res. Conf.*, 1988, p. 195.
52. G. E. Zverda and A. D. Kapustin, *Primen Ultraakustiki, K. Issle Dovaniyu Veschchestsva*, Vol. 15, Moskovsk. Oblastnoi Pedagogisch. Inst., Moscow, 1961, p. 69.
53. P. G. DeGennes, *The Physics of Liquid Crystals*, Clarendon, 1974, Chap. 5.
54. S. Chandrasenkhar, *Liquid Crystal*, Cambridge Monographs on Physics, 1977, Chap. 3.
55. L. M. Blinov, *Electrooptical and Magnetooptical Properties of Liquid Crystals*, Chap. 5.
56. L. K. Vistin, *Sov. Phys. Dud.*, Vol. 15, p. 908, 1971.
57. W. Greubel and W. Wolf, *Appl. Phys. Lett.*, Vol. 19, p. 213, 1971.
58. J. M. Pollack and J. B. Flannery, in J. F. Johnson and R. E. Porter, Eds., *Liquid Crystals and Ordered Fluids*, Vol. 2, Plenum, New York, 1978, p. 557.
59. B. H. Soffer et al., *Mol. Cryst. Liq. Cryst.*, Vol. 70, p. 145, 1981.
60. A. R. Tanguay, Jr. et al., *Opt. Eng.*, Vol. 22, p. 687, 1983.
61. C. S. Wu, Ph.D. Thesis, University of Southern California, 1987.
62. P. G. DeGennes, *The Physics of Liquid Crystals*, Clarendon, Oxford, 1974, Chap. 6.
63. G. Meier, "Applications of LCs," in G. Meier, E. Jackmann, and J. G. Grab Maier, Eds., Springer-Verlag, New York, 1975.
64. L. M. Blinov, *Electro-Optical and Magneto-Optical Properties of Liquid Crystals*, Wiley, New York, 1983, Chap. 6.
65. H. DeVries, *Acta Cryst.*, Vol. 4, p. 219, 1951.
66. R. B. Meyer et al., *J. Phys. Lett.*, Vol. 36, p. L-69, 1975.
67. R. B. Meyer et al., *Mol. Cryst. Liq. Cryst.*, Vol. 40, p. 33, 1977.
68. N. A. Clark and S. T. Lagerwall, *Appl. Phys. Lett.*, Vol. 36, p. 899, 1980.
69. S. T. Lagerwall et al., *Mol. Cryst. Liq. Cryst.*, Vol. 152, p. 503, 1987.
70. S. Garoff and R. B. Meyer, *Phys. Rev.*, Vol. A19, p. 338, 1979.
71. G. Andersson et al., *Appl. Phys. Lett.*, Vol. 51, p. 640, 1987.
72. J. L. Fergason, *SID Dig.*, Vol. 16, p. 68, 1985.
73. N. A. Vaz, *Proc. SPIE*, Vol. 1080, p. 2, 1989.
74. J. W. Doane et al., *Mol. Cryst. Liq. Cryst.*, Vol. 165, p. 511, 1988.
75. G. H. Heilmeier et al., *Appl. Phys. Lett.*, Vol. 13, p. 91, 1968; *Mol. Cryst. Liq. Cryst.*, Vol. 8, p. 293, 1969.
76. D. L. White and G. N. Taylor, *J. Appl. Phys.*, Vol. 45, p. 4718, 1974.
77. T. Uchida et al., *Mol. Cryst. Liq. Cryst.*, Vol. 63, p. 19, 1981.
78. S. Lu and D. H. Davis, *Mol. Cryst. Liq. Cryst.*, Vol. 94, p. 167, 1983.
79. D. W. Berreman et al., U.S. Patent 3,787,110, January 1974.
80. T. Kalfuss and E. Lueder, *Thin Solid Films*, Vol. 61, p. 259, 1979.
81. A. H. Firester, *Solid State Tech.*, December 1988, pp. 63ff.
82. I. DeRyke et al., *1988 Int. Display Res. Conf.* (IEEE pub.), p. 70.
83. T. L. Credelle, *1988 Int. Display Res. Conf.* (IEEE pub.), p. 208.
84. M. Yuki et al., *Proc. Soc. Inf. Disp.*, Vol. 30, p. 143, 1989.
85. R. G. Stewart et al., *Soc. Inf. Disp. 1988 Dig.*, p. 404.
86. D. E. Castleberry et al., *Soc. Inf. Disp. 1988 Dig.*, p. 232.
87. F. Inoue et al., *Proc. Soc. Inf. Disp.*, Vol. 29, p. 265, 1988.
88. H. Matino et al., *Soc. Inf. Disp. Tech. Dig.*, p. 400, 1990.
89. K. Niki et al., *Proc. Soc. Inf. Disp.*, Vol. 29, p. 259, 1988.
90. H. Moriyama et al., *Proc. Soc. Inf. Disp.*, Vol. 31, p. 13, 1990.
91. M. Vijan et al., *Soc. Inf. Disp. Tech. Dig.*, p. 530, 1990.

92. H. Oshima et al., *Soc. Inf. Disp. Tech. Dig.*, p. 408, 1988.
93. Y. Kanemori et al., *Soc. Inf. Disp. Tech. Dig.*, p. 408, 1990.
94. P. W. Ross, *IEEE 1988 Intl. Res. Conf.*, p. 185.
95. H. U. Lauer et al., *Soc. Inf. Disp. Tech. Dig.*, p. 534, 1990.
96. (a) J. Grinberg et al., *IEEE J. Quant. Elec.*, Vol. 17, p. 148, 1981; (b) U. Efron et al., *Opt. Eng.*, Vol. 22, p. 682, 1983; (c) M. S. Welkowsky et al., *Opt. Eng.*, Vol. 26, p. 414, 1987.
97. J. Grinberg et al., *Opt. Eng.*, Vol. 14, p. 217, 1975.
98. D. G. Sikharulidze et al., *Sov. J. Quant.* Elect., Vol. 9, No. 6, p. 747 (1979).
99. I. N. Kompanets et al., *Sov. J. Quant.* Elect., Vol. 9, No. 8, p. 1070, (1979).
100. P. O. Braatz et al., *Intl. Elect. Device Meeting Abstr.*, Vol. 23, p. 540, 1979; U. Efron et al., *J. Appl. Phys.* Vol. 57, p. 1356, (1985).
101. R. C. Lacoe and U. Efron, *Appl. Opt.* (in press).
102. R. C. Lacoe and U. Efron, "One-Dimensional Simulation of the Si–LCLV, Phase I and II," Research Report, Hughes Research Laboratories, 1984.
103. W. P. Brown, S. Wandzura, D. Maydan, and U. Efron, unpublished results, Hughes Research Laboratories, (1986).
104. K. Sayyah et al., *Appl. Opt.*, November 15, 1989.
105. S. G. Latham and M. P. Owen, *GEC J. Res.*, Vol. 4, No. 3, p. 219, 1986.
106. U. Efron et al., Spatial Light Modulators and Applications Topical Meeting, *Opt. Soc. Amer. Tech. Dig.*, 1988, p. 2; U. Efron et al., *Laser Electroopt. Conf.* Vol. 7, paper THW1, 1988.
107. I. N. Kompanets et al., *Sov. J. Quant. Elect.*, Vol. 9, No. 8, p. 1070, 1979.
108. S. P. Berestnev et al., *Radiotekhn. Electron.*, Vol. 6, p. 1212, 1985.
109. M. C. Hebron et al., *Proc. SPIE*, Vol. 825, p. 19, 1987.
110. D. Armitage et al., *Proc. SPIE*, Vol. 825, p. 32, 1987.
111. D. Armitage et al., Topical Meeting on Opticae Computing, Salt Lake City, 1989, Abstracts.
112. A. V. Parfenov el at., *Sov. J. Quant. Elect.*, Vol. 10, No.2, p. 167, 1980.
113. P. Auburg et al., *Appl. Opt.*, Vol. 21, p. 3706, 1982.
114. W. E. L. Haas and G. A. Dir, *Appl. Phys. Lett.*, Vol. 29, p. 325, 1975.
115. U. Efron et al., *Proc. SPIE*, Vol. 1051, 1989.
116. L. Samuelson et al., *Appl. Phys. Lett.* Vol. 34, p. 450, 1979.
117. D. Williams et al., *J. Phys. D, Appl. Phys.*, Vol. 21, p. S156, 1988.
118. K. Johnson et al., *Appl. Opt.*, November 15, 1989.
119. F. J. Kahn et al., *Appl. Phys. Lett.*, Vol. 22, p. 111, 1973.
119a. R. D. Sterling et al., *Soc. Inf. Disp. Tech. Dig.* Vol. 21, p. 327, 1990.
120. A. Sasaki et al., *Proc. Soc. Inf. Display*, Vol. 21, p. 341, 1980.
121. A.G. Dewey et al., *Soc. Inf. Disp. 1983 Dig.*, p. 36.
122. Y. Nagae et al., *Proc. Soc. Inf. Disp.*, Vol. 28, p. 56, 1987.
123. Greyhawk: Spatial Light Modulator System, Model 5075, Technical Specifications, 1989.
124. J. Trias et al., *Proc. Soc. Inf. Disp.*, Vol. 29, p. 275, 1988.
125. D. Haven, *IEEE Trans. Elect. Devices*, Vol. ED30, No. 5, May 1983.
126. B. H. Soffer et al., *Proc. SPIE*, Vol. 218, p. 81, 1980. A. A. Sawchuk et al., *Proc. SPIE*, Vol. 373, p. 69, 1981.
127. J. Millman and C. C. Halkias, *Integrated Electronics*, McGraw-Hill, New York, 1972, Chap. 13, 14.
128. Fatehi et al., *Appl. Opt.*, Vol. 23, p. 2163, 1984.
129. R. A. Athale and S. H. Lee, *Appl. Opt.*, Vol. 20, p. 1424, 1981.
130. V. K. Segupta et al., *Opt. Lett.*, Vol. 3, p. 199, 1978.
131. I. N. Kompanets, *Opt. Comm.*, Vol. 36, p. 415, 1981.

132. (a) H. Tanaka et al., *S.I.D. 1987 Tech. Dig.*, p. 140; (b) D. E. Castleberry et al., *S.I.D. 1988 Tech. Dig.*, p. 232.

133. F. Inoue et al., *S.I.D. 1988 Tech. Dig.*, p. 318.

134. S. Nagata et al., *S.I.D. 1985 Tech. Dig.*, p. 84.

135. J. Koda et al., *Proc. SPIE*, Vol. 760, p. 78, 1987.

136. The term "spatial light modulator," which is interchangeable with "light valve," is often used to emphasize optical processing applications of the device as opposed to the display application.

137. C. Warde and A. D. Fisher, in J. L. Horner, Ed., *Optical Signal Processing*, Academic, New York, 1987, Chap. 7.2.

138. (a) *Proc. SPIE*, Vol. 825, 1987, U. Efron, Ed.; (b) U. Efron, *Proc. SPIE*, Vol. 960, p. 130, 1988.

139. A. G. Ledebuhr, *Soc. Inf. Disp. 1986 Dig.*, p. 379.

140. S. Aruga et al., *Soc. Inf. Disp. 1987 Dig.*, p. 75.

141. Y. Nagae et al., *Soc. Inf. Disp. 1986 Dig.*, p. 368.

142. D. Haven et al., *Soc. Inf. Disp. 1986 Dig.*, p. 372.

143. G. H. Heilemeier et al., *Appl. Phys. Lett.*, Vol. 13, p. 46. 1968.

144. G. H. Heilemeier et al., *Mol. Cryst. Liq. Cryst.*, Vol. 8, p. 293, 1969.

145. J. Wycoski et al., *Phys. Rev. Lett.*, Vol. 20, p. 1024, 1968.

146. J. D. Margerum et al., *Appl. Phys. Lett.*, Vol. 13, p. 5, 1970.

147. T. P. Brody et al., *Trans. IEEE*, Vol. ED31, p. 995, 1973.

148. L. T. Lipton et al., *Soc. Inf. Disp. 1978 Dig.*, 1975.

149. J. L. Fergason, U.S. Patent 3,731, 986, 1973.

150. M. Schadt and W. Helfrich, *Appl. Phys. Lett.*, Vol. 18, p. 127, 1971.

151. J. Grinberg et al., *Opt. Eng.*, Vol. 14, p. 217, 1975.

152. S. Morozumi et al., *Soc. Inf. Disp. 1983 Dig.*, p. 156.

153. J. Goodman, *Introduction to Fourier Optics*, McGraw-Hill, New York, 1968, Chap. 5.

154. G. A. Korn and T. M. Korn, *Mathematical Handbook*, McGraw-Hill, New York, 1968, Chap. 4.

155. R. Rau, *J. Opt. Soc. Amer.*, Vol. 56, p. 1990, 1966.

156. K. Sayyah et al., *J. Appl. Opt.* (in press).

157. E. Marom, *Opt. Eng.*, Vol. 25, p. 274, 1986.

158. S. R. Dashiel and A. A. Sawchuk, *Appl. Opt.*, Vol. 16, p. 1009, 1977.

159. A. Armand et al., *Opt. Lett.*, Vol. 7, p. 451, 1982.

160. B. H. Soffer et al., *Proc. SPIE*, Vol. 218, p. 81, 1980.

161. U. Efron, *Proc. SPIE*, Vol. 960, p. 180, 1988.

162. R. G. Wilson and W. E. Silverton Jr., *Proc. SPIE*, Vol. 178, p. 185, 1979.

163. B. H. Soffer, private communication; U. Efron, B. H. Soffer, and H. J. Caulfield, Proceedings of the NASA Conference on Optical Information Processing for Aerospace Applications, Langley, Hampton NA, 1983.

164. Y. Owechko, W. Byles, and B. H. Soffer, unpublished results, 1986.

165. P. R. Griffith, *Chemical IR Fourier Transform Spectroscopy*, Wiley, 1975, Chap. 1.

166. H. J. Caulfield, "Holographic Spectrascopy," in N. H. Farhat, Ed., *Advances in Holography*, Vol. 2, Dekker, New York, 1976.

167. W. P. Bleha et al., *Opt. Eng.*, Vol. 17, p. 371, 1978.

168. B. H. Soffer and U. Efron, "LC Spatial Light Modulators for EOP/SAR Processing," Final Technical Report N66001-82-C-0613, prepared for NOSC, San Diego, CA, 1982.

169. E. Marom, B. H. Soffer, and U. Efron, *Opt. Lett.*, Vol. 10, p. 43, 1985.

170. B. Clymer and S. A. Collins, Jr., *Opt. Eng.*, Vol. 24, p. 74, 1985.

171. L. A. Pagano-Stauffer et al., *Proc. SPIE*, Vol. 825, p. 141, 1987.
172. J. D. Armitage and A. W. Lohman, *Appl. Opt.*, Vol. 4, p. 399, 1965.
173. A. R. Tanguay, Jr. et al., *Opt. Eng.*, Vol. 22, p. 687, 1983.
174. A. Armand et al., *Opt. Lett.*, Vol. 5, p. 129, 1980.
175. J. W. Goodman et al., *Opt. Lett.*, Vol. 2, p. 1, 1978.
176. H. T. Kung, *Proc. SPIE*, Vol. 241, p. 76, 1981.
177. H. J. Caulfield et al., *Opt. Comm.*, Vol. 40, p. 86, 1981.
178. R. P. Bocker et al., *Appl. Opt.*, Vol. 22, p. 804, 1983.
179. (a) R. P. Lippmann, *IEEE ASSP Magazine*, April 1987, pp. 24–37. (b) D. W. Tank and J. J. Hopfield, *Sci. Amer.*, December 1987, pp. 104–114.
180. J. J. Hopfield and D. W. Tank, *Biol. Cybern.*, Vol. 52, p. 14, 1985.
181. B. H. Soffer et al., *Opt. Lett.*, Vol. 11, p. 118, 1986.
182. D. Z. Anderson and M. C. Erie, *Opt. Eng.*, Vol. 26, p. 434, 1987.
183. Y. Owechko, "Self-pumped Optical Neural Networks," Salt Lake City Topical Meeting on Optical Computing, February 1989.
184. J. A. Anderson et al., *Psych. Rev.*, Vol. 84, p. 413, 1977.
185. J. J. Hopfield, *Proc. Natl. Acad. Sci.*, Vol. 79, p. 2559, 1982.
186. A. D. Fisher et al., *Appl. Opt.*, Vol. 26, p. 5039, 1987.
187. U. Efron and Y. Owechko, Topical Meeting of the OSA on Optical Computing, Salt Lake City, February 1989.
188. B. H. Soffer et al., *Appl. Opt.*, Vol. 25, p. 2295, 1986.
189. J. W. Goodman, *Proc. IEEE*, Vol. 72, p. 850, 1984.
190. B. Clymer and S. A. Collins. Jr., *Opt. Eng.*, Vol. 24, p. 74, 1985.
191. E. Marom and N. Conforti, *Proc. SPIE*, Vol. 700, p. 209, 1986.
192. A. A. Sawchuk et al., *Computer (IEEE)*, Vol. 20, No. 6, p. 50, 1987.
193. K. Johnson et al., *Opt. Eng.*, Vol. 26, p. 389, 1987.
194. A. A. Agranat, *Proc. SPIE*, Vol. 1151, 1989.
195. A. D. Fisher and C. Warde, *Opt. Lett.*, Vol. 8, p. 353, 1983.
196. D. M. Pepper, *Opt. Eng.*, Vol. 21, p. 156, 1982.
197. B. Ya. Zel'dovich, V. I. Popovichev, V. V. Ragulskii, and F. S. Faizullov, *Sov. Phys. JETP*, Vol. 15, p. 109, 1972.
198. J. E. Pearson and S. Hansen, *J. Opt. Soc. Amer.*, Vol. 67, p. 305, 1977.
199. A. D. Fischer and C. Warde, *Opt. Lett.*, Vol. 8, p. 353, 1983.
200. D. M. Pepper and A. Yariv, in R. A. Risher, Ed., *Optical Phase Conjugation*, Academic, New York, 1983, pp. 24–76.
201. E. Marom and U. Efron, *Opt. Lett.*, Vol. 12, p. 504, 1987.
202. O. V. Garibyan, I. N. Kompanets, A. F. Parfyonov, N. F. Filipetsky, V. V. Shkunov, A. N. Sudarkin, A. V. Sukhov, N. V. Tabiryan, A. A. Vasilev, and B. Ya. Zel'dovich, *Opt. Comm.*, Vol. 38, p. 67, 1981.
203. S. T. Wu, U. Efron, J. Grinberg, L. D. Hess, and M. S. Welkowsky, *Proc. Soc. Photo-Opt. Instrum. Eng.*, Vol. 572, p. 94, 1985.
204. U. Efron et al., *Opt. Eng.*, Vol. 24, p. 111, 1985.
205. M. S. Welkowsky et al., *Proc. SPIE*, Vol. 825, p. 193, 1987.
206. S. T. Wu, U. Efron, and T. Y. Hsu, *Opt. Lett.*, Vol. 13, p. 13, 1988.
207. P. O. Braatz, Reflective Matrix Mirror Visible to Infrared Converter Light Valve, patent pending.
208. M. S. Welkowsky et al., *Proc. SPIE*, Vol. 825, p. 193, 1987.
209. I. C. Khoo, *Opt. Eng.*, Vol. 25, p. 198, 1986.

INDEX